The Routledge Handbook of Research Methods for Social-Ecological Systems

The Routledge Handbook of Research Methods for Social-Ecological Systems provides a synthetic guide to the range of methods that can be employed in social-ecological systems (SES) research.

The book is primarily targeted at graduate students, lecturers and researchers working on SES, and has been written in a style that is accessible to readers entering the field from a variety of different disciplinary backgrounds. Each chapter discusses the types of SES questions to which the particular methods are suited and the potential resources and skills required for their implementation, and provides practical examples of the application of the methods. In addition, the book contains a conceptual and practical introduction to SES research, a discussion of key gaps and frontiers in SES research methods, and a glossary of key terms in SES research. Contributions from 97 different authors, situated at SES research hubs in 16 countries around the world, including South Africa, Sweden, Germany and Australia, bring a wealth of expertise and experience to this book.

The first book to provide a guide and introduction specifically focused on methods for studying SES, this book will be of great interest to students and scholars of sustainability science, environmental management, global environmental change studies and environmental governance. The book will also be of interest to upper-level undergraduates and professionals working at the science–policy interface in the environmental arena.

Reinette (Oonsie) Biggs is associated with the Centre for Sustainability Transitions, Stellenbosch University, Stellenbosch, South Africa, and the Stockholm Resilience Centre, Stockholm University, Stockholm, Sweden.

Alta de Vos is associated with the Department of Environmental Science, Rhodes University, Makhanda, South Africa.

Rika Preiser is associated with the Centre for Sustainability Transitions, Stellenbosch University, Stellenbosch, South Africa.

Hayley Clements is associated with the Centre for Sustainability Transitions, Stellenbosch University, Stellenbosch, South Africa.

Kristine Maciejewski is associated with the Centre for Sustainability Transitions, Stellenbosch University, Stellenbosch, South Africa.

Maja Schlüter is associated with the Stockholm Resilience Centre, Stockholm University, Stockholm, Sweden.

The Routledge Handbook of Research Methods for Social-Ecological Systems

Edited by Reinette Biggs, Alta de Vos,
Rika Preiser, Hayley Clements,
Kristine Maciejewski and Maja Schlüter

Routledge
Taylor & Francis Group
LONDON AND NEW YORK

First published 2022
by Routledge
2 Park Square, Milton Park, Abingdon, Oxon OX14 4RN

and by Routledge
605 Third Avenue, New York, NY 10158

Routledge is an imprint of the Taylor & Francis Group, an informa business

British Library Cataloguing-in-Publication Data
A catalogue record for this book is available from the British Library

Library of Congress Cataloging-in-Publication Data
Names: Biggs, Reinette, 1979– editor.
Title: The Routledge handbook of research methods for social-ecological systems / edited by Reinette Biggs, Alta de Vos, Rika Preiser, Hayley Clements, Kristine Maciejewski, Maja Schlüter.
Description: New York: Routledge, 2021. |
Series: Routledge international handbooks | Includes bibliographical references and index. |
Identifiers: LCCN 2020057184 (print) | LCCN 2020057185 (ebook) |
ISBN 9780367898403 (hardback) | ISBN 9781003021339 (ebook)
Subjects: LCSH: Social ecology—Research. | Social sciences—Research—Methodology.
Classification: LCC HD6960 .R68 2021 (print) |
LCC HD6960 (ebook) | DDC 304.2072/1—dc23
LC record available at https://lccn.loc.gov/2020057184
LC ebook record available at https://lccn.loc.gov/2020057185

ISBN: 978-0-367-89840-3 (hbk)
ISBN: 978-1-032-02076-1 (pbk)
ISBN: 978-1-003-02133-9 (ebk)

DOI: 10.4324/9781003021339

Typeset in Bembo
by codeMantra

To the SES researchers of the future

Contents

Methods for analysing systems – directly informing decision-making 399

Editors

Reinette (Oonsie) Biggs is co-director of the Centre for Sustainability Transitions at Stellenbosch University, where she holds the South African Research Chair in Social-Ecological Systems and Resilience. She is also affiliated with the Stockholm Resilience Centre at Stockholm University in Sweden, and currently co-chairs the international Programme on Ecosystem Change and Society (PECS).

Alta de Vos is a senior lecturer at the Department of Environmental Science, Rhodes University, where she teaches several foundational social-ecological systems courses. Her research focuses mainly on protected area resilience, applying a social-ecological systems lens and multiple (often spatial) methods to understand how protected and conservation area networks change over time, and interact with the landscapes in which they exist.

Rika Preiser is a senior researcher at the Centre for Sustainability Transitions, Stellenbosch University. Her research explores the philosophical and conceptual development of complexity and how a deeper understanding of the features and dynamics of complex adaptive systems inform novel ways that participatory and qualitative research methods and practical interventions can be used to foster more ethical social-ecological futures.

Hayley Clements is a researcher at the Centre for Sustainability Transitions, Stellenbosch University. She is also affiliated with the Department of Geosciences and Geography, University of Helsinki. Her research applies social-ecological systems thinking to identify resilient and equitable biodiversity conservation opportunities in a changing world, with a focus on sub-Saharan Africa.

Kristine Maciejewski is an extraordinary senior lecturer at the Centre for Sustainability Transitions at Stellenbosch University, specialising in social-ecological systems. At the time of writing, she was the biodiversity adviser for the USAID-funded Resilient Waters Program that aims to build resilient communities and ecosystems in southern Africa by improving the management of transboundary natural resources.

Maja Schlüter is a professor at the Stockholm Resilience Centre, Stockholm University, where she leads the SES-LINK group (seslink.org). Her research focuses on understanding and theorising about the emergence of social-ecological systems pathways and phenomena through novel ways of combining empirical research with dynamic modelling, and exploring the conceptual foundations of social-ecological systems as intertwined complex adaptive systems.

Figures

Tables

Case studies

Contributors

Géraldine Abrami, G-EAU, Univ Montpellier, AgroParisTech, BRGM, CIRAD, IRD, INRAE, Institut Agro, Montpellier, France.

Razak Abu, Policy Fellow, National Energy Board, Calgary, Canada.

John M. Anderies, School of Sustainability and School of Human Evolution and Social Change, Arizona State University, Tempe, Arizona, USA.

Jacopo Baggio, School of Politics, Security and International Affairs, University of Central Florida, Florida, USA; Sustainable Coastal Systems Cluster, National Center for Integrated Coastal Research, University of Central Florida, Florida, USA.

Olivier Barreteau, G-EAU, Univ Montpellier, AgroParisTech, BRGM, CIRAD, IRD, INRAE, Institut Agro, Montpellier, France.

Xavier Basurto, Nicholas School of the Environment, Duke University, Durham, North Carolina, USA.

Brandon Bestelmeyer, USDA-ARS Jornada Experimental Range, New Mexico State University, Las Cruces, New Mexico, USA.

Joana Carlos Bezerra, Community Engagement Division, Rhodes University, Makhanda, South Africa.

Reinette (Oonsie) Biggs, Centre for Sustainability Transitions, Stellenbosch University, Stellenbosch, South Africa; Stockholm Resilience Centre, Stockholm University, Stockholm, Sweden.

Örjan Bodin, Stockholm Resilience Centre, Stockholm University, Stockholm, Sweden.

Pierre Bommel, CIRAD, UR GREEN 47, MUSE University of Montpellier, Montpellier, France.

Bruno Bonté, G-EAU, Univ Montpellier, AgroParisTech, BRGM, CIRAD, IRD, INRAE, Institut Agro, Montpellier, France.

Jean Boulton, Department of Social and Policy Sciences, University of Bath, Bath, UK.

François Bousquet, CIRAD, UMR SENS, F-34398 Montpellier, France; SENS, University of Montpellier, CIRAD, IRD, Montpellier, France.

Retta Bruegger, Colorado State University Extension, Grand Junction, Colorado, USA.

Tomas Chaigneau, Environment and Sustainability Institute, University of Exeter, Penryn, UK.

Rebecca Chaplin-Kramer, Natural Capital Project, Institute on the Environment, University of Minnesota, Saint Paul, Minnesota, USA; Natural Capital Project, Woods Institute for the Environment, Stanford University, Stanford, California, USA.

Lakshmi Charli-Joseph, Laboratorio Nacional de Ciencias de la Sostenibilidad (LANCIS), Instituto de Ecología, Universidad Nacional Autónoma de México, Mexico City, Mexico.

Hayley Clements, Centre for Sustainability Transitions, Stellenbosch University, Stellenbosch, South Africa.

Jessica Cockburn, Environmental Learning Research Centre, Rhodes University, Makhanda, South Africa.

Anne-Sophie Crépin, Beijer Institute of Ecological Economics, The Royal Swedish Academy of Sciences, Stockholm, Sweden.

Graeme S. Cumming, ARC Centre of Excellence for Coral Reef Studies, James Cook University, Townsville, Australia.

Paul Currie, ICLEI Africa, Cape Town, South Africa; Centre for Sustainability Transitions, Stellenbosch University, Stellenbosch, South Africa.

William's Daré, CIRAD, UR GREEN 47, MUSE University of Montpellier, Montpellier, France.

Stefan Daume, Stockholm Resilience Centre, Stockholm University, Stockholm, Sweden; Beijer Institute of Ecological Economics, The Royal Swedish Academy of Sciences, Stockholm, Sweden.

Alta de Vos, Department of Environmental Science, Rhodes University, Makhanda, South Africa.

Bulgamaa Densambuu, Green Gold and Animal Health Project, Swiss Agency for Development and Cooperation, Ulaanbaatar, Mongolia.

Lisa Deutsch, Stockholm Resilience Centre, Stockholm University, Stockholm, Sweden; Nordic Institute for Latin American Studies, Stockholm University, Stockholm, Sweden.

Scott Drimie, The Southern Africa Food Lab, South Africa; Centre for Sustainability Transitions, Stellenbosch University, Stellenbosch, South Africa.

Raphaëlle Ducrot, G-EAU, Univ Montpellier, AgroParisTech, BRGM, CIRAD, IRD, INRAE, Institut Agro, Montpellier, France.

Graham Epstein, School of Politics, Security, and International Affairs, University of Central Florida, Florida, USA.

Karen Esler, Department of Conservation Ecology and Entomology and Centre for Invasion Biology, Stellenbosch University, Stellenbosch, South Africa.

Maria Fernández-Giménez, Department of Forest and Rangeland Stewardship, Colorado State University, Fort Collins, Colorado, USA.

Ingo Fetzer, Stockholm Resilience Centre, Stockholm University, Stockholm, Sweden; Bolin Centre for Climate Research, Stockholm, Sweden.

Carl Folke, Stockholm Resilience Centre, Stockholm University, Stockholm, Sweden.

Sarah E. Gergel, Department of Forest and Conservation Sciences, University of British Columbia, Vancouver, Canada.

Steven Gray, Department of Community Sustainability, Michigan State University, East Lansing, Michigan, USA.

Jamila Haider, Stockholm Resilience Centre, Stockholm University, Stockholm, Sweden.

Maike Hamann, Centre for Sustainability Transitions, Stellenbosch University, Stellenbosch, South Africa; Humphrey School of Public Affairs, University of Minnesota, Minneapolis, Minnesota, USA; Natural Capital Project, Institute on the Environment, University of Minnesota, Saint Paul, Minnesota, USA.

Dominic A.W. Henry, Statistics in Ecology, Environment and Conservation, Department of Statistical Sciences, University of Cape Town, Cape Town, South Africa; Endangered Wildlife Trust, Johannesburg, South Africa.

Tilman Hertz, Stockholm Resilience Centre, Stockholm University, Stockholm, Sweden.

Tanja Hichert, Hichert and Associates, Somerset West, South Africa; Centre for Sustainability Transitions, Stellenbosch University, Stellenbosch, South Africa.

Lloyd Hill, Department of Sociology and Social Anthropology, Stellenbosch University, Stellenbosch, South Africa.

Marco A. Janssen, School of Sustainability, and School of Complex Adaptive Systems, Arizona State University, Tempe, Arizona, USA.

Justin A. Johnson, Natural Capital Project, Institute on the Environment, University of Minnesota, Saint Paul, Minnesota, USA; Department of Applied Economics, University of Minnesota, Saint Paul, Minnesota, USA.

Stuart Kininmonth, Stockholm Resilience Centre, Stockholm University, Stockholm, Sweden; University of Oslo, Oslo, Norway; University of the South Pacific, Suva, Fiji Islands.

Louis Klein, European School of Governance, Berlin, Germany; International Federation for Systems Research, Austria.

Kasper Kok, Soil Geography and Landscape Group, Wageningen University, Wageningen, the Netherlands.

Steven J. Lade, Stockholm Resilience Centre, Stockholm University, Stockholm, Sweden; Fenner School of Environment and Society, The Australian National University, Canberra, Australia.

Therese Lindahl, Beijer Institute of Ecological Economics, The Royal Swedish Academy of Sciences, Stockholm, Sweden; Stockholm Resilience Centre, Stockholm University, Stockholm, Sweden.

Emilie Lindkvist, Stockholm Resilience Centre, Stockholm University, Stockholm, Sweden.

Marty Luckert, Department of Resource Economics & Environmental Sociology, University of Alberta, Edmonton, Canada.

Kristine Maciejewski, Centre for Sustainability Transitions, School of Public Leadership, Stellenbosch University, Stellenbosch, South Africa.

Colleen Magner, Reos Partners, South Africa.

María Mancilla García, Stockholm Resilience Centre, Stockholm University, Stockholm, Sweden.

Lisa Mandle, Natural Capital Project, Woods Institute for the Environment, Stanford University, Stanford, California, USA.

Amanda Manyani, Centre for Sustainability Transitions, Stellenbosch University, Stellenbosch, South Africa.

Berta Martín-López, Social-ecological Systems Institute, Faculty of Sustainability, Leuphana University, Lüneburg, Germany.

Vanessa A. Masterson, Stockholm Resilience Centre, Stockholm University, Stockholm, Sweden.

Raphaël Mathevet, CEFE, CNRS, University of Montpellier, University Paul Valéry Montpellier III, EPHE, IRD, Montpellier, France; IFP UMIFRE 21 CNRS/MAEE, Pondicherry, India.

Penelope Mograbi, Department of Environmental Science, Rhodes University, Makhanda, South Africa.

Michele-Lee Moore, Stockholm Resilience Centre, Stockholm University, Stockholm, Sweden.

Albert Norström, Stockholm Resilience Centre, Stockholm University, Stockholm, Sweden.

Per Olsson, Stockholm Resilience Centre, Stockholm University, Stockholm, Sweden.

Claudia Pahl-Wostl, Institute of Geography, University of Osnabrück, Osnabrück, Germany.

Laura Pereira, Centre for Sustainability Transitions, Stellenbosch University, Stellenbosch, South Africa; Stockholm Resilience Centre, Stockholm University, Stockholm, Sweden.

Arthur Perrotton, Stockholm Resilience Centre, Stockholm University, Stockholm, Sweden.

Garry Peterson, Stockholm Resilience Centre, Stockholm University, Stockholm, Sweden.

Luigi Piemontese, Stockholm Resilience Centre, Stockholm University, Stockholm, Sweden.

Stephen Polasky, University of Minnesota, Minneapolis, Minnesota, USA.

Gary Polhill, The James Hutton Institute, Aberdeen, UK.

Rika Preiser, Centre for Sustainability Transitions, Stellenbosch University, Stellenbosch, South Africa.

Allyson Quinlan, Resilience Alliance, Ottawa, Canada.

Maureen G. Reed, School of Environment and Sustainability, University of Saskatchewan, Saskatoon, Canada.

Chevonne Reynolds, Fitzpatrick Institute of African Ornithology, DST-NRF Centre of Excellence, University of Cape Town, Cape Town, South Africa; Animal, Plant and Environmental Sciences, University of the Witwatersrand, Johannesburg, South Africa.

Jesse T. Rieb, Department of Natural Resource Sciences, McGill University, Sainte-Anne-de-Bellevue, Canada.

Juan C. Rocha, Stockholm Resilience Centre, Stockholm University, Stockholm, Sweden; Beijer Institute of Ecological Economics, The Royal Swedish Academy of Sciences, Stockholm, Sweden.

Paul Ryan, Australian Resilience Centre, Australia.

Nicolas Salliou, Department of Civil, Environmental and Geomatic Engineering, Institute for Spatial and Landscape Development, Planning of Landscape and Urban Systems (PLUS), ETH Zürich, Zürich, Switzerland.

Caroline Schill, Beijer Institute of Ecological Economics, The Royal Swedish Academy of Sciences, Stockholm, Sweden; Stockholm Resilience Centre, Stockholm University, Stockholm, Sweden.

Maja Schlüter, Stockholm Resilience Centre, Stockholm University, Stockholm, Sweden.

Michael Schoon, School of Sustainability, Arizona State University, Tempe, Arizona, USA.

Kate Schreckenberg, Reader in Development Geography, King's College, London, UK.

Lisen Schultz, Stockholm Resilience Centre, Stockholm University, Stockholm, Sweden.

My Sellberg, Stockholm Resilience Centre, Stockholm University, Stockholm, Sweden.

Odirilwe Selomane, Centre for Sustainability Transitions, Stellenbosch University, Stellenbosch, South Africa.

Ralf Seppelt, UFZ – Helmholtz Centre for Environmental Research, Department of Computational Landscape Ecology, Leipzig, Germany; Institute of Geoscience and Geography, Martin Luther University Halle-Wittenberg, Halle, Germany.

Charlie Shackleton, Department of Environmental Science, Rhodes University, Makhanda, South Africa.

Sheona Shackleton, African Climate and Development Initiative, University of Cape Town, Cape Town, South Africa.

Jesús Mario Siqueiros-Garcia, Laboratorio Nacional de Ciencias de la Sostenibilidad (LANCIS), Instituto de Ecología, Universidad Nacional Autónoma de México, Mexico City, Mexico; Instituto de Investigaciones en Matemáticas Aplicadas y en Sistemas (IIMAS), Universidad Nacional Autónoma de México, Mexico City, Mexico.

Nadia Sitas, Centre for Sustainability Transitions, and Department of Conservation Ecology and Entomology, Stellenbosch University, Stellenbosch, South Africa.

Maria Tengö, Stockholm Resilience Centre, Stockholm University, Stockholm, Sweden.

Ruth H. Thurstan, Centre for Ecology and Conservation, College of Life and Environmental Sciences, University of Exeter, Penryn, UK.

Max Troell, Stockholm Resilience Centre, Stockholm University, Stockholm, Sweden; Beijer Institute of Ecological Economics, The Royal Swedish Academy of Sciences, Stockholm, Sweden.

Sergio Villamayor-Tomas, Institute of Environmental Science and Technology, Autonomous University of Barcelona, Barcelona, Spain.

Nanda Wijermans, Stockholm Resilience Centre, Stockholm University, Stockholm, Sweden.

Olive Zgambo, Centre for Sustainability Transitions, Stellenbosch University, Stellenbosch, South Africa

Foreword

This handbook is long overdue and should be considered an essential tool for any researcher or student interested in traversing the challenging investigative terrain posed by complex social-ecological problems. Complexity science has grown in academic importance as we seek to deal with systemic problems that cannot be relegated to a single discipline or even dealt with by multiple disciplines working collaboratively. The increasingly frequent social-ecological crises brought on by climate change, including massive floods and land erosions, displaced climate refugees, plastics in the ocean, devastating wildfires, antibiotic resistance, wealth disparities and pandemics, are only some of the problems rooted in the intertwined nature of our social and ecological systems. As the world population has increased and trade, communications and travel densely link different societies and ecosystems, temporal and geographical slack has been reduced and such problems take on an urgency that transcends academic engagement. Practitioners, policymakers and citizens demand a voice and look for solutions as well as insight.

All of this means that to research complex problems, new methodologies are required. Furthermore, tackling complexity necessitates a portfolio or broad toolkit of analytic methods and demands the breakdown not only of the traditional barriers between academic disciplines but also of those between different ways of knowing such as lay, local and indigenous expertise.

This is not a trivial challenge. Historically, methodologies have been linked to specific epistemologies which, in turn, are rooted in ontologies defended as revealing 'the truth'. Qualitative methodologies, most commonly associated with the social sciences, depend on unearthing deeper meanings and interpretations. Their goal is to reveal patterns and to produce understanding, and in doing so to reveal the underpinnings of behaviour. Results provide a basis for synthesis and comparison, for theory generation and inductive pattern recognition, but often fall short, by their very nature, of providing the certainty and the potential for falsification which underpins the rigour of testing, statistical results and quantitative methods. When scientific methods from the natural sciences are added to the mix, the tensions often grow, fed by a lack of understanding of the epistemologies which team members bring to the research. The default of research teams is often to create parallel projects, where different disciplines align findings side by side, with little idea of integration. When the challenge is intensified by the demand that researchers integrate lay knowledge or indigenous thought, the task may seem insurmountable, requiring facilitative and transdisciplinary skills outside the training or scope of most academic scientists, whether social or natural.

In the world of post-normal science, we need to rebuild our capacity for synthesis and intuition. We also need to produce knowledge that has been tested enough to earn our confidence, so we can use it as a springboard to action. We need methods that not only allow us this confidence but also open the way to welcoming participation of practitioners as equal partners in our effort to address the complex problems that face us. And we need it urgently. Methodological innovations are evident in this volume and are badly needed; in addition, the volume is structured so that it is possible to envision drawing on elements of different methodologies either in a synthetic manner or sequentially. And to do that we need to know the toolkit and what each approach can provide.

One way to conceptualise this new approach to the research process is as a research journey.[1] The researcher follows a path laid out by the evidence as it emerges. This landscape is defined by at least the two enduring tensions described above. The first tension is between (a) the search for general patterns or broad explanatory studies and theoretical concepts, and (b) academic projects focused on specific problems rooted in specific or grounded observation on the basis of falsifiable hypotheses. This tension can be further defined as one between macro-patterns and micro-studies. The second tension is between (a) data collected through a formalised academic process with the aim of knowledge creation, and (b) co-created knowledge between the researcher and subject(s) (including traditional knowledge and ways of knowing associated with a long-term relationship with place), with the aim of creating transformative action. To address a truly complex social-ecological problem, a researcher (or team of researchers) must explicitly and deliberately move across this research space. However, in order to do so, such researchers need to have at their disposal both the methodological frameworks and the skills to implement the most productive approach for the context at hand.

Of course, a number of factors influence a researcher's trajectory over the research landscape: the researcher's entry point, such as the defined problem, research question, or intellectual and academic training; the research team's composition; emergent aspects of the problem; and the boundaries of the question. The tools and ideas researchers bring to their project affect the research journey's direction at any moment in time. The research journey is an iterative process. As a concept, however, it can help researchers to understand where they are and where they are going, and the important relationship of context to methodological approach. It is important to remember, however, that the journey is not itself a method of inquiry.

The great strength of this handbook is that, for the first time, the wide range of approaches that are currently being tried is presented together in one volume. Moreover, there is a real effort to relate the methods to one another, giving the researcher the capacity to select from a toolkit of methods as the research journey unfolds, allowing the researcher to follow problems through progressively denser and/or thicker contexts, and to revise research questions and methods as understanding of the problem domain changes. This volume represents the first serious effort to formalise complexity research methods and to relate these to one another. It gives both researcher and practitioner the breadth of options to follow to appreciate and engage with that complexity. This is the essence of this book, which represents a great leap forward in our capacity to marry research and action in addressing complex social-ecological challenges.

Frances Westley
13 September 2020

[1] See McGowan, K.A., F. Westley, E.D.G. Fraser, P.A. Loring, K.C. Weathers, F. Avelino, J. Sendzimir, R. Roy Chowdhury, and M-L. Moore. 2014. 'The Research Journey: Travels across the Idiomatic and Axiomatic toward a Better Understanding of Complexity.' *Ecology and Society* 19(3): 37. doi:10.5751/ES-06518-190337, for an elaboration of these ideas.

Preface

The study of social-ecological systems (SES) is an important and exciting research field within the emerging domain of sustainability science. The need for an introductory guide to SES research methods has been apparent for several years, specifically to those working with senior postgraduate students. For instance, at the Stockholm Resilience Centre (SRC) in Sweden, where two of the editors are based, graduate students entering the PhD programme in sustainability science come from a wide diversity of undergraduate backgrounds, including ecology, marine science, political science, economics, anthropology and journalism. Providing this diverse array of students with a common grounding in SES and a basic introduction to SES research methods has been a challenge, and has prompted the development of various short courses since the SRC's inception. These challenges have been echoed by SES researchers based at a wide range of institutions around the world, as well as within key SES research networks, including the international Programme on Ecosystem Change and Society (PECS), and the Resilience Alliance (RA). This handbook specifically aims to meet this need, and is the first broad introduction to SES research methods in the SES field.

The specific idea for this handbook emerged within the context of these broader conversations at an SES winter school in July 2014, hosted by the Southern African Program on Ecosystem Change and Society (SAPECS) in South Africa, where most members of the editorial team are based. One of the key challenges identified right at the outset was how to decide on the list of methods to include in the book. After brainstorming potential ways forward with colleagues in both South Africa and Sweden, we decided to start by developing two background papers. The first (Preiser et al. 2018)[1] provided a conceptual framing for SES as complex adaptive systems (CAS) and the implications this has for choosing research methods for studying SES. The second (De Vos et al. 2019)[2] conducted a systematic review of methods used in SES research as the basis for selecting methods to include in the handbook. The outline and content of the handbook were developed iteratively alongside the papers, and refined through various discussions and presentations, specifically at the SAPECS working group meeting in May 2015, the first international PECS conference in November 2015, a GRAID (Guidance for Resilience in the Anthropocene: Investments for Development) project meeting in March 2016, as well as the Resilience Conference in Sweden in August 2017. In addition, two multi-day workshops focusing specifically on the book were particularly important in crystallising its content: a small workshop in Stockholm in October 2016 helped to finalise the methods to include, and a workshop involving most of the chapter lead

[1] Preiser R., R. Biggs, A. de Vos, and C. Folke. 2018. 'Social-Ecological Systems as Complex Adaptive Systems: Organizing Principles for Advancing Research Methods and Approaches.' *Ecology and Society* 23(4): 46.

[2] De Vos, A., R. Biggs, and R. Preiser. 2019. 'Methods for Understanding Social-Ecological Systems: A Review of Place-based Studies.' *Ecology and Society* 24(4): 16. doi:10.5751/ES-11236-240416.

authors held at Leuphana University in Germany in February 2019 reflected on emerging patterns and key gaps in SES methods across the chapters.

The book has emerged from many inspiring conversations over many years, and it is difficult to identify and acknowledge everyone who has contributed in some way. The comments and suggestions made at various conferences where we presented our ideas have been integral in shaping the book, and we really value these inputs. We thank Christo Fabricius, the organiser of the SAPECS winter school, for helping to create the conditions for the emergence of the initial idea for this book. We had several thoughtful discussions with Carl Folke, who has supported the book since its inception and contributed to the first conceptual paper. We also particularly thank colleagues at the SRC who provided input on the initial idea for the book in late 2014, as well as those who attended the workshop in October 2016, particularly Garry Peterson, Albert Norström, Magnus Nystrom, Andrew Merrie, Tim Daw, Maria Tengö, Miriam Huitric and Lisa Deutsch. The contributors to the book also deserve a special word of thanks. All chapters underwent peer review and multiple rounds of editorial comments to make the book as clear and consistent as possible. The chapter authors showed great fortitude in incorporating all these suggestions; although the chapters are short, they involved a great deal of work to synthesise a diverse set of methods and applications into a broad and succinct summary. Most authors also acted as peer reviewers for other chapters, and a number of chapters benefited from inputs from students. The final product includes 97 different authors, situated at SES research hubs in 16 countries around the world, bringing a wealth of expertise and experience to the final product.

The editorial team has shown great commitment and perseverance to bring this project to fruition. The team emerged somewhat organically over time, initially consisting of Oonsie Biggs, Rika Preiser and Alta de Vos, who led the development of the background papers and the initial outline of the book. Maja Schlüter joined the editorial team during her sabbatical in South Africa in late 2017, bringing substantial experience and expertise on SES research approaches to the table. Kristine Maciejewski and Hayley Clements initially joined as post-doctoral researchers in 2017 and 2018 to help with the enormous amount of coordination work involved in liaising with such a large group of contributors. Ultimately, however, all editors contributed significantly, both intellectually and in terms of practical coordination, and invested substantial amounts of time to see this project through. The team worked in a highly collaborative way, and the combination of our different skills and expertise made for many hours of interesting debates and discussions, and has been a major source of learning for us all.

Editorial consistency across such a large endeavour would not be possible without professional support in the final stages. We sincerely thank Marlene Rose for her tireless attention to detail and dedication in copy-editing, coordinating and compiling all the material for final submission. We thank Rosie Campbell and Ronel van Heerden for redrawing the figures, Tessa Botha and Terry Achieng for helping to check all the references, Patricia Rademeyer for obtaining permissions and Cathy Hill for resizing the pictures. The wonderful front cover design is by artist, scholar and activist Dylan McGarry from South Africa. The cover aims to explore the spaces between ecological relationships, and to offer a symbolic representation of the intimate relationality of human and more-than-human worlds.

A variety of funders have provided critical support for this endeavour. The USAID-funded Resilient Waters Program (project 720-674-18-C-00007) is thanked sincerely for covering the open-access publication fee for the handbook. This is a huge contribution that will greatly enhance the accessibility of the handbook, particularly in developing countries. The DST/NRF-funded South African Research Chair (grant 98766) held by Oonsie Biggs co-funded several editorial meetings, the Leuphana workshop in Germany, and the final copy-editing, redrawing of figures, checking of references and figure permissions. We also thank the SIDA-funded GRAID project for co-funding several of the editorial meetings, several workshops in Sweden and the Leuphana workshop in 2019. In addition, most authors leveraged their own travel support to attend the Leuphana workshop, which was critical to making the workshop possible. During the course of this project, Oonsie's salary has been supported by Swedish Vetenskapsrådet (grant 621-2014-5137); the South African Research Chairs Initiative (grant 98766), which also co-funded Kristine and Hayley; the GRAID project, which also funded Rika and later Kristine; and the Resilient Waters Program, which also co-funded Rika and Kristine. Rika's salary was also co-funded by the Sida-funded Swedbio programme, and Hayley's by a Claude Leon postdoctoral fellowship and a Jennifer Ward Oppenheimer Research Grant. Alta's travel expenses were funded in part through Rhodes University Council Grants (2015–2020), and her participation at the Lüneburg workshop by a South African National Research Foundation (NRF) Knowledge Interchange Grant (UID 118246). Maja's salary and participation in several workshops were funded by a European Research Commission grant under the European Union's Horizon 2020 research and innovation programme (grant agreement no. 682472 – MUSES).

Last but not least, we would like to thank our partners and family for their support, especially during some of the more pressurised phases of the book. Without you we may not have had the energy to pull this together!

Reinette (Oonsie) Biggs
Alta de Vos
Rika Preiser
Hayley Clements
Kristine Maciejewski
Maja Schlüter

Credits list

The editors acknowledge with thanks permission granted to use material from the following sources:

Figure 1.2: Berkes, F., C. Folke, and J. Colding. 2002. *Navigating Social-Ecological Systems: Building Resilience for Complexity and Change.* Cambridge University Press (p. 22, Figure 1.4). Reproduced with permission from Cambridge University Press, through PLSClear Licence #40805.

Figure 1.3: Gunderson, L.H. and C.S. Holling. 2002. *Panarchy: Understanding Transformations in Human and Natural Systems.* Edited by Lance H. Gunderson and C.S. Holling. 2002. Island Press (p. 75; Figure 3–10). Reprinted with permission from Island Press through Copyright Clearance Center.

Figure 1.4: Liu, J., W. Yang, and S. Shuxin Li. 2016. 'Framing Ecosystem Services in the Telecoupled Anthropocene.' *Frontiers in Ecology and the Environment* 14(1), 27–36. (Figure 1). Reprinted with permission from John Wiley and Sons through Copyright Clearance Center.

Figure 1.5: Monroy-Sais, S., A. Castillo, E. García-Frapolli, and G. Ibarra-Manríquez. 2016. 'Ecological Variability and Rule-making Processes for Forest Management Institutions: A Social-Ecological Case Study in the Jalisco Coast, Mexico.' *International Journal of the Commons* 10(2): 1144–1171. Copyright: Content is licensed under a Creative Commons Attribution 3.0 License.

Figure 1.6: Anderies, J. M., M.A. Janssen, and E. Ostrom. 2004. 'A Framework to Analyze the Robustness of Social-Ecological Systems from an Institutional Perspective.' *Ecology and Society* 9(1).

Figure 1.7: Schlüter, M., L.J. Haider, S.J. Lade, E. Lindkvist, R. Martin, K. Orach, N. Wijermans, and C. Folke. 2019. 'Capturing Emergent Phenomena in Social-Ecological Systems – An Analytical Framework.' *Ecology and Society* 24(3). doi:10.5751/ES-11012-240311.

Figure 5.1: Baird, J., R. Plummer, and L. Pickering. 2014. 'Priming the Governance System for Climate Change Adaptation: The Application of a Social-Ecological Inventory to Engage Actors in Niagara, Canada.' *Ecology and Society*, 19(1).

Table 7.2: © Abu, R. 2017. 'Knowledge, Use, and Change in the Saskatchewan River Delta: Assessing the Changing Livelihoods of Cumberland House Métis and Cree Nation.' PhD diss., University of Saskatchewan.

Figure 11.1: © Millennium Ecosystem Assessment 2005.

Figure 15.1: Reason, P., G. Coleman, D. Ballard, M. Williams, M. Gearty, C. Bond, C. Seeley, and E. Maughan McLachlan. 2009. *Insider Voices: Human Dimensions of Low Carbon Technology.* Bath: Centre for Action Research in Professional Practice, University of

Bath. http://people.bath.ac.uk/mnspwr. Originally from Reason, P., and H. Bradbury. 2001. 'Inquiry and Participation in Search of a World Worthy of Human Aspiration.' In *Handbook of Action Research: Participative Inquiry and Practice*, edited by P. Reason and H. Bradbury, 1–14. London: Sage.

Figure 15.2: Reason, P., G. Coleman, D. Ballard, M. Williams, M. Gearty, C. Bond, C. Seeley, and E. Maughan McLachlan. *Insider Voices: Human Dimensions of Low Carbon Technology*. Bath: Centre for Action Research in Professional Practice, University of Bath. 2009. www.peterreason.net/Papers/lowcarbon_insider_voices.pdf

Figure 16.1: Kininmonth et al. 2017. Kininmonth, S., B. Crona, Ö. Bodin, I. Vaccaro, L. Chapman, and C. Chapman. 2017. 'Microeconomic Relationships between and among Fishers and Traders Influence the Ability to Respond to Social-Ecological Changes in a Small-Scale Fishery.' *Ecology and Society* 22(2).

Figure 17.1: Reprinted from Daume, S. 2016. 'Mining Twitter to Monitor Invasive Alien Species – An Analytical Framework and Sample Information Topologies.' *Ecological Informatics* 31: 70–82. doi:10.1016/j.ecoinf.2015.11.014. © 2016, with permission from Elsevier through Copyright Clearance Center.

Figure 18.1: Reprinted from Iniesta-Arandia, I., M. García-Llorente, P.A. Aguilera, C. Montes, and B. Martín-López. 2014. 'Socio-cultural Valuation of Ecosystem Services: Uncovering the Links between Values, Drivers of Change, and Human Well-Being.' *Ecological Economics* 108(December): 36–48. doi:10.1016/j.ecolecon.2014.09.028. © 2014, with permission from Elsevier through Copyright Clearance Center.

Figure 19.1: https://pt.m.wikipedia.org/wiki/Ficheiro:Rio_paraíba_do_sul.jpg

Figure 20.1: Reprinted from Pahl-Wostl, C., and C. Knieper. 2014. 'The Capacity of Water Governance to Deal with the Climate Change Adaptation Challenge: Using Fuzzy Set Qualitative Comparative Analysis to Distinguish between Polycentric, Fragmented and Centralized Regimes.' *Global Environmental Change* 29: 139–154. © 2014, with permission from Elsevier through Copyright Clearance Center.

Figure 22.1: Ostrom, E., R. Gardner, and J. Walker. 1994. *Rules, Games and Common Pool Resources*. Ann Arbor: Michigan University Press. (p.37). Reprinted with permission from University of Michigan Press through Copyright Clearance Center.

Figure 22.2: Monroy-Sais, S., A. Castillo, E. García-Frapolli, and G. Ibarra-Manríquez. 2016. 'Ecological Variability and Rule-Making Processes for Forest Management Institutions: A Social-Ecological Case Study in the Jalisco Coast, Mexico.' *International Journal of the Commons* 10(2): 1144–1171. Copyright: Content is licensed under a Creative Commons Attribution 3.0 License.

Figure 23.1: Maciejewski, K., and G. Cumming. (2015). 'The Relevance of Socio-Economic Interactions for the Resilience of Protected Area Networks.' *Ecosphere*, 6(9), 1–14.

Figure 24.3: Reprinted from Dittrich, A., R. Seppelt, T. Václavík, and A.F. Cord. 2017. 'Integrating Ecosystem Service Bundles and Socio-environmental Conditions – A National Scale Analysis from Germany.' *Ecosystem Services* 28: 273–282 (Figure 4). © 2017, with permission from Elsevier through Copyright Clearance Center.

Figure 26.1: Lade S.J., S. Niiranen, J. Hentati-Sundberg, T. Blenckner, W. Boonstra, K. Orach, M. Quaas, H. Österblom and M. Schlüter. 2015. 'An Empirical Model of the Baltic Sea Reveals the Importance of Social Dynamics for Ecological Regime Shifts.' *Proceedings of the National Academy of Sciences* 112(35): 11120–11125.

Figure 27.1: Densambuu, B., T. Indree, A. Battur, and S. Sainnemekh. 2018. 'State and Transition Models of Mongolian Rangelands.' Agency for Land Management, Geodesy, and Cartography, Ulaanbaatar, Mongolia. © Administration of Land affairs, Geodesy & Cartography, Mongolia.

Figures 28.1 and 28.2: Lindkvist, E., X. Basurto, and M. Schlüter. 2017. 'Micro-level Explanations for Emergent Patterns of Self-governance Arrangements in Small-scale Fisheries – A Modeling Approach'. *PloS ONE* 12(4): e0175532. journals.plos.org/plosone/article?id=10.1371/journal.pone.0175532. Licensed by the respective authors for use and distribution subject to citation of the original source in accordance with the Creative Commons Attribution (CC BY) license. Copyright: © 2017 Lindkvist et al.

Figure 29.1: © Eric Booth 2014 Photo: Eric Booth for the Water Sustainability and Climate Project, under a Creative Commons license. https://wsc.limnology.wisc.edu

Figure 30.1: Reprinted from Deutsch, L., S. Gräslund, C. Folke, M. Huitric, N. Kautsky, M. Troell, and L. Lebel. 2007. 'Feeding Aquaculture Growth through Globalization: Exploitation of Marine Ecosystems for Fishmeal.' *Global Environmental Change* 17(2): 238–249. doi:10.1016/j.gloenvcha.2006.08.004. © 2007. Reprinted with permission from Elsevier through Copyright Clearance Center

Figure 31.1C: Johnson, J.A., S.K. Jones, S.L.R. Wood, R. Chaplin-Kramer, P.L. Hawthorne, M. Mulligan, D. Pennington, and F.A. DeClerck. 2019. 'Mapping Ecosystem Services to Human Well-being: A Toolkit to Support Integrated Landscape Management for the SDGs.' *Ecological Applications* 29(8): e01985. doi:10.1002/eap.1985. Reprinted with permission from John Wiley and Sons through Copyright Clearance Center.

Abbreviations and acronyms

AI	Artificial intelligence
CAS	Complex adaptive systems
EE-MRIO	Environmentally extended multi-regional input-output
EROI	Energy return on investment
GCC	Global commodity chain
GDP	Gross domestic product
GIS	Geographic information system
GPS	Global positioning system
GUI	Graphical user interface
IAD	Institutional analysis and development
ICT	Information and communications technology
LCA	Life Cycle Assessment
MA	Millennium Ecosystem Assessment
MAD	Mutually assured diversity
MEFA	Material- and energy-flow accounting
MEL	Monitoring, evaluation and learning
MRIO	Multi-regional input-output
MuSIASEM	Multi-scale integrated analysis of societal and ecosystem metabolism
NGO	Non-governmental organisation
PTF	Physical trade flow
SES	Social-ecological system(s)
SESMAD	Social-Ecological Systems Meta-Analysis Database

Part 1

1

What are social-ecological systems and social-ecological systems research?

Reinette Biggs,[1,2] Hayley Clements,[1] Alta de Vos,[3] Carl Folke,[2] Amanda Manyani,[1] Kristine Maciejewski,[1] Berta Martín-López,[4] Rika Preiser,[1] Odirilwe Selomane[1] and Maja Schlüter[2]

[1]CENTRE FOR SUSTAINABILITY TRANSITIONS, STELLENBOSCH UNIVERSITY, STELLENBOSCH, SOUTH AFRICA
[2]STOCKHOLM RESILIENCE CENTRE, STOCKHOLM UNIVERSITY, STOCKHOLM, SWEDEN
[3]DEPARTMENT OF ENVIRONMENTAL SCIENCE, RHODES UNIVERSITY, MAKHANDA, SOUTH AFRICA
[4]SOCIAL-ECOLOGICAL SYSTEMS INSTITUTE, FACULTY OF SUSTAINABILITY, LEUPHANA UNIVERSITY, LÜNEBURG, GERMANY

The Anthropocene: the challenge of understanding the world in new ways

The period since the Second World War has been marked by rapid and accelerating changes to many aspects of human society and the environment (Clark, Crutzen, and Schellnhuber 2004; Steffen et al. 2011; Steffen et al. 2015a). There is accumulating evidence and rising concern about the potential consequences these changes hold for key Earth system processes at a global scale, and human well-being and prosperity into the future (Krausmann et al. 2013; Steffen et al. 2015b). The Anthropocene, as this new era of extensive human impact on the Earth has come to be known (Crutzen 2006), manifests in a closely intertwined set of social and ecological changes. Technological advances, increasing human population, rising levels of wealth and consumption, and the institutional arrangements we have developed to govern our economies and societies interplay with one another, and drastically affect the Earth's climate, biological diversity, freshwater and biogeochemical flows, and levels of novel pollutants in the environment (Steffen et al. 2015a). These environmental changes, in turn, contribute to increasingly frequent and severe droughts (Dai 2013; Trenberth et al. 2014), floods (Milly et al. 2002; Nicholls 2004), heatwaves (Guo et al. 2018; Oliver et al. 2018) and the emergence of novel pathogens such as SARS-CoV-2 (Everard et al. 2020; O'Callaghan-Gordo and Antò 2020; Schmeller, Courchamp, and Killeen 2020) that can lead to massive societal disruption and hardship, especially among the poor (Wheeler and Von Braun 2013; Barbier and Hochard 2018).

The intertwined social and ecological changes that underlie the Anthropocene are further reflected in a world that has become highly connected through technology and trade (Green et al. 2019; Keys et al. 2019; Nyström et al. 2019). Nowadays, it is difficult to keep track of the geographic origin of our food, or to account for the various components making up the mobile

DOI: 10.4324/9781003021339-2

phones we use every day. While this connectivity has brought about impressive improvements to many people in terms of the distribution of food and other resources around the world, it has also resulted in conglomerations of markets and resources, making it difficult to trace and hold accountable those polluting rivers and degrading ecosystems. Large and often geographically distant supply chains of resources have increased access to and human consumption of many goods, but simultaneously have had devastating consequences for biodiversity and species habitats, without consumers feeling accountable for or being aware of these impacts (Lenzen et al. 2012; Wilting et al. 2017; Liu et al. 2018). Vast and globally extensive supply chains have also contributed to widespread inequalities between and within countries (Costinot, Vogel, and Wang 2012; Galaz et al. 2018). This multi-dimensional connectivity also means that decisions in one country or part of the world can have far-reaching consequences for other places or countries – economically, socially and ecologically. Small-scale fishers, for example, are now often directly and indirectly connected to distant markets, causing them to be more vulnerable to seemingly unrelated threats and disturbances, such as economic changes in distant economies (Crona et al. 2015; Stoll et al. 2018). Similarly, the interdependence of countries in food supply reduces resilience and increases vulnerability as supply chains are broken (Kummu et al. 2020).

The pressing environmental and social sustainability challenges we face in the 21st century are clearly deeply intertwined. These challenges result from the confluence and interaction of multiple, mutually reinforcing social and ecological processes at multiple scales (Folke et al. 2016), where social processes include economic, political, cultural and technological processes, and ecological processes include biotic (e.g. population dynamics, food web interactions) and abiotic (e.g. nutrient flows, climate patterns) processes. The climate emergency and other environmental changes are underlain by a complex, interacting array of social changes, which themselves are shaped by the environment and environmental disruptions. Similarly, problems of poverty and inequality are often linked to and exacerbated by environmental change and disruption. Ethiopia, for example, has become one of the most food-insecure nations in the world due to complex interactions between environmental degradation, diminishing land holdings, outbreaks of crop and livestock disease, poor infrastructure, political insecurity, and pre- and post-harvest crop losses that have systematically eroded the productive assets of households (Mohamed 2017; Bahru et al. 2019). Factors outside a country also play a role in perpetuating food insecurity in that country, such as discourses about how to address these problems driven by notions of intensification, commercialisation (Jiren et al. 2020) and land acquisition by other countries for their own benefit (e.g. Hules and Singh 2017). The key sustainability challenges of the 21st century cannot be addressed without recognising the systemic, intertwined nature of these problems (Liu et al. 2015).

The recognition that environmental and social sustainability challenges are inherently systemic and intertwined, and the escalating urgency to address these challenges, have driven a paradigm shift in how social and natural systems are studied (Schoon and Van der Leeuw 2015). In most scientific disciplines, humans and nature have been treated as separate entities (Folke et al. 2016). Ecology, for example, has often viewed social systems only as external drivers of ecosystem dynamics (Carpenter et al. 2012; Cumming 2014), whereas economics and other social sciences have considered natural systems simply as resources for extracting capital gains or providing a basis for livelihoods (Gunderson and Holling 2002; Berkes, Colding, and Folke 2003). In recent decades, however, this thinking has been widely contested and is changing, partly influenced by the rise in systems sciences and complexity thinking (see Chapter 2; Preiser et al. 2018). Scholars in different disciplines are increasingly viewing human systems as interdependent, inseparable and intertwined with ecosystems, embedded within and dependent upon the biosphere and the broader Earth system (Folke et al. 2016; Reyers et al. 2018; Schlüter et al.

2019). Furthermore, there is growing recognition of the need for knowledge production processes that account for and engage with the complex interconnections and interplay between the social and the ecological, and the emergent and often unexpected processes, features, problems and opportunities to which they give rise (Preiser et al. 2018).

What are social-ecological systems?

'Social-ecological systems' (SES) is an emerging concept for understanding the intertwined nature of human and natural systems in this new, interconnected and interdependent way. The SES concept developed in the early to mid-1990s through collaboration of scholars working in the interdisciplinary areas of ecological economics and common-pool resource systems (e.g. Berkes 1989; Ostrom 1990; Costanza 1991). Specifically, the volume *Linking Social and Ecological Systems: Management Practices and Social Mechanisms for Building Resilience* combined a systems approach and adaptive management with a focus on dynamic institutions and diverse systems of property rights, with 14 case studies analysing ecological resilience and local and traditional systems engaged in ecosystem management (Berkes and Folke 1998). The concept of SES is based on the notion that 'the delineation between social and natural systems is artificial and arbitrary' (Berkes and Folke 1998), emphasising that people and nature are intertwined. Nature no longer merely sets the space in which social interactions take place; likewise, people are not just an external driver in ecosystem dynamics (Folke et al. 2011; Schoon and Van der Leeuw 2015). Social-ecological systems are therefore not merely social plus ecological systems, but cohesive, integrated systems characterised by strong connections and feedbacks within and between social and ecological components that determine their overall dynamics (Folke et al. 2010; Biggs, Schlüter, and Schoon 2015).

As such, SES are a type of complex adaptive system. These systems comprise many interdependent parts that interact in ways that give rise to emergent, system-wide patterns that cannot be predicted from the properties of the individual system components. Furthermore, these system-wide patterns, in turn, influence the behaviour of the individual system parts and their interactions with other parts, creating a feedback process that shapes the evolution of the system over time and allows it to adapt to changing contexts (Lansing 2003). The continuous interplay between microlevel entities to form emergent macrolevel patterns 'implies that SES are more than the sum of the ecological or the social parts' (Reyers et al. 2018). Furthermore, it means that SES can adapt to changing conditions, learning and self-organising in response to internal or external pressure (Levin et al. 2013). An example of these dynamics is the emergence of adaptive governance, where individuals interact and collaborate, often in response to a crisis, connecting and creating social networks around shared visions and narratives (Folke et al. 2005). As a result, bridging organisations and new institutions emerge and become connected to other levels of governance, influencing them, but also being influenced by them. It has been shown that an entire SES may shift and start to evolve new pathways as a result of this interplay. Examples range from landscape management in Sweden, to large-scale coral reef management in Australia, to a system of global adaptive governance of the regional resources of the Southern Ocean (Schultz et al. 2015).

A recent review by Preiser et al. (2018) identifies six organising principles of complex adaptive systems that help to further inform our understanding of the nature of SES. The first is that such systems are constituted relationally, i.e. the relations and interactions between the components of the system are more important to understanding the properties and behaviour of an SES than the properties of the individual components of the system themselves. This recognition highlights the need to shift from a traditional reductionist scientific approach, which aims to understand a system by breaking it down into its component parts, to a systems-based approach that focuses on system interactions rather than system components (see Chapter 2).

The second principle is that SES have adaptive capacities. The many interrelations in the system create feedback processes that enable an SES to continuously adjust and adapt to changing conditions, brought about either by the system itself or by external forces. Through this process of adaptive change, unique trajectories of development emerge that contain specific historical legacies. These legacies and 'memories', in turn, constrain and shape future development options and possibilities.

The third critical feature of SES is that the dynamic interactions within the system are often non-linear, meaning that small changes can lead to large and surprising effects, or vice versa (Levin et al. 2013). This behaviour is caused by feedback loops that either dampen or amplify system changes and perturbations, and can trigger regime shifts – large, persistent, and often sudden and unexpected reorganisation in the structure and functioning of an SES, such as soil salinisation, ice-sheet collapse or a shift from collaborative institutions to regulate use of common-pool resources to overharvesting (Scheffer et al. 2001; Lade et al. 2013; Biggs, Peterson, and Rocha 2018). Changes leading to regime shifts, and different regime shifts themselves, are often connected across scales, and can lead to cascading regime shifts at different scales (Rocha et al. 2018). Similar non-linear processes underlie transformations – intentional actions that aim to trigger fundamental reorganisation of an SES to create more sustainable and equitable outcomes (Olsson, Galaz, and Boonstra 2014). Enabling such transitions usually involves working at multiple scales to weaken dominant relationships and structures in a system, while at the same time developing new 'shadow' networks and processes that can take their place when a crisis or opportunity for change emerges (Olsson et al. 2006; Pereira et al. 2018).

The fourth feature is that SES do not have clear boundaries. Due to extensive interactions and connections between an SES and its broader environment, it is very difficult to discern which components belong inside the system and which belong to the broader environment. Deciding on system boundaries therefore often depends on the purpose of the study and the perspective of the observer (Cilliers 2001). Linked to this is the fifth feature, namely that SES are context dependent. As the context changes, the system will change and elements in the system may take on a different role or function. Many SES, for example, have through learning and experience developed strategies and institutions that are dormant but that can easily be revived when the context changes, such as in situations of resource scarcity or shocks and stresses (Berkes and Folke 1998).

Finally, SES are characterised by complex causality and emergence. Cause and effect in SES are not unidirectional or linear, but are marked by complex recursive causal pathways. Social-ecological systems therefore cannot be understood nor can their behaviour be predicted based solely on information relating to their individual parts. Many emergent system properties are inherently unpredictable as they involve non-linear effects, learning, evolution, novelty and innovation. Although SES can be influenced, and aspects of these systems can be understood and navigated, these features make prediction and control of SES very difficult, if not impossible.

The recognition that social and ecological systems are inseparable, and function as intertwined complex adaptive systems, offers researchers, policymakers and scholars an alternative entry or viewpoint for studying and engaging with the complex challenges that arise from human–nature interactions (Binder et al. 2013; Preiser et al. 2018; Reyers et al. 2018). In particular, it shifts the focus to understanding how macrolevel system properties emerge from the interactions of microlevel entities and their external environment, rather than separating social and ecological components and studying them in isolation (Levin et al. 2013; Preiser et al. 2018). It also emphasises the dynamic interplay of rapid and gradual change (fast and slow variables), and the critical importance of multi- and cross-scale interactions across space and time in forging different pathways of change and options for the future (Gunderson and Holling 2002; Adger, Arnell and Tompkins 2005; Cash et al. 2006). Finally, the properties of complex adaptive systems emphasise the need to expect and embrace surprise and uncertainty, to be reflexive and acknowledge the limits of what is knowable or controllable in SES. Chapter 2 discusses the properties of complex adaptive systems and their implications for SES research in further detail.

What is SES research?

Understanding SES as complex adaptive systems has profoundly shaped the development of SES research (Levin et al. 2013; Schoon and Van der Leeuw 2015; Preiser et al. 2018; Hertz, Mancilla García, and Schlüter 2020; Mancilla García et al. 2020). Early work on SES (Berkes and Folke 1998) was inspired by insights on people–nature interdependence from anthropology, ecology and geography, among others (e.g. Holling 1973; Bateson 1979; Clark and Munn 1986; Odum 1989; Gunderson, Holling, and Light 1995; Levin 1999; Davidson-Hunt and Berkes 2003). Since the concept of SES was translated into a framework for researching intertwined systems of people and nature (Berkes and Folke 1998), over 13 000 papers have been published on SES, predominantly drawing on the environmental and social sciences, economics and, to a lesser extent, medicine, psychology, and the arts and humanities (Colding and Barthel 2019). Social-ecological systems research now represents a recognised interdisciplinary area (e.g. Ostrom 2009; Colding and Barthel 2019) of sustainability science (Clark and Harley 2020). In several dimensions, it is similar to and overlaps with other people–nature approaches like human–environment systems covering land-system change, vulnerability or environmental literacy (e.g. Turner et al. 1990; Lambin et al. 2001; Turner, Lambin, and Reenberg 2007; Lambin and Meyfroidt 2011; Scholz 2011); coupled human and natural systems (CHANS) and the telecoupling framework (Liu et al. 2007b, 2013); or socio-natural systems that emerged from archaeology (Van der Leeuw 2019). However, a distinguishing feature of the SES approach is its conceptual emphasis on SES as complex adaptive systems (Folke et al. 2016), although the extent to which this is operationalised in empirical research varies.

Social-ecological systems research is largely problem-oriented, with a strong focus on informing sustainability policy and practice (Fischer et al. 2015; Folke et al. 2016), playing a particularly significant part in the development of resilience and adaptive governance approaches (e.g. Gunderson and Holling 2002; Berkes, Colding, and Folke 2003; Dietz, Ostrom, and Stern 2003; Folke et al. 2005; Walker and Salt 2006; Biggs, Schlüter, and Schoon 2015). Notable increases in SES research occurred following calls to move towards 'sustainable development of the Biosphere', and to integrate SES within broader sustainability initiatives, such as the Sustainable Development Goals and Future Earth (Herrero-Jáuregui et al. 2018; De Vos, Biggs, and Preiser 2019). Recent reviews find SES research to be focused on pressing sustainability issues such as climate change, biodiversity loss, livelihoods, poverty, policy, land-use change, water, and social and environmental justice (Herrero-Jáuregui et al. 2018; De Vos, Biggs, and Preiser 2019), and frequently used terms in SES publications include 'policy', 'trade', 'conservation', 'adaptation', 'land-use change', 'water', 'forest loss', 'sustainability', 'urban', 'governance' and 'institutions' (De Vos, Biggs, and Preiser 2019). An analysis of the semantic networks resulting from SES research shows that terms such as 'solution' and 'transformation' are placed towards the centre of the network, being essential to connecting different types of studies (Horcea-Milcu et al. 2020). Baggio, Brown and Hellebrandt (2015), in their citation network analysis of resilience, a concept often closely linked to that of SES, found that

> the social-ecological systems field stands out as an emerging interdisciplinary arena where resilience can effectively act as a bridging concept and facilitate a discussion of

dynamics of complex systems within varied contexts, informed by diverse perspectives, to provide potentially innovative theoretical and applied insights.

Early SES research consisted largely of place-based studies of local resource management (Ostrom 1990; Folke and Berkes 1998). Twenty years on, SES research is still often focused on local-scale studies (Norberg and Cumming 2008; Colding and Barthel 2019; De Vos, Biggs, and Preiser 2019), although network approaches and approaches that capture cross-scale dynamics connecting local, regional and global processes are becoming more common (Galaz et al. 2011; Österblom and Folke 2013; Crona et al. 2015; Cumming et al. 2015; Bodin 2017; Rocha et al. 2018; Selomane et al. 2019). There is also an increasing emphasis on the need for coordination and integration between case studies to enable comparison, synthesis and theory development in SES (Ostrom 2009; Cox 2014; Gurney et al. 2019; Cumming et al. 2020). Several large databases now exist to collect case studies for meta-analyses (Cox 2014; Biggs, Peterson, and Rocha 2018; Partelow 2018), and meta-analysis and large- or cross-scale SES research are increasingly being undertaken (e.g. Cox, Arnold, and Villamayor-Tomas 2010; Hamann, Biggs, and Reyers 2015; Cinner et al. 2016; Ban et al. 2019).

While a focus on the interactions between people and nature is at the core of SES research, the degree to which the 'social' and the 'ecological' are researched as part of a single integrated system varies among studies (Schlüter et al. 2019). Much SES research is still focused either more on the social or more on the ecological elements (Binder et al. 2013; Epstein et al. 2013; Schlüter et al. 2019), and still often investigates one-way links between social and ecological elements, such as the human drivers of ecosystem dynamics or the benefits nature provides to people (Binder et al. 2013; Schlüter et al. 2019). There is a growing recognition, however, that people and nature are interdependent and coevolving, through multiple interactions or feedbacks. Ecosystem services, for example, are increasingly seen as co-produced by both people and nature (Reyers et al. 2013; Palomo et al. 2016); human behaviour and individual and social identities are increasingly understood as relationally constructed and coevolving with the biophysical context (Díaz et al. 2015; Chan et al. 2016; Schill et al. 2019); and the interactions between human well-being or inequality and ecosystems are increasingly recognised as dynamic and reciprocal (Hamann et al. 2018; Masterson et al. 2019). Consequently, research is shifting from focusing on social and ecological elements, to social-ecological relations as the key to SES dynamics (Schlüter et al. 2019; Hertz, Mancilla García, and Schlüter 2020; Mancilla García et al. 2020). Efforts to better capture the dynamic nature of SES have led to an evolution of the original notion of linking social and ecological systems (Berkes and Folke 1998), to research on the 'intertwined' nature of SES (Folke et al. 2016; Schlüter et al. 2019). These changes are evident in the steady increase of articles studying SES as complex adaptive systems (Liu et al. 2007a; Levin et al. 2013), and frameworks for understanding SES (Binder et al. 2013; Schlüter et al. 2019).

The interdisciplinary nature of SES research, its growing need to go beyond the 'sum' of social and ecological research, and the need to focus on cross-scale systemic dynamics have resulted in a high degree of methodological pluralism (De Vos, Biggs, and Preiser 2019). Many methods and approaches used in SES research go beyond traditional social and natural sciences, or any single discipline (Tengö et al. 2014; Folke et al. 2016; De Vos, Biggs, and Preiser 2019). Social-ecological systems research often requires adapting methods, or implementing a sequence of methods, such that they capture both the social and the ecological domains, and their dynamic interdependencies (Fischer et al. 2015; De Vos, Biggs, and Preiser 2019). This has given rise to a variety of models, policies and methods for the practical application of SES research (Rogers et al. 2013; Preiser et al. 2018; Reyers et al. 2018).

Furthermore, given the problem-oriented nature of SES research, there is growing recognition that the application of SES research to policy and practice requires collaborations between researchers and practitioners holding multiple types of knowledge, an approach known as transdisciplinarity (Mauser et al. 2013; Roux et al. 2017).

As a consequence of its inter- and transdisciplinary nature, SES research is predominantly conducted in teams. Student projects are often embedded within larger research projects, which typically involve multiple SES researchers, often with different disciplinary backgrounds (Kelly et al. 2019). Social-ecological systems project teams may also often involve practitioners and various stakeholders relevant to the particular research. This context means that SES research tends to be highly collaborative, requiring particular skills and approaches, as further discussed in Chapter 3. In particular, various knowledge co-production and participatory approaches increasingly characterise transdisciplinary SES research (Gurney et al. 2019) and highlight the need for agile collaboration between SES research and practice, and the development of stronger science–society interfaces to guide research, knowledge co-creation and decision-making (Tengö et al. 2014; Fischer et al. 2015; Reyers et al. 2018).

SES frameworks

All research fields are implicitly or explicitly based on certain conceptions about the nature of the world that they study. Social-ecological systems research as discussed above and throughout this book is motivated by the desire to address pressing sustainability issues facing society, and is framed by an approach grounded in an understanding of SES as complex adaptive systems, where people and nature are intertwined and coevolve (Figure 1.1) (also see Chapter 2). This broad framing underlies different areas of SES research, such as adaptive governance, resilience assessment and transformations towards sustainability. Within these research areas, SES researchers often use one or more frameworks that specify a particular way of conceptualising SES and guide researchers towards the elements, relations and processes of an SES that are considered relevant for a given problem or research question. Ostrom's SES framework (Ostrom 2007), for instance, is a key framework used in the study of common-pool resource management. The different research areas and their frameworks, and the research questions associated with them, in turn, inform the use of particular methods from the diverse array of methods employed in SES research (also see Chapter 3 for further discussion).

Frameworks play a particularly important role in SES research. The SES field does not build on a set of well-established 'laws' or theories that are tested in particular cases; instead, studies are largely guided by the overarching SES approach and various frameworks linked to particular areas of SES research.

Frameworks used in SES research draw on a diverse set of perspectives and theoretical commitments, often originating in a variety of different disciplines. The main purpose of these frameworks is to identify, categorise and organise those factors deemed most relevant to understanding a particular phenomenon (McGinnis 2011). Frameworks aim to guide an investigation or activity by pointing to the concepts, elements, variables, links or processes of an SES that are characteristic or critical, or that help explain or predict particular SES outcomes (e.g. institutional arrangements that facilitate governance for sustainability) or that help affect system change.

Beyond this generic aim, however, purposes and forms of frameworks vary widely. These range from descriptive (conceptual frameworks), to analytical/explanatory (analytical

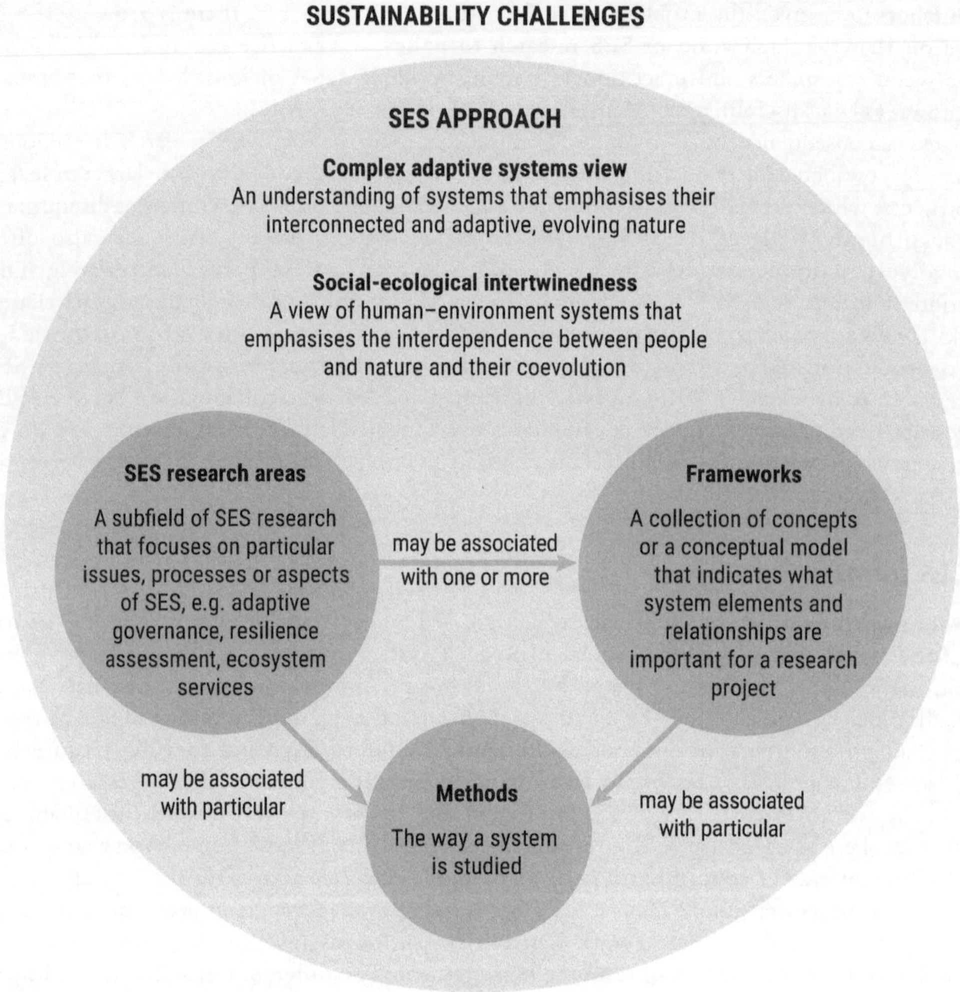

Figure 1.1 The SES approach discussed in this book is grounded in an understanding of SES as intertwined, complex adaptive systems embedded in the biosphere. This understanding directly informs the frameworks and methods employed in different areas of SES research. (© Reinette Biggs)

frameworks), to serving as boundary objects for interdisciplinary collaboration or heuristics for problem solving. In inter- and transdisciplinary processes, frameworks can help bring together insights from different disciplines or highlight incompatibilities between different worldviews. When used as a tool to facilitate collaboration, the process of developing the framework is as important as or even more important than the framework itself. Identifying the elements of the framework and the links between them helps make explicit the assumptions and views of participants about what constitutes the system and how to study it. This increases mutual understanding and helps to develop a shared vocabulary for the study of the problem of interest. It can also facilitate narrowing down an investigation or activity into a feasible endeavour.

Some of the most common frameworks currently in use to study and analyse SES include: the original conceptual framework of linked SES developed by Folke and Berkes (1998) (Figure 1.2); the Panarchy framework depicting system resilience as an outcome of

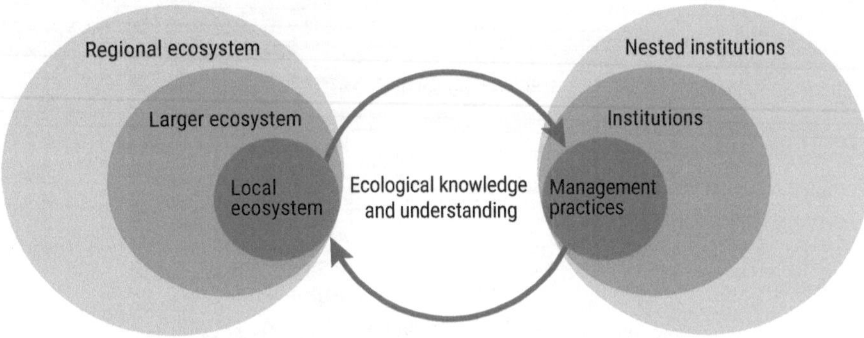

Figure 1.2 Linked SES framework (Folke et al. 2002)

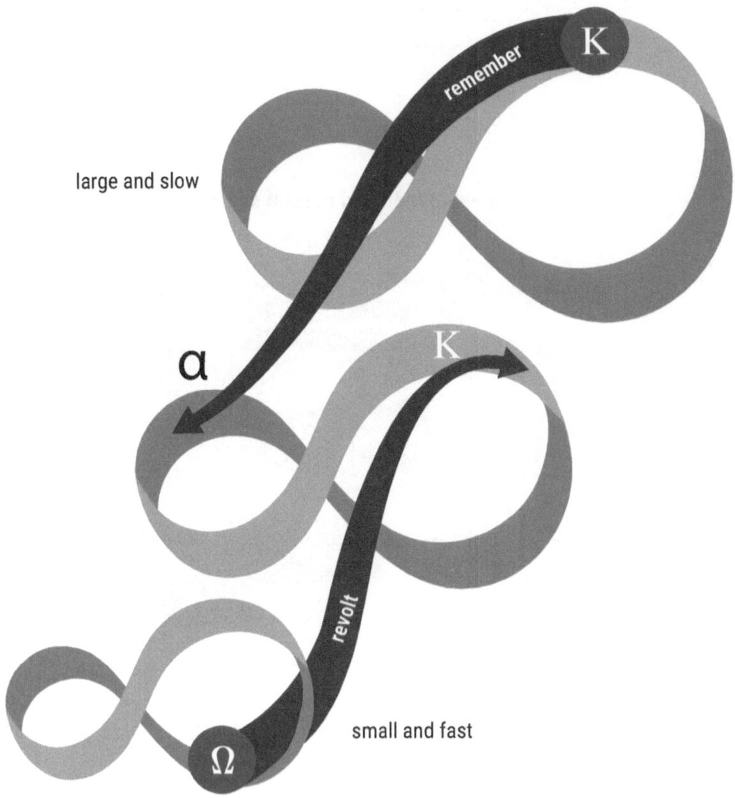

Figure 1.3 The Panarchy framework (Gunderson and Holling 2002)

connected adaptive cycles at different scales (Gunderson and Holling 2002) (Figure 1.3); the telecoupling framework that builds on the CHANS framework, developed by Liu, Yang and Li (2016) (Figure 1.4); the diagnostic framework developed by Ostrom (2007, 2009) to analyse common-pool resource systems (Figure 1.5); the diagnostic framework developed by Anderies, Janssen and Ostrom (2004) to analyse the robustness of SES using institutional analysis (Figure 1.6); and the social-ecological action situation (SE-AS) framework developed

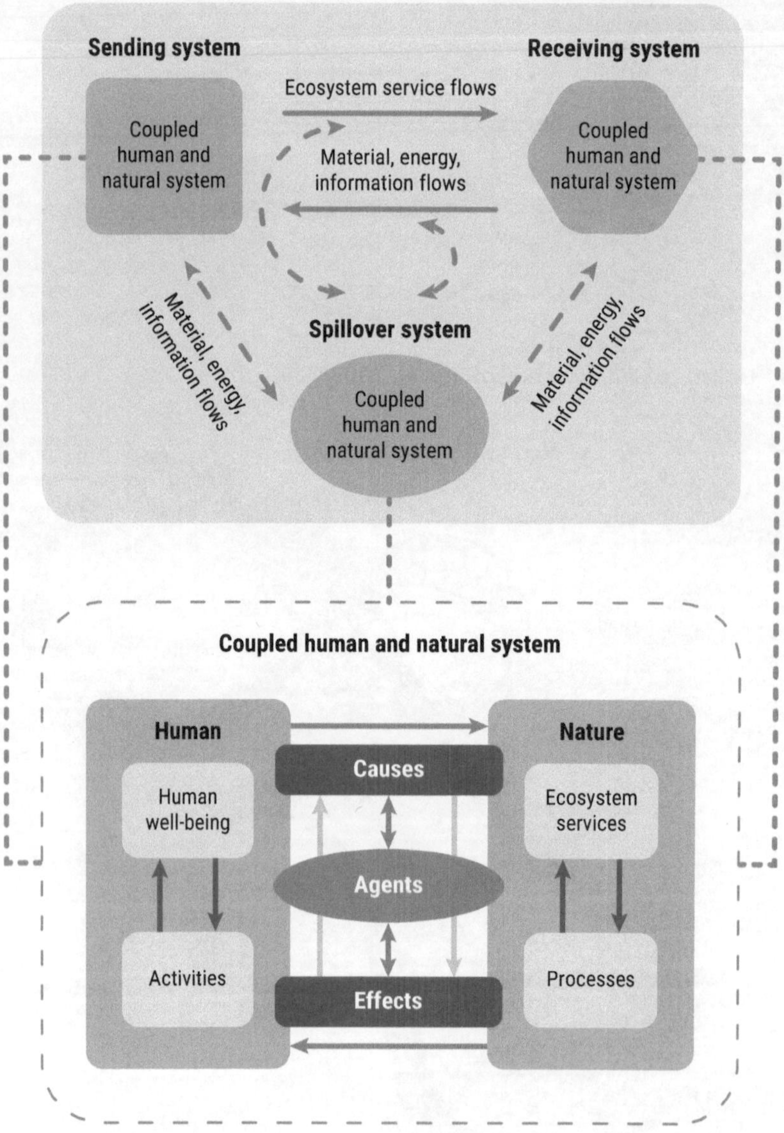

Figure 1.4 The telecoupling framework (Liu, Yang, and Li 2016)

by Schlüter et al. (2019) to analyse the emergence of social-ecological phenomena from social-ecological interactions (Figure 1.7).

The original SES framework (Berkes and Folke 1998) (Figure 1.2) is a conceptual framework that aims to inform the study of local resource management practices and outcomes, and emphasises the links between social and ecological systems and their multi-scale nature. It highlights relations on a highly abstract level, such as the embeddedness of management practices in nested institutions and the links between this multi-level social system and nested ecosystems at different scales. The connection between ecological and institutional systems relies on the ecological knowledge held by local users. The framework focuses on local

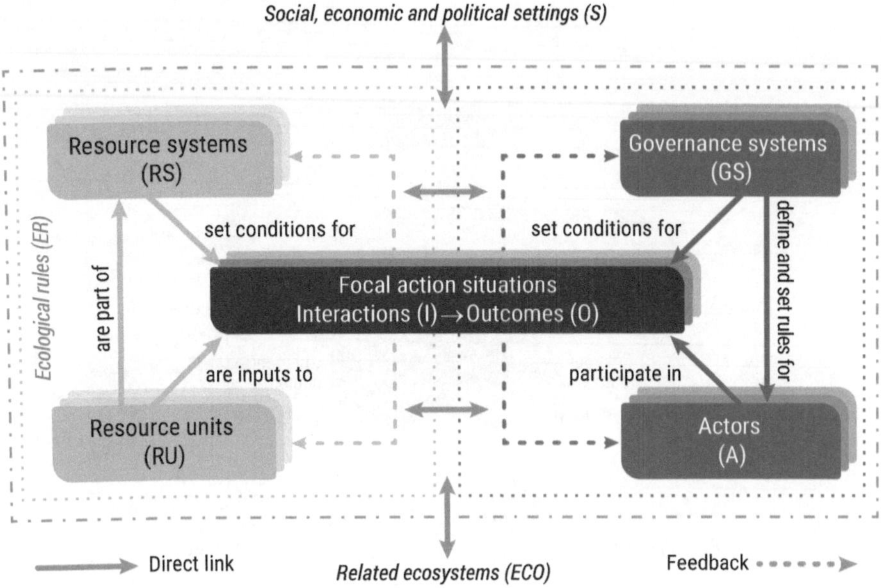

Figure 1.5 Ostrom's SES framework (Ostrom 2007, 2009)

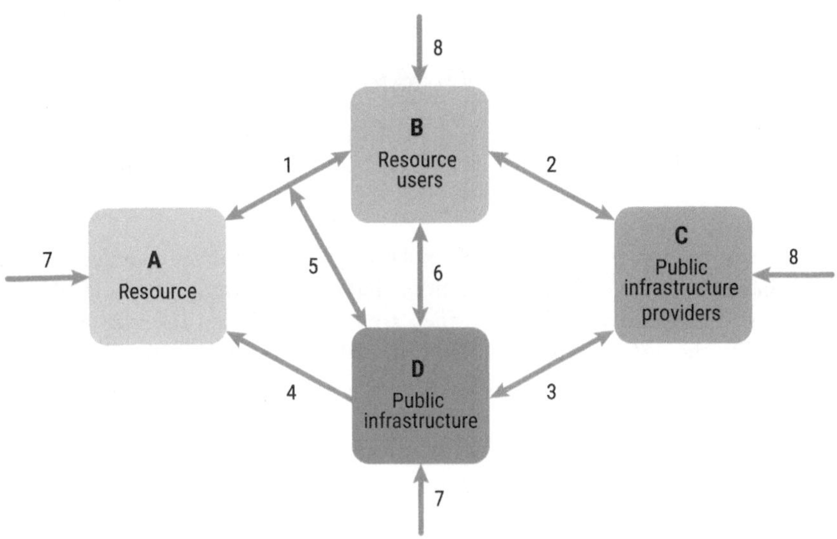

Figure 1.6 The robustness framework (Anderies, Janssen, and Ostrom 2004)

management systems that are able to maintain institutional and ecological resilience in local settings, rather than systems dominated by top-down, conventional, command-and-control resource management (Colding and Barthel 2019). The framework continues to guide conceptual thinking, and is also used in combination with other frameworks such as those focusing on characteristics of resilience (e.g. Galappaththi et al. 2019).

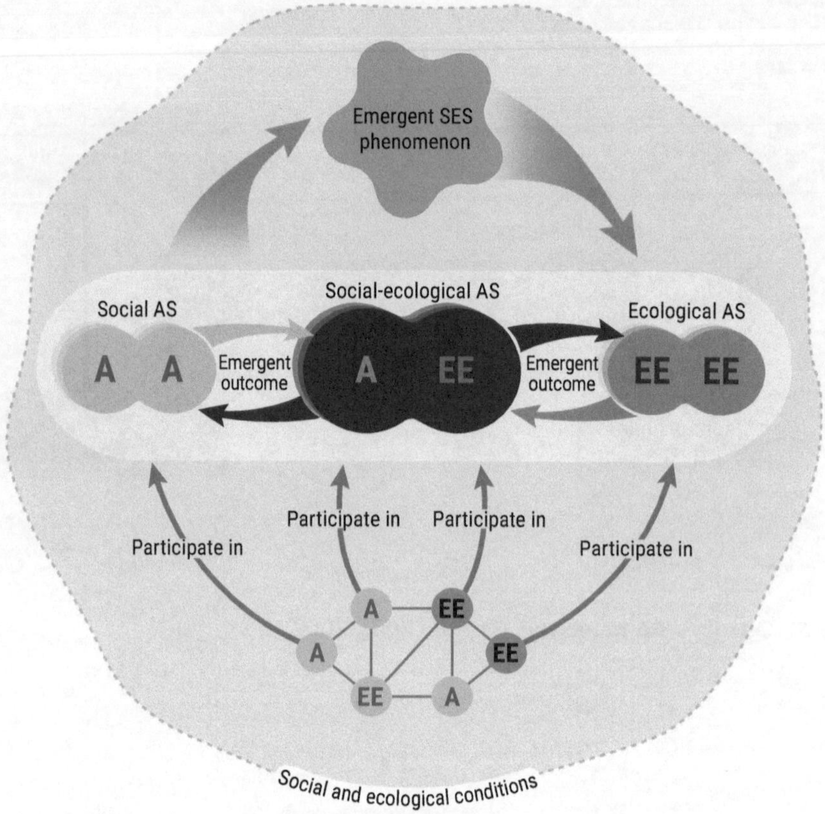

Figure 1.7 The social-ecological action situation (SE-AS) framework (Schlüter et al. 2019) (A: individual and collective actors; EE: ecological elements)

The Panarchy framework (Figure 1.3) explains system resilience as a function of adaptation and change over time, emphasising how changes in a system at one level are affected by the larger-scale systems within which they are embedded, and the smaller-scale systems embedded within them (Gunderson and Holling 2002). The framework is based on a looped 'adaptive cycle' that comprises four stages of change: growth, conservation (consolidation of connections), collapse (creative destruction) and reorganisation (Gunderson and Holling 2002). Panarchy consists of multiple interlinked adaptive cycles at different scales. At each scale, the adaptive cycle operates at different speeds. Larger-scale systems tend to have large, slow cycles that set the conditions for smaller, faster cycles of systems at smaller scales (e.g. in an ecosystem, the interactions between plants and animals, and the species that live there, are determined by climatic conditions and evolution). The faster levels tend to invent, experiment and test, whereas the slower levels stabilise, accumulating knowledge of the past (Folke 2006). In this way, systems develop and adapt to new conditions via a process of creating, testing and maintaining opportunity at one scale, buffered from collapse and regime shifts by the 'slower' processes of change at larger scales (e.g. seedbanks and evolved relationships, reorganised after each disturbance). In recent years, the framework has been used, among other applications, to understand small-island nation recovery and response to extreme weather events (Holdschlag and Ratter 2016), the interplay between legal systems, ecological resilience and adaptive governance in rapidly changing regional water systems (Gunderson et al. 2017; Cosens, Gunderson, and Chaffin 2018), how cross-scale connections can have an impact on social-ecological community resilience (Berkes and Ross 2016) and the spread of the Plague in Europe in the 14th century (Geobey and McGowan 2019).

The 'coupled human and natural systems' (CHANS) framework (Liu et al. 2007a) has a similar descriptive purpose to the Berkes and Folke (1998) framework, although it does not have a standard graphical depiction. Rather, the CHANS framework is associated with several different depictions highlighting that people and nature interact reciprocally across multiple spatial, temporal and organisational scales. The CHANS framework explicitly recognises the relevance of feedbacks, non-linearities, thresholds, time lags, legacy effects, path dependence and emergent properties in shaping human–nature interactions (Liu et al. 2007a).

The telecoupling framework (Figure 1.4) builds on the CHANS framework to incorporate flows of finances, information, energy, goods, organisms and other flows in an integrated way, often between two SES (Hull and Liu 2018) and is evolving to include a third SES, referred to as the 'spillover system' (Liu et al. 2018). The framework's recent applications include uncovering impacts of the following: trade (Friis and Nielsen 2017), telecoupled connections on smallholder farmers (Zimmerer, Lambin, and Vanek 2018), soy demands from China on Amazon forest degradation (Sun, Tong, and Liu 2017), biodiversity conservation (Carrasco et al. 2017) and many others.

Ostrom's SES framework (Ostrom 2007) (Figure 1.5) was developed as an explanatory framework for diagnosing common-pool resource management problems from an institutional and resilience perspective. It is based on the institutional analysis and development (IAD) framework (Ostrom 1990) and 30 years of empirical research on the ability of communities to manage their natural resources sustainably without top-down government regulation. The framework is a collection of social and ecological variables that have proven to be relevant for explaining or predicting when resource users can successfully self-organise to sustainably manage their common-pool resources. Its main unit of analysis is the action situation, which is a social interaction context where resource users interact with one another to produce outcomes (interactions and outcomes) enabled and constrained by rules, ecological settings and attributes of the community. The variables in the framework are organised in four high-level tiers: the resource system, the resource, the users and the governance system. Variables in each tier can be further specified into lower tiers, increasing the specificity of an analysis but at the same time making comparison more difficult. The framework serves two main aims: (a) to guide empirical data collection and analysis by pointing the analyst to those SES variables that may be important for explaining self-organisation and collective action, and (b) to provide a shared vocabulary (i.e. variables) to facilitate cross-case comparisons and support interdisciplinary collaboration. The SESMAD project (sesmad.dartmouth.edu) has collected and defined a list of the most common second- and third-tier variables. This framework has been applied extensively for place-based empirical studies, particularly in marine systems, fisheries, forestry and irrigation (see Partelow 2018 for a review of SES framework applications). While most applications focus on single case studies, there are also several recent studies that use the SES framework for case comparison or meta-analyses (e.g. Gutiérrez, Hilborn, and Defeo 2011; Villamayor-Tomas et al. 2020).

The robustness framework by Anderies and colleagues (2004) (Figure 1.6) is another analytical framework grounded in institutional analysis and the work of Elinor Ostrom. It aims to support the analysis and prediction of the robustness of SES to disturbances, where robustness is defined as the maintenance of system performance when subjected to unpredictable perturbations (Anderies, Janssen, and Ostrom 2004). It focuses on the institutional arrangements that shape the interactions among resource users, resources, public infrastructure providers and public infrastructure. Changes to the links between these critical SES components will affect the robustness of the SES. A disruption in the link between resource users and infrastructure providers, for instance, may critically affect the ability of the SES to respond to disturbance. The framework can help address questions such as how a given institutional arrangement affects the robustness of an SES, or why some SES persist in highly variable environments while others collapse. Recent applications of the framework include analyses of coastal system adaptation to global change (Anderies, Barreteau, and Brady 2019; Naylor et al. 2019) and identifying opportunities for ecosystem-based adaptation (Guerbois et al. 2019).

The social-ecological action situation (SE-AS) framework (Figure 1.7) is a recent development that further develops Ostrom's concept of an action situation to emphasise social-ecological interactions and how they give rise to emergent phenomena such as regime shifts or sustainable ecosystem management (Schlüter et al. 2019). The aim of this analytical framework is to support the development of hypotheses about the intertwined social and ecological processes that may have led to the emergence of particular phenomena. To this end, it introduces two other types of action situations in addition to the social action situations identified by Ostrom (2007): (a) the social-ecological action situations as situations in which humans and non-human elements of an SES (e.g. species, ecosystems, landscapes) interact, and (b) ecological action situations as situations in which ecological or biophysical elements interact (e.g. predation, parasitism, mutualism). A social-ecological phenomenon emerges from interactions of multiple action situations, which influence one another through their emergent outcomes. The collapse of the Baltic cod fishery, for instance, may be explained through the interactions of fishers with cod in fishing action situations, which are influenced not only by the interactions of cod with sprat but also by interactions within the governance system that introduced subsidies. One action situation can influence another, for instance by shaping its rules or changing the attributes of its participants. A possible explanation is thus represented as a configuration or network of action situations that are hypothesised to jointly give rise to a phenomenon, and can be further explored through fieldwork or agent-based modelling. First applications of the framework include supporting the development of global biodiversity targets that capture the interdependencies between biodiversity, ecosystem services and sustainable development (Reyers and Selig 2020), and identifying mechanisms of policy change during transformations to fisheries co-management (Orach and Schlüter, n.d.).

In addition to these SES frameworks, other SES research areas such as ecosystem services and vulnerability are partly undertaken within an SES approach, but substantial areas of work on these topics are also undertaken within other disciplines and approaches. Research on ecosystem services, for example, includes frameworks that span a range from more mechanistic and linear thinking (e.g. the cascade model: Potschin and Haines-Young 2011) to recognising complex dynamics and feedback loops across spatial and temporal scales (e.g. IPBES framework: Díaz et al. 2015). Some of this work such as the concept of 'nature's contributions to people' (Díaz et al. 2018) draws strongly on the SES approach in many respects, while other ecosystem services work is more closely aligned with conventional environmental economics approaches.

Links between research and action

Much SES research is problem focused, and the lines between research and action are often blurred. The goals of SES research are frequently to produce 'usable' knowledge, and often involve liaising or coordinating with policymakers and agencies to promote the integration of scientific knowledge and evidence into policy processes (Turner et al. 2016). Social-ecological systems researchers frequently engage with multiple stakeholders and knowledge types in the course of their research (Lang et al. 2012), and typically produce a broader set of outputs and outcomes than conventional disciplinary research. These researchers therefore often face the challenge of assessing and understanding the complex dynamics of SES, while promoting transdisciplinary processes that facilitate knowledge exchange across disciplines and between academic and non-academic actors (Balvanera et al. 2017).

In generating policy-relevant, solutions-oriented and socially robust knowledge, SES research often involves processes of knowledge co-production. Knowledge co-production is defined as an iterative and collaborative process involving diverse types of expertise, knowledge and actors to produce context-specific knowledge and pathways towards a sustainable future (Norström et al. 2020). Knowledge co-production is premised on the idea that knowledge and action are intertwined (Miller and Wyborn 2018), and can lead to the emergence of potential solutions and open the space for stakeholders to engage with these solutions and co-create strategic plans (Eelderink, Vervoot, and Van Laerhoven 2020). It provides a platform for mutual learning through processes of engagement and negotiation that include different actors and incorporate their diverse and divergent knowledge systems, perspectives, values and interests (Caniglia et al. 2020). These collective and collaborative practices run throughout the entire transdisciplinary research process, from problem formulation, to communication of results or outputs, to design of actions (Wheeler and Root-Bernstein 2020). In these cases, SES research can be seen as 'action-oriented research' with the power to unlock innovative thinking and find the best-fitting strategies to deal with sustainability challenges and problems.

Monitoring, evaluation and learning (MEL) is often incorporated into transdisciplinary SES research programmes to measure the impact of implementation, and identify lessons for future policy and planning (Taylor et al. 2016). When the idea of MEL first emerged, it was conceived within the international development sector as a form of 'evaluation for accountability' or 'summative evaluation' whereby a donor or sponsor is provided with the necessary information to demonstrate that the funded intervention has delivered on its stated ideas and objectives. There has been a gradual shift in the last 20 years to respond to the needs of development funders, planners and practitioners to learn from their previous experiences. This shift recognises that learning plays an essential role in the long-term sustainability of an organisation or initiative. This has given rise to 'evaluation for learning' and has led to an increasing emphasis on the translation of new knowledge into better policy and practice (Morris and Lawrence 2010). There are two main routes to learning that may be applied to SES research: (a) reflexive learning, which is conscious reflection on one's own experience, and (b) learning through exchanging and sharing ideas with other stakeholders. Monitoring and evaluation provides important data and experiences that can contribute to this learning by helping to identify obstacles and highlight possible changes that need to be made as an initiative evolves and develops.

Action-oriented SES research often includes practices of reflexivity and reciprocity. Reflexivity involves critical reflection on the assumptions, values and concepts underlying the research process, as well as a critical analysis of the power asymmetries between different forms of knowledge and actors (Norström et al. 2020; Turnhout et al. 2020). Reflexivity therefore requires the examination of researchers' embeddedness in power-laden transdisciplinary processes, questioning the limitations and potential implications of the researcher's position. Reciprocity entails the empowerment of actors as equal partners during the research process, and the provision of effective and useful outcomes to non-academic actors (Iniesta-Arandia et al. 2016). These practices have been extensively fostered by feminist scholars (Hesse-Biber and Piateli 2012; Iniesta-Arandia et al. 2016).

Social-ecological systems research, especially if it is action-oriented research, typically produces a broader set of outputs than conventional disciplinary research. While academic publications remain important, the process of engagement with stakeholders, and the shared understanding of a problem and system generated through co-production processes, may be as important. In fact, one way of measuring the success of SES research when embedded in transdisciplinary processes is to assess how much the research has supported the co-production of knowledge (from the research questions to the dissemination of findings), and the identification of solutions and design of actions (Mauser et al. 2013; Balvanera et al. 2017). In this sense, stakeholder dialogues and engagement in various management forums may be as important as scientific outputs. Consequently, SES research is often communicated in multiple formats in order to reach a range of audiences and stakeholders (Goring et al. 2014). Examples of these formats are scientific articles and reports, policy briefs, popular science articles and, increasingly, videos and engagement with the arts (see Chapter 33 for further discussion).

Conclusion

Many researchers engage in SES research because they are passionate about addressing the intertwined social and ecological sustainability issues facing society. Social-ecological systems research therefore often requires engaging in larger transdisciplinary initiatives where the lines between research and action are blurred. Any engagement with change processes on the ground requires a thorough understanding of the issues, history and context of the particular place in which one is working and of the diverse actors involved. Engaged SES research is conducted at scales ranging from local places to the global commons, recognising that SES are truly intertwined, across temporal and spatial scales (Cash et al. 2006; Balvanera et al. 2017; Österblom et al. 2017). In this book we discuss SES research methods that span these scales and contexts, focusing on approaches used in SES research rather than primarily in SES governance or management contexts.

This chapter is the first of four introductory chapters comprising Part 1 of the handbook. In this chapter we introduced the concept of SES, and how it is rooted in an understanding of complex adaptive systems and of humanity intertwined with and embedded in the biosphere. Chapter 2 delves deeper into the implications of a complex adaptive systems-based approach for SES research and research methods, while Chapter 3 discusses SES research in practice. With this background in hand, Chapter 4 introduces the structure for different SES research methods covered in the core of the book in Part 2 (Chapters 5–32). Part 3 of the book (Chapter 33) provides a synthesis and reflection on current SES research methods, and potential future areas of development.

Acknowledgements

Reinette Biggs received support from the South African Research Chairs Initiative (SARChI) (grant 98766), the Guidance for Resilience in the Anthropocene: Investments for Development (GRAID) project funded by the Swedish International Development Agency (Sida) and a Young Researchers Grant from the Vetenskapsrådet in Sweden (grant 621-2014-5137). Hayley Clements is funded by a Jennifer Ward Oppenheimer Research Grant and

Kone Foundation. Alta de Vos received support from a Rhodes University Council Grant. Rika Preiser's salary was co-funded by the Swedish International Development Agency's (Sida) GRAID and Swedbio programmes. Maja Schlüter received funding from the European Research Council under the European Union's Horizon 2020 research and innovation programme (grant agreement no. 682472 – MUSES).

References

Adger, W.N., N.W. Arnell, and E.L. Tompkins. 2005. 'Successful Adaptation to Climate Change Across Scales.' *Global Environmental Change* 15(2): 77–86. doi:10.1016/j.gloenvcha.2004.12.005.

Anderies, J.M., O. Barreteau, and U. Brady. 2019. 'Refining the Robustness of Social-Ecological Systems Framework for Comparative Analysis of Coastal System Adaptation to Global Change.' *Regional Environmental Change* 19(7): 1891–1908. doi:10.1007/s10113-019-01529-0.

Anderies, J.M., M.A. Janssen, and E. Ostrom. 2004. 'A Framework to Analyze the Robustness of Social-Ecological Systems from an Institutional Perspective.' *Ecology and Society* 9(1): 18. www.ecologyandsociety.org/vol9/iss1/art18.

Baggio, J.A., K. Brown, and D. Hellebrandt. 2015. 'Boundary Object or Bridging Concept? A Citation Network Analysis of Resilience.' *Ecology and Society* 20(2): 2. doi:10.5751/ES-07484–200202.

Bahru, B.A., C. Bosch, R. Birner, and M. Zeller. 2019. 'Drought and Child Undernutrition in Ethiopia: A Longitudinal Path Analysis.' *PLoS ONE* 14(6): e0217821. doi:10.1371/journal.pone.0217821.

Balvanera, P., T.M. Daw, T. Gardner, B. Martín-López, A.V. Norström, C. Ifejika Speranza, M. Spierenburg et al. 2017. 'Key Features for more Successful Place-based Sustainability Research on Social-Ecological Systems: A Programme on Ecosystem Change and Society (PECS) Perspective.' *Ecology and Society* 22(1): 14. doi:10.5751/ES-08826-220114.

Ban, N.C., G.G. Gurney, N.A. Marshall, C.K. Whitney, M. Mills, S. Gelcich, N.J. Bennett et al. 2019. 'Well-being Outcomes of Marine Protected Areas.' *Nature Sustainability* 2(6): 524–532. doi:10.1038/s41893-019-0306-2.

Barbier, E.B., and J.P. Hochard. 2018. 'The Impacts of Climate Change on the Poor in Disadvantaged Regions.' *Review of Environmental Economics and Policy* 12(1): 26–47. doi:10.1093/reep/rex023.

Bateson, G. 1979. *Mind and Nature: A Necessary Unit.* New York: Bantam Books.

Berkes, F., ed. 1989. *Common Property Resources: Ecology of Community-based Sustainable Development.* London: Belhaven Press.

Berkes, F., J. Colding, and C. Folke, eds. 2003. *Navigating Social-Ecological Systems: Building Resilience for Complexity and Change.* Cambridge: Cambridge University Press.

Berkes, F., and C. Folke, eds. 1998. *Linking Social and Ecological Systems: Management Practices and Social Mechanisms for Building Resilience.* Cambridge: Cambridge University Press.

Berkes, F., and H. Ross. 2016. 'Panarchy and Community Resilience: Sustainability Science and Policy Implications.' *Environmental Science and Policy* 61: 185–193. doi:10.1016/j.envsci.2016.04.004.

Biggs, R., G.D. Peterson, and J. Rocha. 2018. 'The Regime Shifts Database: A Framework for Analyzing Regime Shifts in Social-Ecological Systems.' *Ecology and Society* 23(3): 9. doi:10.5751/ES-10264-230309.

Biggs, R., M. Schlüter, and M.L. Schoon. 2015. 'An Introduction to the Resilience Approach and Principles to Sustain Ecosystem Services in Social-Ecological Systems.' In *Principles for Building Resilience: Sustaining Ecosystem Services in Social-Ecological Systems*, edited by R. Biggs, M. Schlüter, and M.L. Schoon, 1–31. Cambridge: Cambridge University Press.

Binder, C.R., J. Hinkel, P.W.G. Bots, and C. Pahl-Wostl. 2013. 'Comparison of Frameworks for Analyzing Social-Ecological Systems.' *Ecology and Society* 18(4): 26. doi:10.5751/ES-05551-180426.

Bodin, Ö. 2017. 'Collaborative Environmental Governance: Achieving Collective Action in Social-Ecological Systems.' *Science* 357: eaan1114. doi:10.1126/science.aan1114.

Caniglia, C., C. Luederitz, T. von Wirth, I. Fazey, B. Martín-López, K. Hondrila, A. König et al. 2020 (in press). 'A Pluralistic and Integrated Approach to Action-oriented Knowledge for Sustainability.' *Nature Sustainability*. doi:10.1038/s41893-020-00616-z.

Carpenter, S.R., C. Folke, A.V. Norström, O. Olsson, L. Schultz, B. Agarwal, P. Balvanera et al. 2012. 'Program on Ecosystem Change and Society: An International Research Strategy for Integrated Social-Ecological Systems.' *Current Opinion in Environmental Sustainability* 4(1): 134–138. doi:10.1016/j.cosust.2012.01.001.

Carrasco, L.R., J. Chan, F.L. McGrath, and L.T.P. Nghiem. 2017. 'Biodiversity Conservation in a Telecoupled World.' *Ecology and Society* 22(3): 24. doi:10.5751/ES-09448-220324.

Cash, D.W., W.N. Adger F. Berkes, P. Garden, L. Lebel, P. Olsson, and L. Pritchard. 2006. 'Scale and Cross-scale Dynamics: Governance and Information in a Multilevel World.' *Ecology and Society* 11(2): 8. www.jstor:stable/26265993.

Chan, K.M., P. Balvanera, K. Benessaiah, M. Chapman, S. Díaz, E. Gómez-Baggethun, R. Gould et al. 2016. 'Opinion: Why Protect Nature? Rethinking Values and the Environment.' *Proceedings of the National Academy of Sciences* 113(6): 1462–1465. doi:10.1073/pnas.1525002113.

Cilliers, P. 2001. 'Boundaries, Hierarchies and Networks in Complex Systems.' *International Journal of Innovation Management* 5(2): 135–147. doi:10.1142/S1363919601000312.

Cinner, J.E., C. Huchery, M.A. MacNeil, N.A. Graham, T.R. McClanahan, J. Maina, E. Maire et al. 2016. 'Bright Spots among the World's Coral Reefs.' *Nature* 535(7612): 416–419. doi:10.1038/nature18607.

Clark, W.C., P.J. Crutzen, and H.J. Schellnhuber. 2004. 'Science for Global Sustainability.' In *Earth System Analysis for Sustainability*, edited by H.J. Schellnhuber, P.J. Crutzen, W.C. Clark, C. Martin, and H. Hermann, 1–28. Cambridge: MIT Press.

Clark, W.C., and A.G. Harley. 2020 (in press). 'Sustainability Science: Towards a Synthesis.' *Annual Review of Environment and Resources* 45.

Clark, W.C., and R.E. Munn, eds. 1986. *Sustainable Development of the Biosphere*. Cambridge: Cambridge University Press.

Colding, J., and S. Barthel. 2019. 'Exploring the Social-Ecological Systems Discourse 20 Years Later.' *Ecology and Society* 24(1): 2. doi:10.5751/ES-10598-240102.

Cosens, B.A., L.H. Gunderson, and B.C. Chaffin. 2018. 'Introduction to the Special Feature Practicing Panarchy: Assessing Legal Flexibility, Ecological Resilience, and Adaptive Governance in Regional Water Systems Experiencing Rapid Environmental Change.' *Ecology and Society* 23(1): 4. doi:10.5751/ES-09524-230104.

Costanza, R., ed. 1991. *Ecological Economics: The Science and Management of Sustainability*. New York: Columbia University Press.

Costinot, A., J. Vogel, and S. Wang. 2012. 'Global Supply Chains and Wage Inequality.' *American Economic Review* 102(3): 396–401. doi.10.1257/aer.102.3.396.

Cox, M. 2014. 'Understanding Large Social-Ecological Systems: Introducing the Social-ecological System MAD Project.' *International Journal of the Commons* 8(2): 265–276. doi:10.18352/ijc.406.

Cox, M., G. Arnold, and S. Villamayor-Tomas. 2010. 'A Review of Design Principles for Community-based Natural Resource Management.' *Ecology and Society* 15(4): 38. www.jstor.org/stable/26268233.

Crona, B.I., T. van Holt, M. Petersson, T.M. Daw, and E. Buchary. 2015. 'Using Social-Ecological Syndromes to Understand Impacts of International Seafood Trade on Small-scale Fisheries.' *Global Environmental Change* 35: 162–175. doi:10.1016/j.gloenvcha.2015.07.006.

Crutzen, P.J. 2006. 'The "Anthropocene".' In *Earth system Science in the Anthropocene*, edited by E. Ehlers and T. Krafft, 13–18. Berlin: Springer. doi:10.1007/3-540-26590-2_3.

Cumming, G.S. 2014. 'Theoretical Frameworks for the Analysis of Social-Ecological Systems.' In *Social-Ecological Systems in Transition, Global Environmental Studies*, edited by S. Sakai and C. Umetsu, 3–24. Tokyo: Springer.

Cumming, G.S., C.R. Allen, N.C. Ban, D. Biggs, H.C. Biggs, D.H. Cumming, A. de Vos et al. 2015. 'Understanding Protected Area Resilience: A Multi-scale, Social-ecological Approach.' *Ecological Applications* 25(2): 299–319. doi:10.1890/13-2113.1.

Cumming, G.S., G. Epstein, J.M. Anderies, C.I. Apetrei, J. Baggio, Ö. Bodin, S. Chawla et al. 2020. 'Advancing Understanding of Natural Resource Governance Using the Social-Ecological Systems Framework: A Post-Ostrom Research Agenda.' *Current Opinion in Environmental Sustainability* 44: 26–34. doi:10.1016/j.cosust.2020.02.005.

Dai, A. 2013. 'Increasing Drought under Global Warming in Observations and Models.' *Nature Climate Change* 3(1): 52–58. doi:10.1038/nclimate1633.

Davidson-Hunt, I.J., and F. Berkes. 2003. 'Nature and Society through the Lens of Resilience: Toward a Human-in-Ecosystem Perspective.' In *Navigating Social-Ecological Systems: Building Resilience for Complexity and Change*, edited by F. Berkes, J. Colding, and C. Folke, 53–82. Cambridge: Cambridge University Press.

De Vos, A., R. Biggs, and R. Preiser. 2019. 'Methods for Understanding Social-Ecological Systems: A Review of Place-based Studies.' *Ecology and Society* 24(4): 16. doi:10.5751/es-11236-240416.

Díaz, S., S. Demissew, J. Carabias, C. Joly, M. Lonsdale, N. Ash, A. Larigauderie et al. 2015. 'The IPBES Conceptual Framework – Connecting Nature and People.' *Current Opinion in Environmental Sustainability* 14: 1–16. doi:10.1016/j.cosust.2014.11.002.

Díaz, S., U. Pascual, M. Stenseke, B. Martín-López, R.T. Watson, Z. Molnár, R. Hill et al. 2018. 'Assessing Nature's Contributions to People.' *Science* 359(6373): 270–272. doi:10.1126/science.aap8826.

Dietz, T., E. Ostrom, and P.C. Stern. 2003. 'The Struggle to Govern the Commons.' *Science* 302(5652): 1907–1912. www.jstor.org/stable/3835713.

Eelderink, M., J.M. Vervoort, and F. van Laerhoven. 2020. 'Using Participatory Action Research to Operationalize Critical Systems Thinking in Social-Ecological Systems.' *Ecology and Society* 25(1): 16. doi:10.5751/ES-11369-250116.

Epstein, G., J.M. Vogt, S.K. Mincey, M. Cox, and B. Fischer. 2013. 'Missing Ecology: Integrating Ecological Perspectives with the Social-ecological System Framework.' *International Journal of the Commons* 7(2): 432–453. doi:10.18352/ijc.371.

Everard, M., P. Johnston, D. Santillo, and C. Staddon. 2020. 'The Role of Ecosystems in Mitigation and Management of Covid-19 and other Zoonoses.' *Environmental Science and Policy* 111: 7–17. doi:10.1016/j.envsci.2020.05.017.

Fischer, J., T.A. Gardner, E.M. Bennett, P. Balvanera, R. Biggs, S.R. Carpenter, T.M. Daw et al. 2015. 'Advancing Sustainability through Mainstreaming a Social-Ecological Systems Perspective.' *Current Opinion in Environmental Sustainability* 14: 144–149. doi:10.1016/j.cosust.2015.06.002.

Folke, C. 2006. 'Resilience: The Emergence of a Perspective for Social-Ecological Systems Analyses.' *Global Environmental Change* 16: 253–267. doi:10.1016/j.gloenvcha.2006.04.002.

Folke, C., and F. Berkes. 1998. 'Understanding Dynamics of Ecosystem-Institution Linkages for Building Resilience.' *Stockholm, Sweden: Beijer Discussion Paper No. 112*. Stockholm: Beijer Institute of Ecological Economics, The Royal Academy of Sciences.

Folke, C., R. Biggs, A.V. Norström, B. Reyers, and J. Rockström. 2016. 'Social-ecological Resilience and Biosphere-based Sustainability Science.' *Ecology and Society* 21(3): 41. doi:10.5751/ES-08748-210341.

Folke, C., S.R. Carpenter, T. Elmqvist, L.H. Gunderson, C.S. Holling, and B. Walker. 2002. 'Resilience and Sustainable Development: Building Adaptive Capacity in a World of Transformations.' *Ambio* 31(5): 437–440. doi:10.1579/0044-7447-31.5.437.

Folke, C., S.R. Carpenter, B. Walker, M. Scheffer, T. Chapin, and J. Rockström. 2010. 'Resilience Thinking: Integrating Resilience, Adaptability and Transformability.' *Ecology and Society* 15(4): 20. www.ecologyandsociety.org/vol15/iss4/art20.

Folke, C., T. Hahn, P. Olsson, and J. Norberg. 2005. 'Adaptive Governance of Social-Ecological Systems.' *Annual Review of Environment and Resources* 30: 441–473. doi:10.1146/annurev.energy.30.050504.144511.

Folke, C., A. Jansson, J. Rockström, P. Olsson, S.R. Carpenter, F.S. Chapin, A-S. Crépin et al. 2011. 'Reconnecting to the Biosphere.' *Ambio* 40(7): 719. doi:10.1007/s13280-011-0184-y.

Friis, C., and J.Ø. Nielsen. 2017. 'On the System. Boundary Choices, Implications, and Solutions in Telecoupling Land Use Change Research.' *Sustainability* 9(6): 974. doi:10.3390/su9060974.

Galappaththi, E.K., J.D. Ford, E.M. Bennett, and F. Berkes. 2019. 'Climate Change and Community Fisheries in the Arctic: A Case Study from Pangnirtung, Canada.' *Journal of Environmental Management* 250: 109534. doi:10.1016/j.jenvman.2019.109534.

Galaz, V., B. Crona, A. Dauriach, J-B. Jouffray, H. Österblom, and J. Fichtner. 2018. 'Tax Havens and Global Environmental Degradation.' *Nature Ecology and Evolution* 2(9): 1352–1357. doi:10.1038/s41559-018-0497-3.

Galaz, V., F. Moberg, E-K. Olsson, E. Paglia, and C. Parker. 2011. 'Institutional and Political Leadership Dimensions of Cascading Ecological Crises.' *Public Administration* 89(2): 361–380. doi:10.1111/j.1467-9299.2010.01883.x.

Geobey, S., and K.A. McGowan. 2019. 'Panarchy, Ontological and Epistemological Phenomena, and the Plague.' *Ecology and Society* 24(4): 23. doi:10.5751/ES-11089-240423.

Goring, S.J., K.C. Weathers, W.K. Dodds, P.A. Soranno, L.C. Sweet, K.S. Cheruvelil, J.S. Kominoski, J. Rüegg, A.M. Thorn, and R.M. Utz. 2014. 'Improving the Culture of Interdisciplinary Collaboration in Ecology by Expanding Measures of Success.' *Frontiers in Ecology and the Environment* 12(1): 39–47. doi:10.1890/120370.

Green, J.M.H., S.A. Croft, A.P. Durán, A.P. Balmford, N.D. Burgess, S. Fick, T.A. Gardner et al. 2019. 'Linking Global Drivers of Agricultural Trade to on-the-ground Impacts on Biodiversity.' *Proceedings of the National Academy of Sciences* 116(46): 23202–23208. doi:10.1073/pnas.1905618116.

Guerbois, C., U. Brady, A.G. de Swardt, and C. Fabricius. 2019. 'Nurturing Ecosystem-based Adaptations in South Africa's Garden Route: A Common Pool Resource Governance Perspective.' *Regional Environmental Change* 19(7): 1849–1863. doi:10.1007/s10113-019-01508-5.

Guo, Y., A. Gasparrini, S. Li, F. Sera, A.M. Vicedo-Cabrera, M.Z.S. Coelho, P.H.N. Saldiva et al. 2018. 'Quantifying Excess Deaths Related to Heatwaves under Climate Change Scenarios: A Multicountry Time Series Modelling Study.' *PLoS Medicine* 15(7): e1002629. doi:10.1371/journal.pmed.1002629.

Gunderson, L.H., B.A. Cosens, B.C. Chaffin, C.A.T. Arnold, A.K. Fremier, A.S. Garmestani, R.K. Craig et al. 2017. 'Regime Shifts and Panarchies in Regional Scale Social-ecological Water Systems.' *Ecology and Society* 22(1): 31. doi:10.5751/ES-08879-220131.

Gunderson, L.H., and C.S. Holling, eds. 2002. *Panarchy: Understanding Transformations in Human and Natural Systems*. Washington: Island Press.

Gunderson, L.H., C.S. Holling, and S.S. Light, eds. 1995. *Barriers and Bridges to the Renewal of Ecosystems and Institutions*. New York: Columbia University Press.

Gurney, G.G., E.S. Darling, S.D. Jupiter, S. Mangubhai, T.R. McClanahan, P. Lestari, S. Pardede et al. 2019. 'Implementing a Social-Ecological Systems Framework for Conservation Monitoring: Lessons from a Multi-country Coral Reef Program.' *Biological Conservation* 240: 108298. doi:10.1016/j.biocon.2019.108298.

Gutiérrez, N.L., R. Hilborn, and O. Defeo. 2011. 'Leadership, Social Capital and Incentives Promote Successful Fisheries.' *Nature* 470: 386–389. doi:10.1038/nature09689.

Hamann, M., K. Berry, T. Chaigneau, T. Curry, R. Heilmayr, P.J.G. Henriksson, J. Hentati-Sundberg et al. 2018. 'Inequality and the Biosphere.' *Annual Review of Environment and Resources* 43(1): 61–83. doi:10.1146/annurev-environ-102017-025949.

Hamann, M., R. Biggs, and B. Reyers. 2015. 'Mapping Social-Ecological Systems: Identifying "Green-loop" and "Red-loop" Dynamics Based on Characteristic Bundles of Ecosystem Service Use.' *Global Environmental Change* 34: 218–226. doi:10.1016/j.gloenvcha.2015.07.008.

Herrero-Jáuregui, C., C. Arnaiz-Schmitz, M.F. Reyes, M. Telesnicki, I. Agramonte, M.H. Easdale, M.F. Schmitz, M. Aguiar, A. Gómez-Sal, and C. Montes. 2018. 'What Do We Talk About When We Talk About Social-Ecological Systems? A Literature Review.' *Sustainability* 10(8): 2950. doi:10.3390/su10082950.

Hertz, T., M. Mancilla García, and M. Schlüter. 2020. 'From Nouns to Verbs: How Process Ontologies Enhance our Understanding of Social-Ecological Systems Understood as Complex Adaptive Systems.' *People and Nature* 2(2): 328–338. doi:10.1002/pan3.10079.

Hesse-Biber, S., and D. Piatelli. 2012. 'The Feminist Practice of Holistic Reflexivity.' In *Handbook of Feminist Research: Theory and Praxis*, edited by S. Hesse-Biber. Thousand Oaks: Sage.

Holdschlag, A., and B.M. Ratter. 2016. 'Caribbean Island States in a Social-ecological Panarchy? Complexity Theory, Adaptability and Environmental Knowledge Systems.' *Anthropocene* 13: 80–93. doi:10.1016/j.ancene.2016.03.002.

Holling, C.S. 1973. 'Resilience and Stability of Ecological Systems.' *Annual Review of Ecology and Systematics* 4: 1–23. doi:10.1146/annurev.es.04.110173.000245.

Horcea-Milcu, A-I., B. Martín-López, D.P.M. Lam, and D.J. Lang. 2020. 'Research Pathways to Foster Transformation: Linking Sustainability Science and Social-Ecological Systems Research.' *Ecology and Society* 25(1): 13. doi:10.5751/ES-11332-250113.

Hules, M., and S.J. Singh. 2017. 'India's Land Grab Deals in Ethiopia: Food Security or Global Politics?' *Land Use Policy* 60: 343–351, doi:10.1016/j.landusepol.2016.10.035.

Hull, V., and J. Liu. 2018. 'Telecoupling: A New Frontier for Global Sustainability.' *Ecology and Society* 23(4): 41. doi:10.5751/ES-10494-230441.

Iniesta-Arandia, I., F. Ravera, S. Buechler, I. Díaz-Reviriego, M.E. Fernández-Giménez, M.G. Reed, M. Thompson-Hall et al. 2016. 'A Synthesis of Convergent Reflections, Tensions and Silences in Linking Gender and Global Environmental Change Research.' *Ambio* 45(3): 383–393. doi:10.1007/s13280-016-0843-0.

Jiren, T.S., I. Dorresteijn, J. Hanspach, J. Schultner, A. Bergsten, A. Manlosa, N. Jager, F. Senbeta, and J. Fischer. 2020. 'Alternative Discourses Around the Governance of Food Security: A Case Study from Ethiopia.' *Global Food Security* 24: 100338. doi:10.1016/j.gfs.2019.100338.

Kelly, R., M. Mackay, K.L. Nash, C. Cvitanovic, E.H. Allison, D. Armitage, A. Bonn et al. 2019. 'Ten Tips for Developing Interdisciplinary Socio-ecological Researchers.' *Socio-ecological Practice Research* 1(2): 149–161. doi:10.1007/s42532-019-00018-2.

Keys, P.W., V. Galaz, M. Dyer, N. Matthews, C. Folke, M. Nyström, and S.E. Cornell. 2019. 'Anthropocene Risk.' *Nature Sustainability* 2(8): 667–673. doi:10.1038/s41893-019-0327-x.

Krausmann, F., K-H. Erb, S. Gingrich, H. Haberl, A. Bondeau, V. Gaube, C. Lauk, C. Plutzar, and T.D. Searchinger. 2013. 'Global Human Appropriation of Net Primary Production doubled in the 20th Century.' *Proceedings of the National Academy of Sciences* 110(25): 10324–10329. doi:10.1073/pnas.1211349110.

Kummu, M., P. Kinnunen, E. Lehikoinen, M. Porkka, C. Queiroz, E. Röös, M. Troell, and C. Weil. 2020. 'Interplay of Trade and Food System Resilience: Gains on Supply Diversity Over Time at the Cost of Trade Independency.' *Global Food Security* 24: 100360. doi:10.1016/j.gfs.2020.100360.

Lade, S.J., A. Tavoni, S.A. Levin, and M. Schlüter. 2013. 'Regime Shifts in a Social-ecological System.' *Theoretical Ecology* 6(3): 359–372. doi:10.1007/s12080-013-0187-3.

Lambin, E.F., and P. Meyfroidt. 2011. 'Global Land Use Change, Economic Globalization, and the Looming Land Scarcity.' *Proceedings of the National Academy of Sciences* 108(9): 3465–3472. doi:10.1073/pnas.1100480108.

Lambin, E.F., B.L. Turner, H.J. Geist, S.B. Agbola, A. Angelsen, J.W. Bruce, O.T. Coomes et al. 2001. 'The Causes of Land-use and Land-cover Change: Moving Beyond the Myths.' *Global Environmental Change* 11(4): 261–269. doi:10.1016/S0959-3780(01)00007-3.

Lang, D.J., A. Wiek, M. Bergmann, M. Stauffacher, P. Martens, P. Moll, M. Swilling, and C.J. Thomas. 2012. 'Transdisciplinary Research in Sustainability Science: Practice, Principles, and Challenges.' *Sustainability Science* 7(1): 25–43. doi:10.1007/s11625-011-0149-x.

Lansing, J.S. 2003. 'Complex Adaptive Systems.' *Annual Review of Anthropology* 32(1): 183–204. doi:10.1146/annurev.anthro.32.061002.093440.

Lenzen, M., D. Moran, K. Kanemoto, B. Foran, L. Lobefaro, and A. Geschke. 2012. 'International Trade Drives Biodiversity Threats in Developing Nations.' *Nature* 486(7401): 109–112. doi:10.1038/nature11145.Levin, S.A. 1999. 'Fragile Dominion Complexity and the Commons.' *Journal of Ecology* 88(1): 181. doi:10.1046/j.1365-2745.2000.00425-5.x.

Levin, S.A., T. Xepapadeas, A-S. Crépin, J. Norberg, A.D. Zeeuw, C. Folke, T. Hughes et al. 2013. 'Social-Ecological Systems as Complex Adaptive Systems: Modeling and Policy Implications.' *Environment and Development Economics* 18(2): 111–132. doi:10.1017/S1355770X12000460.

Liu, J., T. Dietz, S.R. Carpenter, M. Alberti, C. Folke, M. Alberti, C.L. Redman et al. 2007a. 'Coupled Human and Natural Systems.' *Ambio* 36(8): 639–649. doi:10.1579/0044-7447(2007)36[639:CHANS]2.0.CO;2.

Liu, J., T. Dietz, S.R. Carpenter, M. Alberti, C. Folke, E. Moran, A.C. Pell et al. 2007b. 'Complexity of Coupled Human and Natural Systems.' *Science* 317(5844): 1513–1516. doi:10.1126/science.1144004.

Liu, J., Y. Dou, M. Batistella, E. Challies, T. Connor, C. Friis, J.D.A. Millington et al. 2018. 'Spillover Systems in a Telecoupled Anthropocene: Typology, Methods, and Governance for Global Sustainability.' *Current Opinion in Environmental Sustainability* 33: 58–69. doi:10.1016/j.cosust.2018.04.009.

Liu, J., V. Hull, M. Batistella, R. DeFries, T. Dietz, F. Fu, T.W. Hertel et al. 2013. 'Framing Sustainability in a Telecoupled World.' *Ecology and Society* 18(2): 26. doi:10.5751/ES-05873-180226.

Liu, J., H. Mooney, V. Hull, S.J. Davis, J. Gaskell, T. Hertel, J. Lubchenco et al. 2015. 'Systems Integration for Global Sustainability.' *Science* 347(6225): 963. doi:10.1126/science.1258832.

Liu, J., W. Yang, and S. Li. 2016. 'Framing Ecosystem Services in the Telecoupled Anthropocene.' *Frontiers in Ecology and the Environment* 14(1): 27–36. doi:10.1002/16-0188.1.

Mancilla García, M., T. Hertz, M. Schlüter, R. Preiser, and M. Woermann. 2020. 'Adopting Process-relational Perspectives to Tackle the Challenges of Social-Ecological Systems Research.' *Ecology and Society* 25(1): 29. doi:10.5751/ES-11425-250129.

Masterson, V.A., S. Vetter, T. Chaigneau, T.M. Daw, O. Selomane, M. Hamann, G.Y. Wong et al. 2019. 'Revisiting the Relationships Between Human Well-being and Ecosystems in Dynamic Social-Ecological Systems: Implications for Stewardship and Development.' *Global Sustainability* 2: E8. doi:10.1017/S205947981900005X.

Mauser, W., G. Klepper, M. Rice, Bettina, S. Schmalzbauer, H. Hackmann, R. Leemans, and H. Moore. 2013. 'Transdisciplinary Global Change Research: The Co-creation of Knowledge for Sustainability.' *Current Opinion in Environmental Sustainability* 5(3–4): 420–431. doi:10.1016/j.cosust.2013.07.001.

McGinnis, M.D. 2011. 'An Introduction to IAD and the Language of the Ostrom Workshop: A Simple Guide to a Complex Framework.' *Policy Studies Journal* 39(1): 169–183. doi:10.1111/j.1541-0072.2010.00401.x.

Miller, C.A., and C. Wyborn. 2018. 'Co-production in Global Sustainability: Histories and Theories.' *Environmental Science and Policy* 113: 88–95. doi:10.1016/j.envsci.2018.01.016.

Milly, P.C.D., R.T. Wetherald, K.A. Dunne, K.A, and T.L. Delworth. 2002. 'Increasing Risk of Great Floods in a Changing Climate.' *Nature* 415(6871): 514–517. doi:10.1038/415514a.

Mohamed, A.A. 2017. 'Food Security Situation in Ethiopia: A Review Study.' *International Journal of Health Economics and Policy* 2(3): 86–96. www.sciencepublishinggroup.com/j/hep.

Morris, J., and A. Lawrence. 2010. 'Learning from Monitoring and Evaluation – A Blueprint for an Adaptive Organisation.' *Social and Economic Research Group*. www.forestresearch.gov.uk/research/learning-from-monitoring-and-evaluation-a-blueprint-for-an-adaptive-organisation.

Naylor, L.A., U. Brady, T. Quinn, K. Brown, and J.M. Anderies. 2019. 'A Multiscale Analysis of Social-ecological System Robustness and Vulnerability in Cornwall, UK.' *Regional Environmental Change* 19(7): 1835–1848. doi:10.1007/s10113-019-01530-7.

Nicholls, R.J. 2004. 'Coastal Flooding and Wetland Loss in the 21st century: Changes Under the SRES Climate and Socio-economic Scenarios.' *Global Environmental Change* 14(1): 69–86. doi:10.1016/j.gloenvcha.2003.10.007.

Norberg, J., and G.S. Cumming. 2008. *Complexity Theory for a Sustainable Future*. New York: Columbia University Press.

Norström, A.V., C. Cvitanovic, M.F. Löf, S. West, C. Wyborn, P. Balvanera, A.T. Bednarek et al. 2020. 'Principles for Knowledge Co-Production in Sustainability Research.' *Nature Sustainability* 3: 182–190. doi:10.1038/s41893-019-0448-2.

Nyström, M., J-B. Jouffray, A.V. Norström, B. Crona, P.S. Jørgensen, S.R. Carpenter, Ö. Bodin, V. Galaz, and C. Folke. 2019. 'Anatomy and Resilience of the Global Production Ecosystem.' *Nature* 575(7781): 98–108. doi:10.1038/s41586-019-1712-3.

O'Callaghan-Gordo, C., and J.M. Antó. 2020. 'COVID-19: The Disease of the Anthropocene.' *Environmental Research* 187: 109683. doi:10.1016/j.envres.2020.109683.

Odum, E.P. 1989. *Ecology and our Endangered Life-support Systems*. Sunderland: Sinauer.

Oliver, E.C.J., M.G. Donat, M.T. Burrows, P.J. Moore, D.A. Smale, L.V. Alexander, J.A. Benthuysen et al. 2018. 'Longer and More Frequent Marine Heatwaves over the Past Century.' *Nature Communications* 9(1): 1–12. doi:10.1038/s41467-018-03732-9.

Olsson, P., V. Galaz, and W. J. Boonstra. 2014. 'Sustainability Transformations: A Resilience Perspective.' *Ecology and Society* 19(4): 1. doi:10.5751/ES-06799-190401.

Olsson, P., L.H. Gunderson, S.R. Carpenter, P. Ryan, L. Lebel, C. Folke, and C.S. Holling. 2006. 'Shooting the Rapids: Navigating Transitions to Adaptive Governance of Social-Ecological Systems.' *Ecology and Society* 11(1): 18. www.ecologyandsociety.org/vol11/iss1/art18.

Orach, K., and Schlüter, M. n.d. 'Understanding the Dynamics of Fish Politics: The Role of Diverse Actor Interactions in Transformations Towards Co-management.' *Global Environmental Change* (submitted).

Österblom, H., and C. Folke. 2013. 'Emergence of Global Adaptive Governance for Stewardship of Regional Marine Resources.' *Ecology and Society* 18(2): 4. doi:10.5751/ES-05373-180204.

Österblom, H., C. Folke, J-B. Jouffray, and J. Rockström. 2017. 'Emergence of a Global Science-Business Initiative for Ocean Stewardship.' *Proceedings of the National Academy of Sciences USA* 114: 9038–9043. doi:10.1073/pnas.1704453114.

Ostrom, E. 1990. *Governing the Commons: The Evolution of Institutions for Collective Action*. Cambridge: Cambridge University Press.

Ostrom, E. 2007. 'A Diagnostic Approach for Going beyond Panaceas.' *Proceedings of the National Academy of Sciences* 104(39): 15181–15187. doi:10.1073/pnas.0702288104.

Ostrom, E. 2009. 'A General Framework for Analyzing Sustainability of Social-Ecological Systems.' *Science* 325(5939): 419–422. doi:10.1126/science.1172133.

Palomo I., M. Felipe-Lucía, E.M. Bennet, B. Martín-López, and U. Pascual. 2016. 'Disentangling the Pathways and Effects of Ecosystem Service Co-production.' *Advances in Ecological Research* 54: 245–283. doi:10.1016/bs.aecr.2015.09.003.

Partelow, S. 2018. 'A Review of the Social-Ecological Systems Framework: Applications, Methods, Modifications, and Challenges.' *Ecology and Society* 23(4): 36. doi:10.5751/ES-10594-230436.

Pereira, L., E.M. Bennett, R. Biggs, G.D. Peterson, T. McPhearson, A.V. Norström, P. Olsson, R. Preiser, C. Raudsepp-Hearne, and J. Vervoort. 2018. 'Seeds of the Future in the Present: Exploring Pathways for Navigating Towards 'Good' Anthropocenes.' In *Urban Planet: Knowledge Towards Sustainable Cities*, edited by T. Elmqvist, X. Bai, N. Frantzeskaki, C. Griffith, D. Maddox, T. McPhearson, S. Parnell, P. Romero-Lankao, D. Simon, and M. Watkins, 327–350. Cambridge: Cambridge University Press. https://openaccess.city.ac.uk/id/eprint/19567.

Potschin, M., and R. Haines-Young. 2011. 'Ecosystem Services: Exploring a Geographical Perspective.' *Progress in Physical Geography* 35(5): 575–594. doi:10.1177/0309133311423172.

Preiser, R., R. Biggs, A. de Vos, and C. Folke. 2018. 'Social-Ecological Systems as Complex Adaptive Systems: Organizing Principles for Advancing Research Methods and Approaches.' *Ecology and Society* 23(4): 46. doi:10.5751/ES-10558-230446.

Reyers, B., R. Biggs, G.S. Cumming, T. Elmqvist, A.P. Hejnowicz, and S. Polasky. 2013. 'Getting the Measure of Ecosystem Services: A Social-ecological Approach.' *Frontiers in Ecology and the Environment* 11(5): 268–273. doi:10.1890/120144.

Reyers, B., C. Folke, M-L. Moore, R. Biggs, and V. Galaz. 2018. 'Social-Ecological Systems Insights for Navigating the Dynamics of the Anthropocene.' *Annual Review of Environment and Resources* 43: 267–289. doi:10.1146/annurev-environ-110615-085349.

Reyers, B., and E.R. Selig. 2020. 'Global Targets that Reveal the Social-ecological Interdependencies of Sustainable Development.' *Nature Ecology and Evolution* 4: 1011–1019. doi:10.1038/s41559-020-1230-6.

Rocha, J.C., G.D. Peterson, Ö. Bodin, and S. Levin. 2018. 'Cascading Regime Shifts Within and Across Scales.' *Science* 362(6421): 1379–1383. doi:10.1126/science.aat7850.

Rogers, K.H., R. Luton, H. Biggs, R. Biggs, S. Blignaut, A.G. Choles, C.G. Palmer, and P. Tangwe. 2013. 'Fostering Complexity Thinking in Action Research for Change in Social-Ecological Systems.' *Ecology and Society* 18(2): 31. doi:10.5751/ES-05330-180231.

Roux, D.J., J.L. Nel, G. Cundill, and P.O. Farrell. 2017. 'Transdisciplinary Research for Systemic Change: Who to Learn with, What to Learn about and How to Learn.' *Sustainability Science* 12(5): 711–726. doi:10.1007/s11625-017-0446-0.

Scheffer, M., S.R. Carpenter, J.A. Foley, C. Folke, and B. Walker. 2001. 'Catastrophic Shifts in Ecosystems.' *Nature* 413(6856): 591–596. doi:10.1038/35098000.

Schill, C., J.M. Anderies, T. Lindahl, C. Folke, S. Polasky, J. Camilo Cárdenas, A-S. Crépin et al. 2019. 'A More Dynamic Understanding of Human Behaviour for the Anthropocene.' *Nature Sustainability* 2: 1075–1082. doi:10.1038/s41893-019-0419-7.

Schlüter, M., L.J. Haider, S.J. Lade, E. Lindkvist, R. Martin, K. Orach, N. Wijermans, and C. Folke. 2019. 'Capturing Emergent Phenomena in Social-Ecological Systems – An Analytical Framework.' *Ecology and Society* 24(3): 11. doi:10.5751/ES-11012-240311.

Schmeller, D.S., F. Courchamp, and G. Killeen. 2020. 'Biodiversity Loss, Emerging Pathogens and Human Health Risks.' *Biodiversity Conservation*. doi:10.1007/s10531-020-02021-6.

Scholz, R. 2011. *Environmental Literacy in Science and Society*. Cambridge: Cambridge University Press.

Schoon, M.L., and S. van der Leeuw. 2015. 'The Shift Toward Social-Ecological Systems Perspectives: Insights into the Human-Nature Relationship.' *Natures Sciences Sociétés* 23(2): 166–174. doi:10.1051/nss/2015034.

Schultz, L., C. Folke, H. Österblom, and P. Olsson. 2015. 'Adaptive Governance, Ecosystem Management, and Natural Capital.' *Proceedings of the National Academy of Sciences* 112(24): 7369–7374. doi:10.1073/pnas.1406493112.Selomane, O., B. Reyers, R. Biggs, and M. Hamann. 2019. 'Harnessing Insights from Social-Ecological Systems Research for Monitoring Sustainable Development.' *Sustainability* 11(4): 1190. doi:10.3390/su11041190.

Steffen, W., W. Broadgate, L. Deutsch, O. Gaffney, and C. Ludwig, C. 2015a. 'The Trajectory of the Anthropocene: The Great Acceleration.' *The Anthropocene Review* 2(1): 81–98. doi:10.1177/2053019614564785.

Steffen, W., Å. Persson, L. Deutsch, J. Zalasiewicz, M. Williams, K. Richardson, C. Crumley et al. 2011. 'The Anthropocene: From Global Change to Planetary Stewardship.' *Ambio* 40: 739. doi:10.1007/s13280-011-0185-x.

Steffen, W., K. Richardson, J. Rockström, S.E. Cornell, I. Fetzer, E.M. Bennett, R. Biggs et al. 2015b. 'Planetary Boundaries: Guiding Human Development on a Changing Planet.' *Science* 347(6223): 1259855. doi:10.1126/science.1259855.

Stoll, J.S., B.I. Crona, M. Fabinyi, and E.R. Farr. 2018. 'Seafood Trade Routes for Lobster Obscure Teleconnected Vulnerabilities.' *Frontiers in Marine Science* 5: 239. doi:10.3389/fmars.2018.00239.

Sun, J., Y. Tong, and J. Liu. 2017. 'Telecoupled Land-use Changes in Distant Countries.' *Journal of Integrative Agriculture* 16(2): 368–376. doi:10.1016/S2095-3119(16)61528-9.

Taylor, C., J. Cockburn, M. Rouget, J. Ray-Mukherjee, S. Mukherjee, R. Slotow, D. Roberts, R. Boon, S. O'Donoghue, and E. Douwes. 2016. 'Evaluating the Outcomes and Processes of a Research-Action Partnership: The Need for Continuous Reflective Evaluation.' *Bothalia – African Biodiversity and Conservation* 46(2): 1–16. doi:10.4102/abc.v46i2.2154.

Tengö, M., E. Brondizio, T. Elmqvist, P. Malmer, and M. Spierenburg. 2014. 'Connecting Diverse Knowledge Systems for Enhanced Ecosystem Governance: The Multiple Evidence Base Approach.' *Ambio* 43(5): 579–591. doi:10.1007/s13280-014-0501-3.

Trenberth, K.E., A. Dai, G. van der Schrier, P.D. Jones, J. Barichivich, K.R. Briffa, and J. Sheffield. 2014. 'Global Warming and Changes in Drought.' *Nature Climate Change* 4(1): 17–22. doi:10.1038/nclimate2067.

Turner II, B.L., W.C. Clark, R.W. Kates, J.F. Richards, J.T. Mathews, and W.B. Meyer, eds. 1990. *The Earth as Transformed by Human Action: Global and Regional Changes in the Biosphere over the Past 300 years.* Cambridge: Cambridge University Press.

Turner II, B.L., K.J. Esler, P. Bridgewater, J. Tewksbury, N. Sitas, B. Abrahams, F.S. Chapin III et al. 2016. 'Socio-environmental Systems (SES) research: What Have We Learned and How Can We Use this Information in Future Research Programs.' *Current Opinion in Environmental Sustainability* 19: 160–168. doi:10.1016/j.cosust.2016.04.001.

Turner II, B.L., E.F. Lambin, and A. Reenberg. 2007. 'The Emergence of Land Change Science for Global Environmental Change and Sustainability.' *Proceedings of the National Academy of Sciences* 104(52): 20666–20671. doi:10.1073/pnas.0704119104.

Turnhout, E., T. Metze, C. Wyborn, N. Klenk, and E. Louder. 2020. 'The Politics of Co-production: Participation, Power, and Transformation.' *Current Opinion in Environmental Sustainability* 42: 15–21. doi:10.1016/j.cosust.2019.11.009.

Van der Leeuw, S.E. 2019. *Social Sustainability Past and Present: Undoing Unintended Consequences for the Earth's Survival.* Cambridge: Cambridge University Press.

Villamayor-Tomas, S., C. Oberlack, G. Epstein, S. Partelow, M. Roggero, E. Kellner, M. Tschopp, and M. Cox. 2020. 'Using Case Study Data to Understand SES Interactions: A Model-centered Meta-analysis of SES Framework Applications.' *Current Opinion in Environmental Sustainability* 44: 48–57. doi:10.1016/j.cosust.2020.05.002.

Walker, B.H., and D. Salt, eds. 2006. *Resilience Thinking.* Washington DC, USA: Island Press.

Wheeler, H.C., and M. Root-Bernstein. 2020. 'Informing Decision-making with Indigenous and Local Knowledge and Science.' *Journal of Applied Ecology* 57: 1634–1643. doi:10.1111/1365–2664.13734.

Wheeler, T., and J. von Braun. 2013. 'Climate Change Impacts on Global Food Security.' *Science* 341(6145): 508–513. doi:10.1126/science.1239402.

Wilting, H.C., A.M. Schipper, M. Bakkenes, J.R. Meijer, and M.A.J. Huijbregts. 2017. 'Quantifying Biodiversity Losses Due to Human Consumption: A Global-scale Footprint Analysis.' *Environmental Science and Technology* 51(6): 3298–3306. doi:10.1021/acs.est.6b05296.

Zimmerer, K.S., E.F.B. Lambin, and S.J. Vanek. 2018. 'Smallholder Telecoupling and Potential Sustainability.' *Ecology and Society* 23(1) 30. doi:10.5751/ES-09935-230130.

2

Complexity-based social-ecological systems research: philosophical foundations and practical implications

Rika Preiser,[1] Maja Schlüter,[2] Reinette Biggs,[1,2] María Mancilla García,[2] Jamila Haider,[2] Tilman Hertz[2] and Louis Klein[3]

[1]CENTRE FOR SUSTAINABILITY TRANSITIONS, STELLENBOSCH UNIVERSITY, STELLENBOSCH, SOUTH AFRICA
[2]STOCKHOLM RESILIENCE CENTRE, STOCKHOLM UNIVERSITY, STOCKHOLM, SWEDEN
[3]EUROPEAN SCHOOL OF GOVERNANCE, BERLIN, GERMANY; INTERNATIONAL FEDERATION FOR SYSTEMS RESEARCH, AUSTRIA

Introduction

The social-ecological systems (SES) literature now widely recognises that SES can be characterised as intertwined complex adaptive systems (CAS) (Berkes, Colding, and Folke 2003; Norberg and Cumming 2008, Schoon and Van der Leeuw 2015; Preiser et al. 2018; De Vos, Biggs, and Preiser 2019; Schlüter et al. 2019a). However, understanding the features and behaviour of CAS poses a huge challenge for traditional disciplinary approaches, as researchers are called to study phenomena that are difficult to delineate, define and analyse. This difficulty is related to the fact that CAS have context-sensitive adaptive capacities, which come about as a result of multiple and non-linear recursive causal interactions that cause spill-over effects across different spatial and temporal scales (Levin 2000). Despite the growing scientific understanding that nature and humans in the Anthropocene are intertwined, the tools and technologies we have to measure human influence and effects on natural environments fall short when having to deal with uncertainty and the emergent nature of CAS.

As mentioned in Chapter 1, the field of SES research developed as a response to the growing consensus that there is a need for broader and more integrative approaches to understand and study the interlinked nature of human–environment systems and the challenges to which they give rise (Bammer et al. 2020). There is growing recognition that our traditional scientific frameworks as applied in single disciplines are not adequate for capturing the complexity of global challenges (Wells 2013; Schoon and Van der Leeuw 2015). The speed and scale at which sustainability challenges emerge and change motivate us to find ways to more deeply understand the nature of the problems we face. Building on this recognition, SES research draws on a diverse range of disciplines to form an integrated and multi-disciplinary approach to researching the intertwined nature of social-ecological interactions (Berkes and Folke 1998).

DOI: 10.4324/9781003021339-3

Studying the history of ideas and events that informed our current understanding of SES reveals that foundational SES scholars worked in groups of interdisciplinary research projects or networks such as the Resilience Alliance (RA) network, the Millennium Ecosystem Assessment (MA) and newly created international institutes to address the wide-ranging challenges posed by the aspirations and challenges of achieving sustainability. Interdisciplinary collaboration in the early 1990s and 2000s allowed the ideas of leading complexity scientists (Holland 1995), physicists (Gell-Mann 1994; Prigogine 1996), biologists (Rosen 1991; Kauffman 1993), ecologists (Levin 1998, 2000; Holling 2001) and economists (Ostrom 1990; Arthur 1999) to converge at places like the International Institute of Applied Systems Analysis (IIASA) in Austria, the Santa Fe Institute in New Mexico, USA, and the Beijer Institute of Ecological Economics in Sweden. Under the impetus of these collaborations, a growing community of scientists worldwide was inspired to develop new theoretical frameworks and experimental approaches to explain why human–nature systems should be viewed as living CAS that operate under conditions marked by non-equilibrium. Within resilience thinking, for example, the 'Panarchy' framework (Gunderson and Holling 2002) demonstrates the way in which living systems simultaneously persist and innovate or adapt, and reveals how fast and slow, small and big events and processes can transform ecosystems, organisms and human societies (Holling 2004). From a deeper understanding of how living systems simultaneously foster persistence and renew themselves through evolutionary processes that include cycles of growth and collapse, the intertwined nature of adaptive cycles across scales proved a useful point of departure for positing that SES are complex adaptive systems (Gunderson and Holling 2002; Berkes, Colding, and Folke 2003).

Acknowledging the complex and adaptive nature of living systems, SES research proposes a more integrated approach for studying and engaging with the intertwined nature of human–environment relations. Complex adaptive systems thinking provides a way of bridging the study of social and biophysical sciences to understand the features of these phenomena, the interlinked patterns that emerge and the novelty that is created as a result. It also forms the foundation of many of the new integrative approaches and frameworks in SES research (Berkes and Folke 1998; Holling 2001; Folke et al. 2004; Liu et al. 2007; Levin et al. 2013; Rogers et al. 2013; Schoon and Van der Leeuw 2015; Folke 2016; Preiser et al. 2018; Clark and Harley. 2019; Schlüter et al. 2019a). Key areas of SES research such as resilience, adaptability, transformability and stewardship are all informed by the underlying assumptions that are based on an understanding of the characteristics and dynamics of CAS (Folke et al. 2004; Walker et al. 2004; Levin et al. 2013; Folke 2016). Students who enter the field of SES research take this conceptual foundation for granted. Often, the importance of understanding SES as complex adaptive systems and the implications of doing so, as argued in the late 1990s and early 2000s, are not that deeply emphasised any more.

This chapter provides some background to how ideas about and the understanding of CAS developed, and how the recognition that SES are complex adaptive systems introduces a mindshift in how we understand the nature of the world, and what tools and methods we can use to study and understand SES. This chapter discusses the scientific assumptions that inform our understanding of CAS, what these mean for grounding a CAS-based understanding of reality, and how this shift in focus influences the theories, frameworks and methods that we use to study SES and make decisions about how to act and govern complex adaptive SES.

Origins of the concept of complexity and a complexity-based worldview

The proliferation of ideas relating to the concepts of 'complexity' and 'CAS' is relatively recent in the history of scientific ideas. Publications of classic papers that describe notions of complexity in scientific terms began in the 1940s (Midgley 2003). The work of Waldrop (1993) and Lewin (1993) offered a more popular understanding of complexity that could be applied over a wide range of disciplines. Yet, tracing the conceptual and historical roots of the notion of 'complexity', and the original moment at which one would say that a 'theory of complexity' was born, remains a rather daunting and somewhat impossible task. A search through the ever-growing literature on the study of complex systems reveals that there is no unifying 'theory of complexity' (Chu, Strand, and Fjelland 2003), and that one can trace several conceptual origins rooted in different disciplines that have been combined to form a collective understanding of what we have come to know as 'complexity theory'. A closer investigation into the development of CAS thinking reveals that one should rather speak of 'theories of complexity' (Rasch 1991; Chu, Strand, and Fjelland 2003; Alhadeff-Jones 2008; Morrison 2010), considering the 'range of different theories that deal with the implications related to the notion of complexity' (Alhadeff-Jones 2008, 66). The growth in special CAS research groups, journals and books is so profuse (Allen, Maguire, and McKelvey 2011; Byrne and Callaghan 2014) that one might be led to think that it is possible to speak of a 'complexity turn' (Urry 2005) that is informing new ways of doing science and understanding the nature of real-world problems. Often the term 'complexity science' is used interchangeably with 'complexity theory', 'complex adaptive systems' or even just 'complexity'.

The French complex systems philosopher, Edgar Morin (2008), suggests that one possible definition of complexity could be found by looking at the Latin roots of the word 'complex'. The first meaning of the word 'complexity' comes from the Latin word *complexus*, which means 'what is woven together' (Morin 2008). It seems that even in its original form, the notion of complexity tells us that we should not expect a neatly packaged explanation of where it came from and how it came about. The development of a theory of complexity can consequently be described as a 'weaving together' of discoveries made in different scientific disciplines over a period of time and encompasses a collection of concerns and methods recognisable as an entity (Checkland 1993). In his construction of 'a geography of complexity theory', Thrift (1999, 33) describes 'complexity theory' as being a 'scientific amalgam …, an accretion of ideas, a rhetorical hybrid' that has not developed from one point of diffusion. As a result, a standard account of the development of a 'theory of complexity' is not available. Even so, there are attempts to chronologically trace the developments and possible origins of this paradigm of thought; indeed, there are several such accounts (Waldrop 1993; Heylighen 1997; Rescher 1998; Rasch and Wolfe 2000; Meyers 2009; Ramage and Shipp 2009; Castellani 2018).

Although there is no 'grand theory of complexity', one can recognise a certain 'economy of concepts' (Thrift 1999) that arranges itself around the characteristics of CAS (see Section 'The features and behaviour of CAS'). Checkland (1993) suggests that it might be better to think of all the endeavours that have notions of complexity and the study of complex phenomena as their main purpose as processes that embrace a 'complexity approach' rather than trying to unite these efforts in a 'grand theory' of complexity. Similarly, Cilliers (2007, 4) suggests that it might be more effective to deal with complexity by adopting a 'complexity attitude':

> Once we realize that we are dealing with complex things, and we accept the consequences of this, our approach to what we are doing, irrespective of how we are actually doing it, will change fundamentally.

Checkland's notion of a 'complexity approach' and Cilliers's suggestion of a 'complexity attitude' can be linked to what Morin (2008) calls a 'paradigm of complexity' or what others call 'complexity thinking' (Rogers et al. 2013). Morin's use of the term 'paradigm' is based on Thomas Kuhn's definition: Kuhn (1996) defined 'paradigm' as an overarching collection of beliefs and assumptions that result in the organisation of scientific worldviews and practices. Foucault's notion of *episteme* is also related to the notion of paradigm. Instead of trying to conceptualise a general theory of CAS, the notions of 'approach', 'attitude' and 'paradigm' turn the focus of inquiry around in a radical way. These concepts allow one to expand the idea(l) of complexity to the extent that it becomes 'capable of informing all theories, whatever their field of application or the phenomenon in question' (Morin 1992) might be. Formulating a complexity approach, attitude or paradigm thus allows one to look outwards and alongside other discourses. From this meta-position, the notion of complexity arranges itself in such a way that it does not stay passively outside or alongside other discourses, but actively and dynamically infects and disseminates them. For the purpose of this book, we will rather use the word 'worldview' instead of 'paradigm' so as not to get caught up in technical discussions about when a new way of thinking qualifies as being a new paradigm (or not).

Worldviews and how they shape scientific research and our understanding of the world

A worldview contains ideas about how the world and the universe came about, and expresses what we believe to be real, how we can study these phenomena, and how this informs our values and judgements in deciding how we should act in the world (Dilthey 1954). In the field of philosophy of science, we would say that when we are enquiring into the nature of knowing which phenomena are real or not, we are dealing with the fundamental philosophical questions of **ontology** and **epistemology**. Ontology refers to questions and assumptions that relate to understanding what the nature of reality is (i.e. what exists?). Epistemology concerns itself with questions about how we do or can know what exists (i.e. theories and how to gain knowledge about reality) (Hammond 2005; Rousseau 2017).

In many different scholarly publications on CAS, the study of complexity is often posed as an alternative to the classical scientific or Newtonian worldview. Many scholars pose the acknowledgement of complexity as a shift in worldview (Capra 2005; Mazzocchi 2008; Wells 2013; Boulton, Allen, and Bowman 2015) and frame it as an inquiry into the nature of reality that opens up an alternative approach from the one used by Enlightenment scientists (also known as 'Newtonian thinking' or 'Newtonian science') to ground the basic assumptions of traditional scientific inquiry. To understand what this shift means, it is important to understand the significance of the Newtonian worldview, and the assumptions that inform the modern scientific method. We will first explain this in more detail before we discuss the assumptions that inform a CAS-based worldview.

A short introduction to the Newtonian worldview

The modern Enlightenment or Newtonian worldview is based on insights from 16th- and 17th-century European scholars, and the discoveries they made that informed what is called 'the Scientific Revolution', which includes the 'Copernican Revolution' (Toulmin 2001). Advances in telescopes allowed scholars to observe that the Sun is the centre of the universe (heliocentrism) and it replaced the theory developed many centuries earlier by the Egyptian philosopher Ptolemy, who posed that the Earth was the centre of the universe (geocentrism) (Merchant 1989). These findings, together with the discovery of Newton's laws of physics (Newton 1686), formed the birth of 'modern science', which stated that empirically based knowledge that is universally applicable could be used to study the nature of phenomena in the world.

The Scientific Revolution is considered to be a central episode in the history of science, the historical moment at which the unique way of understanding the world that we call 'modern science' and its institutions emerged (Toulmin 2001). Ever since then, classical mechanics were regarded as the foundation of scientific inquiry and formed the conceptual model of the physical world (Mazzocchi 2008). The metaphor of the Earth or nature being constituted in a mechanical manner informed what philosophers of science call the 'mechanistic view of nature' (Merchant 1989, 2018), positing that all natural phenomena behave in ways that are determined by physical mechanisms (such as the laws of gravity and the laws of motion). This view informed an understanding of the natural world as being in equilibrium (stable and steady), orderly, deterministic and predictable in which the parts of matter, like the parts of machines, were well defined, passive and inert (Arthur 2015).

By claiming that only phenomena that can be observed and quantified can be justified to exist, the Newtonian worldview supports a reductionist ontology which assumes that phenomena are empirically verifiable by dividing them into elementary parts (such as atoms and electrons) that can be studied and observed by means of isolation or analysis (by separating the parts from the whole and by cutting the whole up into its smallest parts). Often the dictum 'the whole is equal to the sum of its parts' is used to qualify a reductionist ontology and implies that all properties of an object can be explained through the individual behaviour of its smallest constituent parts.

The Newtonian method assumes that knowledge can only be considered scientific if the processes of observing, experimenting with and measuring phenomena are based on the conditions of independent verifiability and reproducibility (Joel 1983). Related to these conditions are the principles of empirical verification and deductive reasoning, which establish strict conditions under which a theory can be proven to be true. These principles and conditions underpin the possibility of gaining objective knowledge about phenomena. Objective knowledge is defined as universally verifiable knowledge that is not influenced by contextual variables or subjective interpretations of observations or measurements.

Based on this reductionist ontology and objective epistemology, the Newtonian worldview produced a scientific basis on which to formulate universal knowledge about phenomena that were seen to behave in a deterministic and predictable manner. This allowed scholars to express universal laws and make predictions of how phenomena would behave once initial conditions were known, based on the assumption that the behaviour of the system could be described in terms of linear equations. This ability gave the Newtonian worldview a significant advantage over other worldviews, as it provided the conceptual and methodological tools to justify and verify the truth about what is real. Combined with the advantage of being able to make predictions about behavioural outcomes of material phenomena, the Newtonian worldview provided a means to inform policies, social processes and institutions that favoured and supported the mechanisation, industrialisation, standardisation and formalisation of processes of production and modes of organising societal norms (Toulmin 2001).

The postulates and principles of the Newtonian scientific worldview that were founded in the natural sciences through physics and mathematics formed such a coherent framework for explaining mechanistic and physical phenomena in equilibrium that its assumptions were soon transferred to other fields of study such as the social sciences to revise and reconstruct theories for guiding our understanding of the human condition. In the discipline of economics, for example, the Newtonian worldview inspired theories that viewed the economy as a well-ordered system in equilibrium in which agents are all identical and rational and make independent decisions by analysing trade-offs between personal cost and benefits in order to determine whether the action is worth pursuing for the best possible outcome (Arrow 1968;

Arthur 2015). In the discipline of sociology, Newtonian principles were embraced to inform modernist theoretical positions such as Structuralism. Structuralism provided ontological and methodological foundations that allowed social scientists to uncover and define the abstract structures that underlie all the things that humans do, think, perceive and feel (Blackburn 2008). Structuralism rose to prominence in France in the 1960s and formed a movement that offered a single unified approach to human life that would embrace a wide range of disciplines such as linguistics (De Saussure 1974), psychoanalysis (Lacan 2006), psychology (Piaget 1985), literary theory (structural semiotics) and anthropology (Lévi-Strauss 1963). In the discipline of philosophy, the tradition of analytic philosophy adopted the principles of formal logic and mathematics to establish principles for formulating conceptual clarity and rigour in arguments through the use of language (Tarski 1959). Central figures in this historical development of analytic philosophy were Gottlob Frege (1980), Bertrand Russel (1945), Ludwig Wittgenstein (1953), Saul Kripke (1972) and Karl Popper (1972).

The breakdown of the mechanistic worldview

Although the Newtonian worldview serves as the basis of the modern scientific approach, new discoveries in the field of quantum physics and relativity theory after the 1950s provided significant results that showed that the Newtonian principles were only valid when applied to well-defined problems and explained the behaviour of matter only under certain stable conditions. The breakdown of the mechanistic, unified and stable worldview was further shattered by the inadequacy of the Newtonian paradigm to formalise the behaviour and fundamental nature of quantum particles. In addition, the work of Russian physicist, Ilya Prigogine (1996), revealed that the 'clock-like machine model of nature and society that dominated the better part of three centuries of western thought' (Merchant 2018) broke down when questions were asked about the nature of phenomena on the subatomic level or under conditions of very high or very low temperatures. Prigogine received the Nobel Prize for his study of open systems and argued that classical thermodynamics holds only for systems that are in equilibrium or near equilibrium, such as pendulum clocks, steam engines and solar systems (Prigogine 1996). These are stable systems in which small changes in the system lead to adjustments and adaptations. They are described mathematically by the great Enlightenment scholars who used calculus and linear equations.

But what happens when the input is so large that the system cannot adjust? In these far-from-equilibrium systems, non-linear dynamics take over. In these cases, small inputs can produce new and unexpected effects (Merchant 2018), as we have seen in the COVID-19 global pandemic where it is now assumed that a virus jumped from an infected wild animal to humans, with massive global economic and social consequences. These new discoveries in the physical sciences led to the emergence of a new ontology, and coincided with discoveries that were being made in other scientific fields such as biology, ecology, cybernetics and artificial intelligence. Insights from research in ecology illustrated that, by departing from a mechanistic worldview, the notion of life and living organisms was better understood in terms of a 'systems view of live' (Capra and Luisi 2014). This introduced a new way of thinking, and offered new perceptions, a new language and new concepts for describing the processes and features of living organisms that could also be applied to social phenomena. The birth of general systems theory is often ascribed to the work of the biologist Ludwig von Bertalanffy (1968), who formulated a general theory of systems that could explain the behaviour of all living systems. He was interested in finding those principles that are common or general to all organisms to provide the conceptual language for an 'organismic' science. As he observed,

> there exist models, principles, and laws that apply to generalized systems or their subclasses, irrespective of their particular kind, the nature of their component elements, and the relationships or "forces" between them. It seems legitimate to ask for a theory, not of systems of a more or less special kind, but of universal principles applying to systems in general.
>
> *(Von Bertalanffy 1968, 32)*

In his critique of the mechanistic and reductionist metaphysics of science, Ulanowicz (2009) proposes 'a conversion of mind' concerning the Newtonian paradigm. With his focus on the notion of 'ecology', he suggests that we need a scientific worldview that departs from being 'wholly dependent on physics and chemistry for its explanations'. Building on the work of Gregory Bateson (1972), Ulanowicz (2009) argues that it is crucial to find complementary narratives for the same phenomena in order to accommodate the characteristics of complex and adaptive ecological dynamics. These differences can be used to develop a logical and coherent perception of ecological phenomena in general and to understand the idea of life specifically. Looking through 'Newtonian glasses' (Ulanowicz 2009) does not give the full picture.

What is of importance in tracing the roots of CAS understanding is not the scientific and historical details of these changes, but the fact that a shift took place in the way in which natural scientists and subsequently social scientists and other applied fields of study thought about our relationship to the world and about humankind's status in the world. This rupture in the classical scientific view of the relationship between our descriptions of the world (epistemology) and the nature of what is real (ontology) opened up a new space in which new concepts and theories could take shape and develop. The seed of what is now known as 'the complexity approach' took root in this space, eventually finding its way into informing the pioneering ideas of SES research.

A relational worldview of CAS

A CAS worldview suggests that systems are constituted not only by parts and mechanical kinds of interactions, but that they come about as the result of relations and organisational processes that constitute matter and its interactions. The discoveries in the fields of general systems theory, cybernetics, studies of neural networks, biology and ecology during the 1950s pointed to the fact that not all matter behaves in the same way as planets, deterministic machines or atoms. Living systems, in particular, come about and behave in ways that are the result of underlying emergent and complex sets of causal relations and organisational patterns (Von Foerster 1960; Von Bertalanffy 1968; Bateson 1972; Kauffman 1993; Gell-Mann 1994; Meadows 2008; Capra and Luisi 2014). By drawing on these discoveries, it became clear that the essential properties of living systems are properties of the whole, which none of the constituent parts has. The immune system is a good example of this: we cannot extract it from the body or analyse it under a microscope, because it does not exist in one particular organ, but is a systemic property of the interactions of various organs, processes and body functions.

In her definition of a system, Meadows (2008) argues that a system can be described as 'a set of things – people, cells, molecules, or whatever – interconnected in such a way that they produce their own pattern of behaviour over time'. This basic and succinct description of a system suggests that systemic properties emerge through the dynamic interaction between interconnected elements that cause systems to produce their own patterns of behaviour over time. The interlinked parts of systems produce emergent effects that are different from the combination of effects of each part on their own. In this understanding we see that relations and the emergent causal organisational interactions are acknowledged to have real effects on a systemic level of the whole, and we can therefore say that the relations are ontological (i.e. something real). This view contrasts fundamentally with the Newtonian worldview, which excludes unobservable, immeasurable relations and organisational patterns as having any ontological status.

A CAS-based worldview therefore offers an alternative to the Newtonian ontology, and entails a shift or move beyond viewing the world in a mechanistic sense (Ulanowicz 1999, 2007). When dealing with CAS, the limitations of the grounding principles of the Newtonian worldview are revealed. A CAS approach provides an expanded worldview with assumptions that are more attuned to the nature of living systems. As argued by many CAS scholars, this worldview is based on understanding that the nature of reality is not mechanistic but organic, and allows us to see the world as being made up of interconnected dynamic relations and interactions that are generative and adaptive, unorderly, unpredictable and full of surprises (Wells 2013; Arthur 2015; Merchant 2018). Research in various fields has shown that even when the underlying rules or interactions that constitute a system are extremely simple, the behaviour of the system as a whole can be rich and complex (Cilliers 1998). The processes of organisation in complex systems are not compressible or reversible (Wolfram 2002). Emergent properties (such as life, consciousness, climate) arise from the interactions and relations between the constitutive parts and are destroyed when the system is dissected or isolated. A CAS-based worldview allows us to regard such non-material causes, relations and organisational patterns as being real and regard the emergent nature of phenomena as essential systemic properties.

This kind of worldview is known in philosophy as a process-relational worldview and has been discussed since the pre-Socratics. Heraclitus, for example, is credited with the famous expression 'everything flows'. A process-relational worldview provides conceptual constructs that highlight connections and relational qualities, and focuses attention on processes and relations, as opposed to objects, as the primary constituents of reality (Hertz, Mancilla García, and Schlüter 2020; Mancilla García et al. 2020). Processes can be understood as patterned change over time, and their properties and functions are defined by the set of relations that constitute them. These relations span different fields of study and integrate the social and the ecological, which is why a process-relational worldview is particularly useful for conceptualising SES and integrating CAS theories into the study of SES (Rogers et al. 2013).

Social-ecological intertwinedness

A particular feature that distinguishes SES from other CAS, such as financial systems or ecosystems, is their social-ecological intertwinedness. The notion of intertwinedness captures the co-constitutive nature of social and ecological relationships. Drawing attention to this intertwinedness is necessary to safeguard against reductionist tendencies that treat the social and the ecological as separate realms that exist, and can be studied and understood, independently from each other. Instead, the concept of social-ecological intertwinedness emphasises that SES are co-constituted in ways where one cannot understand the social aspects without making reference to the ecological, and vice versa. Put differently, the conceptualisation of SES as intertwined systems views entities, agency and other SES outcomes as coming into existence through the interaction of social and ecological processes to the extent that it is impossible – and rarely useful – to separate the two.

Intertwinedness is driven by concrete processes in space and time, such as processes of co-adaptation and coevolution. Lansing and Kremer's (1993) account of the emergence of the highly complex crop management system on the island of Bali, for example, is seen as a process of co-adaptation involving farmers, crops, pests, cultural practices and the physical geography of Bali. What these individual components are (and why they do what they do in bringing the emergent phenomena about) is defined by – and can only be understood with reference to – the very process of co-adaptation.

The features and behaviour of CAS

Complex adaptive systems have certain distinguishable properties (features) and behaviours (dynamics) that invite us to explore and discover new ways of studying and governing systems such as SES. Preiser et al. (2018) developed a set of six general organising principles that can be used to qualify and define the way in which complex phenomena present themselves in the world. The six principles present a typology of characteristics that allows us to discern the qualities of CAS and offer suggestions on the practical implications of CAS-based approaches for assessing and applying appropriate methods to study, understand and govern CAS. These principles are:

1. **Constituted relationally:** Complex adaptive systems are constituted relationally, which means that complex behaviour and structures emerge as a result of the recursive and aggregate patterns of relations that exist between the component parts of systems. These relations usually give rise to rich interactions within the system, meaning that any element in the system influences and is influenced by many other ones (Cilliers 1998) either directly, or indirectly via positive (reinforcing) or negative (balancing) feedbacks.
2. **Adaptive:** Complex adaptive systems have adaptive capacities and self-organise and coevolve in relation to contextual changes. Self-organisation describes the process whereby a system can develop complex structures from fairly unstructured beginnings without the intervention of an external designer or the presence of some centralised form of internal control (Ashby 1947). Coevolution describes the recursive patterns or relations of influence that result from ongoing exchanges between components of evolving systems, practices, knowledge, beliefs and values, and the biophysical environment that mutually influence one another (Norgaard 1994; Haider et al. 2020).
3. **Dynamic:** Complex adaptive systems are characterised by dynamic relations. In other words, the relationships in a system are constantly changing in rich and unexpected ways. These relations are mostly non-linear, which means the relationships between any two factors or processes are not necessarily uniform or proportional (Boulton, Allen, and Bowman 2015). Non-linearity can be the result of feedbacks, path dependencies, time lags or multiple time scales, which suppress or magnify processes and interactions, both internally and between the system and its environment. In CAS, non-linear dynamics also arise because the relations between variables constantly change, which renders them uncertain and unpredictable and makes these systems difficult to control (Arthur 1999). Change and not stability is thus the norm in CAS, shifting the focus from analysing stable states to analysing transient processes (the behaviour of the system in between equilibria), and from analysing outcomes to focusing on the trajectories or processes of the system.

4. **Radically open:** Complex adaptive systems are radically open. In other words, it is the activity of the system in relation to the environment that constitutes the system itself (Cilliers 2002). This implies that we cannot clearly discern the boundary between the system and its environment, because the environment co-constitutes the identity of the system. Our definitions of systemic boundaries are thus the product of physical properties (e.g. a watershed boundary that signals a system boundary), mental constructions (i.e. where we choose to draw the line between the system and the environment (Ulrich 2000; Rajagopalan and Midgley 2015)) or the problem or research question we want to address (including the temporal and spatial scales of interest).
5. **Contextual:** Complex adaptive systems are context dependent, meaning that the function(s) of CAS are contingent on context. Changing the context will have an impact on the function of the system. In other words, the environment suppresses or enhances possible systemic functions (Poli 2013). Moreover, the functions that we ascribe to complex systems are contingent on the level of analysis that we employ to understand a system.
6. **Complex causality and emergence:** Complex adaptive systems are characterised by complex causality and emergence. Cause-and-effect interactions in CAS are not unidirectional or linear, but marked by complex recursive causal pathways that are non-linear and dynamic (Rasch and Knodt 1994). Emergence occurs when entities are observed to have systemic properties that are different and non-reducible to the properties of the constituent elements. It is not that the sum is greater than the parts, but rather that the system's effects are different from those of its parts (Urry 2005). Emergent phenomena have causal agency and are real, i.e. they have ontological status (Kauffman 2008).

Table 2.1 summarises the conceptual and practical implications of these features for studying and intervening in SES. The challenge of being able to understand CAS, and the difficulties it poses for research are among the distinguishing characteristics of CAS-based approaches to sustainability (Allen 2001; Bammer et al. 2020). Addressing intertwined, complex sustainability challenges in SES requires expertise in integrating research, and practice to develop a more holistic and comprehensive understanding of the nature of these challenges and to embrace the 'messiness' of working with them (Duit and Galaz 2008; Rogers et al. 2013; Arthur 2015; Boulton, Allen, and Bowman 2015). The implications of these issues for the practical implementation and design of SES research are further discussed in Chapter 3.

Implications of a CAS worldview for doing SES research

Understanding the nature of SES as complex adaptive systems poses new frontiers for studying, governing and influencing SES (Biggs, Schlüter, and Schoon 2015; Bodin 2017; Österblom et al. 2017). Understanding SES as complex adaptive systems has profound implications for our assumptions about what kind of knowledge we can have of SES (ontological implications), how we gain knowledge of SES (epistemological implications) and how we judge whether we have conducted our research in 'good' and just ways (ethical implications).

As discussed earlier in this chapter, many different disciplines have influenced the development of a range of 'theories of complexity'. As a consequence, there also exists a diversity of methods for doing research on CAS (Preiser 2019). Some methods aim to quantify and simulate the behaviour, connections, structures and phases of complex systems by means of mathematical equations, algorithms and computational models (Thurner, Hanel, and Klimek 2018). Other approaches extend the vocabulary of computational complexity to a qualitative engagement with the features of complexity (Audouin et al. 2013).

Table 2.1 General organising principles of complex adaptive systems (CAS), and implications for SES research approaches and methods

Organising principles of CAS	*Conceptual implications for SES research*	*Practical guidelines for SES research*
Constituted relationally	• The nature and structure of relationships in an SES have to be considered explicitly. • Diversity is key and allows for different kinds of SES interactions to take place.	• Assess the nature of relations and structures of networks and connectivities • Foster trust, dialogue, distributed accountability and collaboration across a variety of networks and levels of organisation • Create integrative frameworks and methods to assess relations and connectivity • Actively recognise diversity as a resource in the system • Create transformative spaces where people can learn and foster the experience of being connected to one another and nature in deep and meaningful ways
Adaptive	• The function and structure of SES change with temporal and spatial changes. • Multiple modes of reorganisation are possible when systems undergo change. • Adaptive capacity results from a system's ability to learn and have memory. • Change happens through adaptation, evolution and transformation. • Control is not located in one isolated element of the system, but is spread throughout the nodes and relations of the system.	• Critically reflect on planning and strategy design and implement adaptive co-management practices to foster iterative learning and collaborative processes of engagement • Facilitate continuous innovation based on experiential learning across several iterations of trial and error • Support capacities that allow for self-organising processes • Develop holistic frameworks that cultivate synthesis rather than analysis • Assess resilience and anticipate possible future patterns and pathways
Dynamic	• System behaviour is amplified or dampened by feedback loops, and can lead to tipping points and regime shifts. • Feedback structures are responsible for the changes we experience over time. • Structures and processes are linked across scales. • SES are characterised by inherent unpredictability and uncertainty.	• Map systemic feedbacks across different spatial and temporal scales • Assess which mechanisms build or inhibit systemic agency and resilience • Identify systemic thresholds, traps and indicators that could help detect possible regime shifts • Capture spatial and temporal cross-scale dynamics • Investigate thresholds and tipping points

(*Continued*)

Organising principles of CAS	*Conceptual implications for SES research*	*Practical guidelines for SES research*
Radically open	• Delimiting SES problems and systems is challenging as real-world problems have no natural boundaries. • External variables could have important influences on system behaviour but cannot be included in the models of the system. • Any modelled system is embedded in a larger system.	• Treat projects and geographical locations as if they are not closed and isolated entities • Be aware that unknown variables could have important influences on system behaviour and expect these to have real effects on the system under study • Assess teleconnections and the effects of the flow of energy, matter and information to demonstrate how systems are embedded in other systems
Contextual	• SES are context sensitive. • SES components have multiple functions that change when the context changes. • Context is not a passive backdrop to a system, but an active agent in itself, which enables or inhibits systemic agency. • Many contested problem definitions exist simultaneously and the various stakeholders involved in an SES will have different mental models or beliefs that inform values and understandings of both the causes and the possible actions that could be taken to find possible pathways for action.	• Foster iterative processes of meaning-making that facilitate dialogue to include multiple perspectives from a wide range of stakeholders • Use multiple evidence-based data sources to co-create and integrate different knowledge bases • Develop context-dependent assessments and systemic understanding of challenges
Complex causality and emergence	• Cause-and-effect cannot be traced in linear causal trajectories. • Emergent phenomena arise from multiple recursive patterns and unintended outcomes.	• Engage methods that can illuminate emergence and unexpected outcomes • Adopt a complexity-based frame of mind in considering innovative practices and new decision possibilities • Expect uncertainty and surprises to be part of any engagement with complex SES • Anticipate alternative future pathways and innovations through experimental processes such as scenarios and foresight methods

Source: Adapted from Preiser et al. 2018.

Within the field of SES research, researchers use field research (e.g. Hahn et al. 2006; Gelcich et al. 2010; Herrfahrdt-Pähle et al. 2020), network approaches (e.g. Bodin et al. 2019), and mathematical and computational modelling (Schlüter and Pahl-Wostl 2007; Lade et al. 2017; Martin, Schlüter, and Blenckner 2020), as well as combinations thereof, to shed light on the behaviour of SES, often in exploratory ways. Several methods and tools (like dynamic modelling) are used to explore the different ways an SES may unfold in different contexts (e.g. Lade et al. 2017) and to enhance understanding of key processes and CAS properties such as self-organisation or emergence from local interactions, feedbacks, stochasticity and intertwinedness (Lindkvist, Basurto, and Schlüter 2017; Orach, Duit, and Schlüter 2020). In Chapter 33 we provide a synthesis of the current landscape of SES research methods and their ability to study and account for various CAS features. In the next sections, we discuss the ontological, epistemological and ethical implications of CAS-based approaches to SES research.

Ontological implications: complexity is a real feature of systemic interactions

From a CAS-based worldview, complexity emerges as a real property of systems that exhibit the six principles discussed in the previous section. A CAS-based worldview suggests that reality is constituted by the 'complex interaction between dynamic, open, stratified systems, where particular structures give rise to certain causal powers, tendencies or ways of acting' (Mingers 2000, 1261–1262). Complex adaptive systems therefore do not exist independently from the phenomena and processes that constitute them (Gell-Mann 1994; Holland 1995; Cilliers 1998). As an outcome of dynamic relations and processes, complexity is thus simultaneously a combination of the attributes of the system (ontological) and a function of our present understanding of that system (epistemological) (Cilliers 2008).

This implies that there is no objective position from which to study complex phenomena, as knowledge of CAS is always context sensitive. This knowledge is best generated through methods that seek to understand the bigger picture (holistic) of how certain patterns of behaviour are linked to various contexts, histories and different variables, and how they change over time. This implies that gathering data or information about CAS is often best achieved by methods and research approaches that allow us to record and track the changing nature of phenomena over temporal and spatial scales, and that allow us to see how systems adapt and respond to dynamic interactions such as feedbacks and tipping points. It also suggests that we cannot generate universal objective knowledge about CAS, but have to allow ourselves to delve into the process of observing adaptation, change, diversity and emergent behaviour.

A relational worldview is well supported by the broader theoretical position as upheld in the ideas of critical realism (Mingers 2006). Critical realism is a branch of philosophy that distinguishes between the 'real' world and the 'observable' world. It suggests that although the 'real' (that which exists) cannot be observed, it ultimately exists independently from human perceptions, theories and constructions. Just because we cannot observe it, it does not mean that we should dismiss it (which would revert to a constructivist form of realism).

Critical realism theory can be applied to social sciences as well as natural sciences, and generally informs an understanding of knowledge generation that is grounded in the notion that unobservable events can cause observable events, structures and processes. Values and beliefs about the importance of some natural phenomena or events cannot be observed, for example, but they influence the policies and social norms that inform whether those phenomena are worthy of protection or conservation, or not. A CAS-based worldview does not imply that 'everything goes' nor that we should disregard the validity of certain kinds of knowledge systems. In fact, it rather invites us to consider multiple causes that can interact in seen and unseen ways to influence systems in ways that are unpredictable and not always quantifiable, but have significant influences on CAS dynamics across temporal and spatial scales.

Epistemological implications: embrace methodological pluralism for studying CAS

Complex adaptive systems-based research approaches introduce a different way of thinking about the world and how to understand our place in it. Although it does not provide us with a foolproof, best-practice manual for how to design research projects or change interventions (Preiser and Cilliers 2010), it does provide some general premises that may reduce the tendency to oversimplify reality or analyse systems in ways that generate misleading conclusions. To produce empirically valid and meaningful data and interpretations of the diversity of features and properties of SES, we need to expose and understand the underlying causal relationships, patterns and processes that generate systemic behaviour, patterns and events that govern anthropogenic and non-anthropogenic drivers and social-ecological conditions (Capra 2005; Österblom et al. 2013). Deciding which methods and models are appropriate for a particular purpose and research aim is not obvious, and choices are often made on subjective grounds such as experience, usefulness or even intuition (Mingers 2000; Audouin et al. 2013; Cilliers et al. 2013). This means we often take pragmatic research approaches.

Knowledge of CAS is always partial and our knowledge of certain phenomena can change over time as we gain deeper insight into the features and effects of certain systemic behaviour. Therefore, the best strategy for developing an integrated understanding of SES is to explore a variety of models and frameworks that span a broad spectrum of methodologies and disciplinary divides (Cilliers 2002; Poli 2013; Tengö et al. 2014). A CAS-based epistemology comprises a range of scientific theories and frameworks (Chapman 2016) that can describe, assess and confirm the complex features and dynamics of CAS. By combining different strategies and methods of collecting and interpreting knowledge, insights from different perspectives can be enriched and integrated, and help to contextualise the knowledge claims made by different disciplines (Morin 2008; Bammer et al. 2020). Ulrich (1994, 35) suggests that 'from this new perspective, the implication of the systems idea is not that we must understand the whole system, but rather that we critically deal with the fact that we never do'.

A CAS-based worldview therefore emphasises the benefits and need for methods of inquiry and knowledge-generating practices that draw from a plurality of relevant epistemologies and frameworks (Mitchell 2004; Moon and Blackman 2014; Tengö et al. 2014; Reyers et al. 2015). Engagement with different knowledge types and forms needs to occur in ways that facilitate interactions among researchers from different disciplinary backgrounds (Berkes, Colding, and Folke 2003; Burns and Weaver 2008; Audouin et al. 2013; Klein 2016; Schlüter et al. 2019b) as well as among researchers and stakeholders who should be involved in the research process (Cockburn et al. 2020). Practising methodological pluralism (Norgaard 1994) and epistemological agility (Haider et al. 2018) can therefore be considered as key competencies in SES research. The epistemological limitations of what we can know imply a need for a critical attitude towards tackling problems and issues of decision-making, and provide us with an ethical basis for developing tools of critical reflection.

Ethical implications: actions and decisions can never be made without considering the intertwined nature of complex SES

Complex adaptive systems-based approaches highlight the need for critical reflexivity to align ontological, epistemological and methodological commitments. Social-ecological research is often driven by solutions-oriented and use-inspired outcomes (Clark 2007). In the quest for being pragmatic, a common problem in this kind of 'solutions-oriented' research can be to use readily available approaches and methods that poorly account for CAS assumptions. The use of such methods risks imposing a particular understanding of reality on the research, and closing down other understandings that may allow novel insights and practical approaches to effecting change to emerge. At the same time, it is critical to recognise that choices have to be made; a normative element is therefore always present in our attempts to understand the complex intertwinedness of SES. The normative dimension of our knowledge of CAS means that engagement is needed, not only in generating an understanding of the system itself but also in choosing – and making explicit – the context/framework by which that knowledge is generated. These interrelated tasks typically call for a transdisciplinary approach that includes empirical, pragmatic and normative or value-based knowledge (Max-Neef 2005; Burns, Audouin, and Weaver 2006).

In navigating research, action and decision-making processes in the Anthropocene, the relational interdependencies of SES should always be acknowledged. Complex adaptive systems-based approaches suggest a need to proceed differently and call for more inclusive and integrative modes of engaging with real-world SES problems that acknowledge the intertwinedness of humans and nature, the limits of what is knowable and how we can act to effect change in complex SES. Complex adaptive systems-based approaches call for participatory and collaborative multi-stakeholder processes that foster dialogue and knowledge co-creation, and the development of more systemic awareness (Hammond 2005). In particular, they call for research informed by broader societal framings and understandings of problems, novel forms of collaborative agency with various actors in a system and alternative moral constructs (Woermann 2016). Finally, they imply the need to actively favour the integration of nature, society and technology in policy design and implementation.

Conclusion

This chapter provided insight into the philosophical and conceptual underpinnings of SES research. Knowing that the nature of reality is complex has profound consequences for how we go about understanding and acting in real-world, intertwined SES. A complexity-based understanding of the nature of reality has emerged through insights and influences from a wide variety of different disciplines, and has arisen as an alternative to the Newtonian worldview that has dominated modern science. This shift in our assumptions about the nature of the world, from a mechanistic cause-and-effect understanding to a much more organic understanding with complex causation, has deep consequences for the ways in which we go about doing SES research, our assumptions about what we can know, and how we can act on that knowledge to effect change and address real-world sustainability challenges. The next chapter builds on this understanding to discuss the practical design and execution of SES research.

Acknowledgements

Rika Preiser's salary was co-funded by the Swedish International Development Agency's (Sida) GRAID and Swedbio programmes. Maja Schlüter received funding from the

European Research Council under the European Union's Horizon 2020 research and innovation programme (grant agreement no. 682472 – MUSES). Reinette Biggs received support from the South African Research Chairs Initiative (SARChI) (grant 98766), the GRAID project and a Young Researchers Grant from the Vetenskapsrådet in Sweden (grant 621-2014-5137).

References

Alhadeff-Jones, M. 2008. 'Three Generations of Complexity Theories: Nuances and Ambiguities.' *Educational Philosophy and Theory* 40(1): 66–82.

Allen, P. 2001. 'What is Complexity Science? Knowledge of the Limits to Knowledge.' *Emergence* 3(1): 24–42.

Allen, P., S. Maguire, and B. McKelvey, eds. 2011. *The Sage Handbook of Complexity and Management.* London: Sage.

Arrow, K. 1968. 'Economic Equilibrium.' In *International Encyclopedia of the Social Sciences*, edited by R.K. Merton and D.L. Sills, 376–389. New York: Macmillan.

Arthur, W.B. 1999. 'Complexity and the Economy.' *Science* 284: 107–110. doi:10.1126/science.284.5411.107.

Arthur, W.B. 2015. *Complexity and the Economy.* Oxford: Oxford University Press.

Ashby, W.R. 1947. 'Principles of the Self-organizing Dynamic System.' *Journal of General Psychology* 37: 125–128.

Audouin, M., R. Preiser, S. Nienaber, L. Downsborough, J. Lanz, and S. Mavengahama. 2013. 'Exploring the Implications of Critical Complexity for the Study of Social-Ecological Systems.' *Ecology and Society* 18(3): 12. doi:10.5751/ES-05434-180312.

Bammer, G., M. O'Rourke, D. O'Connell, L. Neuhauser, G. Midgley, J.T. Klein, N.J. Grigg et al. 2020. 'Expertise in Research Integration and Implementation for Tackling Complex Problems: When is it Needed, Where can it be Found and How can it be Strengthened?' *Palgrave Communicatinos* 6, Article 5. doi:10.1057/s41599-019-0380-0.

Bateson, G. 1972. *Steps to an Ecology of Mind.* New York: Ballantine Books.

Berkes, F., J. Colding, and C. Folke. 2003. *Navigating Social-Ecological Systems: Building Resilience for Complexity and Change.* Cambridge: Cambridge University Press. doi:10.1017/CBO9780511541957.

Berkes, F., and C. Folke. 1998. *Linking Social and Ecological Systems.* Cambridge: Cambridge University Press.

Biggs, R., M. Schlüter, and M.L. Schoon. 2015. *Principles for Building Resilience: Sustaining Ecosystem Services in Social-Ecological Systems.* Cambridge: Cambridge University Press. doi:10.1017/CBO9781316014240.

Blackburn, S., ed. 2008. 'Structuralism.' In *Oxford Dictionary of Philosophy* (2nd rev ed.), 353. Oxford: Oxford University Press.

Bodin, Ö. 2017. 'Collaborative Environmental Governance: Achieving Collective Action in Social-Ecological Systems.' *Science* 357(6352). doi:10.1126/science.aan1114.

Bodin, Ö., S.M. Alexander, J. Baggio, M.L. Barnes, R. Berardo, G.S. Cumming, L.E. Dee et al. 2019. 'Improving Network Approaches to the Study of Complex Social-Ecological Interdependencies.' *Nature Sustainability* 2(7): 551–559. doi:10.1038/s41893-019-0308-0.

Boulton, J., P. Allen, and C. Bowman. 2015. *Embracing Complexity: Strategic Perspectives for an Age of Turbulence.* Oxford: Oxford University Press.

Burns, M.E., M.A. Audouin, and A. Weaver. 2006. 'Advancing Sustainability Science in South Africa.' *South African Journal of Science* 102: 379–384.

Burns, M., and A. Weaver (eds.). 2008. *Exploring Sustainability Science: A Southern African Perspective.* Stellenbosch: African SUN MeDIA.

Byrne, D., and G. Callaghan. 2014. *Complexity Theory and the Social Sciences: The State of the Art.* New York: Routledge.

Capra, F. 2005. 'Complexity and Life.' *Theory, Culture & Society* 22(5): 33–44. doi:10.1177/0263276405057046.

Capra, F., and P.L. Luisi. 2014. *The Systems View of Life.* Cambridge: Cambridge University Press. doi:10.1017/CBO9780511895555.

Castellani, B. 2018. *Map of the Complexity Sciences.* Art & Science Factory. www.art-sciencefactory.com/complexity-map_feb09.html.

Chapman, K. 2016. *Complexity and Creative Capacity*. London: Routledge. doi:10.4324/9781315680767.
Checkland, P. 1993. *Systems Thinking, Systems Practice*. Chichester: John Wiley & Sons.
Chu, D., R. Strand, and R. Fjelland. 2003. 'Theories of Complexity. Common Denominators of Complex Systems.' *Complexity* 8(3): 19–30.
Cilliers, P. 1998. *Complexity and Postmodernism: Understanding Complex Systems*. London: Routledge.
Cilliers, P. 2002. 'Why We Cannot Know Complex Things Completely.' *Emergence* 4(1–2): 77–84.
Cilliers, P. 2007. *Thinking Complexity. Complexity and Philosophy Volume 1*. Mansfield: ISCE Publishing.
Cilliers, P. 2008. 'Complexity Theory as a General Framework for Sustainability Science.' In *Exploring Sustainability Science. A Southern African Perspective*, edited by M. Burns and A. Weaver, 39–57. Stellenbosch: African Sun Media.
Cilliers, P., H.C. Biggs, S. Blignaut, A.G. Choles, J.S. Hofmeyr, G.P W. Jewitt, and D.J. Roux. 2013. 'Complexity, Modeling, and Natural Resource Management.' *Ecology and Society* 18(3): 1. doi:10.5751/ES-05382-180301.
Clark, W.C. 2007. 'Sustainability Science: A Room of its Own.' *Proceedings of the National Academy of Sciences of the USA* 104: 1737–1738. doi:10.1073/pnas.0611291104.
Clark, W.C., and A.G. Harley. 2019. *Sustainability Science: Towards a Synthesis*. Sustainability Science Program Working Paper. John F. Kennedy School of Government, Harvard University, Cambridge, MA.
Cockburn, J., M. Schoon, G. Cundill, C. Robinson, J.A. Aburto, S.M. Alexander, J.A. Baggio et al. 2020. 'Understanding the Context of Multifaceted Collaborations for Social-Ecological Sustainability: A Methodology for Cross-case Analysis.' *Ecology and Society* 25(3): 7. doi:10.5751/ES-11527-250307.
De Saussure, F. 1974. *Course in General Linguistics*. London: Fontana.
De Vos, A., R. Biggs, and R. Preiser. 2019. 'Methods for Understanding Social-Ecological Systems: A Review of Place-based Studies.' *Ecology and Society* 24(4): 16. doi:10.5751/ES-11236-240416.
Dilthey, W. 1954. *The Essence of Philosophy*. Translated by S.A. Emery, and W.T. Emery. Chapel Hill: University of North Carolina Press.
Duit, A., and V. Galaz. 2008. 'Governance and Complexity: Emerging Issues for Governance Theory.' *Governance* 21(3): 311–335.
Folke, C. 2016. 'Resilience' (Republished). *Ecology and Society* 21(4): 44. doi:10.5751/ES-09088-210444.
Folke, C., S. Carpenter, B. Walker, M. Scheffer, T. Elmqvist, L. Gunderson, and C. S. Holling. 2004. 'Regime Shifts, Resilience, and Biodiversity in Ecosystem Management.' *Annual Review of Ecology, Evolution, and Systematics* 35: 557–581. doi:10.1146/annurev.ecolsys.35.021103.105711.
Frege, G. 1980. *Philosophical and Mathematical Correspondence*. Chicago: The University of Chicago Press.
Gelcich, S., T.P. Hughes, P. Olsson, C. Folke, O. Defeo, M. Fernandez, S. Foale et al. 2010. 'Navigating Transformations in Governance of Chilean Marine Coastal Resources.' *Proceedings of the National Academy of Sciences* 107(39): 16794–16799. doi:10.1073/pnas.1012021107.
Gell-Mann, M. 1994. *The Quark and the Jaguar: Adventures in the Simple and the Complex*. London: Little, Brown and Company.
Gunderson, L.H., and C.S. Holling. 2002. *Panarchy: Understanding Transformations in Human and Natural Systems*. Washington: Island Press.
Hahn, T., P. Olsson, C. Folke, and K. Johansson. 2006. 'Trust-building, Knowledge Generation and Organizational Innovations: The Role of a Bridging Organization for Adaptive Co-management of a Wetland Landscape around Kristianstad, Sweden.' *Human Ecology* 34(4): 573–592. doi:10.1007/s10745-006-9035-z.
Haider, L.J., W.J. Boonstra, A. Akobirshoeva, and M. Schlüter. 2020. 'Effects of Development Interventions on Biocultural Diversity: A Case Study from the Pamir Mountains.' *Agriculture and Human Values* 37: 683–697. doi:10.1007/s10460-019-10005-8.
Haider, L.J., J. Hentati-Sundberg, M. Giusti, J. Goodness, M. Hamann, V.A. Masterson, M. Meacham et al. 2018. 'The Undisciplinary Journey: Early-career Perspectives in Sustainability Science.' *Sustainability Science* 13: 191–204. doi:10.1007/s11625-017-0445-1.
Hammond, D. 2005. 'Philosophical and Ethical Foundations of Systems Thinking.' *TripleC* 3(2): 20–27. doi:10.31269/triplec.v3i2.20.
Herrfahrdt-Pähle, E., M. Schlüter, P. Olsson, C. Folke, S. Gelcich, and C. Pahl-Wostl. 2020. 'Sustainability Transformations: Socio-political Shocks as Opportunities for Governance Transitions.' *Global Environmental Change* 63: 102097. doi:10.1016/j.gloenvcha.2020.102097.

Hertz, T., M. Mancilla García, and M. Schlüter. 2020. 'From Nouns to Verbs: How Process Ontologies Enhance Our Understanding of Social-Ecological Systems Understood as Complex Adaptive Systems.' *People and Nature* 2(2): 328–338. doi:10.1002/pan3.10079.
Heylighen, F. 1997. 'Publications on Complex, Evolving Systems: A Citation-based Survey.' *Complexity* 2(5): 31–36.
Holland, J. 1995. *Hidden Order: How Adaptation Builds Complexity.* Reading: Addison-Wesley.
Holling, C.S. 2001. 'Understanding the Complexity of Economic, Ecological, and Social Systems.' *Ecosystems* 4(5): 390–405.
Holling, C.S. 2004. 'From Complex Regions to Complex Worlds.' *Ecology and Society* 9(1): 11. www.ecologyandsociety.org/vol9/iss1/art11.
Joel, J.S. 1983. 'Foreword.' *SubStance* 12(3): 5–6.
Kauffman, S. 1993. *The Origins of Order: Self-organization and Selection in Evolution.* New York: Oxford University Press.
Kauffman, S. 2008. *Reinventing the Sacred. The Science of Complexity and the Emergence of a Natural Divinity.* New York: Basic Books.
Klein, L. 2016. 'Towards a Practice of Systemic Change – Acknowledging Social Complexity in Project Management.' *Systems Research and Behavioral Science* 33(5): 651–661. doi:10.1002/sres.242.
Kripke, S. 1972. *Naming and Necessity.* Cambridge: Harvard University Press.
Kuhn, T.S. 1996. *The Structure of Scientific Revolutions* (3rd ed). University of Chicago Press.
Lacan, J. 2006. *Écrits: The First Complete Edition in English.* Translated by B. Fink. New York: W.W. Norton & Co.
Lade, S.J., L.J. Haider, G. Engström, and M. Schlüter. 2017. 'Resilience Offers Escape from Trapped Thinking on Poverty Alleviation.' *Science Advances* 3: e1603043.
Lansing, J.S., and J.N. Kremer. 1993. 'Emerging Properties of Balinese Water Temple Networks: Coadaptation on a Rugged Fitness Landscape.' *American Anthropologist* 95: 97–114. doi: 10.1525/aa.1993.95.1.02a00050.
Lévi-Strauss, C. 1963. *Structural Anthropology.* Translated by C. Jacobson and B.G. Schoepf. New York: Basic Books.
Levin, S.A. 1998. 'Ecosystems and the Biosphere as Complex Adaptive Systems.' *Ecosystems* 1(5): 431–436. doi:10.1007/s100219900037.
Levin, S.A. 2000. 'Multiple Scales and the Maintenance of Biodiversity.' *Ecosystems* 3: 498–506.
Levin, S., T. Xepapadeas, A-S. Crépin, J. Norberg, A. de Zeeuw, C. Folke, T. Hughes et al. 2013. 'Social-Ecological Systems as Complex Adaptive Systems: Modeling and Policy Implications.' *Environment and Development Economics* 18(2): 111–132. doi:10.1017/S1355770X12000460.
Lewin, R. 1993. *Complexity: Life on the Edge of Chaos.* London: Phoenix.
Lindkvist, E., X. Basurto, and M. Schlüter. 2017. 'Micro-level Explanations for Emergent Patterns of Self-governance Arrangements in Small-scale Fisheries – A Modeling Approach.' *PloS ONE* 12: e0175532.
Liu, J., T. Dietz, S.R. Carpenter, M. Alberti, C. Folke, E. Moran, A.N. Pell et al. 2007. 'Complexity of Coupled Human and Natural Systems.' *Science* 317(5844): 1513–1516. doi:10.1126/ science.1144004.
Mancilla García, M., T. Hertz, M. Schlüter, R. Preiser, and M. Woermann. 2020. 'Adopting Process-relational Perspectives to Tackle the Challenges of Social-Ecological Systems Research.' *Ecology and Society* 25(1): 29. doi:10.5751/ES-11425-250129.
Martin, R., M. Schlüter, and T. Blenckner. 2020. 'The Importance of Transient Social Dynamics for Restoring Ecosystems beyond Ecological Tipping Points.' *Proceedings of the National Academy of Sciences of the United States of America.* doi:10.1073/pnas.1817154117.
Max-Neef, M. 2005. 'Foundations of Transdisciplinarity.' *Ecological Economics* 53: 5–16.
Mazzocchi, F. 2008. 'Complexity in Biology.' *EMBO Reports* 9(1): 10–14.
Meadows, D.H. 2008. *Thinking in Systems.* London: Earthscan.
Merchant, C. 1989.*The Death of Nature: Women, Ecology, and the Scientific Revolution.* New York: Harper & Row.
Merchant, C. 2018. *Science and Nature. Past, Present and Future.* New York: Routledge.
Meyers, R., ed. 2009. *Encyclopedia of Complexity and Systems Science.* Berlin: Springer.
Midgley, G., ed. 2003. *Systems Thinking. Volume 1: General Systems Theory, Cybernetics and Complexity.* London: Sage.
Mingers, J. 2000. 'The Contribution of Critical Realism as an Underpinning Philosophy for OR/MS and Systems.' *Journal of the Operational Research Society* 51: 1256–1270.

Mingers, J. 2006. *Realising Systems Thinking: Knowledge and Action in Management Science*. New York: Springer.
Mitchell, S. 2004. 'Why Integrative Pluralism?' *E:CO Emergence: Complexity and Organization* 6: 81–91.
Moon, K., and D.A. Blackman. 2014. 'Guide to Understanding Social Science Research for Natural Scientists.' *Conservation Biology* 28(5): 1167–1177. doi:10.1111/cobi.12326.
Morin, E. 1992. 'From the Concept of System to the Paradigm of Complexity.' *Journal of Social and Evolutionary Systems* 15(4): 371–385.
Morin, E. 2008. *On Complexity*. Cresskill: Hampton Press.
Morrison, K. 2010. 'Complexity Theory, School Leadership and Management: Questions for Theory and Practice.' *Educational Management Administration & Leadership* 38: 374–393.
Newton, I. 1686 [1934]. *Mathematical Principles of Natural Philosophy and His System of the World*. Translated by W. Motte. Berkeley: University of California Press.
Norberg, J., and G. Cumming. 2008. *Complexity Theory for a Sustainable Future*. New York: Columbia University Press.
Norgaard, R.B. 1994. *Development Betrayed: The End of Progress and a Co-evolutionary Revisioning of the Future*. San Francisco: Routledge.
Orach, K., A. Duit, and M. Schlüter. 2020. 'Sustainable Natural Resource Governance under Interest Group Competition in Policy-making.' *Nature Human Behaviour* 4: 898–909. doi:10.1038/s41562-020-0885-y.
Österblom, H., B.I. Crona, C. Folke, M. Nyström, and M. Troell. 2017. 'Marine Ecosystem Science on an Intertwined Planet.' *Ecosystems* 20(1): 54–61. doi:10.1007/s10021-016-9998-6.
Österblom, H., A. Merrie, M. Metian, W.J. Boonstra, T. Blenckner, J.R. Watson, R.R. Rykaczewski et al. 2013. 'Modeling Social-Ecological Scenarios in Marine Systems.' *BioScience* 63(9): 735–744. doi:10.1525/bio.2013.63.9.9.
Ostrom, E. 1990. *Governing the Commons. The Evolution of Institutions for Collective Action*. New York: Cambridge University Press.
Piaget, J. 1985. *The Equilibration of Cognitive Structures: The Central Problem of Intellectual Development*. Chicago: University of Chicago Press.
Poli, R. 2013. 'A Note on the Difference between Complicated and Complex Social Systems.' *Cadmus* 2(1): 142–147.
Popper, K. 1972. *Objective Knowledge: An Evolutionary Approach*. Oxford: Clarendon Press.
Preiser, R. 2019. 'Identifying General Trends and Patterns in Complex Systems Research: An Overview of Theoretical and Practical Implications.' *Systems Research and Behavioral Science* 36: 706–714. doi:10.1002/sres.2619 RESEARCH.
Preiser, R., R. Biggs, A. de Vos, and C. Folke. 2018. 'Social-Ecological Systems as Complex Adaptive Systems: Organizing Principles for Advancing Research Methods and Approaches.' *Ecology and Society* 23(4): 46. doi:10.5751/ES-10558-230446.
Preiser, R., and P. Cilliers. 2010. 'Unpacking the Ethics of Complexity: Concluding Reflections.' In *Complexity, Difference and Identity*, edited by P. Cilliers and R. Preiser, 265–287. Dordrecht: Springer. doi:10.1007/978-90-481-9187-1_13.
Prigogine, I. 1996. *The End of Certainty: Time, Chaos, and the New Laws of Nature*. New York: Free Press.
Rajagopalan, R., and G. Midgley. 2015. 'Knowing Differently in Systemic Intervention.' *Systems Research and Behavioral Science* 32(5): 546–561. doi:10.1002/sres.2352.
Ramage, M., and K. Shipp, eds. 2009. *Systems Thinkers*. London: Springer.
Rasch, W. 1991. 'Theories of Complexity, Complexities of Theory: Habermas, Luhmann, and the Study of Social Systems.' *German Studies Association* 14(1): 65–83.
Rasch, W., and E.M. Knodt. 1994. 'Systems Theory and the System of Theory.' *New German Critique* 61(Special Issue on Niklas Luhmann): 3–7.
Rasch, W., and C. Wolfe, eds. 2000. *Observing Complexity: Systems Theory and Postmodernity*. Minneapolis: University of Minneapolis Press.
Rescher, N. 1998. *Complexity: A Philosophical Overview*. New Brunswick: Transactions.
Reyers, B., J.L. Nel, P.J. O'Farrell, N. Sitas, and D.C. Nel. 2015. 'Navigating Complexity through Knowledge Coproduction: Mainstreaming Ecosystem Services into Disaster Risk Reduction.' *Proceedings of the National Academy of Sciences* 112(24): 7362–7368. doi:10.1073/pnas.1414374112.
Rogers, K.H., R. Luton, H. Biggs, R. Biggs, S. Blignaut, A.G. Choles, C.G. Palmer, and P. Tangwe. 2013. 'Fostering Complexity Thinking in Action Research for Change in Social-Ecological Systems.' *Ecology and Society* 18(2): 31. doi:10.5751/ ES-05330-180231.

Rosen, R. 1991. *Life Itself: A Comprehensive Enquiry into the Nature, Origin, and Fabrication of Life*. New York: Columbia University Press.

Rousseau, D. 2017. 'Systems Research and the Quest for Scientific Systems Principles.' *Systems* 5(2): 25. doi:10.3390/ systems5020025.

Russel, B. 1945. *A History of Western Philosophy*. New York: Simon & Schuster.

Schlüter, M., J. Haider, S.J. Lade, E. Lindkvist, and R. Martin. 2019a. 'Capturing Emergent Phenomena in Social-Ecological Systems: An Analytical Framework.' *Ecology and Society* 24(3): 11. doi:10.5751/ES-11012-240311.

Schlüter, M., K. Orach, E. Lindkvist, R. Martin, N. Wijermans, O. Bodin, and W.J. Boonstra. 2019b. 'Toward a Methodology for Explaining and Theorizing about Social-Ecological Phenomena.' *Current Opinion in Environmental Sustainability* 39: 44–53. doi:10.1016/j.cosust.2019.06.011.

Schlüter, M., and C. Pahl-Wostl. 2007. 'Mechanisms of Resilience in Common-pool Resource Management Systems: An Agent-based Model of Water Use in a River Basin.' *Ecology and Society* 12(2): 4. www.ecologyandsociety.org/vol12/iss2/art4/.

Schoon, M., and S. van der Leeuw. 2015. 'The Shift toward Social-Ecological Systems Perspectives: Insights into the Human–Nature Relationship.' *Natures Sciences Sociétés* 23(2): 166–174. doi:10.1051/nss/2015034.

Tarski, A. 1959. *Introduction to Logic and to the Methodology of Deductive Sciences* (8th ed). New York: Oxford University Press.

Tengö, M., E.S. Brondizio, T. Elmqvist, P. Malmer, and M. Spierenburg. 2014. 'Connecting Diverse Knowledge Systems for Enhanced Ecosystem Governance: The Multiple Evidence Base Approach.' *Ambio* 43(5): 579–591. doi:10.1007/s13280-014-0501-3.

Thrift, N. 1999. 'The Place of Complexity.' *Theory, Culture & Society* 16(3): 31–69.

Thurner, S., R. Hanel, and P. Klimek. 2018. *Introduction to the Theory of Complex Systems*. Oxford: Oxford University Press.

Toulmin, S. 2001. *Return to Reason*. Cambridge: Harvard University Press.

Ulanowicz, R. 1999. 'Life after Newton: An Ecological Metaphysic.' *Biosystems* 50(2): 127–142.

Ulanowicz, R. 2007. 'Ecology: A Dialogue between the Quick and the Dead.' In *Exploring Complexity. Volume 1: Reframing Complexity. Perspectives from the North and South*, edited by F. Capra, A. Juarerro, P. Sotolongo, and J. van Uden. Mansfield: ISCE Publishing.

Ulanowicz, R. 2009. A *Third Window: Natural Life Beyond Newton and Darwin*. West Conshohocken: Templeton Foundation Press.

Ulrich, W. 1994. 'Can We Secure Future-responsive Management through Systems Thinking and Design?' *Interfaces* 24(4): 26–37. doi:10.1287/inte.24.4.26.

Ulrich, W. 2000. 'Reflective Practice in the Civil Society: The Contribution of Critically Systemic Thinking.' *Reflective Practice* 1(2): 247–268. doi:10.1080/713693151.

Urry, J. 2005. 'The Complexity Turn.' *Theory, Culture & Society* 22(5): 1–14.

Von Bertalanffy, L. 1968. *General System Theory: Foundations, Development, Applications*. New York: George Braziller.

Von Foerster, H. 1960. 'On Self-organising Systems and their Environments.' In *Self-Organising Systems*, edited by M.C. Yovits and S. Cameron, 30–50. London: Pergamon Press.

Waldrop, M. 1993. *Complexity: The Emerging Science at the Edge of Order and Chaos*. New York: Simon & Schuster.

Walker, B., C.S. Holling, S.R. Carpenter, and A. Kinzig. 2004. 'Resilience, Adaptability and Transformability in Social-Ecological Systems.' *Ecology and Society* 9(2): 5. doi:10.5751/ES-00650–090205.

Wells, J.L. 2013. *Complexity and Sustainability*. New York: Routledge.

Wittgenstein, L. 1953. *Philosophical Investigations*, edited by G.E.M. Anscombe and R. Rhees. Translated by G.E.M. Anscombe. Oxford: Blackwell.

Woermann, M. 2016. *Bridging Complexity and Post-Structuralism. Insights and Implications*. Switzerland: Springer. doi:10.1007/978-3-319-39047-5.

Wolfram, S. 2002. *A New Kind of Science*. Champaign: Wolfram Media.

3

The practice and design of social-ecological systems research

Alta de Vos,[1] Kristine Maciejewski,[2] Örjan Bodin,[3] Albert Norström,[3] Maja Schlüter[3] and Maria Tengö[3]

[1]DEPARTMENT OF ENVIRONMENTAL SCIENCE, RHODES UNIVERSITY, MAKHANDA, SOUTH AFRICA
[2]CENTRE FOR SUSTAINABILITY TRANSITIONS, STELLENBOSCH UNIVERSITY, STELLENBOSCH, SOUTH AFRICA
[3]STOCKHOLM RESILIENCE CENTRE, STOCKHOLM UNIVERSITY, STOCKHOLM, SWEDEN

Introduction

Studying social-ecological systems (SES) can be a challenging task, as explained in Chapter 2. Phenomena of interest and characteristics of SES research result from both social and ecological processes, and complicated feedback dynamics blur the distinction between cause and effect (Young et al. 2006). Furthermore, multiple causal processes may be operating simultaneously, outcomes are strongly influenced by the system's context and it is difficult to determine system boundaries (Bodin and Prell 2011).

These challenges hold implications for the kinds of research approaches and methods used in SES research (Österblom et al. 2017; Preiser et al. 2018; Hazard et al. 2019). Intertwined SES cannot be understood from within a single discipline, and the context-dependent nature of systems means that generalisable laws are not easy to derive, or even appropriate, to use in SES research. Indeed, given the underlying motivation to inform change towards more sustainable futures, SES research often requires that researchers acknowledge the methodological pluralism that is necessary to understand the different dimensions and especially the interactions between the social and the ecological (Angelstam et al. 2013; Fischer et al. 2015; Preiser et al. 2018).

In response to these demands, the field of SES research draws on a large range of scientific theories and frameworks, and selects from a broad spectrum of methodologies (Fischer et al. 2015; Preiser et al. 2018). Researchers are often required to engage with a variety of stakeholders during the course of their projects. For this they will need to develop inter- and transdisciplinary teamwork skills (Angelstam et al. 2013; Roux et al. 2017) and an ability to engage with multiple knowledge types and values in participatory knowledge co-creation processes (Tengö et al. 2014). To operate in this space, SES researchers need an 'epistemological agility' (Haider et al. 2017) to engage with theories and frameworks from different disciplines that may require fundamentally different theoretical commitments and hold different assumptions, while also developing specific methodological competencies.

How then, do SES researchers develop this epistemological agility, navigate the broad range of theories, frameworks and methods that could potentially inform their study, decide on how

DOI: 10.4324/9781003021339-4

much, and when, they need to engage with other disciplines and stakeholders, and develop their own research identity (Haider et al. 2017; Hazard et al. 2019)? To rise to this challenge, researchers have to recognise that SES research is an iterative and reflexive process for which there is no 'one-size-fits-all' approach (Hazard et al. 2019). Reflexivity here refers to a researcher's examination of how their personal beliefs, judgements, perceptions and worldview may influence research processes they are engaged in, and how these reflections are used to interpret research results within a particular context. Social-ecological systems researchers need to be aware of the assumptions and research traditions underlying the different frameworks, theories and methods they consider using. They have to understand that their research process, and the decisions they make in designing and executing their projects, should be strongly tied to the purpose of their research, which should, in turn, be strongly tied to the societal problem in question.

In this chapter, we attempt to guide the reflexive SES researcher through different components of an SES research project. We first discuss the identification of research problems, the purpose of the research and the SES research areas within which a researcher may be embedded. In many cases, these three factors will determine how researchers engage with frameworks, theories, methods and data; how they combine methods; and the extent to which they will collaborate with other researchers, disciplines and stakeholders (see Chapter 1). We also elaborate on more general issues relating to data generation, participatory research along a gradient of collaboration, and computational and mathematical modelling approaches, as these are common to a number of specific methods discussed in Part 2 of the book (Chapters 5–32). Finally, we consider practical ethics associated with SES research.

Our aim is not to provide a comprehensive overview of all the elements that can possibly be encountered in SES research (which will differ from study to study), but rather to provide general, practical principles for engaging with the SES research process.

Elements of an SES research study

The 'elements' that make up an SES research project are schematically represented in Figure 3.1. In designing and practising SES research, researchers identify problems, work within more specific research areas, draw on different theories and frameworks (which also draw on disciplinary theories), and use and generate different data types. There is no single path through these components and researchers may rely on multiple methods, theories and frameworks, combined and integrated in different ways. Individual SES studies often form part of larger inter- and transdisciplinary SES programmes.

Any project includes the identification of a problem or gap informing the research theme. Problem identification can happen in the 'traditional' way, where the researcher engages with the literature (including their own previous research) and identifies a gap that they would like to address. It may also be that a researcher is an expert in a particular methodology, and looks for (SES) problems to which this methodology can be applied. Alternatively, researchers may identify problems by engaging with policymakers, practitioners and other stakeholders, where the research problem is informed very directly by a societal need or gap. In these cases, problem identification may be an important part of the co-production process.

The researcher can also identify a problem by observing news, events and societal processes, sometimes through systematic, formal techniques such as horizon scanning and other 'futures' methods (Chapter 10). How the problem is identified (and which methods, theory and frameworks are ultimately chosen) will be strongly tied to the purpose of the research and the motivation of the researcher (Hazard et al. 2019). If a researcher is primarily

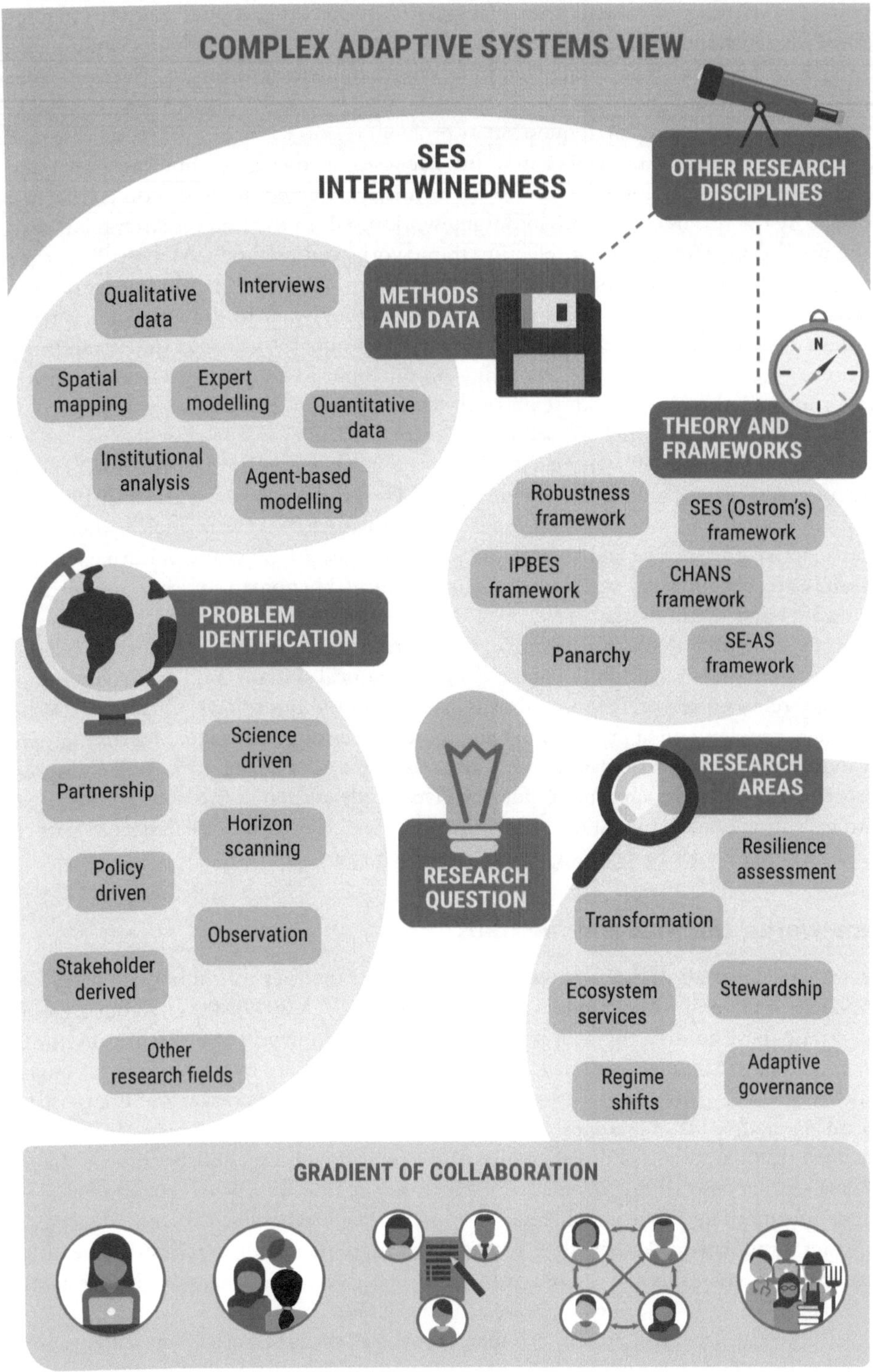

Figure 3.1 Key elements of SES research (© Alta de Vos)

interested in learning a new skill, they may opt for a project that applies that skill to a specific question. A researcher who is interested in affecting transformative change may seek out a mentor or collaborator who is engaged in a transdisciplinary project (see Section 'Working along a gradient of collaboration').

The way in which a research question or problem relates to theory, frameworks, methods and data is not fixed: frameworks may inform research questions, but research questions may also drive the selection of frameworks. Similarly, selected methods could inform or be informed by the research questions and frameworks used. In the SES research field, there are no strict rules or protocols about selecting frameworks and methods. As a result, researchers need to make choices and sometimes tough decisions about the methodological approach they want to take and the theories and frameworks they want to draw on. They will usually have to acknowledge that there are several and equally valuable understandings and approaches that can be adopted, depending on one's purpose and theoretical commitment. These choices and trade-offs can seem overwhelming but in practice they are often guided by the research areas in which a researcher is embedded.

Despite being relatively young (Binder et al. 2013; Pahl-Wostl et al. 2013; Herrero-Jáuregui et al. 2018; Colding and Barthel 2019; De Vos, Biggs, and Preiser 2019), the SES field has sprouted a number of specific research areas such as stewardship, resilience assessment, adaptive governance and transformation in the past 20–30 years. Within each of these areas, researchers typically approach social-ecological phenomena with a set of theories, frameworks and methods that have been developed, tested and debated in the literature. The study of regime shifts, for example, draws strongly on resilience theory (Chapter 14), often uses state-and-transition (Chapter 27) or dynamical systems models (Chapter 26), and frequently relies on remotely sensed data (Chapter 24) and interviews (Chapter 7). Whereas researchers can draw on approaches outside these bodies of knowledge, identifying existing methods used within particular research areas can save a lot of legwork in deciding which approaches may be most appropriate for any given study (although this could blind researchers to gaps in the field). In reality, very little SES research requires researchers to start 'from scratch' when it comes to deciding on their research approach.

Frameworks, theories and methods

Researchers across all disciplines use frameworks and theories to guide their study design and choice of methods (Meyfroidt et al. 2018; Schlüter et al. 2019), although not all research projects use these guiding tools, or use them a priori. Frameworks identify broad sets of SES elements and their linkages, guiding an investigation or activity by pointing to the concepts, elements, variables, links or processes that are characteristic of or critical for SES, or that help explain or predict SES outcomes. As discussed in Chapter 1, frameworks can be descriptive, analytical, or serve as boundary objects for interdisciplinary collaboration or heuristic problem solving (also see Meyfroidt 2016; Meyfroidt et al. 2018). Whereas some frameworks include assumptions about causal relationships between variables, they ultimately cannot posit causal inferences. Rather, they provide the 'ingredients' for theories that hypothesise about causal mechanisms and the relative weight and nature of interactions (Meyfroidt 2016; Meyfroidt et al. 2018; Bodin et al. 2019; Schlüter et al. 2019).

Universal theories that present testable hypotheses across a variety of contexts are not generally possible or desirable in SES research, given the social-ecological intertwinedness and context dependence that characterise the field (Schlüter et al. 2019). Instead, frameworks (particularly conceptual frameworks) are widely used and are often considered to be the 'theoretical element' of SES research studies (Chapter 1). Recently, however, some researchers have started to look to middle-range theories to develop hypotheses that are valid under certain conditions, thus moving towards explaining, and not just describing, phenomena in systems (e.g. Meyfroidt et al. 2018; Bodin et al. 2019). Middle-range theories, still a frontier in SES research (see Chapter 33), can be defined as 'contextual generalizations that describe chains of causal mechanisms explaining a well-bounded range of phenomena, as well as the conditions that trigger, enable, or prevent these causal chains' (Meyfroidt et al. 2018, 53). Middle-range theories thus seek to provide testable hypotheses within very specific and well-defined contexts, striking a balance between ungeneralisable detailed single-case research and universal theories that aim to explain phenomena across and in all general contexts (Meyfroidt et al. 2018; Bodin et al. 2019; Schlüter et al. 2019).

Methods are codified ways of producing knowledge of a focus of interest. They are specific information-generating practices to generate and analyse data (Pahl-Wostl et al. 2013; Stirling 2015). Although frameworks (along with theories and research areas) often guide the selection of methods, not all methods are underpinned or informed by 'formal' frameworks (as discussed above and in Chapter 1). Nevertheless, all methods reflect the underlying values and beliefs of researchers (Poteete, Janssen, and Ostrom 2010) and rely on assumptions that constrain the contexts in which generated knowledge is useful and the degree to which it can be used to make truth claims. Most network analyses (Chapter 23), for example, assume complete knowledge of all connections in a network. Although network analysis is inordinately useful for understanding problems such as disease spread through an agricultural system (e.g. Moore, Grewar, and Cumming 2016) or how the structure and function of natural resource governance systems are linked (Bodin et al. 2019), it may not be a useful method in systems where many of the key actors, or the relationships between them, cannot easily be known. Similarly, many participatory mapping exercises (Chapter 8) assume that participants are able to relate to spatial constructs in their world through reading maps or interacting with a virtual globe. This may not be true in all communities, or there may be some groups more able to engage in this way than others (e.g. Weyer, Bezerra, and De Vos 2019), which may undermine the internal validity of the collected data and the objectives of the participatory process. Finally, models based on economic theory make many simplifying assumptions about the nature of human decision-making (e.g. that the context in which individuals are embedded does not affect their decisions [methodological individualism]). These assumptions may be appropriate in very narrow and controlled contexts where a specific element of the system is under investigation in a specific context (see examples in Chapter 21), but it may be inappropriate to generalise these findings to other contexts or to use these methods to understand complex dynamics in systems.

Combining multiple methods

Given the limited extent to which individual methods can be used to understand SES (Poteete, Janssen, and Ostrom 2010; Preiser et al. 2018), research projects and knowledge co-production processes (see Section 'Working along a gradient of collaboration') often rely on more than one method to achieve their goals (Murray, D'Anna, and MacDonald 2016). At its broadest, multi-method research simply refers to the use of more than one method to understand given phenomena (Anguera et al. 2018) and does not necessarily mean that those methods are integrated or used for triangulation. In the CreativeVoice approach, for example, photos, videos, songs, drawing and paintings are combined to allow more participants to engage with an expressive art form (Rivera Lopez, Wickson, and Hausner 2018).

Similarly, ecological data collection methods (Chapter 6) may be used in combination with interviews and surveys (Chapter 7), sometimes with a view to understanding different aspects of the system rather than understanding the same process or dynamic from a complementary perspective.

In many cases, however, SES researchers explicitly use multi-method designs to integrate different methods by applying mixed-methods and multi-method triangulation approaches. A mixed-methods study is one where the researcher combines a quantitative method and a qualitative method to analyse data (McKim 2017) and integrates these to understand and interpret a particular problem based on the combined strengths of both datasets (Tashakkori and Teddlie 1998; Creswell 2014; Cox 2015). Given the fundamentally different assumptions that typically underlie quantitative and qualitative methods, mixed-methods approaches require careful design and consideration about how and when to 'mix' quantitative and qualitative data (for some suggestions, see Leech and Onwuegbuzie 2009; Teddlie and Tashakkori 2011; Creswell 2014; Cox 2015).

Multi-method triangulation is also usually integrative (Meijer, Verloop, and Beijaard 2002) and specifically refers to approaches where insights regarding a single research problem are strategically drawn from findings generated using different methods (Young et al. 2006; Munafò and Davey Smith 2018). Multi-method triangulation differs from mixed-methods approaches in that it does not necessarily entail integrating quantitative and qualitative methods, but could integrate any set of different methods. Triangulation assumes that researchers can generate a broader span of insights or acquire more support for a potential explanation of a complex phenomenon by relying on different methods. Since every method has its own assumptions, strengths and weaknesses, combining different methods can help to reveal different aspects of phenomena being investigated and can compensate for the limitations of individual methods.

The triangulation of methods can have different objectives and ambitions. A strongly integrative ambition relates to integrating theories, methods and data from different disciplines or knowledge traditions. This approach, known as unification, requires a researcher to consider the theoretical assumptions that underlie methods and, in doing so, start integrating theories (Popa and Guillermin 2017; Persson et al. 2018). Unification can be very rewarding and facilitate new theoretical breakthroughs, particularly when the methods being integrated have similar underlying assumptions (e.g. combining theories of collapse with resilience theory; Cumming and Peterson 2017). However, a unification approach can also lead to indirectly compromising or disregarding the important assumptions that underlie methods (Popa and Guillermin 2017; Persson et al. 2018; Jerneck and Olsson 2020).

A less integrative ambition of triangulation, and one more broadly used and advocated for in SES research (e.g. Norgaard 1989; Popa and Guillermin 2017; Persson et al. 2018; Jerneck and Olsson 2020) builds on pluralism. Pluralism refers to the use of different methods with the aim of investigating a common phenomenon but from different perspectives (e.g. assessing land-use change through remote sensing, interviews, focus groups and participatory mapping; Achieng et al. 2020). Pluralism underscores the autonomy of different methods, along with their associated assumptions and theories. Thus, pluralism is less integrative and less focused on unifying perspectives across disciplinary and knowledge system boundaries. A pluralistic approach is particularly appropriate in knowledge co-production processes where system understanding relies not only on methods from different disciplines but also on different knowledge systems, which may not hold equal power in decision-making. In these processes, it is desirable to 'weave' results from different methods based in different disciplines and knowledge systems, rather than cross-validating one system with another, or unifying them (Tengö et al. 2014).

Data in SES research

As with other choices in SES research, the way in which SES researchers engage with data will depend on the purpose of their research, the motivation of the researchers, the theories and frameworks they employ, their guiding methodology and the available funding. The data used in a particular study will also depend on where the project lies on a 'gradient of intertwinedness': the kind of data used to understand ecological feedbacks resulting from social processes, for example, will be different to the data used in a project setting out to understand social feedbacks resulting from social processes in the context of ecological processes. While multi-method approaches that use different data types are useful in overcoming some of the challenges of SES research, these approaches are also more expensive, conceptually challenging and time consuming, and should not automatically be adopted as the best option in all cases (Creswell 2014; McKim 2017).

In SES research, quantitative and qualitative data can be generated in a number of different ways. In this book, we discuss methods that use the following types of data:

- **Empirical field data** collected through ecological field data collection (Chapter 6), interviews and surveys (Chapter 7), participatory data collection methods (Chapter 8), controlled behavioural experiments (Chapter 21) or methods used for historical assessment (Chapter 25)
- **Co-produced data** generated through participatory data collection methods (Chapter 8), action research (Chapter 15), facilitated dialogues (Chapter 9), futures analysis (Chapter 10), scenario development (Chapter 11), serious games (Chapter 12), participatory modelling (Chapter 13) and resilience assessment (Chapter 14)
- **Textual and archival data** (e.g. photographs and audio) retrieved from published scientific papers and governmental and other policy documents and repositories. These data are stored in libraries, databases and archives and are used in qualitative content analysis (Chapter 19), comparative case study analysis (Chapter 20), institutional analysis (Chapter 22) and historical assessment (Chapter 25)
- **Mined and synthesised data** (often 'big data') using natural-language processing and other pattern-recognition methods (Chapter 17), or through large-scale meta-analysis methods (Chapter 19)
- **Simulated data** produced using dynamical systems models that simulate dynamic systems numerically (Chapter 26), agent-based modelling (Chapter 28), statistical models (Chapter 18), ecosystem service modelling (Chapter 31) or expert modelling (Chapter 16)
- **Downloaded datasets** from public and other databases, including government census data, health and demographic data and remote-sensing and GIS data from a diversity of sectors, compiled case studies and interviews. A collection of platforms commonly used by SES researchers to access useful datasets is given in Table 3.1, and is also provided online (sesmethods.org).

All the data types listed here can include spatial data (Chapter 24). Participatory mapping and GPS-based plots, for example, can be used to co-produce or collect spatial data, remotely sensed spatial data can be synthesised into analysable products, dynamic simulation models can be spatially explicit, and many downloadable SES datasets include spatial data.

Table 3.1 Examples of existing datasets commonly used in SES research, and the platforms where they can be sourced

Type	*Source and website(s)*
Remotely sensed data and global spatial data	• Google Earth Engine: earth-engine/datasets/catalog • Bio-Oracle: bio-oracle.org • Microsoft Earth: microsoft.com/en-us/ai/ai-for-earth-tech-resources • World Resource Institute: datasets.wri.org • NASA's Socioeconomic Data and Applications Centre: sedac.ciesin.columbia.edu/data/sets
Protected areas	• Protected Planet: protectedplanet.net • PADDD (Protected Area Downgrading, Downsizing and Degazettement) tracker: padddtracker.org
Biodiversity data	• Global Biodiversity Information Facility: gbif.org • IUCN Red List of Species and Ecosystems: iucnredlist.org; iucnrle.org • Projecting Responses of Ecological Diversity in Changing Terrestrial Systems (Predicts): predicts.org.uk • The Living Planet Index (LPI): livingplanetindex.org • Nature Map Explorer: explorer.naturemap.earth/map • Map of Life: mol.org
Human impact	• Global Human Footprint: ghsl.jrc.ec.europa.eu/datasets.php • JRC global human settlement layers: ghsl.jrc.ec.europa.eu/datasets.php • Global Human Modification Map (also see 'Remotely sensed data and global spatial data' above): Global_Human_Modification/728308 • Anthropogenic Biomes (also see 'Remotely sensed data and global spatial data'): ecotope.org/anthromes/faq
Development and livelihood data	• World Bank: data.worldbank.org • CIFOR's Poverty and Environment Network (PEN) global dataset: data.cifor.org/dataset • USAID demographic and health surveys (DHS): dhsprogram.com/data • USAID Development Data Library: data.usaid.gov • FLARE network (Forests & Livelihoods: Assessment, Research, and Engagement): forestlivelihoods.org/resources • Comtrade: comtrade.un.org/labs • Food and Agricultural Organization (FAO): fao.org/statistics/databases/en • World Resource Institute data: datasets.wri.org
Case study databases	• Social-Ecological Systems Meta-Analysis Database (SESMAD): sesmad.dartmouth.edu • SES Library: seslibrary.asu.edu/case • Regime Shift Database: regimeshifts.org • Resilience Alliance Thresholds database: resalliance.org/thresholds-db • Engage2020 Action Catalogue: actioncatalogue.eu/about • ISeeChange: Thresholds Database: iseechange.org • Digital Library of the Commons: dlc.dlib.indiana.edu/dlc
Models and methods	• COMSES (Netlogo) models: comses.net • Bayes Net Library: norsys.com/netlibrary/index.htm • SES models: actioncatalogue.eu/about • Engage2020 Action Catalogue: modelingcommons.org
Water	• Water Footprint Network: gwp.org/en/learn/iwrm-toolbox • IWRM Toolbox (see 'Remotely sensed data and global spatial data'): waterfootprint.org/en/resources/waterstat

Type	*Source and website(s)*
Climate	• Climate variability and predictability database (CLIVAR – Climate and Ocean: Variability, Predictability and Change): clivar.org/resources/data • Participatory database on climate change impacts (see 'Remotely sensed data and global spatial data'): iseechange.org
Other	• Google Dataset Search: toolbox.google.com/datasetsearch

A more descriptive list can be found at sesmethods.org

Dynamic modelling in SES research

Dynamic modelling using analytical or computational/simulation approaches is a key approach to studying complex systems. It is particularly suitable for analysing the dynamics of SES, such as the equilibria of a system or the way in which a system unfolds over time from the interplay of social and ecological processes. Modelling allows for experimenting with an SES in ways that are not possible with real systems. Models and simulation experiments can, for example, be used to assess possible intended and unintended consequences of the introduction of a new policy, to unravel the mechanisms that may have generated an outcome of interest, or to assess the uncertainty of outcomes that results from limited knowledge about key processes (such as human decision-making) or the stochastic and emergent nature of SES. Simple mathematical models have been extensively used from the early days of resilience research to study multiple equilibria and regime shifts in ecological systems, and increasingly also SES. Simulation models (e.g. system dynamics or agent-based models) are often used to study the behaviour of particular SES such as a landscape or a fishery that emerges from the interactions of different SES elements.

Models and the modelling process itself can serve many different purposes. They can be used to enhance understanding of SES dynamics; to explore, explain or predict SES outcomes (Edmonds 2017); to facilitate interdisciplinary communication and integration; or to serve as boundary objects in participatory processes (see Chapter 13) (Schlüter, Müller, and Frank 2019; Schlüter et al. 2019). Models can also play different roles in a research or participatory process: they can be used as a thinking tool to support reflection about different beliefs or views of the system or processes of interest and their consequences for system behaviour; as a tool for exploration, e.g. of possible consequences of an intervention; as a tool for eye opening or myth busting; as a tool for explaining social-ecological phenomena of interest by testing possible mechanisms; as an analytical tool to identify generic processes or principles that determine SES behaviour; or as a tool for assessing the effects of policies or the implications of global change (Schlüter, Müller, and Frank 2019). The use of models for policy assessment, for exploring the behaviour of SES as a complex (adaptive) system, and for participatory processes is probably the most common model applications in SES research to date. Although dynamic models can be used to make predictions or to help test and develop theories by formalising and exploring relations between different variables and their outcomes, they are currently very rarely used in this way in SES research.

Models can be theoretical, based on generic processes such as population growth and utility maximisation with the aim to understand the general behaviour of a system, or empirical, based on empirical data and processes with the aim to understand a system's behaviour or response in a particular place. Some models may combine both, for example by using theoretical models where data and knowledge are limited and empirically informed models for other aspects of the system. Stylised or toy models that represent selected aspects of a system in a generic way (often based on insights that hold across different cases but are not necessarily derived from theory) to investigate their impact on SES outcomes are also commonly used. The most common type of modelling used in SES research is dynamical systems modelling, agent-based modelling and state-and-transition modelling. These models differ in how they study the change of the system over time, how and at what level of aggregation they represent a system and how they go about to find solutions or generate the outcomes of the model (for an overview of different common model types, see Schlüter, Müller and Frank 2019). While they can all be used for different purposes and roles, some are more suitable to achieving a given research aim than others.

Contrary to the large-scale simulation models that dominate in Earth systems and climate change research, and to models used in theoretical ecology or economics, many SES models are developed from scratch. This means the researcher or research team goes through all the modelling steps – from collecting and eliciting data and empirical evidence or relevant theoretical models to developing a conceptual model; formalising the model in equations or computer code; testing, verifying and validating the model; running simulations or doing mathematical analyses; analysing the model and communicating the results. These different steps require many different skills, knowledge and methods. The process of building the model, i.e. decisions on what variables/actors and processes/interactions to include, and how to represent functional relationships are major parts of the modelling process. This process is often as valuable as the resulting model as it challenges participants to make explicit their assumptions and understanding about the SES or problem of interest, ideally in a collaborative endeavour that involves a diversity of scientific and non-scientific participants, drawing on different knowledge systems (Schlüter, Müller, and Frank 2019; Schlüter et al. 2019). This is one of the main advantages of co-developing models with stakeholders, as it can facilitate building shared understanding and learning. Similarly, co-developing models with scientists from different disciplinary backgrounds can support a reflective and reflexive approach and thoughtful interdisciplinary integration.

Working along a gradient of collaboration

As explained in Chapter 2, the complex, intertwined nature of SES demands intellectual humility from researchers and a shift in their approach to knowledge development, learning and the ways in which they support policy and change (Audouin et al. 2013; Preiser et al. 2018). Investigating the intertwined human and biophysical dimensions of SES problems requires knowledge on the multi-scale interactions between ecosystems and society (Angelstam et al. 2013; Nash et al. 2017), which demands diverse expertise from many different disciplines and often also requires engagement with key actors and stakeholders engaged in SES governance and management. Collaboration and knowledge co-production with non-academic societal actors are particularly important for SES projects where concrete societal change, and not just knowledge development, is the main objective (e.g. action research (Chapter 15)).

Thus, although SES research can be a solitary practice, it often entails collaboration and working in teams. Teams that only involve academics, but from different disciplines, are commonly referred to as interdisciplinary teams. In these teams, researchers work together to integrate or combine disciplinary knowledge and methods, develop and meet shared goals, and achieve synthetic understanding of a problem or system. In other cases, the sphere of collaboration is expanded to include relevant stakeholders and other non-academics to form transdisciplinary teams (Lang et al. 2012; Angelstam et al. 2013). Research for action, in which supporting and facilitating societal change may be the most important project objectives, is mostly conducted in transdisciplinary teams.

Working in diverse teams has many benefits, the most notable of which is integrating multiple ways of knowing and doing to generate a more comprehensive understanding of complex social-ecological challenges. However, there is also a suite of well-recognised challenges to working in teams (Lang et al. 2012; Kelly et al. 2019). Interdisciplinary work, for example, often requires time and resources to define, understand and combine the disparate concepts and methods upon which diverse disciplines are founded. These intellectual transaction costs can trigger tensions within interdisciplinary teams. Working across disciplines requires practices, attitudes and personality traits such as humility, respectfulness, open-mindedness, patience and the willingness to embrace complexity (Kelly et al. 2019). While these 'soft skills' are consistently identified as critical for collaboration among disciplines and knowledge systems, they are rarely valued or specifically recognised and developed within disciplinary research training (Kelly et al. 2019). Fortunately, the growing number of interdisciplinary projects is generating a growing repository of practical advice to support researchers and institutions wishing to embrace interdisciplinary social-ecological research, as well as scholars with experience in working in interdisciplinary teams.

Working beyond academic boundaries to collaborate with stakeholders such as government officials, community members and civic society organisations often introduces further challenges. Collaborations with stakeholders and other non-academic actors are often seen as pathways to making social-ecological research both more democratic through stakeholder involvement and more impactful by creating ownership and tying it to actions that benefit those stakeholders. In practice, there is a continuum of stakeholder involvement (or participation) (Cvitanovic et al. 2019). At the one end of the spectrum, participation can be limited to very low levels of engagement, where stakeholders merely provide data for a research project (e.g. citizen science) or evaluative feedback on research products, or are informed about research outcomes. While these types of consultative engagements can be useful, they have been criticised for leading to 'token participation' that does not treat non-academics as full partners in the research process, and potentially undermine the extent to which the benefits of participation are conferred.

At the other end of the spectrum of participation is a loosely linked and evolving cluster of approaches that include knowledge co-production and transdisciplinary research (Lang et al. 2012; Wyborn et al. 2019; Norström et al. 2020). These approaches reject the archetypal model of academic knowledge production, where researchers identify problems, carry out research to address them and then communicate this new knowledge to society to be acted upon. The goals of these new forms of participatory approaches are to co-produce knowledge and solutions with, and for, decision-makers and other actors in society. These knowledge co-production processes can facilitate robust solutions to sustainability challenges and their effective and equitable implementation (Norström et al. 2020) by providing a richer, more diverse and more legitimate understanding of the multiple drivers, interdependencies and complexities of SES dynamics and challenges, and of the decision contexts in which research is to be applied (Tengö et al. 2012).

Knowledge co-production holds strong potential for addressing sustainability challenges, but adds new demands and requires new skills to make collaborative research practice successful. A number of critical issues need to be considered and carefully managed to avoid serious pitfalls that end up doing more harm than good. First, who are the actors with whom the researcher wants to engage, and what do they represent? Knowledge co-production needs to explicitly recognise a range of perspectives, knowledge and expertise and build awareness of gender, ethnicity and age-related aspects of who is involved – or left out. This poses a variety of new ethical concerns (Bohle and Preiser 2019; also see Chapter 2, and Section 'Practical and procedural ethics'). The leaders of a co-production process face the task of assembling an appropriately broad or inclusive set of relevant actors, while keeping the process manageable within practical and strategic limits. There may be tensions between actor groups, for example. If that happens, a step-wise approach to participation, where smaller groups are initially convened before a broader group is engaged, may reduce potential points of conflict or allow some facilitating steps to be undertaken.

Second, is the researcher engaging with key stakeholders, key knowledge holders or actors in key positions to implement knowledge? Different objectives or targets of the research may entail different approaches to identifying and actively engaging with actors. This will also have implications for the design of the collaboration – the kinds of interfaces and the types of meetings and activities that are used to facilitate knowledge exchange and learning.

Third, how is the researcher interacting with, and handling, existing power dynamics between actors in collaborative processes? A failure to sufficiently engage with power imbalances can undermine the quality of the engagement and the outcomes, and can derail the entire exercise. These challenges can be addressed by taking the necessary steps to build trust and reveal tensions and expectations between collaborators before the actual phase of knowledge generation begins. Co-produced social-ecological research also needs to understand the historical context of how a particular challenge emerged, who will be affected by the process and its outcome, and how regulatory, institutional and cultural factors will shape the process and the realisation of desired outcomes. Successful co-production is often built on legacies of the past (conceptual insights obtained in previous projects, long-established research sites, earlier relationships with stakeholders) that can all help to provide insights into the context of the current project. Thus, as an early career researcher, it may be valuable to connect with existing networks and relationships.

Finally, as in any team effort, high-quality co-production processes require frequent interaction of participants throughout the process. Interactive exchanges between participants nurture ongoing learning, build trust and increase the salience and legitimacy of generated knowledge.

Practical and procedural ethics

Applying an SES lens to understanding the world has profound ethical implications for how researchers engage with their research projects (also see Chapter 2). A complex adaptive systems perspective and the desire to affect transformative change means that research projects often seek to engage novel forms of collaborative agency through participatory and collaborative multi-stakeholder processes that foster dialogue and knowledge co-creation (Preiser et al. 2018). These approaches and processes have implications for how researchers engage not only in their research projects (or day-to-day 'practical ethics'; Rossman and Rallis 2010) but also with official processes to obtain permission to carry out their studies (procedural ethics).

Social-ecological systems researchers based at universities (as students or staff members) are usually required to obtain ethical clearance and conduct their studies according to the rules of the institution. Most universities have registered ethics committees that are required to adhere to standards set by a higher (state, provincial or national) regulatory body. Clearance from these committees is required before data collection can commence. It must also be shown that the correct process was followed before a degree can be awarded or the results eventually published. The process of applying for ethical clearance for data collection can cause significant delays in getting projects under way. Ethics committees cannot always process applications quickly, and SES research processes, especially where they involve transdisciplinary knowledge co-production, can raise additional questions that can take time to resolve because they are different to conventional research approaches (Cockburn and Cundill 2018). Since SES researchers may have ecological and social components to their research design, they may have to apply for clearance to more than one committee.

Procedural ethics may pose a particular obstacle to transdisciplinary research projects. In most institutions, current research ethics clearance processes are designed for administering research approaches (e.g. surveys) where societal actors only become involved during the data collection phase, and usually as subjects and not participants (Locke, Alcorn, and O'Neill 2013; Cockburn and Cundill 2018). Transdisciplinary research processes are often at odds with procedural ethics, particularly with regard to 'informed consent', which is seen as a one-off activity rather than being negotiated through an ongoing process with research stakeholders (Banks et al. 2013; Locke, Alcorn, and O'Neill 2013; Cockburn and Cundill 2018). Co-production of knowledge and participatory methods can also make it very hard to know who the data belong to and how and where to ethically store the data (Rambaldi et al. 2006; Cockburn and Cundill 2018).

While certain institutional reforms would make for easier procedural practices (Cockburn and Cundill 2018), it is critical to recognise that formal institutional processes and documents will never be able to provide full ethical guidance and standards for SES research projects (Rossman and Rallis 2010; Coburn and Cundill 2018). Procedural ethics can provide some of the principles of ethical engagement, and play an important role in holding individuals and institutions accountable (Cockburn and Cundill 2018). However, researchers themselves are ultimately responsible for the moral considerations and ethical choices they make as part of their daily practice (Rossman and Rallis 2010) throughout their project. In practice, this means that SES researchers need to develop and reflect on their own personal ethics principles for engaging in their project right from the scoping phases, and continue to do so throughout the project (Cockburn and Cundill 2018). The principles they might draw upon will vary according to context, but can include elements such as respect for and the dignity of research participants, transparency and honesty, accountability and responsibility of the researcher, integrity and academic professionalism (Cockburn and Cundill 2018), and sensitivity to power imbalances and the potential impacts of transformative change (Shah et al. 2018). Being an ethical SES researcher requires one to be reflexive (reflecting and reacting) on one's practice not only individually (e.g. by journalling; Meyer and Willis 2019) but also through conversations with advisers, mentors and research participants, and through engaging with reflective peer groups on transdisciplinary research practice, for example (Cockburn and Cundill 2018).

Everyday practical ethics is important in all SES projects, not only for SES researchers engaged in transdisciplinary or participatory projects. In some SES projects, procedural ethics may not be required and the ethical consequences of research may not be immediately obvious. Projects may, for example, have various indirect ecological and social consequences when management decisions flow from interactions between researchers and managers, or when changing the relationship between a community and nature (Schlaepfer, Pascal, and Davis 2011). Big data, natural-language processing and other machine-learning techniques are now becoming commonplace in many global SES studies (Skibins et al. 2012; Di Minin, Tenkanen, and Toivonen 2015). Although these data are in the public domain and thus do not usually require ethical approval, processing and presenting these data can pose severe ethical concerns regarding informed consent, privacy, ownership, objectivity, intellectual property and group-level ethical harms (Mittelstadt and Floridi 2016). Studies that use social media to investigate perceptions of nature, for example, may be biased towards the perception of certain demographics (Di Minin, Tenkanen, and Toivonen 2015). Similarly, remote-sensing data available at global scales could be used to inform policies that are misaligned with local contexts (Veldman et al. 2019). The SES researcher thus has to recognise and take responsibility for the fact that every decision about data collection, analysis, interpretation and presentation has moral dimensions, and that these decisions are ongoing (Rossman and Rallis 2010).

Conclusion

This chapter describes the research landscape that SES researchers navigate when designing and executing their projects. The complex and problem-focused nature of SES research means that researchers have many potential frameworks, theories and methods to choose from, each with their own underlying research traditions and assumptions. Rather than getting overwhelmed by the methodological and epistemological options available, we encourage researchers to define and reflect on the purpose of their research and their ultimate motivations and desired outcomes. With such a reflexive attitude, the researcher can draw further guidance from appropriate frameworks and methods to use in a particular context, based on the particular SES research area within which they are working. They may find themselves in trans- and interdisciplinary collaborations, which will require the development of additional skill sets that are not always well supported by formal graduate training programmes.

Finally, the often collaborative nature of SES research presents significant ethical considerations, particularly as research projects often actively seek to effect actions and change towards sustainability. Regardless of whether they are conducting a mono-, multi-, inter- or transdisciplinary study, the reflective SES researcher recognises that every methodological and theoretical decision is ultimately an ethical one that will have consequences for their understanding of SES, and actions towards sustainability.

Acknowledgements

Alta de Vos received support from a Rhodes University Council Grant. Maja Schlüter received support from the European Research Council under the European Union's Horizon 2020 research and innovation programme (grant agreement no. 682472 – MUSES). Albert Norström received support from the Swedish Research Council FORMAS (grant no. 2017-01326) and the GRAID programme at the Stockholm Resilience Centre.

References

Achieng, T., K. Maciejewski, M. Dyer, and R. Biggs. 2020. 'Using a Social-Ecological Regime Shift Approach to Understand the Transition from Livestock to Game Farming in the Eastern Cape, South Africa.' *Land* 9(4): 97. doi:10.3390/land9040097.

Angelstam, P., K. Andersson, M. Annerstedt, R. Axelsson, M. Elbakidze, P. Garrido, P. Grahn et al. 2013. 'Solving Problems in Social-Ecological Systems: Definition, Practice and Barriers of Transdisciplinary Research.' *Ambio* 42(2): 254–265. doi:10.1007/s13280-012-0372-4.

Anguera, M.T., A. Blanco-Villaseñor, J.L. Losada, P. Sánchez-Algarra, and A.J. Onwuegbuzie. 2018. 'Revisiting the Difference between Mixed Methods and Multimethods: Is It All in the Name?' *Quality and Quantity* 52(6): 2757–2770. doi:10.1007/s11135-018-0700-2.

Audouin, M., R. Preiser, S. Nienaber, L. Downsborough, J. Lanz, and S. Mavengahama. 2013. 'Exploring the Implications of Critical Complexity for the Study of Socialecological Systems.' *Ecology and Society* 18(3): 12. doi:10.5751/ES-05434-180312.

Banks, S., A. Armstrong, K. Carter, H. Graham, P. Hayward, A. Henry, T. Holland et al. 2013. 'Everyday Ethics in Community-Based Participatory Research.' *Contemporary Social Science*. doi:10.1080/21582041.2013.769618.

Binder, C.R., J. Hinkel, P.W.G. Bots, and C. Pahl-Wostl. Claudia. 2013. 'Comparison of Frameworks for Analyzing Social-Ecological Systems.' *Ecology and Society* 18(4): 26. www.ecologyandsociety.org/vol18/iss4/art26.

Bodin, Ö., and C. Prell. 2011. *Social Networks and Natural Resource Management: Uncovering the Social Fabric of Environmental Governance*. Cambridge: Cambridge University Press.

Bodin, S.M.A., J. Baggio, M.L. Barnes, R. Berardo, G.S. Cumming, L.E. Dee et al. 2019. 'Improving Network Approaches to the Study of Complex Social-Ecological Interdependencies.' *Nature Sustainability*. doi:10.1038/s41893-019-0308-0.

Bohle, M., and R. Preiser. 2019. 'Exploring Societal Intersections of Geoethical Thinking.' In *Exploring Geoethics*, edited by M. Bohle, 71–136. New York: Springer. doi:10.1007/978-3-030-12010-8_3.

Cockburn, J., and G. Cundill. 2018. 'Ethics in Transdisciplinary Research: Reflections on the Implications of "Science with Society".' In *The Palgrave Handbook of Ethics in Critical Research*, edited by C.I. Macleod, J. Marx, P. Mnyaka, and G.J. Treharne, 81–97. New York: Springer. doi:10.1007/978-3-319-74721-7_6.

Colding, J., and S. Barthel. 2019. 'Exploring the Social-Ecological Systems Discourse 20 Years Later.' *Ecology and Society* 24(1): 2. doi:10.5751/ES-10598-240102.

Cox, M. 2015. 'A Basic Guide for Empirical Environmental Social Science.' *Ecology and Society* 20(1): 63. doi:10.5751/ES-07400-200163.

Creswell, J.W. 2014. *A Concise Introduction to Mixed Methods Research*. Thousand Oaks: Sage.

Cumming, G.S., and G.D. Peterson. 2017. 'Unifying Research on Social-Ecological Resilience and Collapse.' *Trends in Ecology and Evolution*. doi:10.1016/j.tree.2017.06.014.

Cvitanovic, C., M. Howden, R.M. Colvin, A. Norström, A.M. Meadow, and P.F.E. Addison. 2019. 'Maximising the Benefits of Participatory Climate Adaptation Research by Understanding and Managing the Associated Challenges and Risks.' *Environmental Science and Policy*. doi:10.1016/j.envsci.2018.12.028.

De Vos, A., R. Biggs, and R. Preiser. 2019. 'Methods for Understanding Social-Ecological Systems: A Review of Place-Based Studies.' *Ecology and Society* 24(4): 16. doi:10.5751/es-11236-240416.

Edmonds, B. 2017. 'Different Modelling Purposes.' In *Simulating Social Complexity. Understanding Complex Systems*, edited by B. Edmonds and R. Meyer, 39–58. New York: Springer. doi:10.1007/978-3-319-66948-9_4.

Fischer, J., T.A. Gardner, E.M. Bennett, P. Balvanera, R. Biggs, S. Carpenter, T. Daw et al. 2015. 'Advancing Sustainability through Mainstreaming a Social-Ecological Systems Perspective.' *Current Opinion in Environmental Sustainability* 14: 144–149. doi:10.1016/j.cosust.2015.06.002.

Haider, L.J., J. Hentati-Sundberg, M. Giusti, J. Goodness, M. Hamann, V.A. Masterson, M. Meacham et al. 2017. 'The Undisciplinary Journey: Early-Career Perspectives in Sustainability Science.' *Sustainability Science* (June): 1–14. doi:10.1007/s11625-017-0445-1.

Hazard, L., M. Cerf, C. Lamine, D. Le Magda, and P. Steyaert. 2019. 'A Tool for Reflecting on Research Stances to Support Sustainability Transitions.' *Nature Sustainability* (December). doi:10.1038/s41893-019-0440-x.

Herrero-Jáuregui, C., C. Arnaiz-Schmitz, M. Reyes, M. Telesnicki, I. Agramonte, M. Easdale, M. Schmitz et al. 2018. 'What Do We Talk about When We Talk about Social-Ecological Systems? A Literature Review.' *Sustainability* 10(8). doi:10.3390/su10082950.

Jerneck, A., and L. Olsson. 2020. 'Theoretical and Methodological Pluralism in Sustainability Science.' In *Framing in Sustainability Science: Theoretical and Practical Approaches*, edited by T. Mino and S. Kudo. Singapore: Springer. doi:978-981-13-9061-6.

Kelly, R., M. Mackay, K.L. Nash, C. Cvitanovic, E.H. Allison, D. Armitage, A. Bonn et al. 2019. 'Ten Tips for Developing Interdisciplinary Socio-Ecological Researchers.' *Socio-Ecological Practice Research* 1(2): 149–161. doi:10.1007/s42532-019-00018-2.

Lang, D.J., A. Wiek, M. Bergmann, M. Stauffacher, P. Martens, P. Moll, M. Swilling, and C.J. Thomas. 2012. 'Transdisciplinary Research in Sustainability Science: Practice, Principles, and Challenges.' *Sustainability Science* 7(Supplement 1): 25–43. doi:10.1007/s11625-011-0149-x.

Leech, N.L., and A.J. Onwuegbuzie. 2009. 'A Typology of Mixed Methods Research Designs.' *Quality and Quantity* 43(2): 265–275. doi:10.1007/s11135-007-9105-3.

Locke, T., N. Alcorn, and J. O'Neill. 2013. 'Ethical Issues in Collaborative Action Research.' *Educational Action Research* 21(1): 107–123. doi:10.1080/09650792.2013.763448.

McKim, C.A. 2017. 'The Value of Mixed Methods Research.' *Journal of Mixed Methods Research* 11(2): 202–222. doi:10.1177/1558689815607096.

Meijer, P.C., N. Verloop, and D. Beijaard. 2002. 'Multi-Method Triangulation in a Qualitative Study on Teachers' Practical Knowledge: An Attempt to Increase Internal Validity.' *Quality & Quantity* 36.

Meyer, K., and R. Willis. 2019. 'Looking Back to Move Forward: The Value of Reflexive Journaling for Novice Researchers.' *Journal of Gerontological Social Work*. doi:10.1080/01634372.2018.1559906.

Meyfroidt, P. 2016. 'Approaches and Terminology for Causal Analysis in Land Systems Science.' *Journal of Land Use Science*. doi:10.1080/1747423X.2015.1117530.

Meyfroidt, P., R. Roy Chowdhury, A. de Bremond, E.C. Ellis, K.H. Erb, T. Filatova, R.D. Garrett et al. 2018. 'Middle-Range Theories of Land System Change.' *Global Environmental Change* 53: 52–67. doi:10.1016/j.gloenvcha.2018.08.006.

Minin, E.D., H. Tenkanen, and T. Toivonen. 2015. 'Prospects and Challenges for Social Media Data in Conservation Science.' *Frontiers in Environmental Science* 3(63). doi:10.3389/fenvs.2015.00063.

Mittelstadt, B.D., and L. Floridi. 2016. 'The Ethics of Big Data: Current and Foreseeable Issues in Biomedical Contexts.' *Science and Engineering Ethics*. doi:10.1007/s11948-015-9652-2.

Moore, C., J. Grewar, and G.S. Cumming. 2016. 'Quantifying Network Resilience: Comparison before and after a Major Perturbation Shows Strengths and Limitations of Network Metrics.' *Journal of Applied Ecology* 53(3): 636–645. doi:10.1111/1365-2664.12486.

Munafò, M.R., and G. Davey Smith. 2018. 'Repeating Experiments Is Not Enough.' *Nature*. doi:10.1038/d41586-018-01023-3.

Murray, G., L. D'Anna, and P. MacDonald. 2016. 'Measuring What We Value: The Utility of Mixed Methods Approaches for Incorporating Values into Marine Social-Ecological System Management.' *Marine Policy* 73: 61–68. doi:10.1016/J.MARPOL.2016.07.008.

Nash, K.L., C. Cvitanovic, E.A. Fulton, B.S. Halpern, E.J. Milner-Gulland, R.A. Watson, and J.L. Blanchard. 2017. 'Planetary Boundaries for a Blue Planet.' *Nature Ecology and Evolution* 1(11): 1625–1634. doi:10.1038/s41559-017-0319-z.

Norgaard, R.B. 1989. 'The Case for Methodological Pluralism.' *Ecological Economics* 1(1): 37–57. doi:10.1016/0921-8009(89)90023-2.

Norström, A.V., C. Cvitanovic, M.F. Löf, S. West, C. Wyborn, P. Balvanera, A.T. Bednarek et al. 2020. 'Principles for Knowledge Co-Production in Sustainability Research.' *Nature Sustainability* 3(3): 182–190. doi:10.1038/s41893-019-0448-2.

Österblom, H., B.I. Crona, C. Folke, M. Nyström, and M. Troell. 2017. 'Marine Ecosystem Science on an Intertwined Planet.' *Ecosystems* 20(1): 54–61. doi:10.1007/s10021-016-9998-6.

Pahl-Wostl, C., C. Giupponi, K. Richards, C. Binder, A. de Sherbinin, D. Sprinz, T. Toonen, and C. van Bers. 2013. 'Transition towards a New Global Change Science: Requirements for Methodologies, Methods, Data and Knowledge.' *Environmental Science & Policy* 28: 36–47. doi:10.1016/J.ENVSCI.2012.11.009.

Persson, J., A. Hornborg, L. Olsson, and H. Thorén. 2018. 'Toward an Alternative Dialogue between the Social and Natural Sciences.' *Ecology and Society* 23(4): 14. doi:10.5751/ES-10498-230414.

Popa, F., and M. Guillermin. 2017. 'Reflexive Methodological Pluralism.' *Journal of Mixed Methods Research* 11(1): 19–35. doi:10.1177/1558689815610250.

Poteete, A.R., M.A. Janssen, and E. Ostrom. 2010. *Working Together: Collective Action, the Commons, and Multiple Methods*. Princeton: Princeton University Press.

Preiser, R., R. Biggs, A. de Vos, and C. Folke. 2018. 'Social-Ecological Systems as Complex Adaptive Systems: Organizing Principles for Advancing Research Methods and Approaches.' *Ecology and Society* 23(4): 46.

Rambaldi, G., R. Chambers, M. McCall, and J. Fox. 2006. 'Practical Ethics for PGIS Practitioners, Facilitators, Technology Intermediaries and Researchers.' *Participatory Learning and Action* 54: 106–113. https://pubs.iied.org/pdfs/G02957.pdf.

Rivera Lopez, F., F. Wickson, and V. Hausner. 2018. 'Finding CreativeVoice: Applying Arts-Based Research in the Context of Biodiversity Conservation.' *Sustainability* 10(6): 1778. doi:10.3390/su10061778.

Rossman, G.B., and S.F. Rallis. 2010. 'Everyday Ethics: Reflections on Practice.' *International Journal of Qualitative Studies in Education* 23(4): 379–391. doi:10.1080/09518398.2010.492813.

Roux, D.J., J.L. Nel, G. Cundill, P.O'Farrell, and C. Fabricius. 2017. 'Transdisciplinary Research for Systemic Change: Who to Learn with, What to Learn about and How to Learn.' *Sustainability Science* 12(5): 711–726. doi:10.1007/s11625-017-0446-0.

Schlaepfer, M.A., M. Pascal, and M.A. Davis. 2011. 'How Might Science Misdirect Policy? Insights into the Threats and Consequences of Invasive Species.' *Journal Fur Verbraucherschutz und Lebensmittelsicherheit* 6(Supplement 1): 27–31. doi:10.1007/s00003-011–0690–7.

Schlüter, M., B. Müller, and K. Frank. 2019. 'The Potential of Models and Modeling for Social-Ecological Systems Research: The Reference Frame ModSES.' *Ecology and Society* 24(1): 31. doi:10.5751/ES-10716–240131.

Schlüter, M., K. Orach, E. Lindkvist, R. Martin, N. Wijermans, Ö. Bodin, and W.J. Boonstra. 2019. 'Toward a Methodology for Explaining and Theorizing about Social-Ecological Phenomena.' *Current Opinion in Environmental Sustainability.* doi:10.1016/j.cosust.2019.06.011.

Shah, S.H., L. Rodina, J.M. Burt, E.J. Gregr, M. Chapman, S. Williams, N.J. Wilson, and G. McDowell. 2018. 'Unpacking Social-Ecological Transformations: Conceptual, Ethical and Methodological Insights.' *The Anthropocene Review* 5(3): 250–265. doi:10.1177/2053019618817928.

Skibins, J.C., J.C. Hallo, J.L. Sharp, and R.E. Manning. 2012. 'Quantifying the Role of Viewing the Denali "Big 5" in Visitor Satisfaction and Awareness: Conservation Implications for Flagship Recognition and Resource Management.' *Human Dimensions of Wildlife* 17: 112–128. doi:10.1080/10871209.2012.627531.

Stirling, A. 2015. 'Developing "Nexus Capabilities": Towards Transdisciplinary Methodologies.' doi:10.13140/RG.2.1.2834.9920.

Tashakkori, A., and C. Teddlie. 1998. *Mixed Methodology: Combining Qualitative and Quantitative Approaches.* Thousand Oaks: Sage.

Teddlie, C., and A. Tashakkori. 2011. 'Mixed Methods Research: Contemporary Issues in an Emerging Field.' In *The SAGE Handbook of Qualitative Research*, edited by N.K. Denzin and Y.S. Lincoln, 265–299. Thousand Oaks: Sage.

Tengö, M., E.S. Brondizio, T. Elmqvist, P. Malmer, and M. Spierenburg. 2014. 'Connecting Diverse Knowledge Systems for Enhanced Ecosystem Governance: The Multiple Evidence Base Approach.' *Ambio* 43(5): 579–591. doi:10.1007/s13280-014-0501-3.

Tengö, M., P. Malmer, P. Borraz, C. Cariño, J. Cariño, T. Gonzales, J. Ishizawa et al. 2012. 'Dialogue Workshop on Knowledge for the 21st Century: Indigenous Knowledge, Traditional Knowledge, Science and Connecting Diverse Knowledge Systems.' April. Usdub, Guna Yala, Panama.

Veldman, J.W., J.C. Aleman, S.T. Alvarado, T.M. Anderson, S. Archibald, W.J. Bond, T.W. Boutton, N. Buchmann, E. Buisson, J.G. Canadell, and De Sá Dechoum, M. 2019. 'Comment on "The Global Tree Restoration Potential".' *Science* 366(6463). doi:10.1126/science.aay7976.

Weyer, B., J.C. Bezerra, and A. de Vos. 2019. 'Participatory Mapping in a Developing Country Context: Lessons from South Africa.' *Land* 8(9): 134. doi:10.3390/land8090134.

Wyborn, C., A. Datta, J. Montana, M. Ryan, P. Leith, B. Chaffin, C. Miller, and L. van Kerkhoff. 2019. 'Co-Producing Sustainability: Reordering the Governance of Science, Policy, and Practice.' *Annual Review of Environment and Resources* 44(1): 319–346. doi:10.1146/annurev-environ-101718-033103.

Young, O.R., F. Berkhout, G.C. Gallopin, M.A. Janssen, E. Ostrom, and S. van der Leeuw. 2006. 'The Globalization of Socio-Ecological Systems: An Agenda for Scientific Research.' *Global Environmental Change* 16(3): 304–316. doi:10.1016/j.gloenvcha.2006.03.004.

4

How to use this handbook

Reinette Biggs,[1,2] Hayley Clements,[1] Alta de Vos,[3]
Kristine Maciejewski,[1] Rika Preiser[1] and Maja Schlüter[2]

[1]CENTRE FOR SUSTAINABILITY TRANSITIONS, STELLENBOSCH UNIVERSITY, STELLENBOSCH, SOUTH AFRICA
[2]STOCKHOLM RESILIENCE CENTRE, STOCKHOLM UNIVERSITY, STOCKHOLM, SWEDEN
[3]DEPARTMENT OF ENVIRONMENTAL SCIENCE, RHODES UNIVERSITY, MAKHANDA, SOUTH AFRICA

Introduction

Social-ecological systems (SES) research aims to inform pressing sustainability challenges facing humanity in the 21st century. It is increasingly manifesting in major policy frameworks and initiatives, such as the Sustainable Development Goals (sdgs.un.org) and Future Earth (futureearth.org). Social-ecological systems research is a major subfield within the broader emerging area of sustainability science and provides a key analytical framing to understand the interactions and feedbacks between social and environmental change (see Chapter 1). Social-ecological systems research has helped facilitate increased recognition of the dependence of humanity on ecosystems, inform new environmental management approaches and improve collaboration across disciplines and between science and society.

Research on SES explicitly adopts a complex adaptive systems view (see Chapter 2) and draws on insights and approaches from both the social and the natural sciences. In addition to being adopted by researchers across a variety of fields, SES studies form the focus of a growing number of graduate programmes around the world. However, the conceptual and methodological pluralism that characterises the field commonly contributes to the disorientation for new entrants. Most students enter the field with undergraduate training in either a social or a natural science (although interdisciplinary undergraduate programmes in sustainability science are becoming more common), and are often not familiar with the wide range of methods that can be employed in SES research and the assumptions underlying these methods.

The objectives of this handbook are to provide a brief orientation to the SES research field, and a synthetic guide to the range of methods that can be employed in SES research, highlighting key gaps and frontiers in SES research methods. The book achieves this by providing an introduction to SES research (Chapters 1–3), before presenting a series of chapters introducing different groups or categories of SES research methods (Chapters 5–32). The main aim of the book is to help readers gain an overview of the different methods that could be employed in SES research, the types of questions to which these methods are suited and the potential resources and skills required for their implementation. The goal is to guide and enable readers to identify potential methods that may be suited to their particular questions, activities and context. The book does not aim to provide in-depth information on specific methods, but rather to point readers to more detailed texts

DOI: 10.4324/9781003021339-5

for further information about potential methods of interest. It concludes with a reflection on the diversity of current SES methods, and a discussion of key gaps and frontiers for advancing SES research methods (Chapter 33).

The book is specifically tailored to researchers entering the SES field, and has been written in a style that is accessible to readers from a variety of different disciplinary backgrounds. As a handbook, it is targeted primarily at graduate students, lecturers and researchers working on SES. However, it will also be of interest to those working in the broader areas of sustainability science, environmental management, global environmental change and environmental governance. It may also be useful to upper-level undergraduates, and professionals working at the science–policy interface in the environmental arena.

Overview of the book

The book comprises three parts. Part 1 provides an introduction to SES research and the book itself by means of four introductory chapters. In Chapter 1, we introduced the concept of SES and SES research. We briefly discussed the origins of SES research and the main conceptual frameworks and approaches used in this field. Chapter 2 dug deeper into the underlying theoretical foundations and assumptions of SES research, grounded in an understanding of complex adaptive systems and highlighted the implications this holds for SES research methods. Chapter 3 focused at a practical level on how SES research is conducted. Finally, this chapter introduces the aims, purpose and structure of the handbook and how to navigate the core of the book contained in Part 2.

Part 2 comprises 28 chapters describing different categories of SES methods. The chapters are grouped into three sets: (a) methods for data generation and systems scoping, (b) methods for knowledge co-production and effecting systems change, and (c) methods for analysing systems (Table 4.1). Each chapter provides a concise overview of the set of methods covered in that chapter, including a brief description of the disciplinary origins of the methods covered, key SES research questions that can be addressed using these methods and the limitations and resource implications of implementing the different methods. Each chapter concludes with a discussion of new emerging directions. Each chapter also includes a set of key readings and a case study to illustrate how one or more of the methods discussed in the chapter have been used in practice to answer an SES research question or problem.

While the methods discussed in group (a) include many commonly used methods for collecting data in SES, many SES studies use existing datasets. These are drawn from a wide variety of sources, including government census data, health and demographic data, remote-sensing and GIS data from a diversity of sectors (e.g. agriculture, conservation, health, urbanisation, ecology), compiled case studies and interviews. A collation of platforms commonly used by SES researchers to access useful datasets are given in Table 3.1 (see Chapter 3).

Part 3 comprises a concluding chapter reflecting on the status of SES research methods, key challenges in the field and ways forward. We have also included a glossary with key SES-related terms.

The book is intended as a reference guide and can be accessed from any chapter. Nevertheless, we suggest that most readers, and especially those new to SES research, will find Chapters 1–3 a useful introduction to understanding the framing of and broader context for the methods discussed in the book. This chapter (Chapter 4) provides a guide to navigating the different methods discussed in Part 2, and defines the terminology used in the opening chapter summaries. The final chapter (Chapter 33) provides an overview and reflection on the set of methods currently used in SES research, and highlights gaps and new emerging

Table 4.1 Chapters and methods covered in Part 2 of the handbook

Chapter number and name	*Methods covered*
METHODS FOR DATA GENERATION AND SYSTEMS SCOPING	
5. Systems Scoping	Social-ecological inventories, stakeholder analysis, historical, social and ecological inventories, cultural domain analysis, contextual profiling, policy scoping, literature reviews
6. Ecological Field Data Collection	*Measuring faunal and floral diversity and population structure*: Point counts and survey grids, transects, distance and plotless sampling, quadrats, capture and mark-recapture *Measuring abiotic conditions*: Abiotic environmental sampling, core sampling *Measuring ecological processes*: Telemetry, isotope and genetic analysis
7. Interviews and Surveys	Surveys, in-depth interviews, key informant interviews, life histories, focus group discussions, reflective questioning, conversations and dialogues, arts-based interview methods
8. Participatory Data Collection	Participatory mapping (direct-to-digital participatory mapping, 3D-participatory mapping, participatory GIS), photovoice, transect walks, ranking exercises, focus group discussions, Venn Diagrams, matrix scoring, ecograms, timelines, Q-methodology, community mapping, participatory videography, photo elicitation, seasonal calendars, participatory action research, participatory rural appraisal, participant observation, arts-based methods
METHODS FOR KNOWLEDGE CO-PRODUCTION AND EFFECTING SYSTEM CHANGE	
9. Facilitated Dialogues	Appreciative enquiry, change labs, social innovation labs, the circle, Theory U, T-Labs, scenarios, world café, learning journeys, listening projects, dialogue interviewing
10. Futures Analysis	Scenarios and participatory scenario planning, futures wheels, three horizons framework, design/experiential futures, horizon scanning, Delphi, trend impact analysis, emerging issues analysis, causal layered analysis, appreciative inquiry, gaming (also known as 'gamification' or serious gaming), future workshops, visioning, back-casting, road-mapping
11. Scenario Development	Double uncertainty matrix, Mānoa, scenario archetypes, *La Prospective*, causal layered analysis
12. Serious Games	Serious games
13. Participatory Modelling	*Modelling methods*: System dynamics (group model building, mediated modelling, shared vision planning), agent-based models (ARDI), role-playing games (Wat-A-Game), expert models (Bayesian networks, fuzzy cognitive maps), state-and-transition models, soft system methodologies (rich pictures, concept maps, decision trees, cognitive maps) *Integrated approaches*: Collaborative modelling, companion modelling, participatory system analysis
14. Resilience Assessment	Wayfinder, RAPTA, Resilience Assessment Workbook for Practitioners 2.0, STRESS, operationalising systemic resilience
15. Action Research	Narrative inquiry, learning history, cooperative inquiry

Chapter number and name	*Methods covered*
METHODS FOR ANALYSING SYSTEMS – System components and linkages	
16. Expert Modelling	Bayesian networks, fuzzy cognitive maps
17. Data Mining and Pattern Recognition	Data wrangling, clustering analysis, regression trees, neural networks, sentiment analysis, topic models
18. Statistical Analysis	Descriptive statistics, group comparison, regression models (linear, generalised linear models), multivariate analysis (including clustering, non-metric multidimensional scaling (n-MDS), principal component analysis (PCA), redundancy analysis (RDA), canonical correspondence analysis, factor analysis (FA) and multiple correspondence analysis), time series analysis
19. Qualitative Content Analysis	Discourse analysis, critical discourse analysis, thematic analysis, narrative analysis, critical narrative analysis, interpretative phenomenological analysis
20. Comparative Case Study Analysis	Variable-oriented analysis, archetype analysis (formal concept analysis, qualitative comparative analysis)
21. Controlled Behavioural Experiments	Controlled behavioural experiments
22. Institutional Analysis	Institutional analysis and development framework, SES framework, action situations, networks of action situations, institutional grammar tool, rule typology
23. Network Analysis	Network analysis
24. Spatial Mapping and Analysis	Spatial mapping and analysis, including geography, landscape ecology, remote sensing, statistics, land surveying, brief overview of relevant mapping and analytical approaches
METHODS FOR ANALYSING SYSTEMS – System dynamics	
25. Historical Assessment	Methods related to data obtained from sediment cores, archaeological/zooarchaeological materials, dendrochronology/sclerochronology, land surveys, historical aerial photography, satellite remote sensing, documentary sources, governmental data, interviews and oral histories
26. Dynamical Systems Modelling	Causal loop diagrams, loop analysis, qualitative analysis of differential equations (including bifurcation analysis and stability analysis), numerical simulation of dynamical systems
27. State-and-transition Modelling	State-and-transition modelling
28. Agent-based Modelling	Agent-based modelling
METHODS FOR ANALYSING SYSTEMS – Directly informing decision-making	
29. Decision Analysis Based on Optimisation	Mathematical programming, optimal control theory, game theory, decision theory, cost-benefit analysis, multi-criteria decision analysis

(Continued)

Table 4.1 (Continued)

Chapter number and name	*Methods covered*
30. Flow and Impact Analysis	Physical trade flows, multi-regional input-output analysis, environmentally extended multi-regional input-output analysis, environmental footprints, Life Cycle Assessment, energy return on investment, multi-scale integrated analysis of societal and ecosystem metabolism, global commodity chain analysis
31. Ecosystem Service Modelling	*Decision-support modelling packages*: Integrated valuation of ecosystem services and trade-offs (InVEST), artificial intelligence for ecosystem services (ARIES), Co$ting Nature/WaterWorld *Related technical models and frameworks*: Integrated assessment models, general equilibrium models, Lund–Potsdam–Jena dynamic global vegetation model, Life Cycle Assessment models
32. Livelihood and Vulnerability Analysis	Sustainable livelihood analysis, vulnerability analysis

areas of methodological development. Chapter 33 will be useful to anyone who has read parts of the book or is already quite familiar with SES research methods.

Selection and categorisation of methods

Part 2 of the handbook provides an overview of major categories of methods used in SES research. These categories attempt to pragmatically distil the wide diversity of methods used in SES research into a useful and succinct guide. The categories are based on a process of clustering methods identified in a systematic review of papers reporting on place-based SES research over the past 50 years (De Vos, Biggs, and Preiser 2019). This review identified 311 different methods that were mentioned in abstracts of 4 479 empirical research articles published before 2015. To reduce this list to a set of methods with which one could engage in a meaningful way, we clustered the methods into categories of methods used for similar purposes in SES research. The initial 27 categories (De Vos, Biggs, and Preiser 2019) were derived from a small expert workshop involving several SES researchers from the Stockholm Resilience Centre in addition to the core project team, held in Stockholm in 2016 (see 'Preface' for details).

The categorisation of methods required a number of decisions (De Vos, Biggs, and Preiser 2019). Some method categories contain only a single method (e.g. agent-based modelling) because workshop participants felt that these methods were either very broadly used in SES research, fulfilled a unique function, or are particularly well suited to SES research. Conversely, other categories (e.g. statistical methods) contained a wide diversity of methods and approaches, as they were not seen to play a unique role in SES research and resources for their use are readily available. In categorising the methods, we also kept in mind that certain methods (e.g. statistical methods and GIS) are not commonly mentioned in chapter abstracts, but form important categories of methods used in SES research. We assigned methods to single categories, although many methods could, in reality, fit into several categories.

In line with the objective of the handbook, the initial categories and methods were never intended as a final or definitive list of SES methods, but rather as an entry point for obtaining an overview of the range of methods available for use in SES research. Indeed, the editorial team revised the 27 method categories described by De Vos, Biggs and Preiser (2019) for this handbook. This resulted in a final list of 28 categories, which form the basis of the chapters in Part 2. We added three categories of methods to the original classification: (a) scenarios (previously included under futures analysis, but felt to be a very widely used method in SES research requiring its own chapter); (b) action research (previously subsumed under participatory data collection), and (c) resilience assessment (not included in the original paper because of an analytical decision to not include broader approaches). We also combined some categories (statistics and Bayesian methods were combined, as were optimisation and decision analysis), and changed a few names that were used in the original categorisation (e.g. 'quantitative pattern recognition' was changed to 'data mining and pattern recognition').

Guide to chapter summaries in Part 2

At the beginning of each chapter in Part 2, we provide a summary table of key characteristics of the methods discussed in the chapter to give readers an overview of their disciplinary origins, research approach, temporal and spatial dimensions, and purposes (Table 4.2). We also provide a summary of the systemic SES features that the methods discussed in the chapter are particularly good at addressing. These summaries were compiled by the chapter authors, with input and feedback from the editors. While we aimed for consistency across chapters, the summaries are by their nature somewhat subjective, especially given that different methods may achieve particular goals (e.g. policy/decision support) in very different ways. For chapters that cover many different methods, it was also sometimes challenging to summarise the diversity of ways in which all the individual methods discussed in the chapter are used.

This section provides a description of the elements presented in the summary tables, as defined in this handbook. The goal of the summary tables is to provide readers with a succinct overview of the core focus and origins of the methods contained in each chapter. Importantly, the summaries highlight the most common ways in which the methods discussed in the particular chapter are used, rather than all the possible ways in which they can be used. The summaries should therefore be interpreted as the most common focus or goals of the methods discussed, rather than limiting the interpretation to ways in which the methods discussed in the particular chapter could potentially be used in SES research. Indeed, some of the frontiers in SES research relate precisely to using some of the existing methods in new ways (see Chapter 33).

Research approach

As discussed in Chapter 2, there are several different traditions or ways of doing research across the broad spectrum of disciplines and research fields that exist today. These ways of doing research are based on fundamentally different worldviews regarding the nature of reality, how we can learn about this reality and what we aim to achieve through our research. These philosophical differences are particularly marked across the quantitative–qualitative and the natural–social sciences divides, but there also exist major differences in assumptions and approaches within these broad traditions. In bridging these divides, and drawing on diverse methods from different disciplines, SES research encompasses diverse philosophical traditions and research approaches.

Whereas some methods used in SES research span multiple research approaches, most methods originate clearly within a particular approach. Understanding these origins is important in order to understand the assumptions and potential limitations of particular

Table 4.2 Summary table template of key method characteristics appearing at the start of each chapter in Part 2

SUMMARY TABLE	
DISCIPLINARY BACKGROUND	**KNOWLEDGE TYPE**
The methods in this chapter are derived from or have most commonly been used in: List of relevant disciplines	*The methods in this chapter are primarily used to generate the following types of knowledge:* • Descriptive • Exploratory • Explanatory • Prescriptive
RESEARCH APPROACH	**PURPOSE OF METHOD**
The methods in this chapter originate from or most commonly adopt the following research approaches: • Analytical/objective • Interpretive/subjective • Collaborative/process	*The most common purposes of using the methods in this chapter are:* • Data collection/generation • System understanding • Stakeholder engagement and co-production • Policy/decision support
TEMPORAL DIMENSION	**SYSTEMIC FEATURES AND PROCESSES**
The methods in this chapter are most commonly applied to the following temporal dimensions: • Present (typically within the last 5–10 years) • Recent past (post-1700s) • Pre-industrial revolution (pre-1700s) • Future **SPATIAL DIMENSION** *The methods in this chapter are primarily either or both:* • Non-spatial • Explicitly spatial *The methods in this chapter are most commonly applied at the following spatial scales:* • Local • Regional (provincial/state to continental) • Global • Multiple places/sites around the world	*While most methods can do many things, the methods in this chapter are particularly good (i.e. go-to methods) for addressing the following:* • SES components and linkages • Diversity • Social-ecological dependence and impact • Power relations • Multiple scales and levels or cross-level interactions • Social-ecological interactions over time • Path dependency • Adaptation and self-organisation • Regime shifts • Transformation • Social learning • Collective action and collaborative governance • Evaluating policy options • Exploring uncertainty

methods, and the extent to which they are compatible with or can be combined with other methods (see Chapter 3). In this section of the summary table, we highlight the origins and main orientation of the method or set of methods discussed in the particular chapter in terms of the following categories:

- **Analytical/objective:** Analytical research approaches are grounded in empirical measurements of phenomena that are quantified and described through analysis. The aim is to generate objective descriptions of phenomena that eliminate personal biases, such as a priori assumptions and emotional or subjective interpretations of empirical data. Objective approaches assume that reality (e.g. laws of physics) exists 'out there' independently of the observer, and can be described in universal ways that transcend a particular time or cultural context. This approach underpins most of the natural sciences, but has also been adopted quite widely in the social sciences. While many SES researchers do not subscribe personally to these assumptions (e.g. by acknowledging that there are always personal biases and assumptions introduced in any research process and that a particular context matters), it is important to recognise that many of the methods used in SES research are grounded in an analytical/objective approach, and to be aware of the potential implications of these assumptions.
- **Interpretive/subjective:** Interpretative research approaches focus on the meanings, experiences, feelings and interpretations that people attach to particular phenomena or processes. This approach underpins important branches of the social sciences, especially those studying phenomena such as cultural beliefs, values and practices. Interpretivist approaches assume that the same phenomena can be interpreted and understood in multiple ways, and that any interpretation or experience is highly contingent on a particular time and cultural context. This approach emphasises the researcher's role as interpreter in the process of knowledge creation. One of the implications of this approach is that there can be no framework-independent vantage point from which a phenomenon can be understood, which is at fundamental odds with the starting point for objective research approaches.
- **Collaborative/process:** This research approach specifically aims to co-produce knowledge in collaboration with stakeholders. It aims to elicit, discuss and potentially integrate different types of knowledge (e.g. scientific, indigenous and local, practice-based) and understandings of a situation or phenomenon to facilitate appreciation of diverse viewpoints, and potentially to gain a more comprehensive understanding. This approach assumes that the process of knowledge sharing and joint sense-making is often as important as the specific knowledge being shared or co-created, in that it can build understanding and trust among diverse stakeholders.

Knowledge type

Just as different methods are rooted in different research approaches, they are also typically designed to generate different types of knowledge. Some methods can be used to generate multiple types of knowledge, but most methods are best suited to generating only one or two types of knowledge. Under 'Knowledge type' we distinguish between methods that primarily aim to generate the following types of knowledge:

- **Descriptive:** Methods used to identify and describe system components, connections and processes (e.g. systems scoping, network analysis)

- **Exploratory:** Methods used to explore patterns, connections and systemic behaviour or dynamics without a priori hypotheses (e.g. futures analysis, data mining and pattern recognition)
- **Explanatory:** Methods used to explain phenomena, patterns of connections and processes, i.e. to explain why and how a certain outcome was produced (e.g. statistical analysis, controlled behavioural experiments)
- **Prescriptive:** Methods used to provide specific normative guidelines for policy and practice (e.g. decision analysis based on optimisation, flow and impact analysis)

Purpose of method

Different methods also have different purposes in terms of the research process. Under 'Purpose of method' we refer to what the researcher most commonly aims to achieve by applying the method or sets of methods described in the chapter. More than one purpose is possible. We distinguish between the following main purposes:

- **Data collection/generation:** Methods used to collect or generate empirical data, which can be quantitative or qualitative (e.g. ecological field data collection, interviews and surveys)
- **System understanding:** Methods used to gain understanding of systemic components, processes and behaviours (e.g. network analysis, agent-based modelling)
- **Stakeholder engagement and co-production:** Methods used to engage a range of stakeholders for joint sense-making and knowledge co-creation (e.g. facilitated dialogues, participatory modelling)
- **Policy/decision support:** Methods used to inform or support decision-making, or to facilitate the development of policy (e.g. resilience assessment, ecosystem service modelling)

Temporal and spatial dimensions

For each of the method categories, we give an indication of the temporal and spatial dimensions at which they are most commonly applied. Some methods are commonly applied at multiple temporal and spatial scales. For the temporal dimensions, we distinguish between methods most commonly applied to study the:

- **Present (typically within the last 5–10 years)**
- **Recent past (post-1700)**
- **Pre-industrial revolution (pre-1700)**
- **Future**

In terms of spatial dimensions, we distinguish between methods that are primarily applied in non-spatial versus spatially explicit ways (or in both ways). We also distinguish the most common spatial scales at which the methods are applied (be they non-spatial or spatially explicit):

- **Local**
- **Regional (provincial/state to continental)**
- **Global**
- **Multiple places/sites around the world**

Systemic features and processes

Different methods are suited to understanding different systemic features and processes that people are commonly interested in understanding or affecting when doing SES research. Most methods can do many things, but there are usually a limited number of methods that researchers would turn to as methods of choice if they have a particular interest in understanding or affecting particular features and processes in SES. In this section we highlight systemic features and processes that the methods discussed in a particular chapter are especially good at addressing.

- **SES components and linkages:** Methods that aim to identify the components of an SES and the relations between these components
- **Diversity:** Methods that aim to assess the amount of variation in a system and the implications of this variation. Diversity includes three distinct components: variety (how many different elements), balance (how many of each element) and disparity (how different the elements are from one another). Important SES elements that exhibit diversity include genes, species, landscape patches, cultural groups, livelihood strategies and governance institutions
- **Social-ecological dependence and impact:** Methods that aim to illuminate how social and ecological components of an SES depend on and affect one another
- **Power relations:** Methods that aim to assess the agency someone or an institution or a set of values has over someone else or over resources (dominant or sovereign power). Power can also refer to having the agency or power to act (productive power)
- **Multiple scales and levels or cross-level interactions:** Methods that aim to understand processes at several different discrete scales or levels (multi-scale), or across different scales or levels (cross-scale)
- **Social-ecological interactions over time:** Methods that aim to track and understand how a system changes over time as a result of social-ecological interactions
- **Path dependency:** Methods that aim to track how the particular path or development trajectory of a system is influenced or limited by previous events or decisions
- **Adaptation and self-organisation:** Methods that aim to explore how elements of an SES learn, combine experiences and knowledge, and adjust to changing external drivers and internal processes, and how some form of order arises from local interactions between elements of an SES
- **Regime shifts:** Methods that aim to explore and understand large, persistent changes in the composition, structure and/or function of SES associated with the transgression of critical tipping points
- **Transformation:** Methods that aim to explore or facilitate fundamental reorganisation of SES towards a more sustainable or preferred outcome. Transformations are conceptually similar to regime shifts, but focus on shifts towards more positive futures, often involving radical changes in underlying worldviews, values and governance systems
- **Social learning:** Methods that aim to facilitate or understand how societal learning occurs through social interactions and processes. Social learning goes beyond the individual, enabling new knowledge to become situated within a group, community of practice or society, and can lead to a transformation in values and worldviews
- **Collective action and collaborative governance:** Methods that aim to facilitate, support or analyse processes of collective action or collaborative governance. Collective action refers to joint action by a group of people to achieve a common objective.

Collaborative governance refers to processes and structures of public policy decision-making and management that engage people across public agencies, levels of government, and/or the public, private and civic spheres to carry out a common purpose

- **Evaluating policy options:** Methods that aim to assess intended or unintended outcomes of policies or interventions in an SES in order to directly inform policy choices
- **Exploring uncertainty:** Methods that aim to explore or quantify the potential unknowns or levels of uncertainty about SES dynamics, arising from limited knowledge or the complex adaptive nature of SES

Identifying chapters and methods of interest in Part 2

There are many different ways to navigate Part 2 of this book. You may be interested in getting a sense of the different methods by reading through all the chapters, or you may already have a good idea which particular methods you are interested in learning more about. Alternatively, you may be starting with a specific research problem and purpose but don't have a good idea of appropriate methods that can be used to investigate these questions. The summary tables at the beginning of each chapter can guide you to identify methods that may be appropriate for different purposes by providing information on the key characteristics of a particular method. Tables 4.3 and 4.4 summarise these key characteristics and can serve as a 'roadmap' to identify potential methods and chapters to delve into in more detail. If you want to employ a research approach that is collaborative, for example, you can use Table 4.3 to identify which methods align with this research approach.

While some methods span multiple approaches, others are more restricted in their research approach or the type of knowledge they generate. Some groups of methods can be used to both collect and interpret data, while others only serve one purpose. In this case, you may need to combine multiple methods to achieve your research aim (see Section 'Combining multiple methods'). If your research intends to explore the 'deep' (i.e. pre-industrial) past, you will typically need to use different methods than if your research intends to explore the present or the future. Similarly, different methods are better suited to local studies compared to studies that span the entire globe. Alternatively, perhaps your research will focus on exploring a particular feature of an SES such as regime shifts, transformation or collaborative governance. In this case, Table 4.4 can help you to identify methods that are particularly good at addressing particular features.

In most cases, you may need to consider the various characteristics summarised in Tables 4.3 and 4.4 to identify potential methods of interest for your research, and then delve into those specific chapters to further assess their appropriateness. Each of the chapters provides suggestions of more detailed texts that could be consulted for more information on methods of interest and how they are best implemented.

Combining multiple methods

The methods presented in this handbook may be used as stand-alone methods to address a specific objective or question or, more commonly, can be used in combination or build on one another to achieve a particular purpose. Ecological field data collection methods (Chapter 6), for example, can be used to collect species data in the field. These data can then be analysed using statistical analysis (Chapter 18) to determine species richness or abundance and relationships and interactions between species. This could be combined with interviews

Table 4.3 Summary of the key characteristics of methods discussed in Part 2

	Research approach			Knowledge type				Purpose of method				Temporal dimension				Spatial dimension					
	Analytical/objective	Interpretive/subjective	Collaborative/process	Descriptive	Exploratory	Explanatory	Prescriptive	Data collection/generation	System understanding	Stakeholder engag. and coprod.	Policy/decision support	Present	Recent past	Pre-industrial revolution	Future	Explicitly spatial	Non-spatial	Local	Regional	Global	Multiple places/sites
METHODS FOR DATA GENERATION AND SYSTEMS SCOPING																					
5. Systems Scoping																					
6. Ecological Field Data Collection																					
7. Interviews and Surveys																					
8. Participatory Data Collection																					
METHODS FOR KNOWLEDGE CO-PRODUCTION AND EFFECTING SYSTEM CHANGE																					
9. Facilitated Dialogues																					
10. Futures Analysis																					
11. Scenario Development																					
12. Serious Games																					
13. Participatory Modelling																					
14. Resilience Assessment																					
15. Action Research																					
METHODS FOR ANALYSING SYSTEMS – System components and linkages																					
16. Expert Modelling																					
17. Data Mining and Pattern Recognition																					
18. Statistical Analysis																					
19. Qualitative Content Analysis																					
20. Comparative Case Study Analysis																					
21. Controlled Behavioural Experiments																					
22. Institutional Analysis																					
23. Network Analysis																					
24. Spatial Mapping and Analysis																					
METHODS FOR ANALYSING SYSTEMS – System dynamics																					
25. Historical Assessment																					
26. Dynamical Systems Modelling																					
27. State-and-transition Modelling																					
28. Agent-based Modelling																					
METHODS FOR ANALYSING SYSTEMS – Directly informing decision-making																					
29. Decision Analysis based on Optimisation																					
30. Flow and Impact Analysis																					
31. Ecosystem Service Modelling																					
32. Livelihood and Vulnerability Analysis																					

Table 4.4 Summary of SES features and processes which particular methods are especially good at addressing

	SES components and linkages	Diversity	SE dependence and impact	Power relations	Multiple scales or cross-level interaction	SE interactions over time	Path dependency	Adaptation and self-organisation	Regime shifts	Transformation	Social learning	Collective action and collab. governance	Evaluating policy options	Exploring uncertainty
METHODS FOR DATA GENERATION AND SYSTEMS SCOPING														
5. Systems Scoping	■	■		■	■						■	■		
6. Ecological Field Data Collection	■	■	■											
7. Interviews and Surveys	■	■	■	■										
8. Participatory Data Collection	■		■	■							■			
METHODS FOR KNOWLEDGE CO-PRODUCTION AND EFFECTING SYSTEM CHANGE														
9. Facilitated Dialogues	■	■		■	■	■				■	■	■		■
10. Futures Analysis			■	■		■	■			■	■			■
11. Scenario Development			■			■	■			■	■		■	■
12. Serious Games										■	■	■	■	
13. Participatory Modelling		■	■								■	■	■	
14. Resilience Assessment	■				■			■	■	■	■	■		■
15. Action Research		■		■						■	■	■	■	■
METHODS FOR ANALYSING SYSTEMS – System components and linkages														
16. Expert Modelling	■		■		■				■				■	■
17. Data Mining and Pattern Recognition	■	■							■					
18. Statistical Analysis	■	■	■						■					■
19. Qualitative Content Analysis	■			■						■			■	
20. Comparative Case Study Analysis										■		■		
21. Controlled Behavioural Experiments						■		■				■		
22. Institutional Analysis	■											■		
23. Network Analysis	■			■								■		
24. Spatial Mapping and Analysis	■	■				■			■					
METHODS FOR ANALYSING SYSTEMS – System dynamics														
25. Historical Assessment						■		■	■					
26. Dynamical Systems Modelling	■		■			■	■		■				■	
27. State-and-transition Modelling									■		■		■	
28. Agent-based Modelling	■	■			■	■	■	■					■	■
METHODS FOR ANALYSING SYSTEMS – Directly informing decision-making														
29. Decision Analysis based on Optimisation	■		■										■	
30. Flow and Impact Analysis			■		■	■							■	
31. Ecosystem Service Modelling	■		■										■	
32. Livelihood and Vulnerability Analysis		■	■					■					■	

and surveys (Chapter 7), a method that focuses on gathering people's observations and experiences, to determine how different species are valued and the ways in which they contribute to human well-being. This information can, in turn, feed into futures analysis (Chapter 10) or scenario development (Chapter 11) to explore the integrated future of biodiversity and human well-being, and inform decision-making. When combining different methods, it is critical to keep in mind that not all methods are compatible with one another in terms of their underlying assumptions. Careful consideration is needed when combining different methods, particularly when combining quantitative and qualitative data and approaches (see Chapter 3).

We explored the connections between methods presented in this handbook using network analysis. Figure 4.1 shows a network diagram using the information the authors of each chapter provided about other methods and chapters to which their chapter connects. Nodes represent the chapter numbers, and the edge lines represent the linkages between chapters. Nodes are sized according to the average degree based on the number of edges; the larger the node, the more connected the chapter is to methods in other chapters. The number of arrows moving out of each node is known as the out-degree, and the number of arrows moving in towards the node represents the in-degree. The average degree is the total number of in-degree and out-degree edges.

Based on the network analysis in Figure 4.1, the most highly connected methods are ecological field data collection (Chapter 6), interviews and surveys (Chapter 7), participatory data collection (Chapter 8), statistical analysis (Chapter 18), spatial mapping and analysis (Chapter 24) and agent-based modelling (Chapter 28). Chapters 6–8 are all data-generation methods which feed into many other methods, particularly statistical analysis (Chapter 18) and spatial mapping and analysis (Chapter 24). The results from these analyses, in turn, often feed into more synthetic, participatory decision-support processes such as scenario development (Chapter 11) or resilience assessment (Chapter 14). Agent-based modelling (Chapter 28) is a particularly versatile method that, like all modelling approaches, relies on many other methods to inform model design and analysis, such as systems scoping (Chapter 5), interviews and surveys (Chapter 7), ecological field data collection (Chapter 6) and statistical analysis (Chapter 18). It can also be used to support methods for knowledge co-production such as futures analysis (Chapter 10) and scenario development (Chapter 11), or be combined with other modelling methods such as dynamical systems modelling (Chapter 26).

Some methods are more commonly used in combination than other methods. This may depend on the purpose of the research, or the type of method. Spatial mapping and analysis (Chapter 24), for example, is commonly used in combination with ecological field data collection (Chapter 6), participatory data collection (Chapter 8), participatory modelling (Chapter 13), agent-based modelling (Chapter 28) and ecosystem service modelling (Chapter 31). Sometimes combinations of methods are used to address the limitations of a single method. Combining participatory data collection (Chapter 8) with spatial mapping and analysis (Chapter 24) to understand land-use change, for example, can provide insight into potential biases in people's memories of landscape change and how changes are experienced and affect livelihoods.

Other methods (or more accurately, approaches) constitute steps that combine several methods under a common framing and for a particular goal. Resilience assessment (Chapter 14), for example, is an umbrella process that relies on multiple steps or methods. It is broadly used to address questions about the capacity of an SES to cope with and respond to change. The first step requires methods such as systems scoping (Chapter 5), interviews and

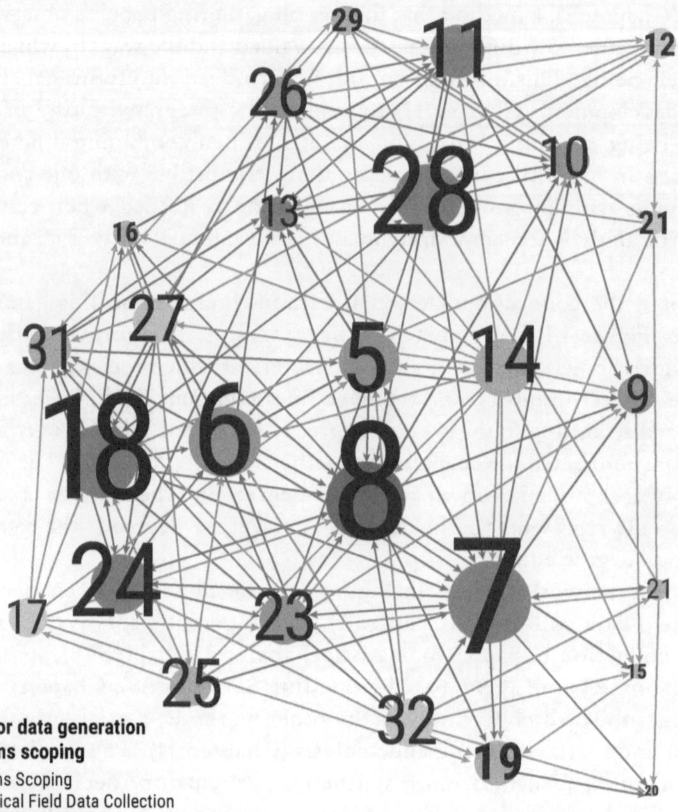

Methods for data generation and systems scoping

5 Systems Scoping
6 Ecological Field Data Collection
7 Interviews and Surveys
8 Participatory Data Collection

Methods for knowledge co-production and effecting system change

9 Facilitated Dialogues
10 Futures Analysis
11 Scenario Development
12 Serious Games
13 Participatory Modelling
14 Resilience Assessment
15 Action Research

System components and linkages

16 Expert Modelling
17 Data Mining and Pattern Recognition
18 Statistical Analysis
19 Qualitative Content Analysis
20 Comparative Case Study Analysis
21 Controlled Behavioural Experiments
22 Institutional Analysis
23 Network Analysis
24 Spatial Mapping and Analysis

System dynamics

25 Historical Assessment
26 Dynamical Systems Modelling
27 State-and-transition Modelling
28 Agent-based Modelling

Directly informing decision-making

29 Decision Analysis based on Optimisation
30 Flow and Impact Analysis
31 Ecosystem Service Modelling
32 Livelihood and Vulnerability Analysis

Figure 4.1 Network analysis of linkages between methods presented in Part 2 of the handbook (© Kristine Maciejewski) surveys (Chapter 7) and participatory data collection (Chapter 8) to collect the data, which are then analysed (as the second step) using network analysis (Chapter 23) or dynamical systems modelling (Chapter 26). Finally, strategic interventions are explored using scenario development (Chapter 11) and futures analysis (Chapter 10). Similarly, livelihood and vulnerability analysis (Chapter 32) and state-and-transition modelling (Chapter 27) also constitute processes with several steps that use different methods.

Conclusion

This handbook is intended as an introduction and guide to the diversity of methods that can be employed in SES research. It provides an overview of the most commonly used methods at this time, but does not constitute an exhaustive or definitive set of SES research methods. Indeed, given the rapid growth of the SES research field, and of science and technology more generally, new methods and novel combinations of methods are constantly being developed, and offer the potential to answer new questions and reshape the field (see Chapter 33).

A diverse array of SES research methods currently in use is covered in this book with the aim of providing SES researchers with a synthetic overview and guide to methods they may consider using in their research. The breadth of the book comes at the expense of detail regarding the assumptions and nuances about the implementation and application of specific methods. The intention is to help readers identify potential methods of interest from among the very large and diverse set of methods that can be employed in SES research, and to point them to more detailed texts on the appropriate use and implementation of specific methods. As in any research endeavour, it is essential to think critically about the appropriateness of any method to be employed and the assumptions underlying that method.

We hope you find this book useful, and that it supports the development of a next generation of SES researchers who can contribute to addressing the pressing sustainability challenges we face. The chapters in this handbook, updates and further materials that may be useful for teaching and research are available at www.sesmethods.org.

Reference

De Vos, A., R. Biggs, and R. Preiser. 2019. 'Methods for Understanding Social-Ecological Systems: A Review of Place-based Studies.' *Ecology and Society* 24(4): 16. doi:10.5751/ES-11236-240416.

Conclusion

Part 2

Methods for data generation and systems scoping

5

Systems scoping

Nadia Sitas,[1] Paul Ryan[2] and Lisen Schultz[3]

[1]CENTRE FOR SUSTAINABILITY TRANSITIONS, AND DEPARTMENT OF CONSERVATION ECOLOGY AND ENTOMOLOGY, STELLENBOSCH UNIVERSITY, STELLENBOSCH, SOUTH AFRICA
[2]AUSTRALIAN RESILIENCE CENTRE, AUSTRALIA
[3]STOCKHOLM RESILIENCE CENTRE, STOCKHOLM UNIVERSITY, STOCKHOLM, SWEDEN

Key methods discussed in this chapter

Social-ecological inventories, stakeholder analysis, historical, social and ecological inventories, cultural domain analysis, contextual profiling, policy scoping, literature reviews

Connections to other chapters

Conducting a systems-scoping exercise can be the precursor to many other analyses outlined in this book and is useful for identifying the key variables and components of the study. A systems-scoping exercise can help to identify what ecological and social data might need to be collected (Chapters 6–8) and whether more participatory and inclusive methods of knowledge co-production are needed (Chapters 9, 12–15 and 22). In addition, a systems-scoping exercise can identify the temporal nature of further research, e.g. whether there needs to be additional historical (Chapter 25) or future-looking analyses (Chapters 10 and 11). By developing a better understanding of the boundaries of the research through systems scoping, the researcher can then get a better idea of which further methods to select, e.g. methods to understand system components (Chapters 16–24), system dynamics (Chapters 25–28) or combinations thereof (Chapters 31 and 32).

Introduction

Which ecological and social variables should be included when trying to understand specific phenomena or issues? How do these variables interact, under what conditions and at which scale? Which knowledge types and actors are important to include? These are some of the typical questions asked when undertaking a systems-scoping exercise in social-ecological systems (SES) research. Many different methods have been used to 'scope' a system, but common to most is a focus on the importance of setting the boundaries of the system under investigation and identifying the system components, their relationships and issues of interest (Walker et al. 2004; Peterson 2005).

Systems scoping has been used in a number of different disciplines, from the health sciences, to social and natural sciences, to humanities, but has a strong history in strategic

DOI: 10.4324/9781003021339-7

SUMMARY TABLE: SYSTEMS SCOPING

DISCIPLINARY BACKGROUND	KNOWLEDGE TYPE
The methods in this chapter are derived from or have most commonly been used in: Environmental Science, Geography, Health Sciences, Management/ Business Studies, Public Administration/ Policy Studies	*The methods in this chapter are primarily used to generate the following types of knowledge:* • Descriptive • Exploratory
RESEARCH APPROACH	**PURPOSE OF METHOD**
The methods in this chapter originate from or most commonly adopt the following research approaches: • Analytical/objective • Interpretive/subjective • Collaborative/process	*The most common purposes of using the methods in this chapter are:* • Data collection/generation • System understanding • Stakeholder engagement and co-production • Policy/decision support
TEMPORAL DIMENSION	**SYSTEMIC FEATURES AND PROCESSES**
The methods in this chapter are most commonly applied to the following temporal dimensions: • Present (typically within the last 5–10 years) • Recent past (post-1700s) **SPATIAL DIMENSION** *The methods in this chapter are primarily either or both:* • Non-spatial • Explicitly spatial *The methods in this chapter are most commonly applied at the following spatial scales:* • Local	*While most methods can do many things, the methods in this chapter are particularly good (i.e. go-to methods) for addressing the following:* • SES components and linkages • Diversity • Power relations • Multiple scales and levels or cross-level interactions • Social learning • Collective action and collaborative governance

planning and associated applied fields such as environmental management and town and regional planning. More specifically in SES research, the goal of systems scoping is to explore a system from a variety of perspectives before establishing a set of key issues (past, present and future), the objectives of a study and a methodology/research design for a more in-depth analysis of particular social-ecological dynamics of interest. In addition, scoping can be used in preparation for a participatory or consultative research process to identify key actors and co-identify issues of concern or interest. Thus, systems scoping can employ a number of different research methods, such as interviews and surveys (Chapter 7), participatory data collection (Chapter 8), network analyses (Chapter 23), institutional analyses (Chapter 22) and facilitated dialogues (Chapter 9).

Social-ecological systems are complex, open systems. To better understand them, one needs to understand not only the system under consideration but also the broader SES it is linked to or embedded in, which is in itself complex. Complex adaptive systems are dynamic, evolve in response to changes in feedbacks and have thresholds and tipping points (see Chapters 1 and 2). A systems-scoping exercise should incorporate approaches that attempt to document the elements that lead to or contribute to these dynamic phenomena. However, to avoid an attempt to include everything, the complexity of the system still needs to be reduced within certain boundaries in order to enable the (co)generation of knowledge (Cilliers 2005; Scholes et al. 2013), while keeping in mind that any choice of this nature means that important interconnections and feedbacks may be missed. Due to the inherent complexity and uncertainty of SES, knowledge of the system can only be gained in relation to a particular framework (or conceptual model) that is employed. However, our conceptualisation of the system will always only be partial. Value-laden choices of what to include, and where the boundaries need to be articulated, are always involved and have consequences for how we understand the system (Cilliers 2005; Heylighen, Cilliers, and Gershenson 2006).

Scoping can refer to a stage in a larger assessment or research process (e.g. in strategic assessments like impact assessments or resilience assessments) or be a stand-alone study that is used to describe a coupled SES (Resilience Alliance 2010; Audouin 2011; Enfors-Kautsky et al. 2018). Systems scoping can also range from a narrow desk-based study to more inclusive participatory processes drawing on diverse actor conceptualisations of a system in order to unearth important issues pertaining to that system (Schultz, Plummer, and Purdy 2011). In some cases, broad participation in a scoping phase is required by law. Scoping should also always be conducted within an ethical framework that is as inclusive as possible (Watson 2010; Cockburn and Cundill 2018). A hidden assumption when conducting a systems-scoping exercise is that those leading or defining the scoping boundaries are aware of the many variables and values inherently present in the system. Here it is important to acknowledge the role of power – not only in terms of method selection but also in determining who and what to be considered or included in the system framing, how diverse values will be accounted for and who might benefit from the research (e.g. researchers, practitioners, funders) – and to highlight the importance of reflexivity in systems-scoping methodologies and approaches (Cote and Nightingale 2012; Audouin et al. 2013; Hankivsky 2014).

Systems scoping is often constrained by institutional and biophysical boundaries, e.g. a water catchment or local government area. When these institutional and biophysical boundaries are spatially different, e.g. when a national border crosses a river, scoping entails making an explicit choice between them (Ison 2008). Systems scoping is also often directed by a conceptual framework, e.g. the Intergovernmental Science Policy Platform for Biodiversity and Ecosystem Services (IPBES) conceptual framework (Díaz et al. 2015) or Ostrom's framework for analysing the sustainability of SES (Ostrom 2009). These conceptual frameworks help to set the boundaries, conditions and variables for a deeper exploratory or analytical exercise based on addressing specific questions or problems.

SES problems and questions

A scoping exercise is often one of the first steps undertaken in research in order to define the boundaries of the research and identify the key relations and dynamics between actors and ecosystems in a given SES. Systems scoping can be applied to a number of different research contexts at different spatial scales. It is very useful in helping to focus a study and outline the important issues that could be considered, and deciding which variables from different domains (e.g. social or environmental) to include. In addition to identifying the key ecological components and important actors within a system, Audouin et al. (2013) identify a number of questions that can underpin a systems scoping. These include:

- Who should be involved in defining the purpose of the study, the problem to be addressed and the skills to be included?
- What values underpin the goals and objectives of the study?
- What assumptions are made in defining the various spatial, temporal and substantive (i.e. issues to be addressed) boundaries of the study?
- What knowledge types (explicit and tacit, informal and formal) are important to include in the process of gaining an understanding of the SES?
- How can the research process, its goals and outcomes be aligned with the needs and values of those most likely to be affected by any recommendations or decisions that might come from the study?
- What is the role of power in shaping relationships and the flows of resources and benefits?

Brief description of key methods

Systems scoping involves the use of a variety of methods in order to understand key dynamics, scales and relationships of social and ecological components in SES. These methods

Table 5.1 Summary of key methods used in systems scoping

Method	*Description*	*References*
Literature review	A literature review is a widespread method used to review published materials that examine recent or current literature on specific topics. As part of a systems-scoping exercise, it can help to define and refine the topic of interest. It can cover a wide range of subjects at various levels of completeness and comprehensiveness, e.g. peer-reviewed literature as well as grey literature.	*Key introductory texts* Grant and Booth 2009 (highlighting the specific types of systematic reviews); Hart 2018 *Applications to SES* Binder et al. 2013; Milkoreit et al. 2018
Policy scoping	Within a systems-scoping process it is often important to identify the relevant policies that can affect the SES and the issues of interest. This can be done as a desk-based process, or through engagement of stakeholders who have knowledge of the particular policies and regulations that might be relevant.	*Key introductory text* Anderies and Janssen 2013 *Applications to SES* Anderies, Janssen, and Ostrom 2004; Kraft and Vig 2006; Garmestani 2014; Orach and Schlüter 2016

Method	*Description*	*References*
Stakeholder analysis	Stakeholder inventories and analyses focus on identifying important actors who have a stake in the research. They are used to gather information from all stakeholders who affect the decision-making process. These methods mostly tend not to account for biophysical components.	*Key introductory text* Reed et al. 2009 *Applications to SES* Grimble and Wellard 1997; Stringer et al. 2006; Prell, Hubacek, and Reed 2009; Petursdottir et al. 2013
Ecological inventory	Ecological inventories aim to document the biophysical landscapes of a specified system and are often compiled through field surveys, remote sensing and/or ecological mapping techniques. Ecological inventories generally omit the social processes influencing the natural system and as such are not often used in isolation in SES research.	*Key introductory text* McRae et al. 2012 *Applications to SES* Wulder et al. 2004; SANBI and UNEP-WCMC, 2016; also see Chapter 6 on ecological field data collection and Chapter 31 on ecological and ecosystem service mapping and modelling
Social inventory	Social inventories are similar to stakeholder analyses/mapping but tend not to focus on the power dimensions of the different actors in the system as much as stakeholder analyses or social network analyses and mapping.	*Key introductory text* Grimble and Wellard 1997 *Applications to SES* Barthel et al. 2005; Colding 2013; Wali et al. 2017
Social-ecological inventories	Social-ecological inventories are used to map key actors engaged in ecosystem management, their values, motives, activities, knowledge, networks and experiences over time, and while doing so to identify and select the most important actors to build trust and work with. Through engagement with key actors, important ecosystem features might be revealed, which could result in identifying additional ecosystem features to consider.	*Key introductory texts* Schultz, Folke, and Olsson 2007; Schultz, Plummer, and Purdy 2011 *Applications to SES* Schultz, Folke, and Olsson 2007; Baird, Plummer, and Pickering 2014
Historical inventory	Historical inventories aim to document important historical events in a system within a specified timeframe and are useful for understanding change and path dependencies in SES. Documenting these events can be conducted either through a literature review or in a more participatory manner, e.g. through interviews or focus groups.	*Key introductory text* Resilience Alliance 2010 *Applications to SES* Ramankutty and Foley 1999; Barthel et al. 2005; Anderson and O'Farrell 2012; Boonstra and De Boer 2014; Zheng et al. 2014
Cultural domain analysis	A cultural domain analysis is used to understand how people in a society think about and define their world. Since all cultures use some system of categories to order experience, the researcher tries to determine what categories are important to people, how these categories are arranged and what values are attached to them.	*Key introductory text* Puri 2011 *Applications to SES* Rodríguez, Pascual, and Niemeyer 2006; Buchmann 2009; Sheil and Liswanti 2006

most commonly include literature reviews, policy scoping, social, ecological, historical and social-ecological inventories, stakeholder analyses and cultural domain analysis. These methods are often used in combination with one another. A social-ecological inventory-based approach can often include social, ecological, cultural and historical characteristics. Table 5.1 provides a summary of key methods used in systems scoping.

Limitations

One of the main limitations in systems-scoping exercises concerns setting the boundaries of the research and how information on stakeholders and spatial and temporal dynamics is intentionally

Case study 5.1: Application of a social-ecological inventory to engage actors in Niagara, Canada

The Niagara region plays an important role in Canada's development, contributing to the economy through industry, agricultural production of fruit and wine, and tourism from the more than 30 million people visiting Niagara Falls annually. The unique biocultural diversity of the area has been recognised with the establishment of the UNESCO Niagara Escarpment Biosphere Reserve in Ontario. In 1998, the Canadian government commissioned a study on the impacts of climate variability on the region. At the time there was no known entity (e.g. government organisation or research institute) focusing on climate change adaptation. Baird, Plummer and Pickering (2014) sought to investigate whether the governance system for climate change adaptation could be primed by undertaking a social-ecological inventory.

The use of a social-ecological inventory seems to have been instrumental in facilitating a multi-sectoral, adaptive co-management governance approach to climate change in the Niagara region (Baird, Plummer, and Pickering 2014). The iterative and dynamic process followed six steps: preparations, preliminary identification, identification of key individuals, interviewing, reviewing and enriching the inventory, and engagement. These steps are outlined in Figure 5.1. This study focused on understanding the role of local knowledge as an important factor in understanding the SES and whether this approach could catalyse co-management approaches to assist with climate change adaptation.

Questions posed to the stakeholders in the research covered issues related to: (a) perceptions of climate change impacts in Niagara, organisational capacity for adaptation and adaptation leadership; (b) specific activities related to climate change and the rationale for these efforts; and (c) networks and relationships with other actors in Niagara related to climate (Baird, Plummer, and Pickering 2014).

The use of a social-ecological inventory-based approach as a precursor to undertaking a climate change adaptation strategy resulted in the following observations: (a) the approach facilitated adaptive co-management mainly through unearthing insights about networks and relationships, which could enable key actors and bridging organisations to be identified, (b) local sources of knowledge were revealed and gaps in knowledge illuminated, and (c) the tailoring of the adaptive co-management process was possible because the approach brought about an awareness of existing actions, desired information and differences in values that could be used in future planning.

or unintentionally included or excluded. Not acknowledging the role of power and politics and how power influences decisions in research can limit the ways in which diversity is taken into account in the conceptualisation of the system (Smith and Stirling 2010; Nayak, Armitage, and Andrachuk 2016). This diversity can include diversity of knowledge based on disciplinary biases, which can minimise the inclusion of diverse disciplinary knowledge or non-disciplinary knowledge and expertise such as local and indigenous knowledge (Van Kerkhoff and Lebel 2006; Tengö et al. 2014). Diversity can also relate to various other social categories such as gender, socio-economic status, ethnicity, religion, age and geographic positionality. Methods for better understanding how these categories intersect and influence the way in which the system and relationships between variables are framed are not always included in scoping processes

The study highlighted future avenues to test the social-ecological inventory approach. These relate to exploring how the approach could be applied in different contexts; exploring different user perceptions of the approach beyond the research community; setting up a database of studies using social-ecological inventories in order to undertake comparative analyses and promoting learning of how the use of social-ecological inventories could be enhanced in different contexts.

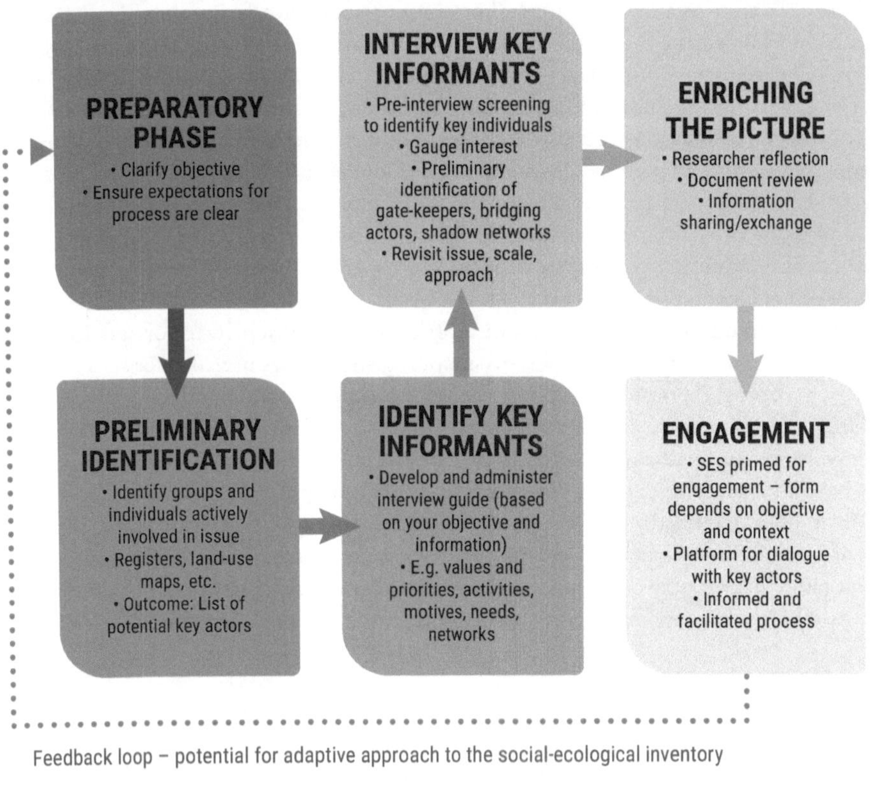

Figure 5.1 Phases of the social-ecological inventory (Baird, Plummer, and Pickering 2014)

(Thompson-Hall, Carr, and Pascual 2016). Power can also mediate the legitimacy of research through excluding key actors or end-users of the research, or by not being explicit about any vested interests in the research going forward (Cash et al. 2003; Clark et al. 2016).

Systems-scoping exercises are normally a descriptive snapshot of the system and its recent past. They are not normally used for explanatory purposes on their own but require additional methods to account for changes and feedbacks in system dynamics across space and time, e.g. scenario and futures assessments (Chapters 10 and 11), social network analyses (Chapter 23) or ecosystem service modelling (Chapter 31). However, systems-scoping exercises can continue to evolve as one learns more about the dynamics and key features of the system that can be used to iteratively focus the study and/or set new boundaries and conditions for research.

Resource implications

The resource implications linked to systems scoping depend on the boundaries and objectives of the specific task. They also depend on acknowledging that there are certain trade-offs and tensions that are important to consider between more technical processes compared to more social, participatory processes. If the aim is to effect change, matching quality information with a legitimate, credible and salient process has a higher capacity for fostering change.

More inclusive, participatory processes for systems scoping require extensive resources (in addition to time as a key resource). These processes typically involve hosting workshops, which includes selecting a suitable venue to use (or hire), travel and transport costs, security measures if necessary, using professional facilitators trained in systems thinking and stakeholder engagement (or undertaking training in order to run these workshops), and materials for running the workshop (see also Chapter 8 and methods for knowledge co-production and effective system change in Chapters 9–15). Depending on the context, it might also be necessary to compensate participants for their participation, either by way of monetary compensation or by providing a meal and transport. Careful attention should always be paid to the ethics of conducting participatory research, especially with vulnerable groups (Watson 2010) (see also Chapter 3).

Systems-scoping exercises often require little in terms of hardware or software when being conducted, but some form of systems-mapping software is useful in order to map out the results of a scoping exercise and capture the relationships between identified variables. These software programs can be simple (preferably open-source) programs such as Visio, MyDraw and draw.io, or more advanced software for simulation or data analysis purposes that would move beyond a scoping exercise for more analytical purposes (e.g. Vensim, Atlas.ti, QGIS, ArcGIS and Google Earth/Maps). Cameras to document specific system features (e.g. a river, place of interest, infrastructure) are useful to provide visuals of specific features for future use, but hand-drawn maps or images are also useful and sometimes easier in under-resourced contexts where access to electricity and other technology is challenging.

New directions

Given the importance of considering interconnected social and ecological components of systems, and that alternative conceptualisations of the system exist, a number of tools are being developed that provide decision-makers with frameworks to guide systems-scoping exercises. These include the Wayfinder tool (Enfors-Kautsky et al. 2018), the STEPS Pathways approach (Leach, Scoones, and Stirling 2007) and the Resilience, Adaptation Pathways and

Transformation Assessment (RAPTA) framework (O'Connell et al. 2015) (Chapter 14). New potential applications include tools that foreground the role of power and politics in systems scoping (e.g. building on Schoon et al. 2015; Berbés-Blazquez, González, and Pascual 2016) and those that try to embed SES perspectives in assessment processes, e.g. IPBES assessments (see the IPBES guide for assessments at ipbes.net/guide-production-assessments).

Transdisciplinary advances for problem framing can be useful in determining the research question and setting the initial boundaries of the research (Pohl and Hadorn 2007; Hadorn et al. 2008). The T-Lab approach provides a novel methodology for designing transformative spaces to negotiate and articulate key issues of interest (Pathways Network 2018; Pereira et al. 2018). Research using cultural historical activity theory (CHAT) can also shed more light on the relationship between the human mind (what people think and feel) and the activity (what people do). CHAT has been used as 'a cross-disciplinary framework for studying how humans purposefully transform natural and social reality, including themselves, as an ongoing culturally and historically situated, materially and socially mediated process' (Roth, Radford, and Lacroix 2012).

Key readings

Audouin, M., M. Burns, A. Weaver, D. le Maitre, P. O'Farrell, R. du Toit, and J. Nel. 2015. 'An Introduction to Sustainability Science and its Links to Sustainability Assessment.' In *Handbook of Sustainability Assessment*, edited by A. Morrison-Saunders, J. Pope, and A. Bond, 321–346. Cheltenham: Edward Elgar.

Enfors-Kautsky, E., L. Järnberg, A. Quinlan, and P. Ryan. 2018. Wayfinder: A Resilience Guide for Navigating Towards Sustainable Futures. GRAID Programme, Stockholm Resilience Centre, Stockholm University. www.wayfinder.earth.

Reed, M.S., A. Graves, N. Dandy, H. Posthumus, K. Hubacek, J. Morris, C. Prell, C.H. Quinn, and L.C. Stringer. 2009. 'Who's in and Why? A Typology of Stakeholder Analysis Methods for Natural Resource Management.' *Journal of Environmental Management* 90(5): 1933–1949.

Schultz, L., R. Plummer, and S. Purdy. 2011. 'Applying a Social-Ecological Inventory: A Workbook for Finding Key Actors and Engaging Them.' www.stockholmresilience.org/publications/artiklar/2011-06-07-applying-a-social-ecological-inventory-a-workbook-for-finding-key-actors-and-engaging-them.html.

Acknowledgements

In addition to the wonderful reviewers of this chapter, the authors thank Michelle Audouin for her insightful comments and the Sida-funded Guidance for Resilience in the Anthropocene: Investments for Development (GRAID) project at the Stockholm Resilience Centre, Sweden for supporting this work.

References

Anderies, J.M., and M.A. Janssen. 2013. 'Robustness of Social-ecological Systems: Implications for Public Policy.' *Policy Studies Journal* 41(3): 513–536.

Anderies, J.M., M.A. Janssen, and E. Ostrom. 2004. 'A Framework to Analyze the Robustness of Social-Ecological Systems from an Institutional Perspective.' *Ecology and Society* 9(1): 18.

Anderson, P.M.L., and P.J. O'Farrell. 2012. 'An Ecological View of the History of the City of Cape Town.' *Ecology and Society* 17(3): 28.

Audouin, M. 2011. *Transdisciplinary Research for Sustainability: User Guide.* Council for Scientific and Industrial Research, Stellenbosch, South Africa.

Audouin, M., R. Preiser, S. Nienaber, L. Downsborough, J. Lanz, and S. Mavengahama. 2013. 'Exploring the Implications of Critical Complexity for the Study of Social-Ecological Systems.' *Ecology and Society* 18(3): 12.

Baird, J., R. Plummer, and K. Pickering. 2014. 'Priming the Governance System for Climate Change Adaptation.' *Ecology and Society* 19(1): 3. doi:10.5751/ES-06152-190103.

Barthel, S., J. Colding, T. Elmqvist, and C. Folke. 2005. 'History and Local Management of a Biodiversity-rich, Urban Cultural Landscape.' *Ecology and Society* 10(2): 10. www.jstor.org/stable/26267721.

Berbés-Blázquez, M., J.A. González, and U. Pascual. 2016. 'Towards an Ecosystem Services Approach that Addresses Social Power Relations.' *Sustainability Science* 19: 134–143. doi:10.1016/j.cosust.2016.02.003.

Binder, C.R., J. Hinkel, P.W.G. Bots, and C. Pahl-Wostl. 2013. 'Comparison of Frameworks for Analyzing Social-Ecological Systems.' *Ecology and Society* 18(4): 26.

Boonstra, W.J., and F.W. de Boer. 2014. 'The Historical Dynamics of Social-Ecological Traps.' *Ambio* 43(3): 260–274.

Buchmann, C. 2009. 'Cuban Home Gardens and their Role in Social-Ecological Resilience.' *Human Ecology* 37(6): 705.

Cash, D.W., W.C. Clark, F. Alcock, N.M. Dickson, N. Eckley, D.H. Guston, J. Jäger, and R.B. Mitchell. 2003. 'Knowledge Systems for Sustainable Development.' *Proceedings of the National Academy of Sciences* 100(14): 8086–8091.

Cilliers, P. 2005. 'Complexity, Deconstruction and Relativism.' *Theory, Culture & Society* 22(5): 255–267.

Clark, W.C., L. van Kerkhoff, L. Lebel, and G.C. Gallopin. 2016. 'Crafting Usable Knowledge for Sustainable Development.' *Proceedings of the National Academy of Sciences* 113(17): 4570–4578.

Cockburn, J., and G. Cundill. 2018. 'Ethics in Transdisciplinary Research: Reflections on the Implications of "Science with Society".' In *The Palgrave Handbook of Ethics in Critical Research*, edited by C.I. Macleod, J. Marx, P. Mnyaka, and G. Treharne, 81–97. London: Palgrave Macmillan.

Colding, J. 2013. 'Local Assessment of Stockholm: Revisiting the Stockholm Urban Assessment.' In *Urbanization, Biodiversity and Ecosystem Services: Challenges and Opportunities*, edited by T. Elqvist, M. Fragkias, J. Goodness, B. Güneralp, P.J. Marcotullio, R.I. McDonald, S. Parnell et al., 313–335. Dordrecht: Springer.

Cote, M., and A.J. Nightingale. 2012. 'Resilience Thinking Meets Social Theory: Situating Social Change in Socio-ecological Systems (SES) Research.' *Progress in Human Geography* 36(4): 475–489.

Díaz, S., S. Demissew, J. Carabias, C. Joly, M. Lonsdale, N. Ash, A. Larigauderie, J.R. Adhikari, S. Arico, and A. Báldi. 2015. 'The IPBES Conceptual Framework – Connecting Nature and People.' *Current Opinion in Environmental Sustainability* 14: 1–16.

Garmestani, A.S. 2014. 'Sustainability Science: Accounting for Nonlinear Dynamics in Policy and Social-Ecological Systems.' *Clean Technologies and Environmental Policy* 16(4): 731–738.

Grant, M.J., and A. Booth. 2009. 'A Typology of Reviews: An Analysis of 14 Review Types and Associated Methodologies.' *Health Information & Libraries Journal* 26(2): 91–108.

Grimble, R., and K. Wellard. 1997. 'Stakeholder Methodologies in Natural Resource Management: A Review of Principles, Contexts, Experiences and Opportunities.' *Agricultural Systems* 55(2): 173–193.

Hadorn, G.H., H. Hoffmann-Riem, S. Biber-Klemm, W. Grossenbacher-Mansuy, D. Joye, C. Pohl, U. Wiesmann, and E. Zemp. 2008. *Handbook of Transdisciplinary Research, Volume 10*. New York: Springer.

Hankivsky, O. 2014. 'Intersectionality 101.' The Institute for Intersectionality Research & Policy, SFU, 1–34.

Hart, C. 2018. *Doing a Literature Review: Releasing the Research Imagination* (2nd ed), 1–381. Thousand Oaks: Sage.

Heylighen, F., P. Cilliers, and C. Gershenson. 2006. 'Complexity and Philosophy.' ArXiv Preprint Cs/0604072.

Ison, R.L. 2008. 'Systems Thinking and Practice for Action Research.' In *The Sage Handbook of Action Research Participative Inquiry and Practice* (2nd ed), edited by P.W. Reason and H. Bradbury, 139–158. London: Sage.

Kraft, M.E., and N.J. Vig. 2006. 'Environmental Policy from the 1970s to the Twenty-First Century.' In *Environmental Policy: New Directions for the Twenty-First Century*, edited by N.J. Vig and M.E. Kraft, 1–33.

Leach, M., I. Scoones, and A. Stirling. 2007. 'Pathways to Sustainability: An Overview of the STEPS Centre Approach.' *STEPS Approach Paper.* Brighton: STEPS Centre. https://steps-centre.org/wp-content/uploads/final_steps_overview.pdf.

McRae, L., B. Collen, S. Deinet, P. Hill, J. Loh, J.E.M. Baillie, and V. Price. 2012. 'The Living Planet Index: Biodiversity, Biocapacity and Better Choices'. World Wild Fund for Nature. https://wwfeu.awsassets.panda.org/downloads/lpr_living_planet_report_2012.pdf.

Milkoreit, M., J. Hodbod, J. Baggio, K. Benessaiah, R. Calderón-Contreras, J.F. Donges, J-D. Mathias, J.C. Rocha, M. Schoon, and S.E. Werners. 2018. 'Defining Tipping Points for Social-Ecological Systems Scholarship – An Interdisciplinary Literature Review.' *Environmental Research Letters* 13(3): 033005.

Nayak, P.K., D. Armitage, and M. Andrachuk. 2016. 'Power and Politics of Social-Ecological Regime Shifts in the Chilika Lagoon, India and Tam Giang Lagoon, Vietnam.' *Regional Environmental Change* 16(2): 325–339.

O'Connell, D., B. Walker, N. Abel, N. Grigg, A. Cowie, and G. Durón. 2015. 'An Introduction to the Resilience, Adaptation Pathways and Transformation Assessment (RAPTA) Framework.' Washington, DC: United Nations Scientific and Technical Advisory Panel. www.stapgef.org/sites/default/files/documents/Summary_RAPTA.pdf.-July-16.pdf.

Orach, K., and M. Schlüter. 2016. 'Uncovering the Political Dimension of Social-Ecological Systems: Contributions from Policy Process Frameworks.' *Global Environmental Change* 40: 13–25.

Ostrom, E. 2009. 'A General Framework for Analyzing Sustainability of Social-Ecological Systems.' *Science* 325(5939): 419–422.

Pathways Network. 2018. 'T-Labs: A Practical Guide – Using Transformation Labs (T-Labs) for Innovation in Social-Ecological Systems.' Brighton: STEPS Centre.

Pereira, L.M., T. Karpouzoglou, N. Frantzeskaki, and P. Olsson. 2018. 'Designing Transformative Spaces for Sustainability in Social-Ecological Systems.' *Ecology and Society* 23(4): 32.

Peterson, G.D. 2005. 'Ecological Management: Control, Uncertainty, and Understanding.' In *Ecological Paradigms Lost: Routes of Theory Change*, edited by K. Cuddington and B.E. Beisner, 371–391. Amsterdam: Elsevier.

Petursdottir, T., O. Arnalds, S. Baker, L. Montanarella, and Á.L. Aradóttir. 2013. 'A Social-Ecological System Approach to Analyze Stakeholders' Interactions within a Large-Scale Rangeland Restoration Program.' *Ecology and Society* 18(2): 29.

Pohl, C., and G.H. Hadorn. 2007. *Principles for Designing Transdisciplinary Research*. Munich: Oekom.

Prell, C., K. Hubacek, and M. Reed. 2009. 'Stakeholder Analysis and Social Network Analysis in Natural Resource Management.' *Society and Natural Resources* 22(6): 501–518.

Puri, R.K. 2011. 'Documenting Local Environmental Knowledge and Change.' In *Conducting Research in Conservation: Social Science Methods and Practice*, edited by H.S. Newing, 146–169.

Ramankutty, N., and J.A. Foley. 1999. 'Estimating Historical Changes in Global Land Cover: Croplands from 1700 to 1992.' *Global Biogeochemical Cycles* 13(4): 997–1027.

Reed, M.S., A. Graves, N. Dandy, H. Posthumus, K. Hubacek, J. Morris, C. Prell, C.H. Quinn, and L.C. Stringer. 2009. 'Who's in and Why? A Typology of Stakeholder Analysis Methods for Natural Resource Management.' *Journal of Environmental Management* 90(5): 1933–1949.

Resilience Alliance. 2010. 'Assessing Resilience in Social-Ecological Systems: Workbook for Practitioners 2.0.' Resilience Alliance. www.resalliance.org/resilience-assessment.

Rodríguez, L.C., U. Pascual, and H.M. Niemeyer. 2006. 'Local Identification and Valuation of Ecosystem Goods and Services from Opuntia Scrublands of Ayacucho, Peru.' *Ecological Economics* 57(1): 30–44.

Roth, W-M., L. Radford, and L. LaCroix. 2012. 'Working with Cultural-Historical Activity Theory.' In *Forum Qualitative Sozialforschung/Forum: Qualitative Social Research* 13(2): Art. 23. http://nbn-resolving.de/urn:nbn:de:0114-fqs1202232.

SANBI and UNEP-WCMC. 2016. 'Mapping Biodiversity Priorities: A Practical, Science-based Approach to National Biodiversity Assessment and Prioritisation to Inform Strategy and Action Planning.' Cambridge: UNEP-WCMC. http://biodiversityadvisor.sanbi.org/wp-content/uploads/2016/06/Mapping-Biodiversity-Priorities-WEB.pdf.

Scholes, R.J., B. Reyers, R. Biggs, M.J. Spierenburg, and A. Duriappah. 2013. 'Multi-Scale and Cross-Scale Assessments of Social-Ecological Systems and their Ecosystem Services.' *Current Opinion in Environmental Sustainability* 5(1): 16–25.

Schoon, M.L., M.D. Robards, K. Brown, N. Engle, C.L. Meek, and R. Biggs. 2015. 'Politics and the Resilience of Ecosystem Services.' In *Principles for Building Resilience: Sustaining Ecosystem Services in Social-Ecological Systems*, edited by R. Biggs, M. Schlüter, and M.L. Schoon, 32–49. Cambridge: Cambridge University Press.

Schultz, L., C. Folke, and P. Olsson. 2007. 'Enhancing Ecosystem Management through Social-Ecological Inventories: Lessons from Kristianstads Vattenrike, Sweden.' *Environmental Conservation* 34(2): 140–152.

Schultz, L., R. Plummer, and S. Purdy. 2011. 'Applying a Social-Ecological Inventory: A Workbook for Finding Key Actors and Engaging Them.' www.stockholmresilience.org/publications/artiklar/2011-06-07-applying-a-social-ecological-inventory-a-workbook-for-finding-key-actors-and-engaging-them.html.

Sheil, D., and N. Liswanti. 2006. 'Scoring the Importance of Tropical Forest Landscapes with Local People: Patterns and Insights.' *Environmental Management* 38(1): 126–136.

Smith, A., and A. Stirling. 2010. 'The Politics of Social-Ecological Resilience and Sustainable Socio-Technical Transitions.' *Ecology and Society* 15(1): 11.

Stringer, L.C., A.J. Dougill, E. Fraser, K. Hubacek, C. Prell, and M.S. Reed. 2006. 'Unpacking "Participation" in the Adaptive Management of Social-Ecological Systems: A Critical Review.' *Ecology and Society* 11(2): 39.

Tengö, M., E.S. Brondizio, T. Elmqvist, P. Malmer, and M. Spierenburg. 2014. 'Connecting Diverse Knowledge Systems for Enhanced Ecosystem Governance: The Multiple Evidence Base Approach.' *Ambio* 43(5): 579–591.

Thompson-Hall, M., E.R. Carr, and U. Pascual. 2016. 'Enhancing and Expanding Intersectional Research for Climate Change Adaptation in Agrarian Settings.' *Ambio* 45(3): 373–382.

Van Kerkhoff, L., and L. Lebel. 2006. Linking Knowledge and Action for Sustainable Development. *Annual Review of Environment and Resources*, 31: 445–477.

Wali, A., D. Alvira, P. Tallman, A. Ravikumar, and M. Macedo. 2017. 'A New Approach to Conservation: Using Community Empowerment for Sustainable Well-Being.' *Ecology and Society* 22(4): 6.

Walker, B., C.S. Holling, S.R. Carpenter, and A. Kinzig. 2004. 'Resilience, Adaptability and Transformability in Social-Ecological Systems.' *Ecology and Society* 9(2): 5.

Watson, C.W. 2010. 'Ethical Issues in Research.' In *Conducting Research in Conservation: Social Science Methods and Practice*, edited by H. Newing. Abingdon: Routledge.

Wulder, M.A., R.J. Hall, N.C. Coops, and S.E. Franklin. 2004. 'High Spatial Resolution Remotely Sensed Data for Ecosystem Characterization.' *BioScience* 54(6): 511–521.

Zheng, Y., A. Byg, B.J. Thorsen, and N. Strange. 2014. A Temporal Dimension of Household Vulnerability in Three Rural Communities in Lijiang, China. *Human Ecology* 42(2): 283–295.

6

Ecological field data collection

Hayley Clements,[1] Karen Esler,[2] Dominic A.W. Henry,[3,4] Penelope Mograbi,[5] Albert Norström[6] and Chevonne Reynolds[7,8]

[1]CENTRE FOR SUSTAINABILITY TRANSITIONS, STELLENBOSCH UNIVERSITY, STELLENBOSCH, SOUTH AFRICA
[2]DEPARTMENT OF CONSERVATION ECOLOGY AND ENTOMOLOGY AND CENTRE FOR INVASION BIOLOGY, STELLENBOSCH UNIVERSITY, STELLENBOSCH, SOUTH AFRICA
[3]STATISTICS IN ECOLOGY, ENVIRONMENT AND CONSERVATION, DEPARTMENT OF STATISTICAL SCIENCES, UNIVERSITY OF CAPE TOWN, CAPE TOWN, SOUTH AFRICA
[4]ENDANGERED WILDLIFE TRUST, JOHANNESBURG, SOUTH AFRICA
[5]DEPARTMENT OF ENVIRONMENTAL SCIENCE, RHODES UNIVERSITY, MAKHANDA, SOUTH AFRICA
[6]STOCKHOLM RESILIENCE CENTRE, STOCKHOLM UNIVERSITY, STOCKHOLM, SWEDEN
[7]FITZPATRICK INSTITUTE OF AFRICAN ORNITHOLOGY, DST-NRF CENTRE OF EXCELLENCE, UNIVERSITY OF CAPE TOWN, CAPE TOWN, SOUTH AFRICA
[8]ANIMAL, PLANT AND ENVIRONMENTAL SCIENCES, UNIVERSITY OF THE WITWATERSRAND, JOHANNESBURG, SOUTH AFRICA

Key methods discussed in this chapter

Measuring faunal and floral diversity and population structure: Point counts and survey grids, transects, distance and plotless sampling, quadrats, capture and mark-recapture
Measuring abiotic conditions: Abiotic environmental sampling, core sampling
Measuring ecological processes: Telemetry, isotope and genetic analysis

Connections to other chapters

Systems scoping can be used to define system boundaries for ecological data collection (Chapter 6). To explore social-ecological interactions, ecological field data are often paired with social data-collection methods such as interviews and surveys (Chapter 7) or participatory data collection (Chapter 8), and analysed using statistical analysis (Chapter 18), network analysis (Chapter 23), spatial analysis (Chapter 24) or models (Chapters 26–28 and 31).

Introduction

Ecology (the study of the relations of organisms to one another and their environment) has a well-established and extensive set of field data-collection methods (Sala et al. 2000; Henderson 2003; Wheater, Bell, and Cook 2011) that have been developed to both advance ecological theory (e.g. optimal foraging theory, theory of island biogeography) and address practical problems (e.g. conservation planning, monitoring ecological restoration). Although ecological field data-collection methods continue to be predominantly developed and used in

DOI: 10.4324/9781003021339-8

SUMMARY TABLE: ECOLOGICAL FIELD DATA COLLECTION	
DISCIPLINARY BACKGROUND	**KNOWLEDGE TYPE**
The methods in this chapter are derived from or have most commonly been used in: Ecology, Zoology, Botany, Conservation Biology	*The methods in this chapter are primarily used to generate the following types of knowledge:* • Descriptive • Exploratory
RESEARCH APPROACH	**PURPOSE OF METHOD**
The methods in this chapter originate from or most commonly adopt the following research approaches: • Analytical/objective	*The most common purposes of using the methods in this chapter are:* • Data collection/generation
TEMPORAL DIMENSION	**SYSTEMIC FEATURES AND PROCESSES**
The methods in this chapter are most commonly applied to the following temporal dimensions: • Present (typically within the last 5–10 years) • Recent past (post-1700s)	*While most methods can do many things, the methods in this chapter are particularly good (i.e. go-to methods) for addressing the following:* • SES components and linkages • Diversity • Social-ecological dependence and impact
SPATIAL DIMENSION	
The methods in this chapter are primarily either or both: • Explicitly spatial *The methods in this chapter are most commonly applied at the following spatial scales:* • Local • Regional (provincial/state to continental) • Multiple places/sites around the world	

the biological sciences (e.g. ecology, zoology, botany), many of these methods are also useful to social-ecological systems (SES) research.

Ecological field methods are largely quantitative and assume that researchers and research methods are entirely objective (i.e. the researcher is an independent observer who does not have an influence on the data that are collected). Studies based on ecological field methods are therefore assumed to be replicable – if a different researcher went to the same site and followed the same method, it should be possible to obtain the same study results (this assumption is common for many scientific methods, but not made by some other methods used in SES research, such as facilitated dialogues (Chapter 9) and scenarios (Chapter 11)). Ecological field methods are often scale sensitive – the results obtained depend on the method grain (minimum spatial resolution of the data collected, e.g. quadrat size) and extent (e.g. study area). Application of these methods therefore assumes that the appropriate spatial scale is being studied, which will depend on the questions being asked and the phenomenon under consideration (Wu et al. 2002). Study scale usually increases from the study of individuals and populations to communities and ecosystems.

SES problems and questions

Ecological field methods can be used to understand how human activities (e.g. resource use, land-use change, conservation actions) influence the ecological elements and processes on which human well-being depends. These ecological elements and processes provide ecosystem services such as climate regulation, food production, sense of place, recreation and relaxation. Ecological field methods can be used to understand the influence of ecological patterns and processes on social outcomes of interest (e.g. how does species composition influence the cultural benefits that people receive from a protected area? (Cumming and Maciejewski 2017)), and vice versa (e.g. what social, institutional and environmental conditions give rise to ecosystems that are in a substantially better condition than expected? (Cinner et al. 2016)). Some studies focus specifically on the interactions between ecological and social processes and outcomes (e.g. how has marine resource harvesting affected the ecosystem, and what are the consequences for local harvesters' livelihoods and economy over time? (Nordlund et al. 2010)).

The consideration of ecological patterns and processes is critical to understanding SES. The risk of a lake shifting from a clear-water state (desirable for drinking, fishing and recreation) to a turbid-water state (i.e. eutrophic, undesirable state) when exposed to anthropogenic nutrient inputs can be influenced by the size, depth and macrophyte density of the lake (Genkai-Kato and Carpenter 2005), for example. This is known as a regime shift. Similarly, the sustainability of fuelwood harvesting, a crucial provisioning ecosystem service in the developing world, is influenced not only by harvester demand and selectivity but also by ecosystem response, such as the regenerative potential of woody vegetation (Swemmer, Mashele, and Ndhlovu 2019). Ecological data are also used to understand and guide natural resource management and are especially effective when contextualised with large-scale data (Edgar et al. 2016). Systems thinking, for example, has promoted ecosystem-based fisheries management (Curtin and Prellezo 2010), recognising that the sustainability of fisheries depends on species-rich and functionally diverse communities that maintain ecological functioning (Nyström et al. 2008).

Ecological field data are by themselves insufficient to answer SES research questions; other methods need to be paired with ecological approaches to provide insight into the interactions of the ecological aspects with social aspects of the system (e.g. interviews – see Chapter 7; land-use maps – see Chapter 24; census data – see Chapter 25). Ecological field data are closely coupled with experimental design and appropriate statistical analysis (Chapter 18); thus, key reference textbooks often combine both field data collection and statistical analysis (e.g. Quinn and Keough 2002; Kent 2011). To understand feedbacks and dynamics in SES, ecological field data can also be used to parameterise models (i.e. inform the relationships that go into the model), such as agent-based models (see Chapter 28) or ecosystem service models (see Chapter 31) (e.g. Perez, Eun-kyeong, and Sengupta 2018). It can also be used in network analyses (see Chapter 23) (e.g. Hong et al. 2013).

Brief description of key methods

Ecological field methods can be used to quantify faunal and floral species richness and abundance (diversity) and population structure across a landscape/seascape or an environmental gradient (e.g. in different land uses, habitats, climates or management regimes). If the study is repeated over time, ecological field methods can be used to assess community and population dynamics (e.g. the process of change and development and the drivers of these dynamics). Other methods quantify abiotic conditions (e.g. water quality, erosion), or focus on ecological processes (e.g. dispersal or predation). A combination of faunal, floral and abiotic methods is often necessary to assess ecosystem function and processes. Table 6.1 provides a summary of key methods used in ecological field data collection.

Limitations

It can be challenging to 'bound the system', or to decide how large a sample area or sample size is necessary to identify trends, or to choose the resolution at which to collect data (e.g. size of quadrat or survey grid; length and number of transects). The appropriate scale of analysis depends on the question or taxa of interest (Wu et al. 2002). At fine spatial scales, for example, the presence of water drives elephant distribution, while at larger scales the presence of available forage drives elephant distribution (De Knegt et al. 2011). Systems scoping can be useful for defining study boundaries (Chapter 5). For some studies, it will also be important to collect data at the correct time of day and/or year (e.g. at night for nocturnal animals, or during breeding season).

Long-term collection of ecological data may be necessary to detect meaningful trends. This can be a limitation due to funding constraints and/or the short-term nature of many research projects that may be mismatched with the longer time scales of many ecological processes. The measurement of trends across space instead of through time ('space-for-time substitution') can sometimes be used to overcome this limitation. Data are collected across a large and diverse area (i.e. space) to understand drivers of species occurrence and abundance (i.e. time) (Edgar et al. 2016). Data can sometimes be collected from sites with different amounts of time elapsing since an event (e.g. disturbance, restoration or protection) to understand how the event has influenced changes in species occurrence and abundance.

Undertaking field data collection over large spatial areas or long temporal periods can be resource and time intensive, particularly for SES research when social data are often being collected concurrently. Consequently, the number of variables measured in the field may be insufficient to identify core variables or drivers of variation (see also Chapter 18). Direct methods such as point counts and quadrat surveys can have practical limitations (e.g. height measurements of extremely tall trees, measuring on steep slopes), whereas indirect methods such as camera traps, GPS tracking and satellite-derived measurements can present technical challenges (e.g. calibration, hardware, software).

Table 6.1 Summary of key methods used in ecological field data collection

Method	*Description*	*References*
MEASURING FAUNAL AND FLORAL DIVERSITY AND POPULATION STRUCTURE		
Point counts and survey grids	Point counts and survey grids are used to inventory and monitor faunal richness and abundance (typically for highly visible and/or audible organisms, such as reef fish or birds). A trained observer records all focal taxa seen or heard from a point-count station over a set time period. This method can also be undertaken by camera trap and acoustic arrays (especially for birds and bats), which are typically positioned across a survey grid.	*Key introductory text* Wheater, Bell, and Cook 2011 *Applications to SES* Daw, Robinson, and Graham 2011; Jouffray et al. 2014; Cinner et al. 2016; Cumming and Maciejewski 2017
Transects, distance sampling and plotless methods	Transects, distance sampling and plotless methods are used to assess faunal or floral richness and abundance along transects, frequently spanning an environmental gradient. The transect can be walked, swum or driven, depending on the taxa of interest. Video transects and aerial surveys can also be used. Flora identified visually is counted and measured, normally along a length of tape where vegetation is recorded at set intervals. Fauna can be identified visually, through acoustics (e.g. cetaceans or bats) or based on tracks and signs. Distances to observed fauna along the transect are used to estimate abundance, based on the statistical assumption that the probability of detecting an animal decreases as its distance from the observer increases. Plotless sampling methods are used to calculate densities of large species that are dispersed (e.g. point-centred quarter method).	*Key introductory texts* Henderson 2003; Kent 2011; Wheater, Bell, and Cook 2011 *Applications to SES* Chanda et al. 2003; Ticktin, Whitehead, and Fraiola 2006; Edgar et al. 2014; Ward et al. 2018
Quadrat sampling	Quadrat sampling is used to assess faunal or floral richness and abundance in an area by surveying randomly or systematically distributed quadrats (sometimes called plots). A quadrat is a frame (traditionally square) used as a standard unit of area. The quadrat is placed multiple times in the study area and a trained observer records all focal taxa seen. The patterns of species occurrence in the sampled area are assumed to be representative of the entire area. It is most appropriate for sessile or slow-moving organisms such as plants or some aquatic animals in intertidal zones.	*Key introductory texts* Sala et al. 2000; Henderson 2003; Kent 2011; Wheater, Bell, and Cook 2011 *Applications to SES* Ticktin, Whitehead, and Fraiola 2006; Andersson, Barthel, and Ahrné 2007; Nordlund et al. 2010; Mandle, Ticktin, and Zuidema 2015

(*Continued*)

Table 6.1 (Continued)

Method	*Description*	*References*
Capture and mark-recapture	Fauna is trapped and can be marked and recaptured to estimate abundance, survival probability and persistence. Methods for trapping vertebrates include using Sherman live trapping (rodents), mistnetting and ringing (birds), pitfall trapping (reptiles and amphibians), phyke or seine netting and electro-fishing (fish). Indirect capture methods such as DNA analysis of scats or camera trapping are also used for capture-recapture methods. Methods for trapping invertebrates include pitfall traps, pan traps (pollinators), hand nets (marine invertebrates), light traps (nocturnal insects) and soil samples.	*Key introductory text* Wheater, Bell, and Cook 2011 *Applications to SES* Sutaria 2009; Mintzer et al. 2013
MEASURING ABIOTIC CONDITIONS		
Abiotic environmental sampling	Abiotic environmental sampling is used to measure abiotic properties of the environment, such as water quality (e.g. pH, nutrient load, presence of heavy metals), soil moisture and respiration, and seismic activity. Abiotic components such as soil can be key determinants of plant species composition, productivity and responses to disturbances such as drought. Landscape functional analysis is a standardised assessment to measure rangeland function using abiotic indicators together with biotic indicators. Abiotic sampling is often performed along an environmental gradient.	*Key introductory texts* Sala et al. 2000; Tongway and Hindley 2004; Tan 2005; Wheater, Bell, and Cook 2011 *Applications to SES* Genkai-Kato and Carpenter 2005; Addison et al. 2013; Read et al. 2016
Core sampling (see also Chapter 25)	Sediment cores are extracted by drilling into the earth's crust with long cylinders, usually in wetlands or the ocean. Tree, ice and coral cores can also be obtained. These cores are dated and analysed to provide insights into the historical climate and biotic conditions. Sediment and ice cores are also used for fossil pollen and charcoal records, and stable isotope analysis.	*Key introductory text* Smith 1987 *Applications to SES* Dearing et al. 2012; Forbes, Gillson, and Hoffman 2018
MEASURING ECOLOGICAL PROCESSES		
Telemetry	Telemetry is used to measure the movement, dispersal or habitat use of species by tracking animals with the use of global positioning system (GPS) collars, radio-tags, a global system for mobile communications (GSM) or even fluorescent dyes, radioactive markers and drones.	*Key introductory text* Hebblewhite and Haydon 2010 *Applications to SES* Johansson et al. 2015; Miguel et al. 2017
Isotope and genetic analysis	Isotope analysis is used to understand a variety of physiological processes (e.g. photosynthetic pathways, water-use efficiency and water nitrogen fixation) and food web studies. Genetic analysis can be used to assess population dynamics (e.g. source-sink dynamics), predation and the consequences of harvesting.	*Key introductory texts* Fry 2006; Kress et al. 2015 *Applications to SES* Villasante 2012; Alexander et al. 2018; Minnie et al. 2018; Kemp et al. 2019

Some field collection methods such as camera trapping, underwater video transects and telemetry can generate extremely large datasets. The analysis of these data can require specialised statistical, programming or modelling skills, as well as adequate computing power. In contrast, some field collection methods may also only generate small sample sizes (e.g. censuses of large predators, or endangered species), which may present challenges for statistical analyses.

Studies can be limited by the lack of a control or counterfactual; for example, it is difficult to say with certainty that a protected area was effective in protecting species unless a similar unprotected site has also been surveyed (Pressey, Visconti, and Ferraro 2015). Experimental design (especially sample replication and site randomisation) that ensures replicability is important to avoid 'pseudoreplication' – a process where artificially inflated replicates compromise the statistical validity of conclusions drawn from the analysis of field data.

Resource implications

Some ecological field methods require highly specialised knowledge and technology to identify species. Field guide books and online species identification apps can help in this regard, but many species look extremely similar, or are only identifiable in certain seasons (e.g. some grasses are only identifiable when they are flowering), under a microscope (e.g. some insects) or require genetic approaches (e.g. coral identification). Involving taxonomic experts and/or people with local knowledge in the data collection can be useful.

Some field methods are expensive, particularly those that make use of modern technology such as camera traps, genetic sampling and acoustic equipment. Ethical clearance also needs to be obtained for data-collection methods that capture animals. The presence of dangerous animals at a field site (e.g. elephants, lions) can constrain data collection or necessitate the use of armed rangers. Handling of dangerous animals such as snakes also requires special training. Fuel and vehicle maintenance can also become a large expense for extensive data collection.

The researcher often needs to obtain a permit before undertaking the study of certain species, or requires permission from the site landholders before being allowed access to field sites. Field studies over large areas may require permission from a diverse range of landholders or other stakeholders (e.g. national park agencies, farmers, forestry companies, local communities).

New directions

While some ecological field data collection techniques have been used widely in SES research (e.g. point counts and quadrats), other methods are less commonly used in interdisciplinary research (e.g. core sampling, isotope and genetic analyses) despite their relevance for advancing the understanding of SES (e.g. Forbes, Gillson, and Hoffman 2018).

Technological development is rapidly expanding the ease and scope of ecological field data collection. Manual methods that are often costly in time and labour can now be augmented with or replaced by more automated techniques that collect objective, repeatable data in remote places, quickly and cost effectively (e.g. drones, camera traps, flux towers, acoustic arrays; see also Chapters 24 and 25). Machine learning can also be used to automate species identification (see Chapter 17). Many of these technological methods are becoming more viable options for research as the technology becomes more affordable and user friendly, and as required software becomes open access.

Some of these new approaches to data collection generate extremely large volumes of data. These big data challenges (see also Chapters 17 and 18) have facilitated the involvement of citizen

Case study 6.1: Land-sharing versus land-sparing to conserve ecosystem services: case studies from Ghana and India

Agriculture currently covers 40% of the planet's ice-free terrestrial surface and the demand for agricultural products is expected to double by 2050 (Godfray et al. 2012). Although there are inevitable trade-offs between meeting this demand and the preservation of ecological functioning and ecosystem service provision (e.g. climate regulation, water purification, pollination), there is scope to reduce this trade-off by understanding how different species respond to different types of production landscapes.

This challenge led Green et al. (2005) to propose the land-sharing versus land-sparing framework, where conservation and food production activities can either occur in the same space (land-sharing, e.g. conservation agriculture) or be separated in space (land-sparing, e.g. intensive farming with a portion of land set aside for protected areas). Green and colleagues suggested that whether land-sharing or land-sparing was better for ensuring species persistence in a landscape would depend on how population densities of species change with agricultural yield. The trade-off between land-sharing and land-sparing can be modelled with a density-yield function, which demonstrates how the population density of an individual species in a landscape (number of individuals per unit area) changes according to the intensity of farming in the landscape (i.e. the yield per unit area of farmed land) (Green et al. 2005; Phalan et al. 2011). Individual species can be designated as 'winners' and 'losers' under the alternative strategies based on the shape of the curve (Figure 6.1).

In Figure 6.1, schematic (A) depicts the land-sharing versus land-sparing model where the area with the squiggly lines represents natural or protected land, the area with the diagonal dotted lines represents high-intensity agriculture and the area with the horizontal lines represents low-intensity agriculture or wildlife-friendly farming. The densities of species in this example are highest in an all-natural (i.e. no agriculture) scenario (B). However, if a set amount of agricultural yield is required from a given land area, either through land-sharing (low-intensity agriculture over the whole area) or through land-sparing (high-intensity agriculture over some of the area and natural land in the remainder of the area), the shape of the relationship between population density and agricultural intensity gives us insight into which strategy is better to conserve a species. The dotted line in (B) represents species that display near-to-natural-level population densities in a land-sharing scenario, making them well suited to this strategy. By contrast, the solid line represents species whose densities decline rapidly under any form of agriculture, even if it is low intensity. For these species, land-sparing is essential to provide the natural areas necessary for them to maintain their populations.

Phalan et al. (2011) collected data on population densities of birds and trees across landscapes in Ghana and India, using standard point count and plot-based survey techniques, respectively. Data on farm yield and profit were collected from household surveys and regressed against the population density of each species to produce the density-yield functions. Phalan et al. (2011) fitted density-yield functions that show how population

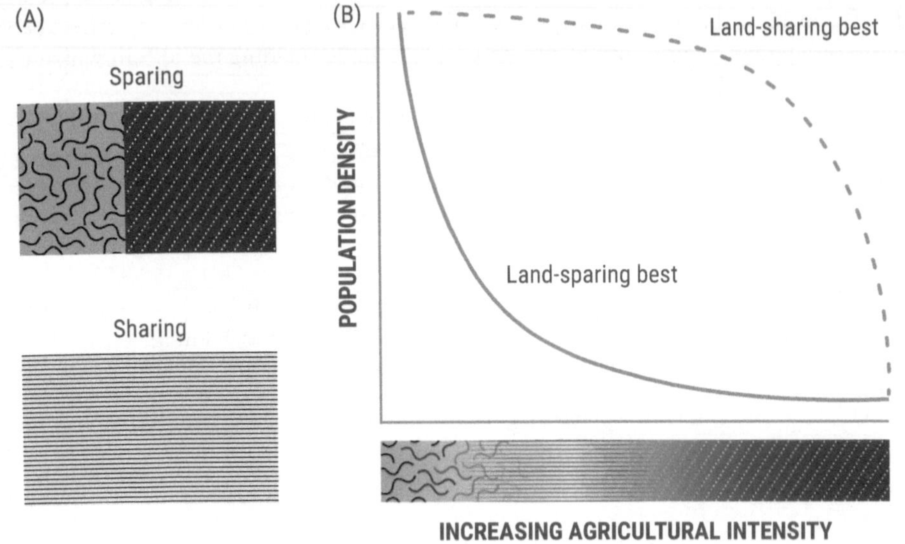

Figure 6.1 A graphic representation of the density-yield function based on the land-sparing versus land-sharing trade-off (© Chevonne Reynolds)

density changes with increasing agricultural yield for 167 bird and 220 tree species in Ghana, and 174 bird and 40 tree species in India for two types of agricultural yield currencies (food production and profit). Both study regions contained forest remnants within a matrix of farmland ranging from low-yielding (i.e. non-intensive) mosaic agriculture to large-scale high-yielding (i.e. intensive) monocultures. The researchers found that more species were negatively affected by agriculture than benefited from it. For both taxa in both countries, land-sparing was a more promising strategy for conserving species while minimising the negative impacts of food production.

Density-yield functions have become widely applied across multiple farming systems, from Uganda's banana–coffee arc to the Eurasian steppes (Hulme et al. 2013; Kamp et al. 2015), providing valuable insights into the potential for people to meet escalating food demand with the least harm to other species. Density-yield functions are therefore excellent examples of how ecological field data can be integrated with socio-economic data to address social-ecological problems. However, one limitation of this approach is the ability to gather agricultural yield data across large scales, which limits the scale at which the trade-off between the two land-use alternatives can be assessed. Novel approaches for quantifying agricultural yield, and for determining population densities of species at larger scales, will be needed if we wish to test this trade-off regionally or nationally.

scientists to assist with analysis (e.g. Edgar et al. 2014), especially of camera- or video-trap data (zooniverse.org). Citizen scientists' enthusiasm has also been harnessed in large atlas projects (e.g. SABAP2), enabling the collection of extensive datasets. Using citizen science to collect ecological data presents a good opportunity to engage people in understanding the SES in which they and other citizens are embedded.

The increasing availability of open-access data has also facilitated the development of global datasets and the ability to do repeat field sampling via globally collaborative projects. The tropical tree database, for example, makes harvested biomass allometry across a suite of global tropical sites freely available (chave.ups-tlse.fr/pantropical_allometry.htm).

Key readings

Henderson, P.A. 2003. *Practical Methods in Ecology*. Malden: Blackwell Publishing.

Kent, M. 2011. *Vegetation Description and Data Analysis: A Practical Approach*. Hoboken: John Wiley and Sons.

Sala, O.E., R.B. Jackson, H.A. Mooney, and R.W. Howarth. 2000 *Methods in Ecosystem Science*. New York: Springer.

Wheater C.P., J.R. Bell, and P.A. Cook. 2011. *Practical Field Ecology: A Project Guide*. Hoboken: John Wiley and Sons.

Acknowledgements

We acknowledge funding from the Claude Leon Foundation and a Jennifer Ward Oppenheimer Research Grant (Hayley Clements), the DSI-NRF Centre for Invasion Biology (Karen Esler), the South African Research Chairs Initiative of the Department of Science and Technology and the National Research Foundation (NRF) of South Africa (grant 84379) (Penelope Mograbi), and an NRF Thuthuka grant (Chevonne Reynolds).

References

Addison, J., J. Davies, M. Friedel, and C. Brown. 2013. 'Do Pasture User Groups Lead to Improved Rangeland Condition in the Mongolian Gobi Desert?' *Journal of Arid Environments* 94: 37–46. doi:10.1016/j.jaridenv.2013.02.009.

Alexander, J., C.T. Downs, M. Butler, S. Woodborne, and C.T. Symes. 2018. 'Stable Isotope Analyses as a Forensic Tool to Monitor Illegally Traded African Grey Parrots.' *Animal Conservation* 1–10. doi:10.1111/acv.12445.

Andersson, E., S. Barthel, and K. Ahrné. 2007. 'Measuring Social and Ecological Dynamics Behind the Generation of Ecosystem Services.' *Ecological Applications* 17(5): 1267–1278. doi:10.1890/06-1116.1.

Chanda, R., O. Totolo, N. Moleele, M. Setshogo, and S. Mosweu. 2003. 'Prospects for Subsistence Livelihood and Environmental Sustainability along the Kalahari Transect: The Case of Matsheng in Botswana's Kalahari Rangelands.' *Journal of Arid Environments* 54(2): 425–445. doi:10.1006/jare.2002.1100.

Cinner, J.E., C. Huchery, M. Aaron MacNeil, N.A.J. Graham, T.R. McClanahan, J. Maina, E. Maire et al. 2016. 'Bright Spots among the World's Coral Reefs.' *Nature* 535(7612): 416–419. doi:10.1038/nature18607.

Cumming, G.S., and K. Maciejewski. 2017. 'Reconciling Community Ecology and Ecosystem Services: Cultural Services and Benefits from Birds in South African National Parks.' *Ecosystem Services* 28: 219–227. doi:10.1016/j.ecoser.2017.02.018.

Curtin, R., and R. Prellezo. 2010. 'Understanding Marine Ecosystem Based Management: A Literature Review.' *Marine Policy* 34(5): 821–830. doi:10.1016/j.marpol.2010.01.003.

Daw, T.M., J. Robinson, and N.A.J. Graham. 2011. 'Perceptions of Trends in Seychelles Artisanal Trap Fisheries: Comparing Catch Monitoring, Underwater Visual Census and Fishers' Knowledge.' *Environmental Conservation* 38(1): 75–88. doi:10.1017/S0376892910000901.

De Knegt, H.J., F. van Langevelde, A.K. Skidmore, A. Delsink, R. Slotow, S. Henley, G. Bucini et al. 2011. 'The Spatial Scaling of Habitat Selection by African Elephants.' *Journal of Animal Ecology* 80(1): 270–281. doi:10.1111/j.1365-2656.2010.01764.x.

Dearing, J.A., X. Yang, X. Dong, E. Zhang, X. Chen, P.G. Langdon, K. Zhang, W. Zhang, and T.P. Dawson. 2012. 'Extending the Timescale and Range of Ecosystem Services through Paleoenvironmental Analyses, Exemplified in the Lower Yangtze Basin.' *Proceedings of the National Academy of Sciences* 109(18): E1111–20. doi:10.1073/pnas.1118263109.

Edgar, G.J., A.E. Bates, T.J. Bird, A.H. Jones, S. Kininmonth, R.D. Stuart-Smith, and T.J. Webb. 2016. 'New Approaches to Marine Conservation Through the Scaling Up of Ecological Data.' *Annual Review of Marine Science* 8(1): 435–461. doi:10.1146/annurev-marine-122414-033921.

Edgar, G.J., R.D. Stuart-Smith, T.J. Willis, S. Kininmonth, S.C. Baker, S. Banks, N.S. Barrett et al. 2014. 'Global Conservation Outcomes Depend on Marine Protected Areas with Five Key Features.' *Nature* 506(7487): 216–220. doi:10.1038/nature13022.

Forbes, C.J., L. Gillson, and M.T. Hoffman. 2018. 'Anthropocene Shifting Baselines in a Changing World: Identifying Management Targets in Endangered Heathlands of the Cape Floristic Region, South Africa.' *Anthropocene* 22: 81–93. doi:10.1016/j.ancene.2018.05.001.

Fry, B. 2006. *Stable Isotope Ecology*. New York: Springer.

Genkai-Kato, M., and S.R. Carpenter. 2005. 'Eutrophication Due to Phosphorus Recycling.' *Ecology* 86(1): 210–219. doi:10.1890/03-0545.

Godfray, H.C.J., J.R. Beddington, I.R. Crute, L. Haddad, D. Lawrence, J.F. Muir, J. Pretty, S. Robinson, S.M. Thomas, and C. Toulmin. 2012. 'Food Security: The Challenge of Feeding 9 Billion People.' *Science* 327: 812–819. doi:10.1126/science.1185383.

Green, R.E., S.J. Cornell, J.P.W. Scharlemann, and A. Balmford. 2005. 'Farming and the Fate of Wild Nature.' *Science* 307(5709): 550–555. doi:10.1126/science.1106049.

Hebblewhite, M., and D.T. Haydon. 2010. 'Distinguishing Technology from Biology: A Critical Review of the Use of GPS Telemetry Data in Ecology.' *Philosophical Transactions of the Royal Society B: Biological Sciences* 365(1550): 2303–2312. doi:10.1098/rstb.2010.0087.

Henderson, P.A. 2003. *Practical Methods in Ecology*. Malden: Blackwell Publishing. doi:10.1111/j.1442-9993.2005.01460.x.

Hong, S.H., B.H. Han, S.H. Choi, C.Y. Sung, and K.J. Lee. 2013. 'Planning an Ecological Network Using the Predicted Movement Paths of Urban Birds.' *Landscape and Ecological Engineering* 9(1): 165–174. doi:10.1007/s11355-012-0194-3.

Hulme, M.F., J.A. Vickery, R.E. Green, B. Phalan, D.E. Chamberlain, D.E. Pomeroy, D. Nalwanga et al. 2013. 'Conserving the Birds of Uganda's Banana-Coffee Arc: Land Sparing and Land Sharing Compared.' *PLoS ONE* 8(2): e54597. doi:10.1371/journal.pone.0054597.

Johansson, Ö., T. McCarthy, G. Samelius, H. Andrén, L. Tumursukh, and C. Mishra. 2015. 'Snow Leopard Predation in a Livestock Dominated Landscape in Mongolia.' *Biological Conservation* 184: 251–258. doi:10.1016/j.biocon.2015.02.003.

Jouffray, J-B., M. Nystrom, A.V. Norstrom, I.D. Williams, L.M. Wedding, J.N. Kittinger, and G.J. Williams. 2014. 'Identifying Multiple Coral Reef Regimes and Their Drivers Across the Hawaiian Archipelago.' *Philosophical Transactions of the Royal Society B: Biological Sciences* 370: 20130268–20130268. doi:10.1098/rstb.2013.0268.

Kamp, J., R. Urazaliev, A. Balmford, P.F. Donald, R.E. Green, A.J. Lamb, and B. Phalan. 2015. 'Agricultural Development and the Conservation of Avian Biodiversity on the Eurasian Steppes: A Comparison of Land-Sparing and Land-Sharing Approaches.' *Journal of Applied Ecology* 52(6): 1578–1587. doi:10.1111/1365-2664.12527.

Kemp, J., A. López-Baucells, R. Rocha, O.S. Wangensteen, Z. Andriatafika, A. Nair, and M. Cabeza. 2019. 'Bats as Potential Suppressors of Multiple Agricultural Pests: A Case Study from Madagascar.' *Agriculture, Ecosystems and Environment* 269: 88–96. doi:10.1016/j.agee.2018.09.027.

Kent, M. 2011. *Vegetation Description and Data Analysis: A Practical Approach*. Hoboken: John Wiley and Sons.

Kress, W.J., C. Garcia-Robledo, M. Uriarte, and D.L. Erickson. 2015. 'DNA Barcodes for Ecology, Evolution, and Conservation Orogen.' *Trends in Ecology and Evolution* 30(1): 25–35. doi:10.1016/j.tree.2014.10.008.

Mandle, L., T. Ticktin, and P.A. Zuidema. 2015. 'Resilience of Palm Populations to Disturbance is Determined by Interactive Effects of Fire, Herbivory and Harvest.' *Journal of Ecology* 103(4): 1032–1043. doi:10.1111/1365-2745.12420.

Miguel, E., V. Grosbois, H. Fritz, A. Caron, M. de Garine-Wichatitsky, F. Nicod, A.J. Loveridge, B. Stapelkamp, D.W. Macdonald, and M. Valeix. 2017. 'Drivers of Foot-and-Mouth Disease in Cattle at Wild/Domestic Interface: Insights from Farmers, Buffalo and Lions.' *Diversity and Distributions* 23(9): 1018–1030. doi:10.1111/ddi.12585.

Minnie, L., A. Zalewski, H. Zalewska, and G.I.H. Kerley. 2018. 'Spatial Variation in Anthropogenic Mortality Induces a Source – Sink System in a Hunted Mesopredator.' *Oecologia* 186(4): 939–951. doi:10.1007/s00442-018-4072-z.

Mintzer, V.J., A.R. Martin, V.M.F. da Silva, A.B. Barbour, K. Lorenzen, and T.K. Frazer. 2013. 'Effect of Illegal Harvest on Apparent Survival of Amazon River Dolphins (Inia geoffrensis).' *Biological Conservation* 158: 280–286. doi:10.1016/j.biocon.2012.10.006.

Nordlund, L., J. Erlandsson, M. de la Torre-Castro, and N. Jiddawi. 2010. 'Changes in an East African Social-Ecological Seagrass System: Invertebrate Harvesting Affecting Species Composition and Local Livelihood.' *Aquatic Living Resources* 23(4): 399–416. doi:10.1051/alr/2011006.

Nyström, M., N.A.J. Graham, J. Lokrantz, and A.V. Norström. 2008. 'Capturing the Cornerstones of Coral Reef Resilience: Linking Theory to Practice.' *Coral Reefs* 27(4): 795–809. doi:10.1007/s00338-008-0426-z.

Perez, L., K. Eun-kyeong, and R. Sengupta. 2018. *Agent-Based Models and Complexity Science in the Age of Geospatial Big Data.* Cham: Springer Nature. doi:10.1007/978-3-319-65993-0.

Phalan, B., M. Onial, A. Balmford, and R.E. Green. 2011. 'Reconciling Food Production and Biodiversity Conservation: Land Sharing and Land Sparing Compared.' *Science* 333(6047): 1289–1291. doi:10.1126/science.1208742.

Pressey, R.L., P. Visconti, and P.J. Ferraro. 2015. 'Making Parks Make a Difference: Poor Alignment of Policy, Planning and Management with Protected-Area Impact, and Ways Forward.' *Philosophical Transactions of the Royal Society of London. Series B, Biological Sciences* 370: 20140280. doi:10.1098/rstb.2014.0280.

Quinn, G.P., and M.J. Keough. 2002. *Experimental Design and Data Analysis for Biologists.* Cambridge: Cambridge University Press.

Read, Z.J., H.P. King, D.J. Tongway, S. Ogilvy, R.S.B. Greene, and G. Hand. 2016. 'Landscape Function Analysis to Assess Soil Processes on Farms Following Ecological Restoration and Changes in Grazing Management.' *European Journal of Soil Science* 67(4): 409–420. doi:10.1111/ejss.12352.

Sala, O.E., R.B. Jackson, H.A. Mooney, and R.W. Howarth. 2000. *Methods in Ecosystem Science.* New York: Springer.

Smith, D.G. 1987. 'A Mini-Vibracoring System.' *Journal of Sedimentary Research* 57(4): 757–758.

Sutaria, D. 2009. 'Species Conservation in a Complex Socio-Ecological System: Irrawaddy Dolphins, Orcaella Brevirostris in Chilika Lagoon, India.' PhD diss., James Cook University.

Swemmer, A.M., M. Mashele, and P.D. Ndhlovu. 2019. 'Evidence for Ecological Sustainability of Fuelwood Harvesting at a Rural Village in South Africa.' *Regional Environmental Change* 19: 403–413. doi:10.1007/s10113-018-1402-y.

Tan, K.H. 2005. *Soil Sampling, Preparation, and Analysis.* Boca Raton: CRC Press.

Ticktin, T., A.N. Whitehead, and H. Fraiola. 2006. 'Traditional Gathering of Native Hula Plants in Alien-Invaded Hawaiian Forests: Adaptive Practices, Impacts on Alien Invasive Species and Conservation Implications.' *Environmental Conservation* 33(3): 185–194. doi:10.1017/S0376892906003158.

Tongway, D., and N. Hindley. 2004. 'Landscape Function Analysis: A System for Monitoring Rangeland Function.' *African Journal of Range and Forage Science* 21(2): 109–113. doi:10.2989/10220110409485841.

Villasante, S. 2012. 'The Management of the Blue Whiting Fishery as Complex Social-Ecological System: The Galician Case.' *Marine Policy* 36(6): 1301–1308. doi:10.1016/j.marpol.2012.02.013.

Ward, D.F.L., S. Wotherspoon, J. Melbourne-Thomas, J. Haapkylä, and C.R. Johnson. 2018. 'Detecting Ecological Regime Shifts from Transect Data.' *Ecological Monographs* 88(4): 694–715. doi:10.1002/ecm.1312.

Wheater, C.P., J.R. Bell, and P.A. Cook. 2011. *Practical Field Ecology: A Project Guide.* Hoboken: John Wiley and Sons.

Wu, J., W. Shen, W. Sun, and P.T. Tueller. 2002. 'Empirical Patterns of the Effects of Changing Scale on Landscape Metrics.' *Landscape Ecology* 17(8): 761–782. doi:10.1023/A:1022995922992.

7

Interviews and surveys

Sheona Shackleton,[1] *Joana Carlos Bezerra,*[2]
Jessica Cockburn,[3] *Maureen G. Reed*[4] *and Razak Abu*[5]

[1]AFRICAN CLIMATE AND DEVELOPMENT INITIATIVE, UNIVERSITY OF CAPE TOWN, CAPE TOWN, SOUTH AFRICA
[2]COMMUNITY ENGAGEMENT DIVISION, RHODES UNIVERSITY, MAKHANDA, SOUTH AFRICA
[3]ENVIRONMENTAL LEARNING RESEARCH CENTRE, RHODES UNIVERSITY, MAKHANDA, SOUTH AFRICA
[4]SCHOOL OF ENVIRONMENT AND SUSTAINABILITY, UNIVERSITY OF SASKATCHEWAN, SASKATOON, CANADA
[5]POLICY FELLOW, NATIONAL ENERGY BOARD, CALGARY, CANADA

Key methods discussed in this chapter

Surveys, in-depth interviews, key informant interviews, life histories, focus group discussions, reflective questioning, conversations and dialogues, arts-based interview methods

Connections to other chapters

This chapter is closely related to the following approaches and methods in this handbook: systems scoping (Chapter 5), participatory data collection (Chapter 8), facilitated dialogues (Chapter 9), futures analysis (Chapter 10), qualitative content analysis (Chapter 19), comparative case study analysis (Chapter 20), institutional analysis (Chapter 22), network analysis (Chapter 23), spatial mapping and analysis (Chapter 24), historical assessment (Chapter 25), and livelihood and vulnerability analysis (Chapter 32).

Introduction

'Every good conversation starts with good listening' – Common saying

Interviews and surveys are a means of gathering information from people who are part of the social-ecological systems (SES) of interest. They involve modes and ways of learning from people through asking questions and recording responses (mainly surveys) and through conversations and listening (mainly interviews). Interviews and surveys are often used at the local level in place-based studies, but are applicable across a variety of scales. While being valid (social science) methods in their own right, interviews and surveys are also foundational to several other approaches and methods in this handbook, especially when exploring the social dimensions of SES and the interactions between the social and the ecological (see Section 'Connections to other chapters').

The use of interviews and surveys originated within the social sciences, including education, psychology and public health. Different interview methods, whether quantitative or qualitative, derive their practices from different ontological and epistemological perspectives about

DOI: 10.4324/9781003021339-9

<table>
<tr><th colspan="2">SUMMARY TABLE: INTERVIEWS AND SURVEYS</th></tr>
<tr><th>DISCIPLINARY BACKGROUND</th><th>KNOWLEDGE TYPE</th></tr>
<tr><td>The methods in this chapter are derived from or have most commonly been used in:
Social Sciences including Anthropology, Economics, Political Science and Sociology; Psychology; Philosophy; Education; Human Geography; Environmental Studies; Public Health; Population Studies</td><td>The methods in this chapter are primarily used to generate the following types of knowledge:
• Descriptive
• Exploratory
• Explanatory</td></tr>
<tr><th>RESEARCH APPROACH</th><th>PURPOSE OF METHOD</th></tr>
<tr><td>The methods in this chapter originate from or most commonly adopt the following research approaches:
• Analytical/objective
• Interpretive/subjective
• Collaborative/process</td><td>The most common purposes of using the methods in this chapter are:
• Data collection/generation
• System understanding
• Policy/decision support</td></tr>
<tr><th>TEMPORAL DIMENSION</th><th>SYSTEMIC FEATURES AND PROCESSES</th></tr>
<tr><td>The methods in this chapter are most commonly applied to the following temporal dimensions:
• Present (typically within the last 5–10 years)
• Recent past (post-1700s)
• Future</td><td rowspan="3">While most methods can do many things, the methods in this chapter are particularly good (i.e. go-to methods) for addressing the following:
• SES components and linkages
• Diversity
• Social-ecological dependence and impact
• Power relations</td></tr>
<tr><th>SPATIAL DIMENSION</th></tr>
<tr><td>The methods in this chapter are primarily either or both:
• Non-spatial
The methods in this chapter are most commonly applied at the following spatial scales:
• Local
• Multiple places/sites around the world</td></tr>
</table>

the nature of reality and how we can study it (Creswell and Plano Clark 2011; Newing 2011). Quantitative survey methods tend to be used by social scientists from a positivist/postpositivist paradigm that includes, for example, economists and some sociologists and psychologists, with the main application being 'prediction' and 'extrapolation' (Moon and Blackman 2014). Qualitative methods, such as in-depth interviews and life histories, are associated with constructivism, interpretivism and critical realism and are used in a wide range of disciplines such as anthropology, education, human geography, psychology, political science and sociology, with the main application being to gain 'understanding' of context-dependent variables and realities.

Since the 1980s, there has been increased recognition of the value of combining qualitative and quantitative methods (Creswell and Plano Clark 2011), especially in applied, holistic research fields such as sustainability science, public health and the geographical and environmental sciences/studies. This approach has been termed 'mixed methods'. Since mixing qualitative and quantitative methods incorporates multiple ways of seeing and making sense of the world, one can argue that, in combination, they provide a better way of understanding complex problems than either method alone (Creswell and Plano Clark 2011). In SES and inter- and transdisciplinary research, qualitative and quantitative interview approaches are often linked with other social and natural science methods, such as geographic information systems (GIS) (see Chapter 24), participatory mapping (see Chapter 8), modelling and ecological surveys (see, for example, Chapter 6), to explore the multiple dimensions of complex sustainability problems and their solutions (see Case study 7.1). These mixed-methods approaches are recognised to be associated with a pragmatic paradigm or pragmatism (Cresswell and Plano Clark 2011).

SES problems and questions

Interviews and surveys can be used to generate data and insights in five main dimensions of SES:

1. Social-ecological dimensions (e.g. human–nature interactions, values and relational values, stewardship, resource use)
2. Institutional dimensions (e.g. governance, management)
3. Social-relational dimensions (e.g. collaboration, social learning, power dynamics)
4. Contextual dimensions (e.g. history, culture, political systems, knowledge systems, socio-economic systems)
5. Individual dimensions (e.g. agency, identity, sense of place, behaviour, perceptions)

Within each of these dimensions, a variety of issues and concepts can be explored and understood through different types of interviews and surveys (Newing 2011). Below are some examples of SES-related research problems or questions that can be addressed using qualitative or quantitative interview methods, or combinations of both types (Table 7.1).

- How do people use resources for livelihoods? (e.g. contribution of dryland natural resources to the livelihoods of communities adjacent to a protected area (Thondhlana, Vedeld, and Shackleton 2012))
- How do people manage and govern resources? (e.g. understanding local institutions for governance (Cundill and Fabricius 2010))
- How do decisions at the global level impact on local resource use? (e.g. exploring the global–local link in environmental decision-making (Charles 2012))

- How do people value resources or ecosystem services? (e.g. tracking how people value cultural ecosystem services associated with water flows in a catchment (Bark, Robinson, and Flessa 2016))
- How do people relate to and engage with places? (e.g. investigating the relationship that rural landholders have with their land and waterways (Baldwin, Smith, and Jacobson 2017))
- How does the environment and changing environmental conditions influence people's identity? (e.g. understanding how a particular place shapes identity (Cundill et al. 2017))
- How do people work together to share resources? (e.g. understanding multi-stakeholder governance for landscape management through social network analysis (this links to Chapter 23 on social network analysis) (Rathwell and Peterson 2012))
- How does human agency mediate people's interactions in the SES in which people are embedded? (e.g. understanding the development of collective agency and capabilities in the management of a biosphere reserve (Pelenc, Bazile, and Ceruti 2015))

Brief description of key methods

Table 7.1 provides a description of each of the quantitative and qualitative methods that support the collection of data required for exploring the issues and different dimensions of SES mentioned above. These different methods can be used to collect a variety of data in diverse contexts. The study research objectives should be used as a guide for their application. Refer to the further readings suggested under 'Applications to SES' in the table for examples on how these methods have been used to address different SES problems and questions.

Table 7.1 Summary of key methods used in interviews and surveys

Method	*Description*	*References*
Surveys and questionnaires	Surveys provide primarily quantitative data, although open-ended questions allow respondents to give their own answers, offering some qualitative data. Data are commonly gathered through questionnaires (structured interviews). These may include a mix of questions that provide factual data such as how long one has stayed in a village, yes or no answers to a range of questions, or the popular five-point Likert scale questions that are usually attached to statements that respondents are required to rate. This can result in both continuous (e.g. years in the village) and discrete (number of yes responses) data. Surveys can be face to face, telephonic or self-administered through drop-off, email or online versions. Within SES research, surveys are often either at household level (related to demographics, livelihood activities and assets, ecosystem service use, shocks and stressors, vulnerabilities) or at individual level (related to perceptions, values, indigenous knowledge, sense of place). Individual-level surveys are often administered to particular target groups such as specific resource users, farmers, managers, women, the elderly or youth.	*Key introductory texts* Angelsen et al. 2011; Newing 2011 *Applications to SES* Shackleton et al. 2008; Thondhlana, Vedeld, and Shackleton 2012; Falayi et al. 2019

Method	*Description*	*References*
Key informant interviews	Key informant interviews are used to collect qualitative data in a purposive manner from people with particular relevance to, or who hold useful knowledge for, the study. Key informant interviews may be used at the beginning (exploratory, scoping) or at the end (confirmatory) of a study. They often provide in-depth, explanatory information on results from a survey. Key informants can be sought to help explain or contextualise data from other sources near the end of a study.	*Key introductory texts* Crabtree and Miller 1999; Newing 2011 *Applications to SES* Elmendorf and Luloff 2001; Cinner et al. 2012; Thondhlana, Vedeld, and Shackleton 2012
Oral/life histories	An oral or life history is a narrative of a person's life as told by the research participant, who has the freedom and power to decide what is important. The method allows the exploration of how and why people live their lives in a particular way, highlighting the causality and temporal dimensions of the past and the present. Interviewers may need more than one session to cover all the topics, which can either be recorded or written down, depending on ethical clearance and participant consent. Also see Chapter 25: Historical assessment.	*Key introductory texts* Hatch and Wisniewski 1995; Atkinson 2002; George and Stratford 2005 *Applications to SES* Sallu, Twyman, and Stringer 2010; Cundill et al. 2017; Abu and Reed 2018; Singh 2018
Focus groups	Focus groups provide a way to gather information from a small group, with the facilitation of the researcher. Focus groups bring participants together to generate new ideas, learning opportunities or consensus about how a group might interpret local phenomena. Focus groups may be employed to confirm or interpret data collected by other means. They can also encourage brainstorming on strategies for collective action and to jointly explore potential solutions. Focus groups are often combined with participatory approaches and those used in methods such as human–environment timelines (see Chapter 8).	*Key introductory texts* Hopkins 2007; Longhurst 2016 *Applications to SES* Ibarra et al. 2014; Rivera et al. 2014; Andrachuk and Armitage 2015; Masunungure and Shackleton 2018
In-depth interviews (structured, semi-structured, unstructured)	In-depth interviews generate qualitative data through extensive individual interviews and can be structured, with set questions; semi-structured, with guiding questions; and unstructured, with themes that the interviewer would like to cover. Both in-depth interviews and life histories can be used to bring culturally derived understandings into the assessment of long-term social, ecological, economic and cultural changes in SES.	*Key introductory text* Legard, Keegan, and Ward 2003 *Applications to SES* Tenza et al. 2017; Abu, Reed, and Jardine 2019
Conversations and informal interviews	In some social-ecological research contexts, a more informal form of interview may be necessary, as opportunities for gathering information or insights sometimes arise unexpectedly. The casual nature of these interactions helps to build trust and participants might feel more comfortable to share valuable data. Whereas researchers may not be able to prepare in detail for unstructured conversations, the general principles guiding interviews should be adhered to as far as possible. Daily reflections diaries can be used to capture some of these data.	*Key introductory texts* Turner 2010; Gideon and Moskos 2012 *Applications to SES:* Barthel, Folke, and Colding 2010

(Continued)

Table 7.1 (Continued)

Method	*Description*	*References*
Guided reflection	In guided reflections or 'reflective interviews', the interviewer asks questions that will trigger participants to reflect rather than simply answer a question. Reflection, which is crucial to learning, requires critical thinking about behaviour, beliefs and values and how the person reflecting might change their behaviour in light of the reflection. This is often used in workshop settings, for knowledge co-production and for exploring actions or solutions.	*Key introductory texts* Mezirow 1990; Lee and Barnett 1994 *Application to SES* Sriskandarajaha et al. 2010; Cockburn et al. 2018
Arts-based and visual approaches to interviews	Arts-based and visual tools like photographs, drawings, body movement, etc. can be used to enrich an interview by engaging people in a generative and meaningful conversation. These tools can be particularly relevant in change-oriented research where it is important to draw people in at an emotional level, or when working with illiterate research participants or children. These creative approaches can be instrumental in overcoming barriers such as language and race. Also see Chapter 8: Participatory data collection and Chapter 19: Qualitative content analysis.	*Key introductory texts* Douglas 1985; Collier and Collier 1986; Kara 2015 *Applications to SES* Trell and Van Hoven 2010; Berbés-Blázquez 2012; Pearson et al. 2018; Steelman et al. 2018

Limitations

Understanding the dynamics of SES through (individual or group) interview-based methods is challenging. As Case study 7.1 points out, these are best suited when they are combined with other sources to better understand ecological and social thresholds and how these interrelate. Some might argue that people's opinions and perceptions are subjective and therefore require corroboration from other sources such as instrumental or other observation-based approaches. However, this view can also be criticised because local people not only perceive change, they also experience and observe it. Hence, their observations may also be used to corroborate or correct scientific observations that contribute to an SES analysis. This debate points to an important element of interview-based methods: researchers play a significant role throughout the research process – from how interview questions are posed to how answers are framed.

Researchers must be ethical and transparent in presenting their processes of analysis and interpretation, including with the research participants with whom they work. For any type of interview, the researcher is interacting with another person or other people and it is because of this interaction that ethics and power gain crucial significance. The ethics involved in the build-up to interviews – such as seeking ethical clearance, acquiring official or unofficial permissions to conduct the interviews, and explaining the aim and outputs of the research – will set the tone for how the interview process will progress. The researcher–participant relationship can be exploitative or reciprocal (England 1994). The position of the researcher will be crucial in determining how this relationship develops, especially in qualitative interviews.

Furthermore, responses elicited from participants are context dependent. Historical legacies and contemporary realities, such as colonialisation, globalisation, gender-based inequalities, language barriers and racism, may form intersecting axes of marginalisation and privilege that affect researcher–participant interactions, data collection, the analysis methods selected and the interpretation of results. Therefore, respectful and reciprocal relations with participants and communities must accompany these methods to enrich understanding of social-ecological change, along with opportunities to deepen interpretation through additional locally appropriate research methods.

We must also recognise that these methods are themselves embedded in larger frameworks and assumptions that also require critical reflection. In considering how to bring different knowledge sources together, Johnson et al. (2016, 3) point out:

> [i]n shaping a dialogue with Indigenous sciences, the explicit universalism of science and the need for more than locally or contextually tailored solutions to problems, confronts a need to build frameworks for understanding that are themselves pluralist, open and engaged across (linguistic, cultural, epistemological, spatial and temporal) difference.

This observation suggests that the methods will only serve as well as the frameworks within which they are embedded.

Resource implications

Both surveys and interviews can be time-demanding and budget-consuming processes. They also require specific skills related to asking good questions, designing instruments and performing both quantitative and qualitative analysis. When surveys are used to generalise to a larger population, usually a large sample size (a few hundred questionnaires) is required depending on the size of the population under consideration. If the questionnaire is administered face to face, then trained interviewers are needed as the researchers may not have the time to do this themselves. It can be costly to hire these people or train less experienced interviewers, although the inclusion of local youth as interviewers, for example, can have multiple benefits. For smaller sample sizes, the researcher may be able to conduct the interviews but may require an interpreter and/or a translator, which will add to costs. There is also the cost of data capture if additional support is needed.

Qualitative interviews can be equally resource demanding. For instance, it may be possible to do only one oral history per day, requiring lengthy periods in the field. In addition, most interviews are recorded and need to be transcribed, which again may require additional assistance and funding. For both qualitative and quantitative interviews, software packages for analysis are required. These can be expensive, although increasingly there are free options (e.g. R, which can be used for both qualitative and quantitative data). Furthermore, training in the use of the software may be needed. These practical issues relating to time and budget need to be carefully thought through in the research design process.

New directions

There are multiple and emerging strategies to flatten the power dynamics of more conventional interview methods that have historically favoured researchers over participants. Indigenous methodologies have introduced the conversational method (Kovach 2010), modified sharing circles (Lavallée 2009) and storytelling (Fernández-Llamazares and Cabeza 2017), among others, as approaches that can give more power to participants during an interview process. There may also be creative ways to establish interviews as exercises in collaborative learning, such as through photo elicitation (Clark-Ibáñez 2004; Steelman et al. 2018), co-creating mental maps, or the use of technology such as tablets or phone applications. Each of these has the potential to give participants more agency in directing the course of an interview than standard structured or semi-structured questions. Good practice also suggests that data analysis

Case study 7.1: Combining oral histories with other methods to link indigenous and scientific knowledge

This case study considers how diverse forms of knowledge can be brought together to understand long-term change in the SES of the Saskatchewan River Delta, Canada (Abu and Reed 2018; Abu, Reed, and Jardine 2019). The Saskatchewan River Delta is North America's largest freshwater delta, covering an area of about 10 000 km^2. However, since the 1960s, three upstream dams have been built, which have resulted in rapid and ongoing ecological changes in the delta and socio-cultural changes for the indigenous peoples who reside there. These changes have had a wide variety of impacts, including altering patterns of water availability, changing transportation modes and patterns, diminishing fish and wildlife habitat, and decreasing harvesters' ability to access traditional food.

Abu, Reed and Jardine (2019) used a 'two-eyed seeing' approach to collect and analyse changes in the delta and impacts of the dams. Suggested to scientists by an indigenous (Mi'kmaq) elder in Canada, two-eyed seeing is a metaphor that suggests 'seeing together' from indigenous and Western scientific lenses. The idea is to harness the strengths of each to appreciate the differences brought by each eye and to use both in order to gain a wider and deeper view to better understand complex and interrelated phenomena. Two-eyed seeing thus offers a respectful and practical means to bring Western science and indigenous knowledge systems together by providing strategies for checking the accuracy and filling in the knowledge gaps of each, without one knowledge system subsuming the other.

To demonstrate how to engage in two-eyed seeing, the study drew on, compared and evaluated three sources of evidence – indigenous knowledge, archival records and instrumental observations (i.e. information collected using scientific instruments such as water gauges, GIS and laboratory tests). The archival records and instrumental observations were the Western science portion of the 'two-eyed seeing'. The indigenous knowledge consisted of oral history and semi-structured interviews conducted with eight elders and 34 resource users – fishers, hunters, trappers and plant harvesters (see Table 7.2, compiled by Abu (2017)). The oral history interviews with elders provided first-hand knowledge of key historical events, especially prior to the 1960s, which were not documented in archival records. The semi-structured interviews also provided evidence of social impacts of ecological change, including reductions in harvesters' ability to access traditional food and changes in the taste of fish and meat, which instrumental observations did not detect.

By combining indigenous knowledge from oral histories and semi-structured interviews with archival records and instrumental observations, two-eyed seeing provided a

and representation should be approved by participants. Some authors suggest co-authorship as a means to recognise research participants as equal partners in knowledge co-creation (e.g. Castleden, Morgan, and Lamb 2012; Adams et al. 2014). Hence, introducing new methods is not simply about inserting methods into traditional research frameworks; it must be accompanied by the introduction of new ways of conceiving and implementing overall research design and follow-up.

Table 7.2 Evidence that indigenous knowledge, archival records and instrumental observations provide on social-ecological change in the Saskatchewan River Delta, Canada (© Razak Abu)

Social-ecological change	Indigenous knowledge from field interviews	Archival records	Instrumental observations
Altered seasonal flow	'Our highest water was in June and July... Now the way it is, they reversed that; the high water is in January.' (Participant 9)	Seasonal flows have changed from natural high summer, low winter to low summer, high winter levels. (Godwin 1968)	Reduced summer flows and elevated winter flows at gauge at The Pas in the post-dam era.
Northern pike production	'There's always that northern pike where you can't get rid of, lots of them jackfish.' (Participant 5)	Pike and suckers have thrived in the deteriorating water conditions. (Waldram 1989)	Pike harvest has declined to near zero.
Changes in berry season	'Like I remember one summer [we] were picking Saskatoons in July instead of June. And then we were picking raspberries in August instead of July.' (Participant 14)	No data	No data

more complete description of long-term social-ecological change than any single knowledge system could have done alone. Moreover, when coupled with a commitment to community-based research, two-eyed seeing made it possible for non-indigenous researchers to use Western science and indigenous knowledge in an appropriate way that demonstrated respect for both knowledge traditions.

Key readings

Angelsen, A., H.O. Larsen, J.F. Lund, C. Smith-Hall, and S. Wunder, eds. 2011. *Measuring Livelihoods and Environmental Dependence: Methods for Research and Fieldwork.* London: Earthscan.

Creswell, J.W., and V.L. Plano Clark. 2011. *Designing and Conducting Mixed Methods Research* (2nd ed). Thousand Oaks: Sage.

Moon, K., and D. Blackman. 2014. 'A Guide to Understanding Social Science Research for Natural Scientists.' *Conservation Biology* 28(5): 1167–1177.

Newing, H. 2011. 'Qualitative Interviews and Focus Groups.' In *Conducting Research in Conservation: A Social Science Perspective*, edited by H. Newing, 98–118. Oxon: Routledge.

Turner, D.W. III. 2010. 'Qualitative Interview Design: A Practical Guide for Novice Investigators.' *The Qualitative Report* 15(3): 754–760. https://nsuworks.nova.edu/cgi/viewcontent.cgi?article=1178&context=tqr.

Acknowledgements

The writing of this chapter was supported by funding from a South African National Research Foundation incentive funding grant (Sheona Shackleton), a Rhodes University post-doctoral fellowship (Jessica Cockburn), and research funds from the Canada Excellence Research Chair in Water Security, the Natural Sciences and Engineering Research Council of Canada (grant no. 445292), the Social Sciences and Humanities Research Council of Canada (grant no. 430-2013-000347) and the Arctic Institute for North America.

References

Abu, R. 2017. 'Knowledge, Use, and Change in the Saskatchewan River Delta: Assessing the Changing Livelihoods of Cumberland House Métis and Cree Nation.' PhD diss., University of Saskatchewan.

Abu, R., and M.G. Reed. 2018. 'Adaptation through Bricolage: Indigenous Responses to Long-term Socio-Ecological Change in the Saskatchewan River Delta, Canada.' *The Canadian Geographer / Le Géographe Canadien* 62(4): 437–451. doi:10.1111/cag.12469.

Abu, R., M.G. Reed, and T. Jardine. 2019. 'Using Two-eyed Seeing to Bridge Western Science and Indigenous Knowledge Systems and Understand Long-term Change in the Saskatchewan River Delta, Canada.' *International Journal of Water Resources Development*. doi:10.1080/07900627.2018.1558050.

Adams, M.S., J. Carpenter, J.A. Housty, D. Neasloss, P.C. Paquet, C. Service, J. Walkus, and C.T. Darimont. 2014. 'Toward Increased Engagement Between Academic and Indigenous Community Partners in Ecological Research.' *Ecology and Society* 19(3): 5. doi:10.5751/ES-06569-190305.

Andrachuk, M., and D. Armitage. 2015. 'Understanding Socio-Ecological Change and Transformation Through Community Perceptions of System Identity.' *Ecology and Society* 20(4): 26. doi:10.5751/ES-07759-200426.

Angelsen, A., H.O. Larsen, J.F. Lund, C. Smith-Hall, and S. Wunder, eds. 2011. *Measuring Livelihoods and Environmental Dependence: Methods for Research and Fieldwork*. London: Earthscan.

Atkinson, R. 2002. 'The Life Story Interview.' In *Handbook of Interview Research: Context and Method*, edited by J.F. Gubrium and J.A. Holstein, 121–140. London: Sage. doi:10.4135/9781412986205.

Baldwin, C., T. Smith, and C. Jacobson. 2017. 'Love of the Land: Socio-Ecological Connectivity of Rural Landholders.' *Journal of Rural Studies* 51: 37–52. doi:10.1016/j.jrurstud.2017.01.012.

Bark, R.H., C.J. Robinson, and K.W. Flessa. 2016. 'Tracking Cultural Ecosystem Services: Water Chasing the Colorado River Restoration Pulse Flow.' *Ecological Economics* 127: 165–172. doi:10.1016/j.ecolecon.2016.03.009.

Barthel, S., C. Folke, and J. Colding. 2010. 'Socio-Ecological Memory in Urban Gardens – Retaining the Capacity for Management of Ecosystem Services.' *Global Environmental Change* 20: 255–265. doi:10.1016/j.gloenvcha.2010.01.001.

Berbés-Blázquez, M. 2012. 'A participatory Assessment of Ecosystem Services and Human Wellbeing in Rural Costa Rica using Photo-voice.' *Environmental Management* 49(4): 862–875. doi:10.1007/s00267-012-9822-9.

Castleden, H., V.S. Morgan, and C. Lamb. 2012. '"I Spent the First Year Drinking Tea": Exploring Canadian University Researchers' Perspectives on Community-based Participatory Research Involving Indigenous Peoples.' *Canadian Geographer / Le Géographe canádien* 56(2): 160–179. doi:10.1111/j.1541-0064.2012.00432.x.

Charles, A. 2012. 'People, Oceans and Scale: Governance, Livelihoods and Climate Change Adaptation in Marine Socio-Ecological Systems.' *Current Opinion in Environmental Sustainability* 4(3): 351–357. doi:10.1016/j.cosust.2012.05.011.

Cinner, J.E., T.R. McClanahan, M.A. MacNeil, N.A.J. Graham, T.M. Daw, A. Mukminin, D.A. Feary et al. 2012. 'Co-management of Coral Reef Socio-Ecological Systems.' *Proceedings of the National Academy of Sciences* 109(14): 5219–5222. doi:10.1073/pnas.1121215109.

Clark-Ibáñez, M. 2004. 'Framing the Social World with Photo-elicitation Interviews.' *American Behavorial Scientist* 47(12): 1507–1527. doi:10.1177/0002764204266236.

Cockburn, J., C.T.G. Palmer, H. Biggs, and E. Rosenberg. 2018. 'Navigating Multiple Tensions for Engaged Praxis in a Complex Socio-Ecological System.' *Land* 7(4): 129. doi:10.3390/land7040129.

Collier, J., and M. Collier. 1986. *Visual Anthropology: Photography as a Research Method.* Albuquerque: University of New Mexico Press.

Crabtree, B.F., and W.L. Miller, eds. 1999. *Doing Qualitative Research* (2nd ed). Thousand Oaks: Sage.

Creswell, J.W., and V.L. Plano Clark. 2011. *Designing and Conducting Mixed Methods Research* (2nd ed). Thousand Oaks: Sage.

Cundill, G., J.C. Bezerra, A. de Vos, and N. Ntingana. 2017. 'Beyond Benefit Sharing: Place Attachment and the Importance of Access to Protected Areas for Surrounding Communities.' *Ecosystem Services* 28(Part B): 140–148. doi:10.1016/j.ecoser.2017.03.011.

Cundill, G., and C. Fabricius. 2010. 'Monitoring the Governance Dimension of Natural Resource Co-Management.' *Ecology and Society* 15(1): 15. doi:10.5751/ES-03346-150115.

Douglas, J. 1985. *Creative Interviewing.* Beverly Hills: Sage.

Elmendorf, W.F., and A.E. Luloff. 2001. 'Using Qualitative Data Collection Methods when Planning for Community Forests.' *Journal of Arboriculture* 27(3): 139–151. https://pennstate.pure.elsevier.com/en/publications/using-qualitative-data-collection-methods-when-planning-for-commu.

England, K.V.L. 1994. 'Getting Personal: Reflexivity, Positionality, and Feminist Research.' *The Professional Geographer* 46: 80–89. doi:10.1111/j.0033-0124.1994.00080.x.

Falayi, M., S.E. Shackleton, G.C. Kemp, and C.M. Shackleton. 2019. 'Changes in the Use and Value of Local Environmental Resources over a 15-Year Period in a Rural Village, South Africa.' *Forests, Trees and Livelihoods* 28(2): 90–107. doi:10.1080/14728028.2019.1568309.

Fernández-Llamazares, Á., and M. Cabeza. 2017. 'Rediscovering the Potential of Indigenous Storytelling for Conservation Practice.' *Conservation Letters* 11(3): 1–12. doi:10.1111/conl.12398.

George, K., and E. Stratford. 2005. 'Oral History and Human Geography.' In *Qualitative Research Methods in Human Geography* (2nd ed), edited by I. Hay, 106–115. Oxford: Oxford University Press.

Gideon, L., and P. Moskos. 2012. 'Interviewing.' In *Handbook of Survey Methodology for the Social Sciences*, 109–118. New York: Springer.

Godwin, R.B. 1968. *A Study of the Probable Water Level of Cumberland Lake.* Regina: Saskatchewan Water Resources Commission, Hydrology Division, Investigation and Planning Branch.

Hatch, J.A., and R. Wisniewski. 1995. 'Life History and Narrative: Questions, Issues, and Exemplary Works.' In *Life History and Narrative*, edited by J.A. Hatch and R. Wisniewski, 113–135. Washington, DC: Falmer Press.

Hopkins, P.E. 2007. 'Thinking Critically and Creatively about Focus Groups.' *Area* 39: 528–553.

Ibarra, A.M.S., V.A. Luzadis, M.J.B. Cordova, M. Silva, T. Ordoñez, E.B. Ayala, and S.J. Ryan. 2014. 'A Socio-Ecological Analysis of Community Perceptions of Dengue Fever and *Aedes aegypti* in Machala, Ecuador.' *BMC Public Health* 14: 1135. doi:10.1186/1471-2458-14-1135.

Johnson, J.T., R. Howitt, G. Cajete, F. Berkes, R.P. Louis, and A. Kliskey. 2016. 'Weaving Indigenous and Sustainability Sciences to Diversify Our Methods.' *Sustainability Science* 11(1): 1–11. https://link.springer.com/article/10.1007/s11625-015-0349-x.

Kara, H. 2015. *Creative Research Methods in the Social Sciences: A Practical Guide.* Bristol: Policy Press.

Kovach, M. 2010. 'Conversational Method in Indigenous Research.' *First Peoples Child and Family Review* 5(1): 40–48. http://journals.sfu.ca/fpcfr/index.php/FPCFR/article/view/172.

Lavallée, L.F. 2009. 'Practical Application of an Indigenous Research Framework and Two Qualitative Indigenous Research Methods: Sharing Circles and Anishnaabe Symbol-based Reflection.' *International Journal of Qualitative Methods* 8: 21–40. doi:10.1177/160940690900800103.

Lee, G.V., and B.G. Barnett. 1994. 'Using Reflective Questioning to Promote Collaborative Dialogue.' *Journal of Staff Development* 15(1): 16–21.

Legard, R., J. Keegan, and K. Ward. 2003. 'In-depth Interviews.' In: *Qualitative Research Practice: A Guide for Social Science Students and Researchers*, edited by J. Ritchie and J. Lewis. London: Sage.

Longhurst, R. 2016. 'Semi-structured Interviews and Focus Groups.' In *Key Methods in Geography*, edited by C. Nicholas, M. Cope, T. Gillespie, and S. French, 117–132. Thousand Oaks: Sage.

Masunungure, C., and S.E. Shackleton. 2018. 'Exploring Long-term Livelihood and Landscape Change in Two Semi-arid Sites in Southern Africa: Consequences for Vulnerability.' *Land* 7(2): 50. www.mdpi.com/2073-445X/7/2/50.

Mezirow, J. 1990. 'How critical reflection triggers transformative learning.' In *Fostering Critical Reflection in Adulthood*, edited by J. Mezirow, 1–20. San Francisco: Jossey-Bass Publishers.

Moon, K., and D. Blackman. 2014. 'A Guide to Understanding Social Science Research for Natural Scientists.' *Conservation Biology* 28(5): 1167–1177. www.academia.edu/27283236/A_Guide_to_Understanding_Social_Science_Research_for_Natural_Scientists.

Newing, H. 2011. 'Qualitative Interviews and Focus Groups.' In *Conducting Research in Conservation: A Social Science Perspective*, edited by H. Newing, 98–118. Oxon: Routledge.

Pearson, K.R., M. Bäckman, S. Grenni, A. Moriggi, S. Pisters, and A. de Vrieze. 2018. 'Arts-based Methods for Transformative Engagement: A Toolkit'. Wageningen: SUSPLACE. www.sustainableplaceshaping.net/arts-based-toolkit.

Pelenc, J., D. Bazile, and C. Ceruti. 2015. 'Collective Capability and Collective Agency for Sustainability: A Case Study'. *Ecological Economics* 118(Supplement C): 226–239. doi:10.1016/j.ecolecon.2015.07.001.

Rathwell, K.J., and G.D. Peterson. 2012. 'Connecting Social Networks with Ecosystem Services for Watershed Governance: A Socio-Ecological Network Perspective Highlights the Critical Role of Bridging Organizations'. *Ecology and Society* 17(2): 24. doi:10.5751/ES-04810-170224.

Rivera, A., S. Gelcich, L. García-Florz, J.L. Alcázar, and J.L. Acuña. 2014. 'Co-management in Europe: Insights from the Gooseneck Barnacle Fishery in Asturias, Spain.' *Marine Policy* 50(Part A): 300–308. doi:10.1016/j.marpol.2014.07.011.

Sallu, S.M., C. Twyman, and L.C. Stringer. 2010. 'Resilient or Vulnerable Livelihoods? Assessing Livelihood Dynamics and Trajectories in Rural Botswana.' *Ecology and Society* 15(4): 3. www.ecologyandsociety.org/vol15/iss4/art3.

Shackleton, S., B. Campbell, H. Lotz-Sisitka, and C.M. Shackleton. 2008. 'Links Between the Local Trade in Natural Products, Livelihoods and Poverty Alleviation in a Semi-Arid Region of South Africa.' *World Development* 36(3): 505–526. doi:10.1016/j.worlddev.2007.03.003.

Singh, C. 2018. 'Using Life Histories to Understand Temporal Vulnerability to Climate Change in Highly Dynamic Contexts.' *SAGE Research Methods Cases.* doi:10.4135/9781526440358.

Sriskandarajaha, N., R. Bawden, C. Blackmore, K.G. Tidball, and A.E.J. Wals. 2010. 'Resilience in Learning Systems: Case Studies in University Education.' *Environmental Education Research* 16: 559–573. doi:10.1080/13504622.2010.505434.

Steelman, T.A., E. Andrews, S. Baines, L. Bharadwaj, E.R. Bjornson, L. Bradford, K. Cardinal et al. 2018. 'Identifying Transformational Space for Transdisciplinarity: Using Art to Access the Hidden Third.' *Sustainability Science* 14(3): 771–790. doi:10.1007/s11625-018-0644-4.

Tenza, A., I. Pérez, J. Martínez-Fernández, and A. Giménez. 2017. 'Understanding the Decline and Resilience Loss of a Long-lived Socio-Ecological System: Insights from System Dynamics.' *Ecology and Society* 22(2): 15. doi:10.5751/ES-09176-220215.

Thondhlana, G., P. Vedeld, and S. Shackleton. 2012. 'Natural Resource Use, Income and Dependence among San and Mier Communities Bordering Kgalagadi Transfrontier Park, Southern Kalahari, South Africa.' *International Journal of Sustainable Development and World Ecology* 19(5): 460–470. doi: 10.1080/13504509.2012.708908.

Trell, E-M., and B. van Hoven. 2010. 'Making Sense of Place: Exploring Creative and (Inter)Active Research Methods with Young People.' *Fennia – International Journal of Geography* 188(1): 91–104. https://fennia.journal.fi/article/view/2522.

Turner, D.W. III. 2010. 'Qualitative Interview Design: A Practical Guide for Novice Investigators.' *The Qualitative Report* 15(3): 754–760. https://nsuworks.nova.edu/cgi/viewcontent.cgi?article=1178&context=tqr.

8

Participatory data collection

Alta de Vos,[1] *Rika Preiser*[2] *and Vanessa A. Masterson*[3]

[1]DEPARTMENT OF ENVIRONMENTAL SCIENCE, RHODES UNIVERSITY, MAKHANDA, SOUTH AFRICA
[2]CENTRE FOR SUSTAINABILITY TRANSITIONS, STELLENBOSCH UNIVERSITY, STELLENBOSCH, SOUTH AFRICA
[3]STOCKHOLM RESILIENCE CENTRE, STOCKHOLM UNIVERSITY, STOCKHOLM, SWEDEN

Key methods discussed in this chapter

Participatory mapping (direct-to-digital participatory mapping, 3D-participatory mapping, participatory GIS), photovoice, transect walks, ranking exercises, focus group discussions, Venn diagrams, matrix scoring, ecograms, timelines, Q-methodology, community mapping, participatory videography, photo elicitation, seasonal calendars, participatory action research, participatory rural appraisal, participant observation, arts-based methods

Connection to other chapters

Participatory data collection methods can be used by many of the data analysis methods covered in this book, particularly those that require qualitative data. Indeed, the lines between data collection and analysis or modelling may be blurred at times. In particular, participatory modelling (Chapter 13), fuzzy cognitive mapping (Chapter 16), companion modelling (Chapter 12), as well as futures analysis (Chapter 10), scenario development (Chapter 11) and facilitated dialogues (Chapter 9) often involve the use of many of the methods described in this chapter, and can also be considered 'participatory data collection' processes in their own right. Many action research (Chapter 15), qualitative content analysis (Chapter 19), spatial mapping and analysis (Chapter 24), and livelihood and vulnerability analysis (Chapter 32) projects may also use the methods listed in this chapter.

Introduction

Social-ecological systems (SES) research acknowledges that how systems are framed depends on the observer, making it possible to have multiple valid descriptions or conceptualisations of a system (Preiser et al. 2018). These framings imply that, to understand how and why systems change, researchers often (but not always, see Hurlbert and Gupta 2015) need to employ participatory approaches and knowledge co-production. Knowledge co-production is the 'collaborative process of bringing a plurality of knowledge sources and types together

DOI: 10.4324/9781003021339-10

SUMMARY TABLE: PARTICIPATORY DATA COLLECTION	
DISCIPLINARY BACKGROUND	**KNOWLEDGE TYPE**
The methods in this chapter are derived from or have most commonly been used in: Sociology, Anthropology, Psychology, Art, Development Studies	*The methods in this chapter are primarily used to generate the following types of knowledge:* • Descriptive • Exploratory
RESEARCH APPROACH	**PURPOSE OF METHOD**
The methods in this chapter originate from or most commonly adopt the following research approaches: • Interpretive/subjective • Collaborative/process	*The most common purposes of using the methods in this chapter are:* • Data collection/generation • Stakeholder engagement and co-production
TEMPORAL DIMENSION	**SYSTEMIC FEATURES AND PROCESSES**
The methods in this chapter are most commonly applied to the following temporal dimensions: • Present (typically within the last 5–10 years) • Recent past (post-1700s) • Future	*While most methods can do many things, the methods in this chapter are particularly good (i.e. go-to methods) for addressing the following:* • SES components and linkages • Social-ecological dependence and impact • Power relations • Social learning
SPATIAL DIMENSION	
The methods in this chapter are primarily either or both: • Non-spatial • Explicitly spatial *The methods in this chapter are most commonly applied at the following spatial scales:* • Local • Multiple places/sites around the world	

to address a defined problem and build an integrated or systems-oriented understanding of that problem' (Armitage et al. 2011, 996). As researchers and practitioners are increasingly acknowledging, local people hold important knowledge for addressing SES problems. They also have a stake in co-defining these problems in the first place, given that their livelihoods and well-being are most at risk of impact (Fischer and Eastwood 2016; Nel et al. 2016). Participatory approaches provide a mechanism to facilitate feedbacks and social learning (Stringer et al. 2006) and participatory data collection methods (participatory methods specifically used to generate or co-create data) are often used to establish shared interest and overlapping understanding across multiple knowledge domains (Roux et al. 2017). Participatory research focuses on a process of sequential actions, in which local people are part of, instead of the subject of, research processes (Cornwall and Jewkes 1995). Local knowledge and perspectives form the basis for research and planning (Cornwall and Jewkes 1995). What sets participatory research apart from non-participatory social research processes is its focus on navigating power imbalances between researchers and research participants, and among individuals in communities (Campbell 2002; Pain 2004; Van Riet and Boettiger 2009).

Participatory methods originate from many different fields in the social sciences, most notably development studies, anthropology, organisational studies, psychology and public health. Many of the methods discussed in this chapter also form part of well-established methodologies such as participatory rural appraisal (PRA), participatory learning and action (PLA), participatory action research (PAR) (see Chapter 13) and participatory mapping. All these methods have in common a focus on the co-production of knowledge with local knowledge- and stakeholders, prioritising the engagement process over the collection of data, and acknowledging and integrating multiple knowledge types.

Most participatory research emphasises 'knowledge for action' and requires a critical, reflective approach from the researcher. It demands an openness to see reality from the participants' point of view, and requires dynamic sharing of knowledge and perspectives between researchers and participants (Williams and Hardison 2013). Participatory methods often facilitate knowledge exchange and knowledge co-production among different knowledge systems, recognising that different knowledge systems are all internally valid and have their own strengths (Tengö et al. 2012, 2014). Tengö et al. (2012) identify the following essential principles for exchange across knowledge systems: trust, respect, reciprocity, equity, transparency, free, prior and informed consent. Williams and Hardison (2013) and Rambaldi et al. (2006) call for safeguards related to the rights of communities to their knowledge and knowledge ownership, including proper implementation of informed consent related to the sharing of their knowledge, and capacity building about the potential risks associated with sharing knowledge.

Many of the approaches used in social-ecological research today are linked to participatory rural appraisal and participatory learning and action approaches first adopted in the field of development in the 1960s and 1970s. Following a recognition that desired development outcomes would be achieved more effectively by working with the intended beneficiaries, researchers sought to modify existing social science methodologies in ways that could better incorporate affected people's local views. The early result was rapid rural appraisal, typically comprising a short visit by multiple experts who would set out to get a quick understanding of the system by interviewing local experts and consulting archival sources. This approach was heavily criticised, mostly because important decisions were being made based on limited, biased views. By the 1980s, participatory rural appraisal had developed (Chambers 1994) to emphasise more in-depth research engagement with communities and the inclusion of a greater diversity of voices within communities. Both participatory rural appraisal and participatory learning and action approaches have since been adopted in many different research fields, not only in appraisals and a rural setting.

The diversity of participatory methods has increased into the 21st century, with many innovations to better acknowledge and include the diverse needs and knowledge systems of local and indigenous people in co-defining problems (Smith et al. 2017), as well as inclusive ways of incorporating new technologies (e.g. participatory modelling, companion modelling, 3D-PGIS) (see Rambaldi et al. 2007; Barreteau et al. 2014).

SES problems and questions

Participatory data collection methods have not been specifically created to address SES questions, but many are inherently well suited to the SES domain and the incorporation of knowledge, preferences and values into decision-making about natural resources (Lynam et al. 2007). The richness and holistic nature of participatory data collection methods allow for the understanding of the nature of feedbacks between people and nature.

Participatory data collection methods are mostly employed at a local scale and are well suited to place-based research. These data collection methods are particularly useful in projects where the engagement process is important to the content and legitimacy of the outcome, and where mobilising local knowledge and perceptions is expedient and ethically responsible. When used responsibly, participatory methods can be useful tools to manage or dampen power dynamics (Reed 2008; Hill et al. 2012; Villamor et al. 2014), allowing voices that may otherwise have been silent or undervalued to be heard and legitimised (Stirling 2008).

Participatory methods are often used in research that concerns values and perceptions related to natural resources, their management and governance, and in projects that seek integration across knowledge systems. Certain participatory data collection methods, such as timelines, can be very useful for finding out how and why systems might change, particularly when combined with methods such as participatory modelling (see Chapter 13) or agent-based modelling (see Chapter 28).

Typical questions for participatory research could include the following:

- Where are important natural resources located, and how have they changed over time? (Levine and Feinholz 2015)
- How do different people within communities use natural resources differently? (Kalibo and Medley 2007)
- Where are priority areas for environmental restoration, and which restoration methods are most appropriate? (Cockburn et al. 2018; Weyer, Bezerra, and De Vos 2019)
- How and why have landscapes changed over time? (Sieber, Medeiros, and Albuquerque 2011)
- What are the potential impacts or unintended consequences of development interventions for different people within a community? (Mehryar et al. 2017)
- How do power and hierarchies influence people's access to ecosystem services? (Weyer, Bezerra, and De Vos 2019)

- How can local visions of the future be used to inform natural resource management? (Palomo et al. 2011)
- How can local and indigenous knowledge inform local restoration or conservation actions? (Ramirez-Gomez, Brown, and Tjon Sie Fat 2013)

Brief description of key methods

Methods used in participatory data collection research all have in common a focus on facilitating data co-creation and navigating power relations among individuals in communities, and between researchers and communities. To facilitate the inclusion of multiple types of knowledge and knowledge systems and to engage a diversity of non-academic abilities and capacities across a variety of cultural contexts, many participatory data collection methods make use of visual methods (e.g. participatory photography, arts-based methods), whereas others employ modes of spatial and temporal mapping (e.g. community/participatory mapping, seasonal calendars, 3D-PGIS).

While not strictly a participatory method, it is important to also mention ethnographic research and participant observation here, as these tools for deep inductive and exploratory qualitative research are often used alongside participatory methods. Ethnographic research makes use of participant observation over long periods of time, research diaries and interviews to triangulate insights and produce rich and thick descriptions of phenomena. In this way ethnographic research approaches demand 'participation' of the researcher in the day-to-day life of study communities. For an introduction to ethnographic research, see LeCompte and Schensul (2010). For an example of how these methods have been applied to SES research, see Moerlein and Carothers (2012), Frey and Berkes (2014) and Laborde et al. (2016). Table 8.1 provides a summary of participatory data collection methods used in SES research.

Table 8.1 Summary of key participatory data collection methods used in SES research

Method	*Description*	*References*
Participatory photography	Visual methods can reduce power imbalances between researchers and the researched. One-on-one interviews of photo elicitation are considered participatory when they focus on photographs taken by the interviewee, which allows participants to retain control over what information they share with researchers. Photovoice is a related but distinct form of participatory visual storytelling, where participants take their own photographs related to a theme and share them in a group setting. Photovoice facilitates learning processes by capturing and sharing complex issues through a visual narrative and allows for the co-construction of knowledge through group participatory processes of collective reflexivity and meaning-making. There is often emphasis on advocacy and the use of photographs to convey a message, e.g. to decision-makers.	*Key introductory texts* Wang and Burris 1994, 1997; Harper 2002; Pink 2011 *Applications to SES* Beilin 2005; Mitchell and De Lange 2011; Berbés-Blázquez 2012; Maclean and Woodward 2013; Kong et al. 2015; Robinson et al. 2016; Masterson et al. 2018

(Continued)

Table 8.1 (Continued)

Method	*Description*	*References*
Timelines, matrices, Venn diagrams, ranking exercises	These tools are often used in a focus group setting and are considered 'classic' participatory rural appraisal techniques. The timeline technique comprises a drawn line, on which participants highlight key events, changes or visions of the future. Matrix-scoring exercises are often used to discern cause and effect. Venn diagrams are used to highlight key social interactions, whereas ranking exercises involve prioritising resources, livelihoods and other elements in order of importance.	*Key introductory texts* Newing et al. 2011; Schreckenberg et al. 2016 *Applications to SES* Bunce et al. 2010; Malinga et al. 2013; Sinare, Gordon, and Enfors-Kautsky 2016; Masterson et al. 2017
Transect walks	A transect walk is a systematic walk by the research team and community members along a defined path (transect) across the community/project area together with community members to explore certain SES contexts and conditions by observing, asking and listening. The result is a transect map. Transect walks are usually conducted during the early phase of fieldwork.	*Key introductory texts* Newing et al. 2011; Schreckenberg et al. 2016 *Applications to SES* Kalibo and Medley 2007; Malmborg et al. 2018
Focus group discussions	Many of the other participatory methods mentioned in this table can happen in a focus group setting, but focus group discussions do not necessarily have to involve particular exercises. They can simply be group discussion on how people relate to the environment and how they adapt, with an emphasis on understanding their views and values.	*Key introductory text* Newing et al. 2011 *Applications to SES* Nyirenda and Drive 2015; Sinare, Gordon, and Enfors-Kautsky 2016; Sylvester, Segura, and Davidson-Hunt 2016
Q-methodology	Q-methodology originated in the field of psychology and is useful when researchers wish to understand and describe subjectivity. Many variants of Q-methodology require participants to sort statements on a pre-configured grid, according to their preferences, followed by group discussion.	*Key introductory texts* Newing et al. 2011; Watts and Stenner 2012 *Applications to SES* Milcu et al. 2014; Forrester et al. 2015; Murray, D'Anna, and MacDonald 2016; West, Cairns, and Schultz 2016; Armatas, Venn, and Watson 2017; Rust 2017
Seasonal calendars	Seasonal calendars are a tool to map seasonal changes in resources, events, institutions and customs, usually in a public, participatory setting.	*Key introductory texts* Newing et al. 2011; Schreckenberg et al. 2016 *Applications to SES* Aburto et al. 2013

Method	*Description*	*References*
Community mapping	Community mapping concerns the mapping of resources or important places in relation to one another.	*Key introductory text* Schreckenberg et al. 2016 *Applications to SES* Belay 2012; Villamor et al. 2014
Participatory GIS	Participatory geographic information systems (GIS) differ from community mapping primarily on account of the accuracy of spatial representation. Participatory GIS involves either the use of digitised maps in a GIS, or the use of virtual earth and mapping technologies (most commonly Google Earth) to map information such as preferences or use directly onto a digital platform. 3D-PGIS requires the construction of a 3D model of the landscape by community members, typically out of papier-mâché, ribbons and pins. The focus in 3D-PGIS is on the process of building the model, and ownership of the resultant data and model belongs to the community who created it.	*Key introductory texts* Rambaldi et al. 2007; Bryan 2015; Brown 2017 *Applications to SES* Rambaldi et al. 2007; Raymond et al. 2009; Olson, Hackett, and DeRoy 2016; Ramirez-Gomez et al. 2017; Samuelsson et al. 2018
Arts-based methods	Arts-based methods refer to a broad group of participatory methods that specifically make use of at least one of a diversity of art genres (e.g. performance, writing, photography, mosaicking, collage, sculpture, painting) in collecting data. The co-creation of artistic pieces with participants may elicit knowledge, values and emotions. Creating art and performance pieces together can facilitate discussion on a shared platform that is more familiar to many, and even produce a 'boundary object' which might be understood differently by different participants but initiates discussion. Many of these methods, e.g. applied theatre, are devised in close collaboration with the communities they target and are often directed at social change.	*Key introductory texts* Liamputtong and Rumbold 2008; Bagnoli 2009; Heras and Tàbara 2014 *Applications to SES* Walker 2012; Lemelin et al. 2013; Heras and Tàbara 2014; Brown et al. 2017; Johansson and Isgren 2017
Participant observation and ethnographic research	Participant observation is not participatory in the same way as other methods in this chapter, in that it does not require participation from the research participants as such. Rather, it involves the researchers immersing themselves in community life and in a sense becoming participants in the community, thus blurring the lines between researcher and research 'subject'. Ethnography is a type of qualitative research that involves immersing yourself in a particular community or organisation to observe their behaviour and interactions up close. Ethnography is a flexible research method that allows one to gain a deep understanding of a group's shared culture, conventions and social dynamics. However, it also involves some practical and ethical challenges.	*Key introductory text* LeCompte and Schensul 2010 *Applications to SES* Moerlein and Carothers 2012; Frey and Berkes 2014; Laborde et al. 2016

Limitations

Participatory data collection explicitly requires interaction with people. The level of engagement and trust required in these exercises can carry substantial ethical risks, which need to be considered prior to commencement. These ethical risks particularly concern unmet expectations and equality of voice. Many kinds of engagements may count as 'participatory' and it may not always be clear to research participants what is expected of them, or what the research might entail. This may lead to the generation of mistrust between researchers and research participants when expectations are not met. It is also important to consider which methods are appropriate for a setting and the cultural context and contemporary issues faced by communities. Researchers need to be cognisant of the power of participatory methods (especially those with roots in advocacy such as forum theatre or photovoice) for learning and illuminating issues that inspire a desire for change. They should therefore be responsible when initiating dialogues with communities (Wang and Burris 1994; Belay 2012).

How problems are framed has important consequences for achieving or undermining justice in participatory research (Stirling 2008; Scoones et al. 2018). It can be easy to exclude the most vulnerable voices in participatory research, which may also result in shallow depictions of the local reality (Schreckenberg et al. 2016). Without careful consideration of who are included in participatory research and in what way, participatory methods may perpetuate existing gender and cultural biases.

Participatory methods are also associated with more practical and logistical limitations. These methods may be biased in favour of areas and people who are easily accessible and who have the capability to participate in the research activities, and areas that are more similar to the cultural norms and language of the researcher (Campbell 2002). Accessibility to participants, along with other elements of participatory research, may be sensitive to seasonality (Schreckenberg et al. 2016). Accounting for accessibility, however, has significant implications for the time and financial resources needed to conduct this kind of research. In addition to the time required for the researchers to get to participants, participatory activities can place a large demand on participants' time (Campbell 2002; Pain 2004; Schreckenberg et al. 2016; Brown and Kyttä 2018).

Since participatory data collection methods always concern the co-production of knowledge, the questions of who the data belong to and how the data should be stored and shared can be difficult ones to answer. This is particularly true in studies and projects that make use of participatory mapping techniques and produce maps as an outcome (Rambaldi et al. 2006), or visual ethnography methods that produce photographs (Pink 2011).

As with other social research, it is key that free, prior and informed consent is sought from participants. The research protocol must be approved by a research ethics council before research begins. Different countries and institutions have different procedures and requirements that may apply to research with or on vulnerable groups, so it is critical that researchers consult their university administration or ethics office before embarking on research of this nature. It is also important to consult specific codes of research ethics drawn up by indigenous people (e.g. Callaway 2017), where these exist.

Resource implications

Many participatory mapping methods require very little in the way of hardware or software, relying on flipcharts and markers, or even drawings in the sand. Voice recorders are often the only hardware that accompanies researchers in the field. Some participatory methods require specific equipment. Photovoice, for example, requires the use of cameras or smartphones and the opportunity to print or project photographs. Methods such as 3D-PGIS require constructing material, such as papier-mâché, for building models. Mapping with virtual globes (such as Google Earth, which is free to download) requires a computer, pointer, projector and screen. Although an Internet connection is desirable, it is not critical.

Participatory GIS requires basic GIS knowledge, and researchers administrating this method should be familiar with making maps. Digitising hard-copy maps and creating maps may require the use of ESRI's ArcGIS or QGIS. Analysing participatory data may require significantly more technical skill and knowledge of statistical packages, GIS tools, tools and platforms to conduct thematic analysis, including qualitative computer packages such as Atlas.ti or InVivo (see Chapter 19).

Whereas materials to run participatory processes may not be very expensive, participatory methods are nevertheless resource intensive. Getting participants involved in research and using mapping software or cameras, for example, mean the researcher has an added responsibility of managing and negotiating the type of capacity, skills and training people would need to engage in the research on an equitable basis. Projects that make use of participatory methods can be very expensive, as a project should be able to afford the cost of interpreters or translators, skilled facilitators and drivers as well as accessing sometimes very inaccessible locations. Some field sites may require the use of specific vehicles and specialist drivers, or the need for extra security. Setting up participatory processes can be very time consuming and may also require the use of community resources. Time is one of the most important resource considerations for participatory research, especially if there are multiple iterations of a process.

New directions

Participatory data collection methods are now widely used in SES research. Participatory mapping has already benefited from the introduction of technology, particularly the use of virtual globes like Google Earth. The more recent development of virtual reality promises to deepen the experience of participatory data collection. Platforms such as Ushahidi, Kobo Collect and Open Data Kit have greatly improved the affordability and ease of participatory monitoring. These technologies potentially allow for participatory mapping methods to be used over larger spatial extents, widening the degree to which different people may 'participate'. However, they also open up new ethical concerns around data privacy and ownership.

In recent years, participatory data collection methods have been adapted and applied increasingly in approaches that go beyond 'participation' towards the co-production of knowledge (e.g. Tengö et al. 2014, 2017; Scoones et al. 2018) and achieving greater social and epistemic justice (Roux et al. 2017). Notably, the multiple evidence base approach (Tengö et al. 2014, 2017) emphasises the self-representation of knowledge and perspectives, and the internal validation of knowledge systems. Similarly, the STEP centre's pathways approach (Leach, Scoones, and Stirling 2010) emphasises the importance of research pathways that use methods and methodologies in a way that favours the rights, interests and values of marginalised and excluded people.

Participatory methods are increasingly being used in action research focused on transformation (Chapter 15). This includes processes described elsewhere in this book, such as facilitated dialogues (Chapter 9) and scenario development (Chapter 11). Arts-based methods are also increasingly being incorporated in participatory data collection processes as a way of deepening conversations with research participants towards learning and transformation (Bennett et al. 2016).

Case study 8.1: Understanding the role of sense of place in landscape dynamics in South Africa

Masterson, Mahajan and Tengö (2018) sought to understand the ways in which people in the former Transkei, in the Eastern Cape province of South Africa, perceive social and ecological changes in the landscape over time and how these changes have influenced their well-being. The former Transkei homeland (a homeland refers to an area established during the apartheid era to which black South Africans were forcibly removed under the policy of separate development) was historically a place of small-scale farming, supported by remittance wages from migrant family members. However, the area has witnessed a long-term decline in cultivation and animal husbandry coupled with bush encroachment. Today, the region remains underdeveloped with high rates of outmigration to urban areas and a heavy reliance on social welfare grants.

The research team (see Masterson 2016; Masterson, Mahajan, and Tengö 2018) explored rural residents' experiences of and responses to declining subsistence agriculture and continued labour migration, through a lens of sense of place. To gain a deep understanding of local perspectives on well-being and to overcome cultural differences and language barriers, the researchers settled on photovoice as their main methodology.

Figure 8.1 The kraal (© M. Bili 2013) 'That is the kraal [cattle byre], but you can see there inside, there's no manure, which clearly shows that they have no livestock there at that home. But even if you haven't got livestock, it's important to have a kraal at home, because that is a place that you need when you perform your rituals. As people we have different perspectives. Some people are in the cities and have jobs and earn much money. For those they see having livestock as something unimportant.' – M. Bili, photovoice exercise, 2013, Gqunqe, South Africa

In this study, four demographic groups of between three and six individuals used small digital cameras to capture aspects of the local landscape that were important to them. Each group convened multiple times (between four and six times) over five weeks. After basic training in visual literacy, individuals in the groups took photographs that represented their lived experience in the rural village landscape. Each participant selected the most important images, which were printed. These photographs were then either narrated or captioned by the photographers. This process formed the basis of in-depth discussions in the group, all of which were recorded with the free, prior and informed consent of the participants. All four groups chose to make a poster based on an issue that the group had identified, and displayed this in public places around the villages. Photographs, captions and translated transcriptions of the discussions formed the data for a thematic qualitative analysis by the researchers.

In this case study, photovoice provided an inventory of human–nature relationships highlighting the diversity of ways in which ecosystems influenced people's well-being, despite a low economic reliance on these resources. Participants' photographs illustrated the often hidden cultural and non-monetised connections that people have to an agricultural lifestyle. This was important for maintaining subjective aspects

Figure 8.2 Collecting firewood (© N. Zibonele 2013) 'This is our way of living. You have to go to the forest and come back with a headbundle of firewood to use at home. That's where women go to get firewood. It's good and bad to have forests. There in the forest as *umama* going to collect firewood, you can meet a rapist hiding there. There are times when you will feel happy in the forest – when it's very hot and you enjoy the shade of the trees and rest there. But to be alone in the forest is not safe.' – N. Zibonele, photovoice exercise, 2013, Nobuswana, South Africa

(*Continued*)

of well-being and was a strong motivator to continue farming practices despite the changing livelihood portfolios of local people (Figure 8.1).

Photographs and discussions also pointed to the role of nature in the well-being of different groups in the community. This was particularly because the method of taking photographs was able to engage individuals who might be overlooked in focus groups or written surveys. Young women, for example, do not have many opportunities to represent themselves in this patriarchal cultural context, but through photovoice they could represent their concerns and be the experts on their lived experience, illustrating gender-specific perspectives on the landscape that affected, for example, women's sense of safety (Figure 8.2).

Through their photographs and discussions, participants also communicated their knowledge of the dynamics of the SES. Photographs of landscape elements demonstrated their culturally contingent sense-making of the complexity of the interacting drivers of field abandonment, bush encroachment and migration. Most importantly, photovoice encouraged mobilisation of tacit knowledge and reflection on these insights through group discussion.

Key readings

Lynam, T., W. de Jong, D. Sheil, T. Kusumanto, and K. Evans. 2007. 'A Review of Tools for Incorporating Community Knowledge, Preferences, and Values into Decision Making in Natural Resources Management.' *Ecology and Society* 12(1): 5.

Newing, H. 2010. *Conducting Research in Conservation: Social Science Methods and Practice*. Abingdon: Routledge.

Rambaldi, G., R. Chambers, M. McCall, and J. Fox. 2006. 'Practical Ethics for PGIS Practitioners, Facilitators, Technology Intermediaries and Researchers.' *Participatory Learning and Action* 54(1): 106–113.

Schreckenberg, K., C.A. Torres-Vitolas, S. Willcock, C. Shackleton, C.A. Harvey, and D. Kafumbata. 2016. 'Participatory Data Collection for Ecosystem Services Research: A Practitioner's Manual.' *ESPA Working Paper Series* 3.

Acknowledgements

The authors would like to thank the participant photographers in the photovoice study for their generous hospitality and willingness to co-create this research. They are also grateful to Faith Mabusela for assistance in the field. Alta de Vos is supported by a Rhodes Council Grant. Rika Preiser is funded by the USAID Resilient Waters Program (contract no. 72067418C00007). Vanessa Masterson acknowledges support from the Swedish Research Council (VR 2017-06337).

References

Aburto, J., G. Gallardo, W. Stotz, C. Cerda, C. Mondaca-Schachermayer, and K. Vera. 2013. 'Territorial User Rights for Artisanal Fisheries in Chile – Intended and Unintended Outcomes.' *Ocean and Coastal Management* 71: 284–295.

Armatas, C., T. Venn, and A. Watson. 2017. 'Understanding Social-Ecological Vulnerability with Q-Methodology: A Case Study of Water-based Ecosystem Services in Wyoming, USA.' *Sustainability Science* 12(1): 105–121.

Armitage, D., F. Berkes, A. Dale, E. Kocho-Schellenberg, and E. Patton. 2011. 'Co-management and the Co-Production of Knowledge: Learning to Adapt in Canada's Arctic.' *Global Environmental Change* 21(3): 995–1004.

Bagnoli, A. 2009. 'Beyond the Standard Interview: The Use of Graphic Elicitation and Arts-based Methods.' *Qualitative Research* 9(5): 547–570.

Barreteau, O., F. Bousquet, M. Étienne, V. Souchère, and P. d'Aquino. 2014. 'Companion Modelling: A Method of Adaptive and Participatory Research.' In *Companion Modelling*, edited by M. Étienne, 13–40. Dordrecht: Springer.

Beilin, R. 2005. 'Photo-elicitation and the Agricultural Landscape: 'Seeing' and 'Telling' about Farming, Community and Place.' *Visual Studies* 20(1): 56–68.

Belay, M. 2012. 'Participatory Mapping, Learning and Change in the Context of Biocultural Diversity and Resilience.' PhD diss., Rhodes University.

Bennett, N.J., J. Blythe, S. Tyler, and N.C. Ban. 2016. 'Communities and Change in the Anthropocene: Understanding Social-Ecological Vulnerability and Planning Adaptations to Multiple Interacting Exposures.' *Regional Environmental Change* 16(4): 907–926.

Berbés-Blázquez, M. 2012. 'A Participatory Assessment of Ecosystem Services and Human Wellbeing in Rural Costa Rica Using Photo-voice.' *Environmental Management* 49(4): 862–875.

Brown, G. 2017. 'A Review of Sampling Effects and Response Bias in Internet Participatory Mapping (PPGIS/PGIS/VGI).' *Transactions in GIS* 21(1): 39–56.

Brown, G., and M. Kyttä. 2018. 'Key Issues and Priorities in Participatory Mapping: Toward Integration or Increased Specialization?' *Applied Geography* 95: 1–8.

Brown, K., N. Eernstman, A.R. Huke, and N. Reding. 2017. 'The Drama of Resilience: Learning, Doing, and Sharing for Sustainability.' *Ecology and Society* 22(2): 8.

Bryan, J. 2015. 'Participatory Mapping.' In *The Routledge Handbook of Political Ecology*, edited by T.A. Perreault, G. Bridge, and J. McCarthy, 249–262. New York: Routledge.

Callaway, E. 2017. 'South Africa's San People Issue Ethics Code to Scientists.' *Nature News* 543(7646): 475–476.

Campbell, J. 2002. 'A Critical Appraisal of Participatory Methods in Development Research.' *International Journal of Social Research Methodology* 5(1): 19–29.

Chambers, R. 1994. 'The Origins and Practice of Participatory Rural Appraisal.' *World Development* 22(7): 953–969.

Cockburn, J., C. Palmer, H. Biggs, and E. Rosenberg. 2018. 'Navigating Multiple Tensions for Engaged Praxis in a Complex Social-Ecological System.' *Land* 7(4): 129.

Cornwall, A., and R. Jewkes. 1995. 'What is Participatory Research?' *Social Science and Medicine* 41(12): 1667–1676.

Fischer, A., and A. Eastwood. 2016. 'Coproduction of Ecosystem Services as Human–Nature Interactions – An Analytical Framework.' *Land Use Policy* 52: 41–50.

Forrester, J., B. Cook, L. Bracken, S. Cinderby, and A. Donaldson. 2015. 'Combining Participatory Mapping with Q-Methodology to Map Stakeholder Perceptions of Complex Environmental Problems.' *Applied Geography* 56: 199–208.

Frey, J., and F. Berkes. 2014. 'Can Partnerships and Community-based Conservation Reverse the Decline of Coral Reef Social-Ecological Systems?' *International Journal of the Commons* 8(1): 26–46.

Harper, D. 2002. 'Talking about Pictures: A Case for Photo Elicitation.' *Visual Studies* 17(1): 13–26.

Heras, M., and J. D. Tàbara. 2014. 'Let's Play Transformations! Performative Methods for Sustainability.' *Sustainability Science* 9(3): 379–398.

Hill, R., C. Grant, M. George, C.J. Robinson, S. Jackson, and N. Abel. 2012. 'A Typology of Indigenous Engagement in Australian Environmental Management: Implications for Knowledge Integration and Social-Ecological System Sustainability.' *Ecology and Society* 17(1): 23.

Hurlbert, M., and J. Gupta. 2015. 'The Split Ladder of Participation: A Diagnostic, Strategic, and Evaluation Tool to Assess When Participation is Necessary.' *Environmental Science and Policy* 50: 100–113.

Johansson, E.L., and E. Isgren. 2017. 'Local Perceptions of Land-use Change: Using Participatory Art to Reveal Direct and Indirect Socioenvironmental Effects of Land Acquisitions in Kilombero Valley, Tanzania.' *Ecology and Society* 22(1): 3.

Kalibo, H.W., and K.E. Medley. 2007. 'Participatory Resource Mapping for Adaptive Collaborative Management at Mount Kasigau, Kenya.' *Landscape and Urban Planning* 82(3): 145–158.

Kong, T.M., K. Kellner, D.E. Austin, Y. Els, and B.J. Orr. 2015. 'Enhancing Participatory Evaluation of Land Management through Photo Elicitation and Photovoice.' *Society and Natural Resources* 28(2): 212–229.

Laborde, S., A. Fernández, S.C. Phang, I.M. Hamilton, N. Henry, H.C. Jung, A. Mahamat et al. 2016. 'Social-Ecological Feedbacks Lead to Unsustainable Lock-in in an Inland Fishery.' *Global Environmental Change* 41: 13–25.

Leach, M., I. Scoones, and A. Stirling. 2010. 'Governing Epidemics in an Age of Complexity: Narratives, Politics and Pathways to Sustainability.' *Global Environmental Change* 20(3): 369–377.

LeCompte, M., and J. Schensul. 2010. *Designing and Conducting Ethnographic Research: An Introduction* (2nd ed). New York: Rowman Altamira.

Lemelin, R.H., E.C. Wiersma, L. Trapper, R. Kapashesit, M.S. Beaulieu, and M. Dowsley. 2013. 'A Dialogue and Reflection on Photohistory: Engaging Indigenous Communities in Research through Visual Analysis.' *Action Research* 11(1): 92–107.

Levine, A. S., and C.L. Feinholz. 2015. 'Participatory GIS to Inform Coral Reef Ecosystem Management: Mapping Human Coastal and Ocean Uses in Hawaii.' *Applied Geography* 59: 60–69.

Liamputtong, P., and J. Rumbold. 2008. *Knowing Differently: Arts-based and Collaborative Research Methods*. New York: Nova Science Publishers.

Lynam, T., W. de Jong, D. Sheil, T. Kusumanto, and K. Evans. 2007. 'A Review of Tools for Incorporating Community Knowledge, Preferences, and Values into Decision Making in Natural Resources Management.' *Ecology and Society* 12(1): 5.

Maclean, K., and E. Woodward. 2013. 'Photovoice Evaluated: An Appropriate Visual Methodology for Aboriginal Water Resource Research.' *Geographical Research* 51(1): 94–105.

Malmborg, K., H. Sinare, E. Enfors-Kautsky, I. Ouedraogo, and L.J. Gordon. 2018. 'Mapping Regional Livelihood Benefits from Local Ecosystem Services Assessments in Rural Sahel.' *PLoS ONE* 13(2): e0192019.

Masterson, V.A. 2016. 'Sense of Place and Culture in the Landscape of Home: Understanding Social-Ecological Dynamics on the Wild Coast, South Africa.' PhD diss., Stockholm University.

Masterson, V., S. Mahajan, and M. Tengö. 2018. 'Photovoice for Mobilizing Insights on Complex Social-Ecological Dynamics – Case Studies from Kenya and South Africa.' *Ecology and Society* 23(3): 13.

Mehryar, S., R. Sliuzas, A. Sharifi, D. Reckien, and M. van Maarseveen. 2017. 'A Structured Participatory Method to Support Policy Option Analysis in a Social-Ecological System.' *Journal of Environmental Management* 197: 360–372.

Milcu, A.I., K. Sherren, J. Hanspach, D. Abson, and J. Fischer. 2014. 'Navigating Conflicting Landscape Aspirations: Application of a Photo-Based Q-Method in Transylvania (Central Romania).' *Land Use Policy* 41: 408–422.

Mitchell, C., and N. de Lange. 2011. 'Community-based Participatory Video and Social Action in Rural South Africa.' In *The SAGE Handbook of Visual Research Methods*, edited by E. Margolis and L. Pauwels, 171–183. Los Angeles: Sage Publications.

Moerlein, K.J., and C. Carothers. 2012. 'Total Environment of Change: Impacts of Climate Change and Social Transitions on Subsistence Fisheries in Northwest Alaska.' *Ecology and Society* 17(1): 10.

Murray, G., L. D'Anna, and P. MacDonald. 2016. 'Measuring What We Value: The Utility of Mixed Methods Approaches for Incorporating Values into Marine Social-Ecological System Management.' *Marine Policy* 73: 61–68.

Nel, J.L., D.J. Roux, A. Driver, L. Hill, A.C. Maherry, K. Snaddon, C.R. Petersen, L.B. Smith-Adao, H. van Deventer, and B. Reyers. 2016. 'Knowledge Co-production and Boundary Work to Promote Implementation of Conservation Plans.' *Conservation Biology* 30(1): 176–188.

Newing, H., C.M. Eagle, R.K. Puri, and C. W. Watson. 2011. *Conducting Research in Conservation: Social Science Methods and Practice*. Abingdon: Routledge.

Nyirenda, V.R., and J. Drive. 2015. 'Role of Relational Social Capital in Transforming Conservation Inequalities and Conflicts to Sustainable Solutions in Developing Countries Chansa Chomba.' *International Journal of Sustainable Development* 18(3): 229–246.

Olson, R., J. Hackett, and S. DeRoy. 2016. 'Mapping the Digital Terrain: Towards Indigenous Geographic Information and Spatial Data Quality Indicators for Indigenous Knowledge and Traditional Land-use Data Collection.' *The Cartographic Journal* 53(4): 348–355.

Pain, R. 2004. 'Social Geography: Participatory Research.' *Progress in Human Geography* 28(5): 652–663.

Palomo, I., B. Martín-López, C. López-Santiago, and C. Montes. 2011. 'Participatory Scenario Planning for Protected Areas Management under the Ecosystem Services Framework: The Doñana Social-Ecological System in Southwestern Spain.' *Ecology and Society* 16(1): 23.

Pink, S. 2011. 'Images, Senses and Applications: Engaging Visual Anthropology.' *Visual Anthropology* 24(5): 437–454.

Preiser, R., R. Biggs, A. de Vos, and C. Folke. 2018. 'Social-Ecological Systems as Complex Adaptive Systems: Organizing Principles for Advancing Research Methods and Approaches.' *Ecology and Society* 23(4): 46.

Rambaldi, G., R. Chambers, M. McCall, and J. Fox. 2006. 'Practical Ethics for PGIS Practitioners, Facilitators, Technology Intermediaries and Researchers.' *Participatory Learning and Action* 54(14): 106–113.

Rambaldi, G., J. Muchemi, N. Crawhall, and L. Monaci. 2007. 'Through the Eyes of Hunter-Gatherers: Participatory 3D Modelling among Ogiek Indigenous Peoples in Kenya.' *Information Development* 23(2–3): 113–128.

Ramirez-Gomez, S.O.I, G. Brown, and A. Tjon Sie Fat. 2013. 'Participatory Mapping with Indigenous Communities for Conservation: Challenges and Lessons from Suriname.' *The Electronic Journal of Information Systems in Developing Countries* 58(1): 1–22.

Ramirez-Gomez, S.O.I., P. Verweij, L. Best, R. van Kanten, G. Rambaldi, and R. Zagt. 2017. 'Participatory 3D Modelling as a Socially Engaging and User-useful Approach in Ecosystem Service Assessments among Marginalized Communities.' *Applied Geography* 83: 63–77.

Raymond, C.M., B.A. Bryan, D.H. MacDonald, A. Cast, S. Strathearn, A. Grandgirard, and T. Kalivas. 2009. 'Mapping Community Values for Natural Capital and Ecosystem Services.' *Ecological Economics* 68(5): 1301–1315.

Reed, M.S. 2008. 'Stakeholder Participation for Environmental Management: A Literature Review.' *Biological Conservation* 141(10): 2417–2431.

Robinson, C.J., K. Maclean, R. Hill, E. Bock, and P. Rist. 2016. 'Participatory Mapping to Negotiate Indigenous Knowledge used to Assess Environmental Risk.' *Sustainability Science* 11(1): 115–126.

Roux, D.J., J.L. Nel, G. Cundill, P. O'Farrell, and C. Fabricius. 2017. 'Transdisciplinary Research for Systemic Change: Who to Learn With, What to Learn About and How to Learn.' *Sustainability Science* 12(5): 711–726.

Rust, N.A. 2017. 'Can Stakeholders Agree on How to Reduce Human–Carnivore Conflict on Namibian Livestock Farms? A Novel Q-Methodology and Delphi Exercise.' *Oryx* 51(2): 339–346.

Samuelsson, K., M. Giusti, G.D. Peterson, A. Legeby, S.A. Brandt, and S. Barthel. 2018. 'Impact of Environment on People's Everyday Experiences in Stockholm.' *Landscape and Urban Planning* 171: 7–17.

Schreckenberg, K., C.A. Torres-Vitolas, S. Willcock, C. Shackleton, C.A. Harvey, and D. Kafumbata. 2016. 'Participatory Data Collection for Ecosystem Services Research: A Practitioner's Manual.' *ESPA Working Paper Series* 3.

Scoones, I., A. Stirling, D. Abrol, J. Atela, and L. Charli-Joseph. 2018. 'Transformations to Sustainability.' *STEPS Working Paper* 104. Brighton: STEPS Centre.

Sieber, S.S., P.M. Medeiros, and U.P. Albuquerque. 2011. 'Local Perception of Environmental Change in a Semi-Arid Area of Northeast Brazil: A New Approach for the Use of Participatory Methods at the Level of Family Units.' *Journal of Agricultural and Environmental Ethics* 24(5): 511–531.

Sinare, H., L.J. Gordon, and E. Enfors-Kautsky. 2016. 'Assessment of Ecosystem Services and Benefits in Village Landscapes – A Case Study from Burkina Faso.' *Ecosystem Services* 21: 141–152.

Smith, B.M., P. Chakrabarti, A. Chatterjee, S. Chatterjee, U.K. Dey, L.V. Dicks, B. Giri, S. Laha, R.K. Majhi, and P. Basu. 2017. 'Collating and Validating Indigenous and Local Knowledge to Apply Multiple Knowledge Systems to an Environmental Challenge: A Case-Study of Pollinators in India.' *Biological Conservation* 211: 20–28.

Stirling, A. 2008. 'Opening Up and Closing Down.' *Science, Technology, and Human Values* 33(2): 262–294.

Stringer, L.C., A.J. Dougill, E. Fraser, K. Hubacek, C. Prell, and M.S. Reed. 2006. 'Unpacking "Participation" in the Adaptive Management of Social-Ecological Systems: A Critical Review.' *Ecology and Society* 11(2): 39.

Sylvester, O., A.G. Segura, and I.J. Davidson-Hunt. 2016. 'Wild Food Harvesting and Access by Household and Generation in the Talamanca Bribri Indigenous Territory, Costa Rica.' *Human Ecology* 44(4): 449–461.

Tengö, M., E.S. Brondizio, T. Elmqvist, P. Malmer, and M. Spierenburg. 2014. 'Connecting Diverse Knowledge Systems for Enhanced Ecosystem Governance: The Multiple Evidence Base Approach.' *Ambio* 43(5): 579–591.

Tengö, M., R. Hill, P. Malmer, C.M. Raymond, M. Spierenburg, F. Danielsen, T. Elmqvist, and C. Folke. 2017. 'Weaving Knowledge Systems in IPBES, CBD and Beyond – Lessons Learned for Sustainability.' *Current Opinion in Environmental Sustainability* 26–27: 17–25.

Tengö, M., P. Malmer, P. Borraz, C. Cariño, J. Cariño, T. Gonzales, J. Ishizawa et al. 2012. Dialogue Workshop on Knowledge for the 21st Century: Indigenous Knowledge, Traditional Knowledge, Science and Connecting Diverse Knowledge Systems. Usdub, Guna Yala, Panama, 10–13 April 2012. Workshop Report. Stockholm: Stockholm Resilience Centre.

Van der Riet, M., and M. Boettiger. 2009. 'Shifting Research Dynamics: Addressing Power and Maximising Participation through Participatory Research Techniques in Participatory Research.' *South African Journal of Psychology* 39(1): 1–18.

Villamor, G.B., I. Palomo, C.A.L. Santiago, E. Oteros-Rozas, and J. Hill. 2014. 'Assessing Stakeholders' Perceptions and Values towards Social-Ecological Systems Using Participatory Methods.' *Ecological Processes* 3(1): 22.

Walker, G. 2012. 'Climate Change Oppression: Media Production as the Practice of Freedom.' *The Journal of Sustainable Development* 9(1): 97–106.

Wang, C., and M.A. Burris. 1994. 'Empowerment through Photo Novella: Portraits of Participation.' *Health Education and Behaviour* 21: 171–186.

Wang, C., and M.A. Burris. 1997. 'Photovoice: Concept, Methodology, and Use for Participatory Needs Assessment.' *Health, Education and Behaviour* 24(3): 369–386.

Watts, S., and P. Stenner. 2012. *Doing Q Methodological Research: Theory, Method and Interpretation*. London: Sage.

West, S., R. Cairns, and L. Schultz. 2016. 'What Constitutes a Successful Biodiversity Corridor? A Q-study in the Cape Floristic Region, South Africa.' *Biological Conservation* 198: 183–192.

Weyer, D., J. Bezerra, and A. de Vos. 2019. 'Participatory Mapping in a Developing Country Context: Lessons from South Africa.' *Land* 8(9): 134.

Williams, T., and P. Hardison. 2013. 'Culture, Law, Risk and Governance: Contexts of Traditional Knowledge in Climate Change Adaptation.' *Climatic Change* 120(3): 531–544.

Methods for knowledge co-production and effecting system change

9

Facilitated dialogues

Scott Drimie,[1,2] *Colleen Magner,*[3] *Laura Pereira,*[2,5]
Lakshmi Charli-Joseph,[4] *Michele-Lee Moore,*[5] *Per Olsson,*[5]
Jesús Mario Siqueiros-Garcia[4,6] *and Olive Zgambo*[2]

[1]THE SOUTHERN AFRICA FOOD LAB, SOUTH AFRICA
[2]CENTRE FOR SUSTAINABILITY TRANSITIONS, STELLENBOSCH UNIVERSITY, STELLENBOSCH, SOUTH AFRICA
[3]REOS PARTNERS, SOUTH AFRICA
[4]LABORATORIO NACIONAL DE CIENCIAS DE LA SOSTENIBILIDAD (LANCIS), INSTITUTO DE ECOLOGÍA, UNIVERSIDAD NACIONAL AUTÓNOMA DE MÉXICO, MEXICO CITY, MEXICO
[5]STOCKHOLM RESILIENCE CENTRE, STOCKHOLM UNIVERSITY, STOCKHOLM, SWEDEN
[6]INSTITUTO DE INVESTIGACIONES EN MATEMÁTICAS APLICADAS Y EN SISTEMAS (IIMAS), UNIVERSIDAD NACIONAL AUTÓNOMA DE MÉXICO, MEXICO CITY, MEXICO

Key methods discussed in this chapter

Appreciative enquiry, change labs, social innovation labs, the circle, Theory U, T-Labs, scenarios, world café, learning journeys, listening projects, dialogue interviewing

Connections to other chapters

Facilitated dialogues often draw on action research to enable deeper learning while also providing opportunity for action research to interrogate dialogue-into-action (Chapter 15). They also often make use of visioning or imagining the future to liberate participants to transcend the constraints of the present and so explore potential pathways to a better outcome. There is a strong link to the discussions on futures analysis in Chapter 10 and scenario development in Chapter 11.

Introduction

There is a growing body of experience surrounding the design of social dialogue to create spaces that can be used to enable transformation (Pohl et al. 2010; Fazey et al. 2018; Naumann et al. 2018; Pereira et al. 2018b; Schäpke et al. 2018; Pereira et al. 2019, 2020). As one key intention, they 'seek to generate social-ecological innovations aimed at challenging and changing existing roles and routines, power dynamics, relations among groups and networks, resource flows, as well as meaning and values (and culture) across different contexts and scales' (Schäpke et al. 2018, 91).

 DOI: 10.4324/9781003021339-12

SUMMARY TABLE: FACILITATED DIALOGUES	
DISCIPLINARY BACKGROUND	**KNOWLEDGE TYPE**
The methods in this chapter are derived from or have most commonly been used in: Sustainability Science, Public Leadership, Sociology, Geography	*The methods in this chapter are primarily used to generate the following types of knowledge:* • Exploratory • Explanatory
RESEARCH APPROACH	**PURPOSE OF METHOD**
The methods in this chapter originate from or most commonly adopt the following research approaches: • Interpretive/subjective • Collaborative/process	*The most common purposes of using the methods in this chapter are:* • Stakeholder engagement and co-production • Policy/decision support
TEMPORAL DIMENSION	**SYSTEMIC FEATURES AND PROCESSES**
The methods in this chapter are most commonly applied to the following temporal dimensions: • Present (typically within the last 5–10 years) • Recent past (post-1700s) • Future **SPATIAL DIMENSION** *The methods in this chapter are primarily either or both:* • Non-spatial *The methods in this chapter are most commonly applied at the following spatial scales:* • Local • Regional (provincial/state to continental) • Multiple places/sites around the world	*While most methods can do many things, the methods in this chapter are particularly good (i.e. go-to methods) for addressing the following:* • SES components and linkages • Diversity • Power relations • Multiple scales and levels or cross-level interactions • Social-ecological interactions over time • Transformation • Social learning • Collective action and collaborative governance • Exploring uncertainty

In response to the growing number of complex social-ecological challenges today – those shining a spotlight on unsustainable trends and the realities of inequality and injustice – many are calling for transformation (Feola 2015; Blythe et al. 2018). For scholars, engaging in or supporting attempts to deliberately push for or navigate transformation necessitates moving into a new kind of transdisciplinarity; that is, action-oriented knowledge generation that is co-produced and interwoven with multiple knowledge(s) (Tengö et al. 2014; Pereira et al. 2019). The complexity of global challenges and their increasing interdependence demand solutions that transcend individual actors, specialists, sectors or disciplines and draw on a wide range of perspectives. Dialogue between actors in a complex social-ecological systems (SES) can contribute to the institutional change required for deep social innovation and transformation of that system (Mair and Hehenberger 2014). However, this demands a specific type of facilitation, one that enables dialogue, sense-making, reflection and reflexive learning, while supporting the reframing of issues in ways that allow solutions – or at the very least, attempt to experiment and transform – to be co-created and co-realised (Sharpe et al. 2016). In this way, we use 'dialogue' as an umbrella term to include a suite of facilitated processes that might also be used during processes for knowledge co-production, participatory action research, lab processes, future search processes and more.

Interventions that are driven by 'top-down' directives are widely recognised as problematic since they often leave stakeholders feeling that the project outcomes are imposed from the outside. Strategies like these are often met with resistance and fail because they are inappropriate in addressing contextual sensibilities and do not convey a sense of ownership from stakeholders during the decision-making process (Freeth and Drimie 2016). In contrast, facilitated dialogues are carefully designed processes. They aim to support multi-stakeholder groups in addressing complex SES problems through the creation of 'safe' or 'safe enough' spaces where SES innovations can be fostered and developed (Pereira et al. 2019, 2020).

The most common dictionary definition of a dialogue is simply a conversation between two or more people. A much deeper and more distinct meaning is provided by David Bohm (Bohm 1996), who interrogated the source of the word: it is derived from the Greek root *dia*, which means 'through', and *logos*, which means 'the word' or 'meaning'. By drawing on this deeper interpretation of the etymological origins of the word, we see that the definition can be expanded to include an emphasis on questions, inquiry, co-creation and listening to uncover one's own assumptions and those of others while suspending judgement and pursuing a collective search for truth. These qualities shape and frame the conditions for dialogue to take place in a more reflective and dynamic manner. Greater inquiry into the viewpoints of others helps to develop greater understanding of those others and creates an opportunity to adopt new ways of thinking. When assumptions are explored, participants can challenge their own ideas and recognise bias and thought patterns that influence – and possibly inhibit – lively engagement. Successful dialogue often requires a skilled facilitator who is capable of navigating the tensions between different viewpoints carried by the participants in the dialogue (Drimie et al. 2018).

Dialogue is an emergent and generative communicative interaction between actors, moving well beyond an exchange of information to include the building of relationships. These relationships are both a means to address a challenge and an end in themselves. In other words, dialogue is inherently relational and as such requires a critical mass of different actors who need to relate to one another, sometimes under tense circumstances. A diversity of perspectives in dialogue processes potentially allows for the emergence of innovation when different people from different backgrounds, with their own ideas and creativity, interact to address challenges (Drimie et al. 2018; Pereira et al. 2020). This needs to be thoughtfully constructed in order to reach some kind of coherent outcome (Pereira et al. 2020).

Deciding on what a successful outcome might be is a critical first step before a dialogue is convened. The specific details are not important. What is important is being clear about the intentions of the dialogue before bringing people together and deciding which tools to use. The following questions are important. Why are we bringing this group of people together? What purpose lies behind this specific process of change? What is this whole thing about?

Most, if not all, the tools presented in Table 9.1 are underpinned by the essential principle of clarity of purpose. Building on this, the tools focus on enabling open communication, honest speaking and genuine listening. They allow people to take responsibility for their own learning and ideas. They create a safe space or 'container' for people to voice their assumptions, question their previous perceptions, judgements and worldviews, and change the way they think. The tools can generate new ideas or solutions that go beyond what anyone had thought of before, and create a different level of understanding of people and problems. They allow for more contextual and holistic ways of seeing.

SES problems and questions

To give an example of where facilitated dialogue has been used to address challenges in SES, a recent paper on social-ecological transformations analysed nine projects ranging from the Xochimilco Wetland in Mexico and Mopani farmers in South Africa, to Argentinian seeds and agro-ecology in Soweto, South Africa, to peri-urban South Asia, Mombasa in Kenya and Cabo Delgado in Mozambique (Pereira et al. 2019). Facilitated dialogue involves a range of methods since these dialogues for transformation are intervention processes that require thorough planning but are still flexible enough to allow 'emergence and the unexpected to occur' (Pereira et al. 2018b). Ideally, the form a facilitated dialogue will take depends on the local context and the people involved (Feola and Butt 2017). The following are some of the conditions under which a facilitated dialogue may be deemed a useful method for generating new solutions and experiments that aim to have a transformative impact (Westley et al. 2013; Westley et al. 2015; Ely and Marin 2016):

- There is a complex SES challenge to address, in which impact is difficult to achieve, sense-making is needed, but agreement has emerged that 'business as usual' is no longer an option (Moore et al. 2014; Westley et al. 2015; Olsson et al. 2017).
- A diverse group of participants with the potential for transformative agency exists and can move any new idea or process forward and source the necessary resources (Moore and Westley 2011; Westley et al. 2013).
- There is an identifiable action-oriented outcome as the end goal of the process, as opposed to just a product or 'thing' (Pereira et al. 2019).
- There is a motivated convenor who is willing to invest the resources needed for the process (Westley et al. 2015; Drimie et al. 2018).
- There has been no successful implementation of alternative innovations that counter the dominant way of doing things (Westley, McGowan, and Tjörnbo 2017).
- There exist noticeable shifts in the culture or economic or political context that can serve as potential windows of opportunity for facilitated dialogue to take place and be effective (Olsson, Folke, and Hahn 2004; Gelcich et al. 2010; Westley et al. 2013).

Facilitated dialogue presents opportunities for researchers to explore the intersection of action and analysis, where they navigate the fine line between actively intervening in processes to enable change and also being able to provide a critical analysis of the types of changes that are occurring. Some researchers are finding themselves to be 'transformative space-makers'

(Marshall, Dolley, and Priya 2018). By facilitating dialogues, researchers have been able to open up a space for productive collaboration and interaction between diverse stakeholders with the intention that there may be actionable outcomes with which policy and other decision-making actors can engage (Tengö et al. 2017; Marshall, Dolley, and Priya 2018).

Brief description of key methods

As facilitated processes, dialogues typically combine a number of different participatory methods over several workshops and experiments, based on the requirements of the group and the complex question being examined. The tools that are presented in Table 9.1 draw largely from an important book reflecting on dialogue entitled *Mapping Dialogue: Essential Tools for Social Change* (Bojer et al. 2008). This is a key resource for more detailed descriptions of each tool.

A key skill in facilitating dialogue lies in the ability to combine different methods to achieve the desired space that is most conducive to realising the intended objectives of the interaction. A skilled facilitator needs to be able to take the group on a journey, reaching specific landmarks along the way. The choice of method and the order in which it is undertaken lie at the core of this and are largely intuitive.

Table 9.1 Summary of key methods used in facilitated dialogues

Method	*Description*	*References*
Appreciative inquiry	Appreciative inquiry is an approach and process that turns problem solving on its head. Instead of finding the best ways to solve a pressing problem, it places the focus on identifying the best of what's already present in an organisation or community and finding ways of enhancing this to pursue dreams and possibilities of what could be.	*Key introductory texts* Barrett and Fry 2005; Cooperrider, Whitney, and Stavros 2008 *Applications to SES* Whitney and Trosten-Bloom 2003; appreciative-inquiry.org; imaginechicago.org
Theory U	Theory U is a multi-stakeholder dialogic change process that places emphasis on 'presencing' the internal consciousness and how it affects the ways in which we engage with the complex systems around us. It is designed to generate collective insight, shared commitment and the creative capacities needed to address complex problems.	*Key introductory texts* Senge et al. 2004; Kahane 2004; Scharmer 2008 *Applications to SES* The Sustainable Food Laboratory SFL: sustainablefoodlab.org
Social innovation labs and transformation labs (T-Labs)	Social innovation labs combine insights from group psychology and whole-system processes, along with social innovation theory, with techniques and tools from design thinking to create a new framework for both invention and institutionalisation of social innovations. T-Labs advanced this thinking, adding an emphasis on SES dynamics while exploring the plurality of pathways that contribute to sustainability transformations.	*Key introductory texts* Westley et al. 2015; The Pathways Network 2018; Pereira et al. 2020 *Applications to SES* Charli-Joseph et al. 2018; Van Zwanenberg et al. 2018

Method	*Description*	*References*
The circle	At its most essential level, the circle is a participatory format that allows a group of people to slow down, practise deep listening and truly think together. When practised fully, it can be a physical embodiment of the root of the word dialogue: 'meaning flowing through'.	*Key introductory texts* Baldwin 1994; fromthefourdirections.org; peerspirit.com *Applications to SES* Kufunda Village (a learning centre focusing on rural community development in Zimbabwe): kufunda.org
Scenarios	Scenarios are possible and plausible pictures of the future that can be developed in a variety of ways (see Chapter 11). Participatory scenarios are created through a series of conversations in which a group of people invent and consider several varied stories about how the future may unfold. Ideally, these stories should be carefully researched and full of detail, be able to expose new understandings and hold some surprises. Scenarios can be powerful tools for challenging current assumptions about the world. In doing so, they lift the barriers that constrain our own creativity and understanding about the future.	*Key introductory texts* Schwartz 1991; Senge et al. 2004; Van der Heijden 2005; Kahane 2004, 2012 *Applications to SES* The Future of Food: southernafricafoodlab.org/transformative-scenario-planning; Pereira et al. 2018a; Freeth and Drimie 2016
World café	The world café is an intentional way to create a living network of conversations about questions that matter. It is a methodology that enables (12 to 1 200) people to think together and intentionally create new, shared meaning and collective insight.	*Key introductory text* Brown, Isaacs, and The World Café Community 2005 *Applications to SES* collectivewisdominitiative.com/papers/pioneers_dialogue/13_world.pdf
Learning journeys	Learning journeys are about getting away from behind one's desk, out of one's comfort zone, conference rooms and hotels. They are physical journeys from one place to another, intended to explore and experience the world first-hand.	*Key introductory text* reospartners.com/tools/learning-journeys *Applications to SES* Pereira et al. 2020; reospartners.com/projects/bhavishya-alliance-for-child-nutrition; reospartners.com/wp-content/uploads/old/bhavishya.pdf
Listening projects and dialogue interviewing	Listening projects and dialogue interviewing are methods that create an opportunity for asking meaningful questions, listening with an open mind and connecting to what another person is saying, to help that person uncover knowledge they may not even have known they had.	*Key introductory texts* dialogonleadership.org; listeningproject.info *Applications to SES* Drimie et al. 2018; alertademocratica.org/en; reospartners.com/moving-through-tough-terrain-the-role-of-hope/; reospartners.com/tools/dialogue-interviews

Case study 9.1: Food systems in the Western Cape, South Africa

A process of knowledge co-production and initiating social innovation was instituted as part of a research project in the Western Cape province of South Africa. The main aim of the process was to convene a facilitated dialogue called a transformation lab (T-Lab) to discuss challenges in the provincial food system seen from a range of diverse perspectives, and to originate potentially transformative innovations intended to shift the system onto a more desirable trajectory. This example shows how a number of different methods were merged to create a dynamic process with concrete results.

A transformative process values actor interaction. By connecting alternative food system actors and proponents, the dialogue created bridges by, for example, linking chefs to producers, restaurateurs to informal traders and academics to actual work on the ground. These processes were an opportunity for these actors to reimagine the ways in which food is produced, processed and consumed and potentially to become more embedded, sustainable and strategically aligned to influence the dominant food system. The facilitated dialogue consisted of two workshops.

Workshop 1

The first workshop was a 'safe space' for participants from across food systems, particularly in the Western Cape, with an interest or stake in these systems. The aim was to determine:

- The viability of linking alternative food actors into the mainstream without losing the integrity that makes them small-scale/alternative
- How to build relationships that enable alternative food systems to grow

Participants included chefs, researchers, artists, food activists, producers, retailers, food innovators, an anthropologist, a food scientist and an artisanal baker. Four researchers from the Centre for Sustainability Transitions at Stellenbosch University, the Southern Africa Food Lab and the Stockholm Resilience Centre facilitated the dialogue. Beforehand, a rapid survey was sent out to participants who had confirmed their attendance. The survey consisted of five open-ended questions focused on activities that the actors are involved in within the food system, their expectations of the dialogue and areas that they considered important intervention points. The intention was for these feedbacks to shape the process.

Workshop 2

The second dialogue was designed as a consolidation workshop and included both former and new participants (Figure 9.1). All participants from the first workshop were invited, as well as new contacts in wider networks. Participants included permaculture specialists, food and land activists, restaurateurs, urban farmers, a representative from the informal traders' association, researchers, an anthropologist and an indigenous food innovator.

As food systems are so complex, with a multitude of actors and underlying issues and outcomes, the dialogue built on a systems approach that integrates thinking, reviewing, reflecting and doing. The aim was to translate concrete coalitions and ideas into action through building relationships and commitment for the actors to drive change.

The second dialogue focused on strengthening the trust between participants in the emerging coalition of change, which would enable them to continue to define and implement breakthrough solutions. It sought to build on what was identified in the first workshop: ideas and actions pivoting on the intersections between niche, artisanal and fledgling projects intended to provide alternatives to the dominant food system so as to contribute to its disruption over time.

The consolidation workshop was based on three distinct movements that unfolded over two days. These were:

1. Sensing the system
2. Letting go (old ways of working)
3. Letting come (emerging innovation)

Two facilities were also available to support participants as they immersed themselves in activities:

- **The ideas room** was a physical space available to all participants at any time for deeper reflection. It contained coloured pens, wax crayons, playdough, water, seeds and images to form a food exhibition.
- **Collective food preparation** was also an important part of the process. This was done in a way that built an understanding of combining different foods, flavours and textures through experimentation and eating.

Figure 9.1 (A) Participants of a food systems dialogue in Stellenbosch during a plenary discussion, July 2016, and (B) participants during a speed-dating discussion at a T-Lab on transforming the Western Cape food system, May 2019 (© Laura Pereira)

Limitations

Many dialogue processes take a significant amount of time in terms of design, preparation and execution. Given that these processes can span multiple generations, there may not be 'immediate results' or 'evidence of transformation'. During and beyond the dialogue, there is a need for real-world experimentation, but the concept of 'experimenting' in areas of complex social-ecological challenges has impacts on people and the planet, often in ways that are unanticipated (Moore et al. 2018). In addition, in terms of SES work, facilitated dialogue may not necessarily help to identify critical tipping points (i.e. it will depend on the knowledge that emerges during the facilitated processes) or to reveal complex causality (see Schlüter et al. 2019). This raises questions of ethics and accountability within the process, particularly for researchers and facilitators who may 'see' those dynamics more clearly but are too top-down in their facilitation roles.

The facilitation processes discussed in Table 9.1 are not recipes for dialogue that should be applied universally, nor are they prescribed as specific tools for specific situations. Rather, each tool should be understood in terms of the context, story and impetus behind how these processes were developed. Similarly, tools can provide safety and comfort as they may help one to function in a complex world. A challenge is that a tool can become like a lens that affects how our surroundings are seen. If we wear only one lens all the time, our perception of the very thing we are trying to change may become distorted.

Resource implications

In terms of resource implications, at a minimum a design of a process, a facilitator and a physical space are required. Key participants involved in the exchange need to be convened with a facilitator in a place conducive to meaningful conversation and exchange as appropriate and within context. An effective facilitator with experience and integrity is necessary. Convening dialogue processes are active interventions in a system, designed for instigating change. Researchers should therefore allow time for careful reflection about what the potential implications of the dialogue process might be. In addition, all ethical considerations should be taken into account. The reflections from researchers who are increasingly using these methods for more action-oriented research are outlined in Pereira et al. (2019). It is strongly recommended to design these collaborative spaces in conjunction with key stakeholders and experienced facilitators. In addition to being a time-consuming process to set up and conduct, dialogues can also be very emotionally draining for the convenors as they seek to hold the space through all the highs and lows of group dynamics. It is therefore important to have a good team that can hold things together. Convening dialogues is not for everybody, but can be a really rewarding experience for those prepared to develop facilitation skills.

New directions

Variations on conventional approaches to collaboration are emerging, based on the acknowledgement that collaboration is not the only nor necessarily the best option in all situations. Choosing to collaborate is a pragmatic choice, often when high levels of complexity exist and where a clear way forward is not apparent. In conventional terms, collaboration assumes that the group first strives for agreement on what the problem is that they are trying to solve and identify a common purpose for what they need to do. In contrast, 'stretch collaboration' (Kahane 2017) is based on three propositions. The first stretch proposes that collaboration requires both conflict and connection rather than finding harmony. Second, stretch collaboration suggests that when the future is highly volatile and contested, we need to experiment our way forward. This is in contrast to conventional collaboration, which focuses on identifying a solution as quickly as possible and creating a plan to achieve it. The third stretch suggests that the group needs to change both themselves and others, and step into the game. In contrast, conventional collaboration tries to advocate for influencing others' actions – to change 'them'. Given the complex and uncertain nature of many SES challenges, along with the social conflict that transformations may unearth, we suggest that it is essential to take the insights from stretch collaboration forward into dialogue processes.

Key readings

Bojer, M.M., H. Roehl, M. Knuth, and C. Magner. 2008. *Essential Tools for Social Change.* Chagrin Falls: Taos Institute Publications.

Fuller Transformation Collaborative. 2019. *The Art of Systems Change: Eight Guiding Principles for a Green and Fair Future,* edited by B. Banerjee, K. Claborn, L. Gaskell, L. Glew, J. Griffin, P. Hovmand, S.L. Mahajan et al. Washington: World Wide Fund for Nature.

Kahane, A. 2017. *Collaborating with the Enemy: How to Work with People You Don't Agree with or Like or Trust.* Oakland: Berrett-Koehler Publishers.

Schwartz, T. 2018. 'What it Takes to Think Deeply About Complex Problems.' *Harvard Business Review* 9 May. https://hbr.org/2018/05/what-it-takes-to-think-deeply-about-complex-problems.

Westley, F., S. Laban, C. Rose, K. Robinson, K. McGowan, O. Tjörnbo, and M. Tovey. 2015. *Social Innovation Lab Guide.* Waterloo: University of Waterloo.

Acknowledgements

The authors would like to thank all those who have participated in their facilitated sessions over the year and from whom they have learnt so much. This work is based on research supported in part by the National Research Foundation of South Africa (grant 115300).

References

Baldwin, C. 1994. *Calling the Circle: The First and Future Culture.* New York: Bantam Books.

Barrett, F.J., and R.E. Fry. 2005. *Appreciative Inquiry: A Positive Approach to Building Cooperative Capacity.* Chagrin Falls: Taos Institute Publications.

Blythe, J., J. Silver, L. Evans, D. Armitage, N.J. Bennett, M.L. Moore, T.H. Morrison, and K. Brown. 2018. 'The Dark Side of Transformation: Latent Risks in Contemporary Sustainability Discourse.' *Antipode.* doi:10.1111/anti.12405.

Bohm, D. 1996. *On Dialogue.* New York: Routledge.

Bojer, M.M., H. Roehl, M. Knuth, and C. Magner. 2008. *Essential Tools for Social Change.* Chagrin Falls: Taos Institute Publications.

Brown, J., D. Isaacs, and The World Café Community. 2005. *The World Café: Shaping Our Futures Through Conversations That Matter.* San Francisco: Berrett-Koehler Publishers.

Charli-Joseph, L., J.M. Siqueiros-Garcia, H. Eakin, D. Manuel-Navarrete, and R. Shelton. 2018. 'Promoting Agency for Social-Ecological Transformation: A Transformation-Lab in the Xochimilco Social-Ecological System.' *Ecology and Society* 23(2): 46. doi:10.5751/ES-10214-230246.

Cooperrider, D.L., D. Whitney, and J.M. Stavros. 2008. *Appreciative Inquiry Handbook* (2nd ed). San Francisco: Berrett-Koehler Publishers.

Drimie, S., R. Hamann, A.P. Manderson, and N. Mlondobozi. 2018. 'Creating Transformative Spaces for Dialogue and Action: Reflecting on the Experience of the Southern Africa Food Lab.' *Ecology and Society* 23(3): 2. doi:10.5751/ES-10177-230302.

Ely, A., and A. Marin. 2016. 'Learning about 'Engaged Excellence' across a Transformative Knowledge Network.' *IDS Bulletin* 47(6). doi:10.19088/1968-2016.200.

Fazey, I., N. Schäpke, G. Caniglia, J. Patterson, J. Hultman, B. van Mierlo, F. Säwe et al. 2018. 'Ten Essentials for Action-Oriented and Second Order Energy Transitions, Transformations and Climate Change Research.' *Energy Research & Social Science* 40: 54–70. doi:10.1016/j.erss.2017.11.026.

Feola, G. 2015. 'Societal Transformation in Response to Global Environmental Change: A Review of Emerging Concepts.' *Ambio* 44(5): 376–390. doi:10.1007/s13280-014-0582-z.

Feola, G., and A. Butt. 2017. 'The Diffusion of Grassroots Innovations for Sustainability in Italy and Great Britain: An Exploratory Spatial Data Analysis.' *Geographical Journal* 183(1): 16–33. doi:10.1111/geoj.12153.

Freeth, R., and S. Drimie. 2016. 'Participatory Scenario Planning: From Scenario "Stakeholders" to Scenario "Owners."' *Environment: Science and Policy for Sustainable Development* 58(4): 32–43. doi:10.1080/00139157.2016.1186441.

Gelcich, S., T.P. Hughes, P. Olsson, C. Folke, O. Defeo, M. Fernández, S. Foale et al. 2010. 'Navigating Transformations in Governance of Chilean Marine Coastal Resources.' *Proceedings of the National Academy of Sciences of the United States of America* 107(39): 16794–99. doi:10.1073/pnas.1012021107.

Kahane, A. 2004. *Solving Tough Problems: An Open Way of Talking, Listening, and Creating New Realities.* San Francisco: Berrett-Koehler Publishers.

Kahane, A. 2012. *Transformative Scenario Planning: Working Together to Change the Future.* Oakland: Berrett-Koehler Publishers.

Kahane, A. 2017. *Collaborating with the Enemy: How to Work with People You Don't Agree with or Like or Trust.* Oakland: Berrett-Koehler Publishers.

Mair, J., and L. Hehenberger. 2014. 'Front-stage and Backstage Convening: The Transition from Opposition to Mutualistic Coexistence in Organizational Philanthropy.' *Academy of Management Journal* 57(4): 1174–1200. doi:10.5465/amj.2012.0305.

Marshall, F., J. Dolley, and R. Priya. 2018. 'Transdisciplinary Research as Transformative Space Making for Sustainability: Enhancing Propoor Transformative Agency in Periurban Contexts.' *Ecology and Society* 23(3): 8. doi:10.5751/ES-10249-230308.

Moore, M.L., P. Olsson, W. Nilsson, L. Rose, and F.R. Westley. 2018. 'Navigating Emergence and System Reflexivity as Key Transformative Capacities: Experiences from a Global Fellowship Program.' *Ecology and Society* 23(2): 38. doi:10.5751/ES-10166-230238.

Moore, M.L., O. Tjörnbo, E. Enfors, C. Knapp, J. Hodbod, J.A. Baggio, A. Norström, P. Olsson, and D. Biggs. 2014. 'Studying the Complexity of Change: Toward an Analytical Framework for Understanding Deliberate Social-Ecological Transformations.' *Ecology and Society* 19(4): 54. doi:10.5751/ES-06966-190454.

Moore, M.L., and F. Westley. 2011. 'Surmountable Chasms: Networks and Social Innovation for Resilient Systems.' *Ecology and Society* 16(1): 5. doi:10.5751/ES-03812-160105.

Naumann, S., M. Davis, M-L. Moore, and K. McCormick. 2018. 'Utilizing Urban Living Laboratories for Social Innovation.' In *Urban Planet*, edited by T. Elmqvist, X. Bai, N. Frantzeskaki, C. Griffith, D. Maddox, and T. McPhearson, 197–217. Cambridge: Cambridge University Press. doi:10.1017/9781316647554.012.

Olsson, P., C. Folke, and T. Hahn. 2004. 'Social-Ecological Transformation for Ecosystem Management: The Development of Adaptive Co-Management of a Wetland Landscape in Southern Sweden.' *Ecology and Society* 9(4): 2.

Olsson, P., M-L. Moore, F.R. Westley, and D.D.P. McCarthy. 2017. 'The Concept of the Anthropocene as a Game-Changer: A New Context for Social Innovation and Transformations to Sustainability.' *Ecology and Society* 22(2): 31. doi:10.5751/ES-09310-220231.

Pereira, L., S. Drimie, O. Zgambo, and R.O. Biggs. 2020. 'Planning for Change: Transformation Labs for an Alternative Food System in the Western Cape.' *Urban Transformations.* doi:10.1186/s42854-020-00016-8.

Pereira, L., N. Frantzeskaki, A. Hebinck, L. Charli, J. Scott, M. Dyer, H. Eakin et al. 2019. 'Transformative Spaces in the Making: Key Lessons from Nine Cases in the Global South.' *Sustainability Science* 1–18. doi:10.1007/s11625-019-00749-x.

Pereira, L.M., T. Hichert, M. Hamann, R. Preiser, and R. Biggs. 2018a. 'Using Futures Methods to Create Transformative Spaces: Visions of a Good Anthropocene in Southern Africa.' *Ecology and Society* 23(1): 19. doi:10.5751/ES-09907-230119.

Pereira, L.M., T. Karpouzoglou, N. Frantzeskaki, and P. Olsson. 2018b. 'Designing Transformative Spaces for Sustainability in Social-Ecological Systems.' *Ecology and Society* 23(4): 32. doi:10.5751/ES-10607-230432.

Pohl, C., S. Rist, A. Zimmermann, P. Fry, G.S. Gurung, F. Schneider, C.I. Speranza et al. 2010. 'Researchers' Roles in Knowledge Co-production: Experience from Sustainability Research in Kenya, Switzerland, Bolivia and Nepal.' *Science and Public Policy* 37(4): 267–281. doi:10.3152/030234210X496628.

Schäpke, N., F. Stelzer, G. Caniglia, M. Bergmann, M. Wanner, M. Singer-Brodowski, D. Loorbach, P. Olsson, C. Baedeker, and D.J. Lang. 2018. 'Jointly Experimenting for Transformation? Shaping Real-world Laboratories by Comparing Them.' *GAIA – Ecological Perspectives for Science and Society* 27(1): 85–96. doi:10.14512/gaia.27.S1.16.

Scharmer, O. 2008. *Theory U: Leading from the Future as It Emerges*. San Francisco: Berrett-Koehler Publishers.

Schlüter, M., K. Orach, E. Lindkvist, R. Martin, N. Wijermans, Ö. Bodin, and W.J. Boonstra. 2019. 'Toward a Methodology for Explaining and Theorizing about Social-Ecological Phenomena.' *Current Opinion in Environmental Sustainability*. Amsterdam: Elsevier. doi:10.1016/j.cosust.2019.06.011.

Schwartz, P. 1991. *The Art of the Long View: Planning for the Future in an Uncertain World*. New York: Currency Doubleday.

Senge, P., O.C. Scharmer, J. Jaworski, and B.S. Flowers. 2004. *Presence: Human Purpose and the Field of the Future*. Cambridge: Society for Organizational Learning.

Sharpe, B., A. Hodgson, G. Leicester, A. Lyon, and I. Fazey. 2016. 'Three Horizons: A Pathways Practice for Transformation.' *Ecology and Society* 21(2): 47. doi:10.5751/ES-08388-210247.

Tengö, M., E.S. Brondizio, T. Elmqvist, P. Malmer, and M. Spierenburg. 2014. 'Connecting Diverse Knowledge Systems for Enhanced Ecosystem Governance: The Multiple Evidence Base Approach.' *Ambio* 43(5): 579–591.

Tengö, M., R. Hill, P. Malmer, C.M. Raymond, M. Spierenburg, F. Danielsen, T. Elmqvist, and C. Folke. 2017. 'Weaving Knowledge Systems in IPBES, CBD and beyond – Lessons Learned for Sustainability.' *Current Opinion in Environmental Sustainability* 26–27: 17–25. doi:10.1016/J.COSUST.2016.12.005.

The Pathways Network. 2018. 'T-Labs: A Practical Guide – Using Transformation Labs (T-Labs) for Innovation in Social-Ecological Systems.' Brighton: STEPS Centre.

Van der Heijden, K. 2005. *Scenarios: The Art of Strategic Conversation* (2nd ed). Hoboken: John Wiley & Sons. www.wiley.com/en-za/Scenarios:+The+Art+of+Strategic+Conversation, +2nd+Edition-p-9780470023686.

Van Zwanenberg, P., A. Cremaschi, M. Obaya, A. Marin, and V. Lowenstein. 2018. 'Seeking Unconventional Alliances and Bridging Innovations in Spaces for Transformative Change: The Seed Sector and Agricultural Sustainability in Argentina.' *Ecology and Society* 23(3): 11. doi:10.5751/ES-10033-230311.

Westley, F., S. Laban, C. Rose, K. Robinson, K. McGowan, O. Tjörnbo, and M. Tovey. 2015. *Social Innovation Lab Guide*. Waterloo: University of Waterloo.

Westley, F.R., K. McGowan, and O. Tjörnbo. 2017. *The Evolution of Social Innovation: Building Resilience through Transitions*. Cheltenham: Edward Elgar Publishing.

Westley, F.R., O. Tjörnbo, L. Schultz, P. Olsson, C. Folke, B. Crona, and Ö. Bodin. 2013. 'A Theory of Transformative Agency in Linked Social-Ecological Systems.' *Ecology and Society* 18(3): 27. doi:10.5751/ES-05072-180327.

Whitney, D., and A. Trosten-Bloom. 2003. *The Power of Appreciative Inquiry: A Practical Guide to Positive Change*. San Francisco: Berrett-Koehler Publishers.

10

Futures analysis

Tanja Hichert,[1,2] *Reinette Biggs*[2,3] *and Alta de Vos*[4]

[1]HICHERT AND ASSOCIATES, SOMERSET WEST, SOUTH AFRICA
[2]CENTRE FOR SUSTAINABILITY TRANSITIONS, STELLENBOSCH UNIVERSITY, STELLENBOSCH, SOUTH AFRICA
[3]STOCKHOLM RESILIENCE CENTRE, STOCKHOLM UNIVERSITY, STOCKHOLM, SWEDEN
[4]DEPARTMENT OF ENVIRONMENTAL SCIENCE, RHODES UNIVERSITY, MAKHANDA, SOUTH AFRICA

Key methods discussed in this chapter

Scenarios and participatory scenario planning, futures wheels, three horizons framework, design/experiential futures, horizon scanning, Delphi, trend impact analysis, emerging issues analysis, causal layered analysis, appreciative inquiry, gaming (also known as 'gamification' or serious gaming), future workshops, visioning, back-casting, road-mapping

Connections to other chapters

Futures analysis methods enable the imagination and generation of alternative images of futures that are yet to exist. 'Utilising' and/or working with these alternatives, preferred and otherwise, and the signs that signal them, connects futures analysis very strongly to scenario development and participatory scenario planning (discussed in detail in Chapter 11). In this regard, futures analysis also connects to participatory modelling and planning (Chapter 13), serious games (Chapter 12) and facilitated dialogues (Chapter 9).

Introduction

Futures analysis methods can help people to think constructively and systematically about the future and advance our understanding of change and uncertainty in complex social-ecological systems (SES). This is important because there is not one single predictable future but multiple ones, depending on the complex, unpredictable interplays and interactions of actors, institutions, ecological processes and other elements of the system and its dynamics. Actively developing ideas, images and/or stories about different futures can enable us to make different choices and take different actions in the present in relation to, for example, risk mitigation, adaptation, resource allocation and strategy development, which can help build more sustainable and just futures.

Futures analysis methods are mostly derived from the field of futures studies – also commonly known as strategic foresight. While futures studies is established in academic arenas (see rossdawson.com/futurist/university-foresight-programs), it is not widespread

 DOI: 10.4324/9781003021339-13

SUMMARY TABLE: FUTURES ANALYSIS

DISCIPLINARY BACKGROUND	KNOWLEDGE TYPE
The methods in this chapter are derived from or have most commonly been used in: Futures Studies. For quantitative forecasting: Mathematical Modelling, Simulation Modelling, Statistical Modelling, Operations Research	*The methods in this chapter are primarily used to generate the following types of knowledge:* • Descriptive • Exploratory
RESEARCH APPROACH	**PURPOSE OF METHOD**
The methods in this chapter originate from or most commonly adopt the following research approaches: • Analytical/objective • Interpretive/subjective • Collaborative/process	*The most common purposes of using the methods in this chapter are:* • System understanding • Stakeholder engagement and co-production • Policy/decision support
TEMPORAL DIMENSION	**SYSTEMIC FEATURES AND PROCESSES**
The methods in this chapter are most commonly applied to the following temporal dimensions: • Present (typically within the last 5–10 years) • Recent past (post-1700s) • Future **SPATIAL DIMENSION** *The methods in this chapter are primarily either or both:* • Non-spatial • Explicitly spatial *The methods in this chapter are most commonly applied at the following spatial scales:* • Local • Regional (provincial/state to continental) • Global • Multiple places/sites around the world	*While most methods can do many things, the methods in this chapter are particularly good (i.e. go-to methods) for addressing the following:* • Social-ecological dependence and impact • Power relations • Social-ecological interactions over time • Path dependency • Transformation • Social learning • Exploring uncertainty

or universally accepted as a field of research (Bengston, Kubik, and Bishop 2012). Key features of futures studies are:

- It deals with wicked (complex, interconnected) problems.
- It emphasises mutually assured diversity (MAD), meaning that foresight practitioners and participants should be conscious of having multiple identities and bring this to their work, e.g. being 'a scientific researcher', together with being 'a citizen of a particular country', together with being 'someone who holds a particular worldview', together with being 'a parent of young children'.
- It adopts a sceptical perspective that questions dominant axioms and assumptions.
- It is futureless in the sense that the results are mostly useful in the present (Sardar 2010, 177).

Futures studies generally emphasise the exploration of multiple futures to make sense of the present in order to better understand and potentially influence the future. Many futures analysis methods are well suited to SES work as futures studies generally adopt a systems approach and emphasise 'using' multiple futures to make sense of the present in order to better understand the future.

Three levels of futures studies can be identified: forecast, foresight and anticipation (Poli 2017). Forecast is the first level and is past oriented. It is quantitative and based on statistical calculations, such as time series regression, in which the past is extrapolated into the future. Forecasts are thus a product of probabilistic prediction based on the past dynamics of the system and have limited use when trying to understand complex, volatile and novel change.

The second level is foresight, which is future oriented. It works with the inherent ability of humans to imagine a future that does not exist, tell stories about it and thereby engage tacit knowledge to make assumptions explicit, and make sense of and prepare for what has yet to happen (Wilkinson 2017). Forecasting is often used in conjunction with foresight to provide an assessment of the most likely future – a baseline depicting a future which assumes that all other things remain equal.

Anticipation, as the third level of futures studies (not as in anticipatory systems, as it is sometimes used as a technical term – Poli 2017), is a present-oriented approach and consists of a 'forward-looking attitude' combined with the 'use' of that attitude that results in action (Poli 2017). In other words, anticipation focuses on understanding behavioural change based on an idea, or image, of the future. Anticipatory behaviour, or an anticipation capability, 'uses' the future in present-day decision-making processes. This behaviour is more robust than purely reactive behaviour. Strictly speaking, the future can only exist in the present as anticipation. 'Studying' the future – even though it does not exist – involves learning about and understanding how different ways of framing the future generate different perceptions of the present and hence alters preferences and choices.

Futures analysis methods (and tools) are applied differently from project to project and are nearly always highly customised. These methods are also constantly evolving. It therefore becomes important that the SES researcher knows why, and for what purpose, they want to incorporate futures analysis methods in their work. Is it to increase awareness of change? To create visions of a preferred future? To make sense of the present? Having a purpose for using futures analysis will ensure an appropriate choice of methods. Futures analysis methods can be used at any stage of an SES study, in conjunction with, and to augment, other SES analyses.

SES problems and questions

Futures analysis methods are useful for helping to understand change. This includes understanding the patterns, systems and underlying driving forces that cause change to happen, making sense of the present, surfacing and challenging assumptions about the present and present worldviews, embracing uncertainty and pursuing alternatives rather than being 'locked' into forecasts.

For these reasons, SES research often turns to futures analysis methods to interpret, reiterate and recalibrate predictions made by more deterministic models and methods. Gao et al. (2016), for example, used scenario analysis to better understand deep uncertainty around outputs of land-use change models. Whereas scenario analysis (which is covered in Chapter 11) is probably the futures analysis method most used in this way, other futures analysis methods are also sometimes used. Strategic foresight is starting to become more popular in long-term conservation planning, for example (Cook et al. 2014).

Many futures analysis studies in the SES field concern questions of uncertainty and risk related to global change processes, particularly climate change (e.g. Bohensky et al. 2011), land-use change (e.g. Gao et al. 2016) and changes in socio-political and economic systems (e.g. Bohensky et al. 2011). In most cases, futures analysis is not merely used to identify elements of risk and uncertainty. Since many futures analyses are largely participative, they are also used to explore options for navigating risks and uncertainty, usually through co-production of knowledge with key stakeholders.

The combination of participatory, narrative approaches and the ability of futures analysis to interact with more quantitative predictions make these methods particularly well suited to questions raised in the management and prioritisation of social-ecological landscapes, and for ultimately developing policies to manage these systems (Francis, Levin, and Harvey 2011). To name two examples: (a) scenario approaches and back-casting (Table 10.1) have been used to set conservation targets in social-ecological protected landscapes (Levin et al. 2015), and (b) visioning, scenarios and predictive forecasting are often combined to set realistic objectives and targets and to explore the consequences of potential management decisions on ecosystems under multiple potential scenarios of change (Francis, Levin, and Harvey 2011).

Futures analysis methods are used to understand not only what might be driving change in a system but also the cascading consequences of drivers of change, i.e. the many direct and indirect implications (Bengston 2016). Methods such as futures wheels and three horizons are well suited to exploring deeper drivers of change and possible future trajectories, which may be particularly relevant in resolving conflict and understanding management options in SES in flux. Case study 10.1 is an example of how futures analysis tools may be used in this way.

Futures analysis methods (especially Delphi and horizon scans) are sometimes used to understand the kind of questions relevant to SES research (e.g. Shackleton et al. 2011) and how research approaches that seek to understand SES (e.g. adaptive co-management, Plummer and Armitage 2007) might be changing and developing.

Brief description of key methods

Table 10.1 contains brief descriptions of some futures analysis methods that are useful for SES research, with references for further reading. The methods are categorised according to their main purpose: increasing awareness of change, exploring impacts of change, exploring alternative futures, exploring preferred futures, and informing strategies and action.

Table 10.1 Summary of key methods used in futures analysis, categorised according to purpose

Method	*Description*	*References*
INCREASING AWARENESS OF CHANGE		
Horizon scanning	Horizon scanning focuses on identifying new and emerging issues, typically called 'weak signals', as well as existing trends. It can also serve as a future-oriented sense-making exercise. Horizon scanning entails a systematic information/intelligence gathering and analysing activity. Discussing the impacts of scanning output on an issue is often conducted in a workshop format. Output from a horizon-scanning exercise often serves as input for scenarios, with the objective of systematically looking for the 'driving forces' that shape the future of the topic or issue being examined. Horizon scanning usually covers a wide range of domains including social, technological, economic, environmental and political domains.	*Key introductory texts* Hines et al. 2018; UNDP 2018 *Applications to SES* Shackleton et al. 2011; Bengston 2013; Sutherland et al. 2020
Emerging issues analysis	Emerging issues analysis (EIA) is similar to horizon scanning in that it seeks to identify initial sources of change, usually by monitoring fringe thinking, niches or outliers. Emerging issues are not mainstream realities in the present, but could become emerging patterns, major drivers or the source of a new trend. It can be very effective to combine EIA with the three horizons framework.	*Key introductory text* Molitor 2003 *Applications to SES* Bennett et al. 2016 (although the activity of searching for 'seeds of a Good Anthropocene' is not explicitly called (crowdsourced) emerging issues analysis, it is essentially that, and is an excellent example)
Delphi	The Delphi method is also often referred to as 'expert panels', although a true Delphi specifically involves iterative assessment of what selected anonymous experts think future developments for a given topic may be, and not just a random survey of experts. Several rounds are conducted and experts are allowed to alter their input after being exposed to previous rounds. The aim is to clarify consensus. A Delphi can be conducted either online or via interviews. Specialised software is also available.	*Key introductory texts* Linstone and Turoff 1975; Glenn and Gordon 2004 *Applications to SES* Plummer and Armitage 2007

Method	*Description*	*References*
Trend impact analysis	Trend impact analysis (TIA) focuses on the potential impacts and implications of trends and megatrends on the topic or issue under consideration. Megatrends are the great forces in societal and natural development that are likely to affect the future in all areas over the next 10–15 years. A framework measuring likelihood, scale and speed of arrival is often used. TIA ranges from highly sophisticated exercises, e.g. the government of Singapore and World Economic Forum's risk mapping, to brainstorming sessions about issues that are very difficult to measure and track, such as the growth in the popularity of veganism.	*Key introductory text* Glenn and Gordon 2004 *Applications to SES* Nair, Wen, and Ling 2014
EXPLORING IMPACTS OF CHANGE		
Futures wheels	Futures wheels is a group brainstorming method that explores and maps multiple levels of consequences of trends, events, emerging issues and/or future possible decisions. It is a graphic visualisation of direct and indirect, positive and negative future consequences of a particular change or development.	*Key introductory text* Glenn and Gordon 2004 *Applications to SES* Bengston 2016; Bengston, Dockry, and Shifley 2018; Pereira et al. 2018; Hichert, Biggs, and Preiser 2019
Three horizons framework	The three horizons framework is a conceptual model to aid people's thinking about current assumptions, emerging changes and possible and desired futures. It is a graphical approach developed to explore the change in importance of issues over time and connect the future to the present. It is an adaptable tool that is often used as an intuitive, accessible introduction to futures thinking and to make sense of emerging changes. At its most basic it is a systems model about the way things change over time. It is particularly good for working with complexity, developing future consciousness and recognising transformative change, while exploring how to manage transitions.	*Key introductory texts* Curry and Hodgson 2008; h3uni.org/practices/foresight-three-horizons *Applications to SES* Sharpe et al. 2016; Pereira et al. 2018; Hichert, Biggs, and Preiser 2019
EXPLORING ALTERNATIVE FUTURES		
Design/experiential futures	Experiential futures refer to a set of approaches to make alternative futures present and 'feel' real, i.e. the aim is to get people to experience the future. It is essentially a future brought to life materially or performatively, or both. It is all about engaging with futures using design (often prototyping), performance, film and materiality – objects and things – as well as media and modalities that have not traditionally been used.	*Key introductory text* Candy 2014 *Applications to SES* Hichert, Biggs, and Preiser 2019

(*Continued*)

Table 10.1 (Continued)

Method	*Description*	*References*
Scenario development	Scenario development (see Chapter 11) is regarded as a structured process (but it is useful to keep in mind that there are many different approaches and processes for creating scenarios (Bishop, Hines, and Collins 2007)) where a group creates narratives or images of alternative futures of an issue, and how those futures might unfold and affect the issue. These stories are then used to inform decisions, plans and policies with a view to improving them.	*Key introductory text* Wack 1985 *Applications to SES* Carpenter, Bennett, and Peterson 2006
EXPLORING PREFERRED FUTURES		
Appreciative inquiry	Appreciative inquiry originated in organisation development. It is a large-group collaborative change method structured as a set of iterative cyclical questions that concentrate on the positive. The aim is to identify what is working well, identify the energy for change and build on that. The underlying assumption is that people are more comfortable co-creating an unknown future if they can take parts of today's successes forward and build upon them.	*Key introductory text* Bushe 2013 *Applications to SES* Van der Merwe, Biggs, and Preiser 2018
Future workshop	This is a three-day structured workshop process which originated in Europe in the early 1960s. The aim is to create visions of a preferred future. It starts off by analysing problems and challenges in the present, followed by brainstorming possible solutions. The best potential solutions are chosen democratically and are formulated into viable projects. The workshop ends with a plan of action.	*Key introductory text* Jungk and Mullert 1987 *Applications to SES* No known example
Visioning	Visioning is any participatory activity or exercise designed to come up with compelling visions/narratives/images of preferred, often transformative, futures. These preferred futures are always normative as opposed to possible or plausible futures generated by scenarios. The aim is to inspire, engage and enable people to act towards creating a preferred future.	*Key introductory text* Ziegler 1991 *Applications to SES* Pereira et al. 2018 Hamann et al. 2020
Causal layered analysis	Causal layered analysis is a four-level analysis examining the litany ('headlines'), systems, worldviews and myths/metaphors associated with an issue. It is used to identify different perspectives about the future and is good for 'surfacing' underlying, sometimes sensitive issues. Changing the deepest levels of myths/metaphors about an issue is one way of developing preferred futures.	*Key introductory text* Inayatullah 2008 *Applications to SES* Heinonen et al. 2017
INFORMING STRATEGIES AND ACTION		
Road-mapping	Road-mapping is a 'vision-into-action' technique which is often used for technology planning to help turn ideas into products or services. It maps potential pathways, with timelines and actions, from the present to the preferred future to help make it possible to reach that future.	*Key introductory texts* Garcia and Bray 1997; Jackson 2013 *Applications to SES* No known example

Method	*Description*	*References*
Back-casting	Much like road-mapping, back-casting works backward from the preferred future to the present. It is a set of imaginary steps detailing how a preferred future was reached or brought about. These steps then form the basis of actions to be taken, decisions and policies to be made, and resources needed to create that preferred future.	*Key introductory text* Jackson 2013 *Applications to SES* Palomo et al. 2011
Gaming, 'gamification' and/or serious games	Gaming in futures-oriented activities – the terms 'gamification' or serious games are also commonly used – simulates real-world situations and predicaments and engages participants by means of, often goal-directed, play. It includes foresight card decks, board games, immersive role-playing experiences, futures labs and various types of online games such as Foresight Engine. According to Bengston (2016), an important rationale for the use of gaming methods in futures research is that active learning methods are often most effective.	*Key introductory text* Bengston 2019 *Applications to SES* Vervoort et al. 2010

Limitations

Because there is no set 'formula', other than quantitative forecasting, for when and how to use which futures analysis method, it can be difficult to successfully incorporate these methods into SES research as many of the methods require skilled facilitation. However, there is huge potential for adding value to SES research when employing these methods in collaboration with skilled future/foresight practitioners, which can then also lead to skills transfer.

None of the futures analysis methods aims to predict the future – something which is impossible. Despite the emphasis of horizon scanning and emerging issues analysis on detecting weak signals and early signs of change, neither of these methods is able to foresee tipping points or entirely novel events. They are, however, very useful in fostering greater awareness of, and learning about, sudden, surprising change, volatility and systemic turbulence. The methods are good for sense-making and sensitisation.

The outputs of some of the methods, especially the more creative ones, like causal layered analysis, or those working with outlier emerging issues, like emerging issues analysis, are often not regarded as authoritative and may suffer from a lack of credibility. Many of the methods are fairly new to the public sector, development sector and civil society, and people may be unfamiliar and uncomfortable with engaging in these processes. In contrast, many of these methods have either been developed in or used for many decades by the military and business sectors.

Futures analysis methods and the field of futures studies are actively evolving. In many instances, new developments are practitioner led, with the result that new methods are not well documented and knowledge about new developments can be quite fragmented. Fortunately, the field has several professional associations (such as the Association of Professional Futurists (apf.org), the World Futures Studies Federation (wfsf.org) and the US Public Sector Foresight Network (publicsectorforesight.org)) that share knowledge among members and make some of their publications available. Notable among these is the Foresight Competency Model and *The Future of Futures* e-book, both published by the Association of Professional Futurists.

Resource implications

Many of the methods mentioned here require experienced, skilled facilitators with some background and knowledge of futures studies. With training, knowledge transfer and self-learning, however, the methods can be successfully applied by SES researchers themselves.

In general, and excluding potential future developments, no specialised hardware or software is necessary, although online curating tools such as Factr (factr.com), Pearltrees (pearltrees.com) and Pocket (getpocket.com) are very useful for organising feedstock for horizon scanning, trend impact analysis and emerging issues analysis. Scanning can be very time consuming – sifting signals from noise – and generates large amounts of surplus data. It is, therefore, advisable to devise a scanning system and obtain institutional support before committing to it. Doing a Delphi will require access to experts and it can be a very time-consuming exercise for them.

All the methods are participatory, except when trend impact analysis and emerging issues analysis are conducted as single-researcher desk-based, or online crowdsourced, exercises.

Case Study 10.1: Using futures analysis methods to generate visions of a 'Good Anthropocene'

In November 2016, the Centre for Complex Systems in Transition (now the Centre for Sustainable Transitions) (CST) at Stellenbosch University, South Africa, wanted to 'solicit, explore, and develop a suite of alternative visions for "Good Anthropocenes" – positive futures that are socially and ecologically desirable, just, and sustainable' (CST-GRAID 2017, 4). The project's aim was to create good stories about the future. The critical question for the project was: how can we imagine radically alternative positive futures for southern Africa arising from small-scale, experimental sustainability initiatives employing new ways of thinking or doing?

The first step could be regarded as a highly defined horizon-scanning process. Rather than throwing a wide net to identify emerging signals of change, scientists, researchers, sustainability practitioners and their networks compiled an initial database of 100 'seed' projects (goodanthopocenes.net), each catalogued and categorised. These are all small-scale, experimental projects and initiatives – new social institutions, technologies or frameworks for understanding the world – that are not yet mainstream. Examples include projects on urban gardening and renewable energy, as well as technological advances like gene therapy (Hamann et al. 2020).

The project team and 23 participants (split into four mixed groups) made up of a roughly equal mix of scientists, artists, social entrepreneurs and social/policy researchers created the visions of a 'Good Anthropocene' from the seeds in a three-day workshop. Each working group was responsible for building a positive scenario of the Good Anthropocene by combining three very different 'seeds' from the database: two southern African seeds and one technology 'wildcard' seed.

Participants first considered each of their three assigned seeds, one by one, and explored the impacts and implications of each. Futures wheels were used as the backbone of this exploration (Figure 10.1). To begin the impact mapping, each

The participatory nature implies access to a suitable venue, often for a few days, and standard workshopping material (see Hichert, Biggs, and Preiser 2019). All the participative methods, but causal layered analysis in particular, benefit from participants with multiple cultural perspectives and different disciplinary backgrounds.

Engaging in futures and foresight work is often an exciting, inspiring and revelatory experience for participants (Pereira et al. 2018), so it is wise to allow enough time to do it properly.

New directions

Schultz (2012) suggests that new directions for futures analysis methods involve exploring humanity's inner spaces, such as fears, hopes and beliefs, psychological stance and cognitive bias with methods such as integral futures and verge (which draws on ethnographic futures concepts). Simultaneously there is 'a clear and welcome trend towards decentralised, massively distributed and inclusive futures work. Global computing and interconnected

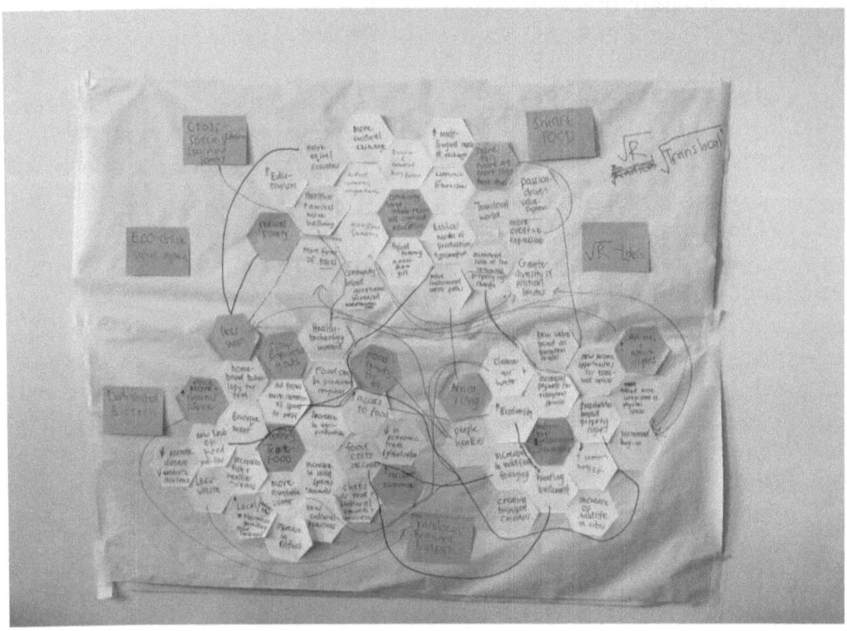

Figure 10.1 Three futures wheels placed in proximity to one another, together with cross-impact matrix cards, so that it is possible to conduct the influence mapping exercise (© Gys Loubser)

seed was imagined in its future, mature form – as a mainstream 'new normal' rather than a fringe activity. Participants then mapped the primary, secondary and tertiary impacts and consequences cascading out from each seed. To consider how the three seeds and their impacts would interconnect, the afternoon of Day 1

(Continued)

was used to create cross-impact matrices to explore outcomes as each of the seeds affected the others. As a final step on the first day, each working group created an influence map connecting the interactions between seeds, their impacts and the sub-impacts. After standing back and getting a sense of the emerging story, each group presented their new scenario via an artistic image (any medium), three fictional statistics and a social commentary/news headline (CST-GRAID 2017; Hamann et al. 2020).

Day 2 focused on building out the visionary narratives from the previous day's work. The goal was 'bold, vivid, hopeful scenario narratives – Visions of a Good Anthropocene in southern Africa' (CST-GRAID 2017; Hamann et al. 2020). The fact that the 'seeds' at the heart of each vision are existing pilot projects and initiatives demonstrating ways in which humans might be able to live in a 'Good Anthropocene' adds local specificity and enhances the credibility of each story. Because seed representatives were workshop participants, the experiences of active change agents enriched the vision stories, in addition to those details added by the artists and the scientists.

Back-casting is the tool most often used to connect visionary statements of audacious goals to action in the present. CST chose to use the three horizons tool – often chosen as a framing tool or to make sense of the changes emerging from scanning – for back-casting. Each group considered their vision's place on Horizon Three, and then connected it to the present of Horizon One by 'looking for, and talking about, systemic changes, amplifications, clashes and potential inflection points' (CST-GRAID 2017; Hamann et al. 2020) in the Horizon Two transition space that bridges to the vision on Horizon Three (Figure 10.2).

The final activity for the participants was to share their visions in a creative, expressive and immersive manner. This is where experiential futuring comes in. Participants were given complete freedom as to how to do this, and examples included role-playing, dance, visual art, objects and theatrical performance (Figure 10.3).

Everything that informed the choice of methodology resulted from conversations about picking tools to get really far away from the generic futures, the ambient futures embedded in the social context, popular media and regular academic literature. The project team wanted to push past these 'everyday', more common images of the future. They wanted to avoid so-called 'used futures' and develop positive transformative stories that felt fresh and local. This drove their choice of the tools and methods mentioned above, which resulted in inspiring output that maximises difference from current conditions.

communication support digital exploration of our possible futures with new levels of creativity, rigor and participation' (Schultz 2012, 7).

This involves the application of online ICT tools such as Futurescaper (futurescaper.com) and Sensemaker (cognitive-edge.com/sensemaker), and global foresight games such as Foresight Engine (iftf.org/what-we-do/foresight-tools/collaborative-forecasting-games/foresight-engine). Evolving artificial intelligence and big data analytics will change futures analysis methods in fundamental ways going forward.

Also worth watching are new methods, such as the Mānoa mash-up, emanating from the non-Western world (Pereira et al. 2018; Hichert, Biggs, and Preiser 2019) as there is a shift

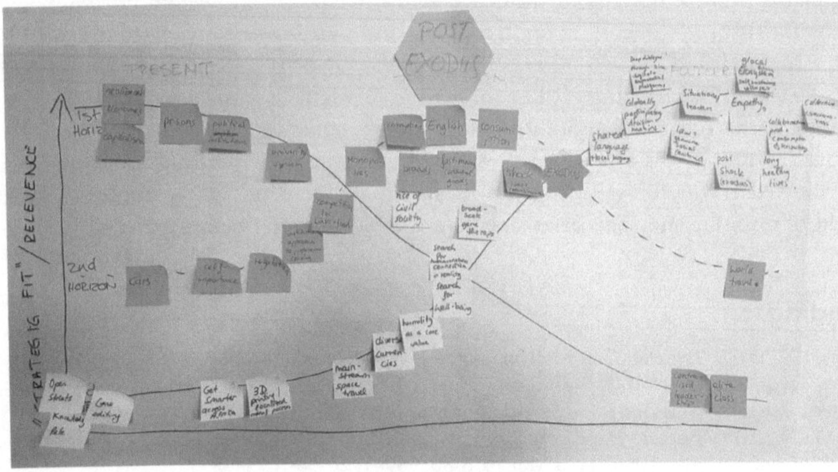

Figure 10.2 A populated three horizons framework (© Gys Loubser)

Figure 10.3 Sharing a vision of a future in an immersive, creative way (© Gys Loubser)

away from the formalisation of futures thinking in Europe and the USA to vibrant communities of practice, albeit small, in Asia and Africa.

In the afterword to *The Future of Futures* e-book, Curry (2012, 46) mentions some characteristics that will shape the future of futures thinking and foresight:

- More distributed and more networked, more at home with the social media tools
- More data
- Take a 'complexity turn': 'While the futures academy has already engaged with complex adaptive systems and emergence, these have been slower to inform futures methods.'

- 'Rediscover some of its roots in philosophy, building (or re-building) a knowledge base that places more emphasis on how we know what we say we know when we make claims for futures work. The epistemology and ontology of futures work will become increasingly visible.'
- Different ways of knowing will become more evident in futures practice
- Become better informed about its history and its contexts
- Futures thinking was born into a world of growth, the emerging consumer economy and the Cold War. It will grow up in an age of 'descent' where futures addressing a world of resource shortage or even collapse will no longer be regarded as dystopian.

Key readings

Bengston, D.N. 2017. *Ten Principles for Thinking about the Future: A Primer for Environmental Professionals. Gen. Tech. Rep. NRS-175.* Newtown Square: US Department of Agriculture, Forest Service, Northern Research Station. www.fs.usda.gov/treesearch/pubs/55548.

Bengston, D.N. 2019. 'Futures Research Methods and Applications in Natural Resources.' *Society and Natural Resources* 32(10): 1099–1113. doi:10.1080/08941920.2018.1547852.

Bishop, P.C., and A. Hines. 2012. *Teaching about the Future*. London: Palgrave Macmillan.

Hichert, T., R. Biggs, and R. Preiser. 2019. *Generating Visions of Good Anthropocenes: The Mānoa Mash-up Scenarios Methodology*. CST Toolkit. University of Stellenbosch. www0.sun.ac.za/cst/publication/generating-visions-of-good-anthropocenes-the-manoa-mash-up-scenarios-methodology.

UNDP Global Centre for Public Service Excellence. 2018. *Foresight Manual: Empowered Futures for the 2030 Agenda*.

Acknowledgements

This chapter benefited from the valuable insights, comments and contributions by David N. Bengston, PhD, environmental futurist and social scientist of the US Forest Service's Northern Research Station in St Paul, Minnesota. Thank you, David!

References

Bengston, D.N. 2013. *Horizon Scanning for Environmental Foresight: A Review of Issues and Approaches*. Gen. Tech. Rep. NRS-121. Newtown Square: US Department of Agriculture, Forest Service, Northern Research Station. doi:10.2737/NRS-GTR-121.

Bengston, D.N. 2016. 'The Futures Wheel: A Method for Exploring the Implications of Social-Ecological Change.' *Society and Natural Resources* 29(3): 374–379.

Bengston, D.N. 2019. 'Futures Research Methods and Applications in Natural Resources.' *Society and Natural Resources* 32(10): 1099–1113. doi:10.1080/08941920.2018.1547852.

Bengston, D.N., M.J. Dockry, and S.R. Shifley. 2018. 'Anticipating Cascading Change in Land Use: Exploring the Implications of a Major Trend in US Northern Forests.' *Land Use Policy* 71: 222–229. www.fs.usda.gov/treesearch/pubs/55563.

Bengston, D.N., G.H. Kubik, and P.C. Bishop. 2012. 'Strengthening Environmental Foresight: Potential Contributions of Futures Research.' *Ecology and Society* 17(2): 10.

Bennett, E.M., M. Solan, R. Biggs, T. McPhearson, A.V. Norström, P. Olsson, L. Pereira et al. 2016. 'Bright Spots: Seeds of a Good Anthropocene.' *Frontiers in Ecology and the Environment* 14(8): 441–448.

Bishop, P., A. Hines, and T. Collins. 2007. 'The Current State of Scenario Development: An Overview of Techniques.' *Foresight* 9(1): 5–25. doi:10.1108/14636680710727516.

Bohensky, E., J.R. Butler, R. Costanza, I. Bohnet, A. Delisle, K. Fabricius, M. Gooch et al. 2011. 'Future Makers or Future Takers? A Scenario Analysis of Climate Change and the Great Barrier Reef.' *Global Environmental Change* 21(3): 876–893.

Bushe, G.R. 2013. 'The Appreciative Inquiry Model.' In *Encyclopedia of Management Theory*, Volume 1, edited by E.H. Kessler, 41–44. Thousand Oaks: Sage.

Candy, S. 2014. 'Experiential Futures.' *The Futurist* 48(5): 34–37.

Carpenter, S.R., E.M. Bennett, and G.D. Peterson. 2006. 'Scenarios for Ecosystem Services: An Overview.' *Ecology and Society* 11(1): 29. http://ecologyandsociety.org/vol11/iss1/art29.

Cook, C.N., S. Inayatullah, M.A. Burgman, W.J. Sutherland, and B.A. Wintle. 2014. 'Strategic Foresight: How Planning for the Unpredictable Can Improve Environmental Decision-making.' *Trends in Ecology and Evolution* 29(9): 531–541.

CST-GRAID. 2017. *Report on the Anthropocene Visioning Workshop, 15–18 November 2016, Cape Town, South Africa.* GRAID Project Workshop. Centre for Complex Systems in Transition, Stellenbosch University, South Africa.

Curry, A., ed. 2012. The Future of Futures. Houston: Association of Professional Futurists.

Curry, A., and A. Hodgson. 2008. 'Seeing in Multiple Horizons: Connecting Futures to Strategy.' *Journal of Futures Studies* 13(1): 1–20.

Francis, T.B., P.S. Levin, and C.J. Harvey. 2011. 'The Perils and Promise of Futures Analysis in Marine Ecosystem-based Management.' *Marine Policy* 35(5): 675–681.

Gao, L., B.A. Bryan, M. Nolan, J.D. Connor, X. Song, and G. Zhao. 2016. 'Robust Global Sensitivity Analysis under Deep Uncertainty via Scenario Analysis.' *Environmental Modelling and Software* 76: 154–166.

Garcia, M.L., and O.H. Bray. 1997. *Fundamentals of Technology Roadmapping.* Albuquerque: Sandia National Laboratories.

Glenn, J.C., and T.J. Gordon, eds. 2004. Futures Research Methodology Version 3.0.

Hamann, M., R. Biggs, L. Pereira, R. Preiser, T. Hichert, R. Blanchard, H.W. Coetzee et al. 2020. 'Scenarios of Good Anthropocenes in Southern Africa.' *Futures* 118: 102526.

Heinonen, S., M. Minkkinen, J. Karjalainen, and S. Inayatullah. 2017. 'Testing Transformative Energy Scenarios Through Causal Layered Analysis Gaming.' *Technological Forecasting and Social Change* 124: 101–113.

Hichert, T., R. Biggs, and R. Preiser. 2019. Generating Visions of Good Anthropocenes: The Mānoa Mash-up Scenarios Methodology. CST Toolkit. University of Stellenbosch. www0.sun.ac.za/cst/publication/generating-visions-of-good-anthropocenes-the-manoa-mash-up-scenarios-methodology.

Hines, A., D.N. Bengston, M.J. Dockry, and A. Cowart. 2018. 'Setting Up a Horizon Scanning System: A US Federal Agency Example.' *World Futures Review* 10(2): 136–151.

Inayatullah, S. 2008. 'Six Pillars: Futures Thinking for Transforming.' *Foresight* 10(1): 4–21.

Jackson, M. 2013. Practical Foresight Guide Chapter 3 – Methods. www.shapingtomorrow. com/media-centre/pf-ch03.pdf.

Jungk, R., and N. Mullert. 1987. *Future Workshops: How to Create Desirable Futures.* London: Institute for Social Inventions.

Levin, P.S., G.D. Williams, A. Rehr, K.C. Norman, and C.J. Harvey. 2015. 'Developing Conservation Targets in Social-Ecological Systems.' *Ecology and Society* 20(4): 6.

Linstone, H.A., and M. Turoff. 1975. The Delphi Method: Techniques and Applications. https://web.njit.edu/~turoff/pubs/delphibook/delphibook.pdf.

Molitor, G. 2003. *The Power to Change the World: The Art of Forecasting.* Potomac: Public Policy Forecasting.

Nair, S., W.K. Wen, and C.M. Ling. 2014. 'Bangkok Flood Risk Management: Application of Foresight Methodology for Scenario and Policy Development.' *Journal of Futures Studies* 19(2): 87–112.

Palomo, I., B. Martín-López, C. López-Santiago, and C. Montes. 2011. 'Participatory Scenario Planning for Protected Areas Management under the Ecosystem Services Framework: The Doñana Social-Ecological System in Southwestern Spain.' *Ecology and Society* 16(1): 23.

Pereira, L., T. Hichert, M. Hamann, R. Preiser, and R. Biggs. 2018. 'Using Futures Methods to Create Transformative Spaces: Visions of a Good Anthropocene in Southern Africa.' *Ecology and Society* 23(1): 19.

Plummer, R., and D.R. Armitage. 2007. 'Charting the New Territory of Adaptive Co-management: A Delphi Study.' *Ecology and Society* 12(2): 10. www.ecologyandsociety.org/vol12/iss2/art10.

Poli, R. 2017. *Introduction to Anticipation Studies.* Dordrecht: Springer.

Sardar, Z. 2010. 'The Namesake: Futures; Futures Studies; Futurology; Futuristic; Foresight – What's in a Name?' *Futures* 42(3): 177–184.
Schultz, W. 2012. 'The History of Futures.' In *The Future of Futures*, edited by A. Curry. Houston: Association of Professional Futurists.
Shackleton, C.M., B.J. Scholes, C. Vogel, R. Wynberg, T. Abrahamse, S.E. Shackleton, F. Ellery, and J. Gambiza. 2011. 'The Next Decade of Environmental Science in South Africa: A Horizon Scan.' *South African Geographical Journal* 93(1): 1–14.
Sharpe, B., A. Hodgson, G. Leicester, A. Lyon, and I. Fazey. 2016. 'Three Horizons: A Pathways Practice for Transformation.' *Ecology and Society* 21(2): 47.
Sutherland, W.J., M.P. Dias, L.V. Dicks, H. Doran, A.C. Entwistle, E. Fleishman, D.W. Gibbons et al. 2020. 'A Horizon Scan of Emerging Global Biological Conservation Issues for 2020.' *Trends in Ecology & Evolution* 35(1): 81–90.
UNDP Global Centre for Public Service Excellence. 2018. Foresight Manual: Empowered Futures for the 2030 Agenda.
Van der Merwe, S.E., R. Biggs, and R. Preiser. 2018. 'Building Social Resilience in Socio-Technical Systems through a Participatory and Formative Resilience Assessment Approach.' *Systemic Change Journal* 1(1): 1–34.
Vervoort, J.M., K. Kok, R. van Lammeren, and T. Veldkamp. 2010. 'Stepping into Futures: Exploring the Potential of Interactive Media for Participatory Scenarios on Social-Ecological Systems.' *Futures* 42(6): 604–616.
Wack, P. 1985. 'Scenarios: Uncharted Waters Ahead' and 'Scenarios: Shooting the Rapids.' *Harvard Business Review*, September–October and November–December.
Wilkinson, A. 2017. Strategic Foresight Primer. European Political Strategy Centre.
Ziegler, W. 1991. 'Envisioning the Future.' *Futures* 23(5): 516–527.

11

Scenario development

Tanja Hichert,[1,2] *Reinette Biggs,*[2,3] *Alta de Vos*[4] *and Garry Peterson*[3]

[1]HICHERT AND ASSOCIATES, SOMERSET WEST, SOUTH AFRICA
[2]CENTRE FOR SUSTAINABILITY TRANSITIONS, STELLENBOSCH UNIVERSITY, STELLENBOSCH, SOUTH AFRICA
[3]STOCKHOLM RESILIENCE CENTRE, STOCKHOLM UNIVERSITY, STOCKHOLM, SWEDEN
[4]DEPARTMENT OF ENVIRONMENTAL SCIENCE, RHODES UNIVERSITY, MAKHANDA, SOUTH AFRICA

Key methods discussed in this chapter

Double uncertainty matrix, Mānoa, scenario archetypes, *La Prospective*, causal layered analysis

Connections to other chapters

Scenario development connects to various other methods, but in particular to futures analysis (Chapter 10) as it is also regarded as a futures analysis method. It warrants a separate chapter in addition to the other futures analysis methods (described in Chapter 10) because of its importance and widespread use. Scenario development also connects to facilitated dialogues (Chapter 9), serious games (Chapter 12), participatory modelling and planning (Chapter 13), dynamical systems modelling (Chapter 26) and agent-based modelling (Chapter 28).

Introduction

Scenario development has a rich history and has been widely used for more than five decades in the corporate and military sectors, from where it originates (Bradfield et al. 2005). Over the past three decades, scenario approaches have been increasingly used in social-ecological systems (SES) research in many different contexts (e.g. for exploring the integrated future of biodiversity, land-use change, ecosystem services, and changes in value systems, global markets and the climate), at scales from local communities to the entire planet (Peterson, Cumming, and Carpenter 2003; Carpenter, Bennett, and Peterson 2006; Oteros-Rozas et al. 2015). Scenario development in global assessments is used to focus scientific investigation, integrate different models and data, and improve decision-making (Kok et al. 2017; Rosa et al. 2017), whereas local-scale scenarios often involve participatory processes that enhance stakeholder engagement and legitimacy in decision-making (Oteros-Rozas et al. 2015).

At their most basic, scenarios are a group of stories, often called narratives, which together describe a range of possible and coherent future worlds for a given system (Curry 2012). Central to scenario development is the concept of exploring multiple, alternative futures. Scenario development never attempts to predict the future, and scenarios never offer a single view of the future (Kosow and Gaßner 2008). Scenario development always focuses on multiple futures.

DOI: 10.4324/9781003021339-14

SUMMARY TABLE: SCENARIO DEVELOPMENT	
DISCIPLINARY BACKGROUND	**KNOWLEDGE TYPE**
The methods in this chapter are derived from or have most commonly been used in: Military Strategy, Operations Research, Futures Studies, Strategic Foresight, Management Science, Business Administration, Strategic Planning	*The methods in this chapter are primarily used to generate the following types of knowledge:* • Exploratory
RESEARCH APPROACH	**PURPOSE OF METHOD**
The methods in this chapter originate from or most commonly adopt the following research approaches: • Analytical/objective • Interpretive/subjective • Collaborative/process	*The most common purposes of using the methods in this chapter are:* • System understanding • Stakeholder engagement and co-production • Policy/decision support
TEMPORAL DIMENSION	**SYSTEMIC FEATURES AND PROCESSES**
The methods in this chapter are most commonly applied to the following temporal dimensions: • Present (typically within the last 5–10 years) • Recent past (post-1700s) • Future	*While most methods can do many things, the methods in this chapter are particularly good (i.e. go-to methods) for addressing the following:* • Social-ecological dependence and impact • Social-ecological interactions over time • Path dependency • Transformation • Social learning • Evaluating policy options • Exploring uncertainty
SPATIAL DIMENSION	
The methods in this chapter are primarily either or both: • Non-spatial • Explicitly spatial *The methods in this chapter are most commonly applied at the following spatial scales:* • Local • Regional (provincial/state to continental) • Global • Multiple places/sites around the world	

Descriptions and analyses of scenarios typically include terms such as 'plausible', 'possible' and 'preferable' futures (Bishop, Hines, and Collins 2007; Alcamo 2008; Kosow and Gaßner 2008). Importantly, these different futures are not extrapolations or variations around a base case. Rather, each scenario in a set offers a structurally different view of the future, based on different assumptions about groups of key variables that might shape systems change into the future (Bishop, Hines, and Collins 2007; Alcamo 2008; Kosow and Gaßner 2008).

Since scenario development uses narratives to connect sometimes unrelated drivers of change or projections, it can create stories around disconnected and random events. Scenario development can also overemphasise porous system boundaries to enable clear stories to emerge. The real world is full of complex, cross-scale drivers and feedbacks, therefore creating understandable, cohesive yet rich stories is very challenging. However, by incorporating diverse types of knowledge, being flexible and telling stories, scenario construction can inject an appreciation of the variety of possible futures into decision-making, planning and science.

Scenario development is used for decision-making, sense-making and to change mindsets (Bishop, Hines, and Collins 2007). Similar to the futures analysis methods listed in Chapter 10, the overarching aim of scenario development is to 'work' with and 'learn' from reflecting on the future in order to make better decisions and choices in the present because our actions in the present can influence the future (Glenn and Gordon 2009; Kosow and Gaßner 2008). There are many different scenario development approaches, and SES scenarios often involve a combination of qualitative and quantitative methods and tools. Scenarios draw on our scientific understanding of historical patterns, current conditions, physical and social processes, and relationships. They also draw on the imagination to conceive, articulate and evaluate alternative pathways of development and the impacts and interactions of these with the environment (Kosow and Gaßner 2008). This understanding is then used to illuminate links between issues, the relationships between global and regional development, and the role of human actions in shaping the future (Raskin and Kemp-Benedict 2004).

SES problems and questions

Scenario planning methods typically share a combination of the following main objectives (Wright, Bradfield, and Cairns 2013):

- To increase our understanding of the consequences of interactions and interrelations in a system, including causal processes (e.g. what are the impacts of different development futures on various ecosystem services and human well-being? (MA 2005))
- To challenge prevailing or entrenched thinking and reframe perceptions, potentially changing the mindsets of decision-makers (participants) (e.g. what current small-scale initiatives or innovations could potentially radically reshape the future way in which the world works? (Bennett et al. 2016; Pereira et al. 2018))
- To improve decision-making with a view to better strategy and policy development (e.g. what policies or strategies are robust in a variety of different potential futures? (Rosa et al. 2017))

An additional objective is to facilitate stakeholder engagement and knowledge co-production in order to draw on and integrate different knowledge types and improve the legitimacy of decision-making (e.g. how can scientific, and indigenous and local knowledge be integrated to better understand the future? (Sandker et al. 2009; Vervoort et al. 2013)).

Social-ecological scenarios can help to clarify, distinguish and explore social-ecological feedbacks, uncertainties and potential surprises, and enable exploration of the dynamics and sustainability of SES. The Millennium Ecosystem Assessment (MA) scenarios, for example, involved a combination of narrative storylines and detailed quantitative models to explore the future of a wide range of ecosystem services and their implications for human well-being, at local, regional and global scales (MA 2005). Scenarios can also be used to explore important cross-scale processes and feedbacks that link local, regional and global scales (Rosa et al. 2017).

Scenarios can enable inclusive, participatory, dialogue-stimulating processes that are essential for exploring the normative dimensions of sustainable development. Participatory scenario processes are extensively used in SES research and can enhance stakeholder engagement in decision-making processes (Oteros-Rozas et al. 2015). These processes allow for the integration of stakeholder views and increase the relevance, acceptance and legitimacy of scenario findings (Kok, Biggs, and Zurek 2007). Participatory scenarios can also provide an important avenue for integrating practitioner, indigenous and local knowledge with scientific knowledge, which can fill important information gaps, increase agency and empower stakeholders (Tengö et al. 2014; IPBES 2016; Bourgeois et al. 2017; Falardeau, Raudsepp-Hearne, and Bennett 2018). Scenario development can therefore contribute to dealing with the methodological challenges of sustainability science by linking local and global perspectives and accounting for temporal inertia and urgency. This can be done by linking long-term goals; highlighting complex linkages, multiple stressors and inconsistencies; and revealing links among disciplines, themes and issues (Swart, Raskin, and Robinson 2004).

Scenario development can help to identify novel lines of enquiry and challenge existing assumptions about how the world works (Ramirez et al. 2015). The Mānoa mash-up method, for example, explores how a variety of technological and social innovations may come together to change the current major driving forces of change in the world (Pereira et al. 2018; Raudsepp-Hearne et al. 2019). Multiple alternative futures can be used to 'stress-test' or screen strategic objectives, plans, policies and projects to see whether they will perform well under different possible futures. It is often found that plans and policies become obsolete or fail if the future does not turn out as expected (Enfors et al. 2008). The exercise to determine how future changes might affect the ability to deliver a set of strategic objectives, plans or policies is called 'wind tunnelling'.

Different types of scenario processes can address the needs of alternative policy and decision contexts (IPBES 2016, Figure 5.1). The policy process can be considered as consisting of agenda-setting, policy-design, implementation and evaluation contexts. While bridging different perspectives is particularly important in agenda-setting contexts, evaluating alternatives is particularly important in policy-design contexts. Exploratory scenarios examine a range of plausible futures based on potential trajectories of key drivers and can contribute significantly to understanding system dynamics, high-level problem identification and agenda setting. In contrast, intervention scenarios focus on informing policy design and implementation by evaluating alternative policy or management options (IPBES 2016).

Brief description of key methods

Many different approaches to scenario development exist, including participatory, expert-driven, qualitative, quantitative and hybrid approaches. 'Participatory' here refers specifically to collaborative scenario processes that involve directly affected stakeholders. Participatory scenario development can be exclusively qualitative, but often involves hybrid approaches that use a combination of qualitative and quantitative approaches (Oteros-Rozas et al. 2015). In hybrid approaches, scenario storylines are often initially developed using qualitative approaches and then used to parameterise one or more models to explore the outcomes and check for plausibility, which may then feed back into revising the scenario storyline (Alcamo, Van Vuuren, and Ringler 2005). These story and simulation approaches may involve a wide variety of different modelling approaches, including integrated assessment models, agent-based models (Chapter 28), state-and-transition models (Chapter 27) and dynamical systems modelling (Chapter 26), and draw on various participatory modelling approaches (Chapter 13). Protocols that deal with the challenges of converting 'narrative' to 'numbers' in these hybrid scenarios include methods such as cross-impact balance and simulation (CIBAS) (Kosow 2011) and fuzzy cognitive mapping (FCM) (Kok 2009). In contrast to participatory scenarios, expert-driven scenarios are created with only expert input. Although these scenarios may involve qualitative aspects, they tend to focus on quantitative model-based exercises.

Within these broader categories, there is a range of methods for developing scenarios (Glenn and Gordon 2009). By far the most popular and well-known approach is the '2×2 double uncertainty matrix'. Other methods include scenario archetypes, Mānoa scenario building, *La Prospective*, which involves morphological scenarios (also known as field anomaly relaxation or FAR), causal layered analysis and new developments where methods are combined (Table 11.1). These methods range from 'harder', more technically oriented approaches to 'softer', more intuitive methods and from those focused more on structure to those focused more on values (Curry 2012).

Table 11.1 Summary of key scenario development approaches most commonly used in SES research

Approach	*Description*	*References*
Quantitative scenarios	Quantitative scenarios rely on quantitative simulation models to generate plausible outcomes under different simulated conditions.	*Key introductory text* Popper 2008 *Applications to SES* Nelson 2005
Participatory scenarios	Participatory scenarios can use a variety of exclusively qualitative methods, but more often use a hybrid approach (see below) to engage different stakeholders in the scenario process.	*Key introductory text* Popper 2008 *Applications to SES* Palomo et al. 2011; Oteros-Rozas et al. 2015
Hybrid scenarios	Hybrid scenarios, also called the story and simulation approach (described above), broadly refer to the combination of qualitative and quantitative scenario methodologies.	*Key introductory texts* Kemp-Benedict 2004; Kosow 2011 *Applications to SES* Alcamo, Van Vuuren, and Ringler 2005
2×2 double uncertainty matrix	The 2×2 double uncertainty matrix (also known as the GBN method) is the most common and best-known method. By brainstorming and analysing key drivers of change – 'driving forces' of the issue of concern – participants choose two highest impact, highly uncertain drivers, often after clustering, and extrapolate them to their opposite extremes (polarities) to provide four cells representing the kernels of four alternative futures. These are then elaborated into alternative stories or images.	*Key introductory text* Schwartz 1991 *Applications to SES* Hamann et al. 2012

(*Continued*)

Table 11.1 (Continued)

Approach	*Description*	*References*
Archetypes or using existing scenario sets	The archetype method makes use of generic images of the future (or existing scenario sets), typically categorised as Growth, Collapse/New Beginnings, Discipline and Transform. These archetypal, generic images of the future are presented to participants who are then asked to add details to the scenarios, using data and specifics of their issue of concern. Participants can consider how they would redefine, reinvent or otherwise transform their objectives, activities, plans or policies to succeed under each scenario. The original archetypes were derived from a content analysis of futures research and other forecasts available in the 1970s.	*Key introductory texts* Dator 2009, 2017 *Applications to SES* Carpenter et al. 2015; Sitas et al. 2019
The Mānoa method	The Mānoa method for constructing scenarios is used to maximise difference from the present. It is usually used to create exploratory, possible scenarios but can be adapted for creating preferred, normative scenarios. The Mānoa method relies on futures wheels, which is a futures analysis method (see Chapter 10) used for identifying cascading waves of change and dynamics in the system under consideration.	*Key introductory text* Schultz 2015b *Applications to SES* Pereira et al. 2018; Hichert, Biggs, and Preiser 2019
La Prospective	*La Prospective* is a French approach to scenario planning, encompassing quantitative and qualitative techniques and relying on computer-based tools to analyse structural conditions and stakeholder positions. It has a morphological element (also known as field anomaly relaxation) combined with a participatory aspect, and has recently been adapted to develop 'futures literacy' (the capability of 'using' the future to change the present, to change the future) to empower grassroots level stakeholders.	*Key introductory texts* Godet 1986; Bourgeois et al. 2017 *Applications to SES* Del Mar Delgado-Serrano et al. 2015
Causal layered analysis	Causal layered analysis is a futures analysis method (Chapter 10) that can also be used to generate scenarios. It translates different ways of knowing into four layers: (a) 'litany' (the way in which trends and issues are presented in the public domain), (b) 'systems' (causal and institution-based understanding), (c) 'worldview', and (d) 'metaphor'. Scenarios are developed by working through the layers to worldview and metaphor, then 'inflecting' (fundamentally changing) them. The scenarios emerge by reinterpreting the layers through the lens of the deepest level inflection.	*Key introductory text* Inayatullah 2004 *Applications to SES* Heinonen et al. 2017
Combining methods	Constructing scenarios is practice based and therefore constantly evolving and being adapted for specific purposes. This means methods are increasingly being combined with one another and other futures/foresight tools, methods and approaches (Chapter 10). This includes combining axes of uncertainty (2×2 double uncertainty matrix) with visioning, causal layered analysis with gaming, and visionary scenarios with the three horizons framework and back-casting.	*Applications to SES* Heinonen et al. 2017; Falardeau 2018; Pereira et al. 2018; Hichert, Biggs, and Preiser 2019; Iwaniec et al. 2020

Limitations

Scenario development exists between the rigour of process-based models and the flexibility of stories. As such, some of the strengths of scenario planning are also limitations when viewed from other perspectives. Because scenario development often uses approximations rather than 'exact' inputs, it risks being less extreme or variable than reality itself. Scenarios cannot, and should not, be used for forecasting and/or prediction purposes. In participatory scenarios, a lack of experienced facilitators and flaws in participant selection owing to a lack of diversity, dysfunctional power relations or an inability to take cognitive bias into account can compromise the scenario development process and lead to weak or biased scenarios that do not meet the objectives of the exercise. Participatory processes can be difficult to conduct in situations where there is a lack of trust, extreme power inequality and a lack of resources. In these situations, other approaches may be more fruitful.

Hybrid scenarios – the story and simulation approach – can be problematic owing to differences between qualitative and quantitative methods, levels of stakeholder expertise and attempts to integrate different types of knowledge into the analysis (Wiebe et al. 2018). Converting narrative statements embedded in scenario storylines into quantifiable parameters for numerical modelling can be difficult if the variables addressed in the quantitative versus qualitative processes are quite different. In these situations, the quantitative variables tend to be emphasised at the expense of the intangible or difficult-to-quantify variables (Booth et al. 2016). Davenport et al. (2018) propose a simple analytical framework based on five categories of capital assets as part of a protocol for overcoming the conversion problem in hybrid scenario analysis.

These limitations are easier to address when scenario planning is embedded in ongoing, long-term research processes that already bridge different groups and knowledge systems. These long-term relationships can improve the quality of scenarios, reduce the difficulty of creating them and ensure that they can be more easily connected to policy and decision-making. When scenarios are embedded in an ongoing process, their use and impact are also easier to evaluate.

Resource implications

Successful hybrid participatory scenario planning exercises, especially those dealing with social-economic systems, rely on seasoned facilitators, process designers, modellers, project coordinators and enough time to integrate diverse types of knowledge, often on an iterative basis. This all adds up to expense. Oteros-Rozas et al. (2015) make good recommendations on how to deal with these challenges by, among others things, building on existing networks and scenario sets.

The quantitative modelling aspect requires access to specialised modelling skills, as well as the hardware and software elements associated with them, whereas the qualitative, participatory aspect relies on skilled facilitators and appropriate stakeholders as participants in workshop settings.

New directions

Participatory scenario planning methods are constantly evolving and new practices in the broader futures and scenarios fields will continue to spill over into SES research. A number of emerging methods attempt to better incorporate chaos, complexity and tipping points

Case study 11.1: The development of the Millennium Ecosystem Assessment scenarios

The goal of the Millennium Ecosystem Assessment (MA) was to provide decision-makers and stakeholders with scientific information on the links between ecosystem change and human well-being. The MA was ground-breaking in its scope and social-ecological approach (Carpenter et al. 2009). The scenarios component of the MA set out to use creative thinking and storytelling, combined with quantitative modelling of drivers and trends, to develop potential global futures of ecosystem change and human well-being.

The development of the MA scenarios at the global level involved three core phases: (a) organisation, (b) scenario storyline and quantification, and (c) synthesis, review and dissemination (Alcamo, Van Vuuren, and Ringler 2005). The first phase involved establishing a scenario guidance team, a scenario panel, conducting interviews with scenario end-users, determining the objectives and focus of the scenarios, and devising the focal questions of the scenarios. For the MA, the guiding question was identified as: 'What are the consequences of plausible changes in development paths for ecosystems and their services over the next 50 years and what will be the consequences of those changes for human well-being?' (Alcamo, Van Vuuren, and Ringler 2005). This broader question was then focused around four key themes relating to economic and human development, local and regional safety and protection, the use of technologies, and adaptive management and learning about the consequences of management interventions for ecosystem services.

In the second phase, the MA's storyline teams constructed a zero-order draft of scenario storylines based on the key guiding questions and a review and evaluation of past scenario efforts (Raskin 2005). At the same time, a team of modellers was organised to quantify the scenarios. Five global models covering global change processes and provisioning ecosystem services, and two models describing changes in biodiversity, were chosen. Test calculations were carried out. After several iterations, the zero-order storylines were adjusted and cross-checked for internal consistency. These zero-draft scenarios allowed the modelling team, in consultation with the storyline team, to set quantitative levels of key driving forces consistent with the storylines (Nelson 2005). These driving forces included both indirect drivers (demographic, economic, cultural and religious, socio-political, science and technology drivers) and direct drivers (climate variability and change, plant nutrient use, land conversion, biological invasions and diseases) of ecosystem change. Based on the subsequent model outcomes, and a number of feedback workshops with the MA board and stakeholder groups, the storylines were further iterated and focused into first-order storylines. These iterated storylines were then used to inform the revision of model inputs, after which the models were rerun to analyse impacts on ecosystem services and human well-being.

In the final phase, the global scenarios (both qualitative storylines and quantitative modelling calculations) were distributed for general review. Feedback from this process, which included presentations, workshops and the MA review process, were incorporated into the final published version of the storylines and their associated models. Four scenarios emerged from the analysis (MA 2005, Figure 1). Three of the four scenarios suggested that significant changes in policies and institutions can mitigate some of the negative consequences of anthropogenic pressures on the planet. Despite limitations, the MA scenarios influenced many policy processes, conventions and businesses at global and regional scales, and represent a landmark change in the way global change assessments engage with futures methods (Reid and Mooney 2016).

A particular challenge for the MA was the multi-scale nature of the relationships between ecosystem services and human well-being. This was addressed through assessments at different spatial scales. Many of the MA regional assessments developed their own scenarios, which were linked to the global scenarios by incorporating some of the global storylines into regional processes, having members of the global team participate in regional assessments and developing regional storylines for the global archetypes (Alcamo, Van Vuuren, and Ringler 2005).

The MA developed four global-scale scenarios that had a landmark impact on understanding long-term social-ecological change (Figure 11.1). The 'Global orchestration' scenario depicted a highly connected world with well-developed global markets and supranational institutions that deal with environmental problems. In the 'Order from strength' scenario, the world is disconnected, fragmented and individualised, with an emphasis on security and the protection of regional economies. In the 'Adapting mosaic' scenario, discredited global institutions have been replaced by stronger local institutions aimed at improving local ecosystem management. In the 'TechnoGarden' scenario, ecosystem services are often delivered by engineered and highly managed systems, some of which have unintended consequences such as the loss of local culture, customs and traditional knowledge.

Figure 11.1 shows the net change in the number of ecosystem services enhanced or degraded under each of the four scenarios, for each of the four ecosystem service categories. A 100% degradation or enhancement score means all the services in that category would be degraded or enhanced by 2050.

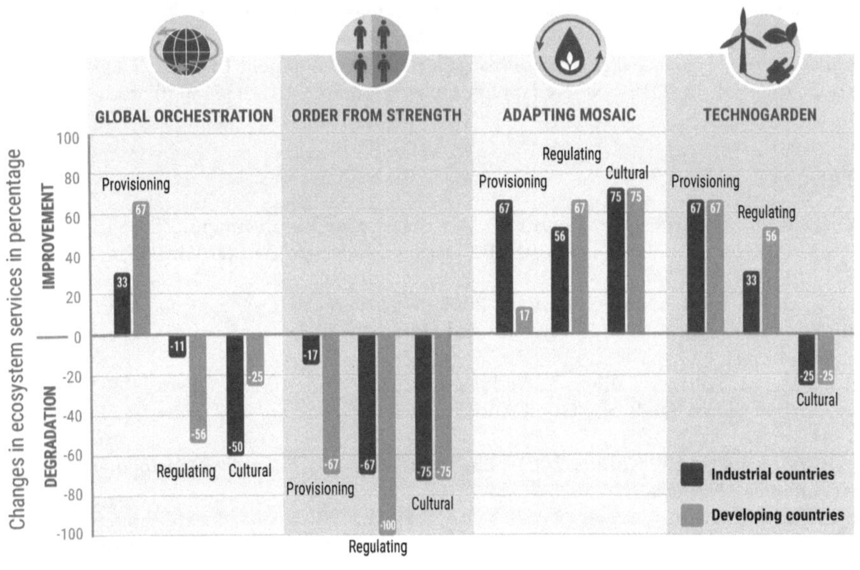

Figure 11.1 Net change in the number of ecosystem services enhanced or degraded under each of the four scenarios, for each of the four ecosystem service categories (© Millennium Ecosystem Assessment 2005)

(black swan type surprises or wildcards) (e.g. Hamann et al. 2012). In the field of futures studies, concepts such as 'crazy futures' (Schultz 2015a), 'preposterous futures' (Voros 2017) and 'post-normal times' (Sardar and Sweeney 2016) are expected to influence existing methods and inspire new ones.

Innovative and creative practices around combining different scenario methods and/or the mashing up of scenarios and foresight methods are also expected to grow. This includes incorporating activities and practices such as gaming, design, art and virtual reality.

Experiential scenarios (Candy 2010), science-fiction prototyping (Merrie et al. 2018) and scenarios based on crowdsourcing, using tools such as SenseMaker (sensemaker.cognitive-edge.com) and Futurscaper (futurescaper.com) (Raford 2012), are emerging methods and practices that hold particular promise for the SES field.

Finally, expanding the number of people able to participate in scenario processes can have great potential. Most participatory scenarios involve relatively small groups of people. However, there have been experiments with massive online scenarios that have the potential to include many more voices (McGonigal 2011). Developing new scenario methods that enable large-scale participation could be very useful in incorporating teleconnections and plurality, and creating more radical futures to explore the interconnected world of the Anthropocene.

Key readings

Alcamo, J. 2008. 'Environmental Futures: The Practice of Environmental Scenario Analysis.' *Developments in Integrated Environmental Assessment, Volume 2*. Amsterdam: Elsevier. doi:10.1016/S1574-101X(08)00406-7.

Bishop, P., A. Hines, and T. Collins. 2007. 'The Current State of Scenario Development: An Overview of Techniques.' *Foresight* 9(1): 5–25.

Curry, A. 2012. 'The Scenarios Question.' In *The Future of Futures*, edited by A. Curry. Houston: Association of Professional Futurists.

Glenn, J.C., and T.J. Gordon. 2009. *Futures Research Methods 3.0*. Millennium Project. www.millennium-project.org/publications-2/futures-research-methodology-version-3-0.

Kosow, H., and R. Gaßner. 2008. *Methods of Future and Scenario Analysis: Overview, Assessment, and Selection Criteria, Volume 39*. In Studies from German Development Institute/Deutsches Institut für Entwicklungspolitik (DIE). edoc.vifapol.de/opus/volltexte/2013/4381/pdf/Studies_39.2008.pdf.

References

Alcamo, J. 2008. 'Environmental Futures: The Practice of Environmental Scenario Analysis.' *Developments in Integrated Environmental Assessment, Volume 2*. Amsterdam: Elsevier. doi:10.1016/S1574-101X(08)00406-7.

Alcamo, J., D. van Vuuren, and C. Ringler. 2005. *Methodology for Developing the MA Scenarios*. Millennium Ecosystem Assessment. Ecosystems and Human Well-being: Scenario Assessment. Washington: Island Press.

Bennett, E.M., M. Solan, R. Biggs, T. McPhearson, A.V. Norström, P. Olsson, S.R. Carpenter et al. 2016. 'Bright Spots: Seeds of a Good Anthropocene.' *Frontiers in Ecology and the Environment* 14(8): 441–448.

Bishop, P., A. Hines, and T. Collins. 2007. 'The Current State of Scenario Development: An Overview of Techniques.' *Foresight* 9(1): 5–25.

Booth, E.G., J. Qiu, S.R. Carpenter, J. Schatz, X. Chen, C.J. Kucharik, S.P. Loheide II et al. 2016. 'From Qualitative to Quantitative Environmental Scenarios: Translating Storylines into Biophysical Modeling Inputs at the Watershed Scale.' *Environmental Modelling and Software* 85: 80–97. doi:10.1016/j.envsoft.2016.08.008.

Bourgeois, R., E. Penunia, S. Bisht, and D. Boruk. 2017. 'Foresight for All: Co-elaborative Scenario Building and Empowerment.' *Technological Forecasting and Social Change* 124: 178–188.

Bradfield, R., G. Wright, G. Burt, G. Cairns, and K. van der Heijden. 2005. 'The Origins and Evolution of Scenario Techniques in Long Range Business Planning.' *Futures* 37(8): 795–812.

Candy, S. 2010. *The Futures of Everyday Life: Politics and the Design of Experiential Scenarios*. PhD diss., University of Hawaii at Mānoa.

Carpenter, S.R., E.M. Bennett, and G.D. Peterson. 2006. 'Scenarios for Ecosystem Services: An Overview.' *Ecology and Society* 11(1): 29. www.ecologyandsociety.org/vol11/iss1/art29/.

Carpenter, S.R., E.G. Booth, S. Gillon, C.J. Kucharik, S. Loheide, A.S. Mase, M. Motew et al. 2015. 'Plausible Futures of a Social-Ecological System: Yahara Watershed, Wisconsin, USA.' *Ecology and Society* 20(2): 10. doi:10.5751/ES-07433-200210.

Carpenter, S.R., H.A. Mooney, J. Agard, D. Capistrano, R.S. deFries, S. Díaz, T. Dietz et al. 2009. 'Science for Managing Ecosystem Services: Beyond the Millennium Ecosystem Assessment.' *Proceedings of the National Academy of Sciences* 106(5): 1305–1312.

Curry, A. 2012. 'The Scenarios Question.' In *The Future of Futures*, edited by A. Curry. Houston: Association of Professional Futurists.

Dator, J. 2009. 'Alternative Futures at the Mānoa School.' *Journal of Futures Studies* 14(2): 1–18.

Dator, J. 2017. 'Some Sources of the Generic Four Alternative Images of the Futures of the Mānoa School.' *Design Develop Transform Conference*, Antwerp. https://ddtconference.files.wordpress.com/2016/06/dator-sourcesfourfutures.pdf.

Davenport, M., M. Delport, J.N. Blignaut, T. Hichert, and G. van der Burgh. 2018. 'Combining Theory and Wisdom in Pragmatic, Scenario-based Decision Support for Sustainable Development.' *Journal of Environmental Planning and Management* 62(4): 692–716.

Del Mar Delgado-Serrano, M., E. Oteros-Rozas, P. Vanwildemeersch, C. Ortíz-Guerrero, S. London, and R. Escalante. 2015. 'Local Perceptions on Social-Ecological Dynamics in Latin America in Three Community-based Natural Resource Management Systems.' *Ecology and Society* 20(4): 24. doi:10.5751/ES-07965-200424.

Enfors, E.I., L.J. Gordon, G.D. Peterson, and D. Bossio. 2008. 'Making Investments in Dryland Development Work: Participatory Scenario Planning in the Makanya Catchment, Tanzania.' *Ecology and Society* 13(2): 42.

Falardeau, M., C. Raudsepp-Hearne, and E.M. Bennett. 2018. 'A Novel Approach for Co-producing Positive Scenarios that Explore Agency: Case Study from the Canadian Arctic.' *Sustainability Science* August: 1–16.

Glenn, J.C., and T.J. Gordon, eds. 2009. *Futures Research Methodology Version 3.0*. The Millennium Project. www.millennium-project.org/publications-2/futures-research-methodology-version-3-0.

Godet, M. 1986. 'Introduction to La Prospective.' *Futures* 18: 134–157. doi:10.1016/0016-3287(86)90094-7.

Hamann, M., V. Masterson, R. Biggs, M. Tengö, B. Reyers, L. Dziba, and M.J. Spierenburg. 2012. 'Social-Ecological Scenarios for the Eastern Cape Province, South Africa 2012–2050.' https://sapecs.org/wp-content/uploads/2013/08/Eastern-Cape-Scenarios-Report-Aug-2012_-final.pdf.

Heinonen, S., M. Minkkinen, J. Karjalainen, and S. Inayatullah. 2017. 'Testing Transformative Energy Scenarios Through Causal Layered Analysis Gaming.' *Technological Forecasting and Social Change* 124: 101–113.

Hichert, T., R. Biggs, and R. Preiser. 2019. *Generating Visions of Good Anthropocenes: The Mānoa Mash-up Scenarios Methodology*. CST Toolkit. University of Stellenbosch. www0.sun.ac.za/cst/publication/generating-visions-of-good-anthropocenes-the-manoa-mash-up-scenarios-methodology/.

Inayatullah, S. 2004. 'Causal Layered Analysis: Theory, Historical Context, and Case Studies.' In *The Causal Layered Analysis (CLA) Reader*, edited by S. Inayatullah. Taipei: Tamkang University.

IPBES. 2016. *Summary for Policymakers of the Methodological Assessment of Scenarios and Models of Biodiversity and Ecosystem Services of the Intergovernmental Science-Policy Platform on Biodiversity and Ecosystem Services*. Edited by S. Ferrier, K.N. Ninan, P. Leadley, R. Alkemade, L.A. Acosta, H.R. Akçakaya, L. Brotons et al. Secretariat of the Intergovernmental Science-Policy Platform on Biodiversity and Ecosystem Services, Bonn, Germany. www.ipbes.net/sites/default/files/downloads/pdf/spm_deliverable_3c_scenarios_20161124.pdf.

Iwaniec, D.M., E.M. Cook, M.J. Davidson, M. Berbés-Blázquez, M. Georgescu, E.S. Krayenhoff, A. Middel, D.A. Sampson, and N.B. Grimm. 2020. 'The Co-production of Sustainable Future Scenarios.' *Landscape and Urban Planning* 197: 103744.

Kemp-Benedict, E. 2004. 'From Narrative to Number: A Role for Quantitative Models in Scenario Analysis.' In *iEMSs 2004 International Congress: Complexity and Integrated Resources Management*, edited by C. Pahl-Wostl, S. Schmidt, and T. Jakeman, 765–770. Osnabrück: International Environmental Modelling and Software Society.

Kok, K. 2009. 'The Potential of Fuzzy Cognitive Maps for Semi-Quantitative Scenario Development, with an Example from Brazil.' *Global Environmental Change* 19: 122–133. doi:10.1016/j.gloenvcha.2008.08.003.

Kok, K., R. Biggs, and M. Zurek. 2007. 'Methods for Developing Multiscale Participatory Scenarios: Insights from Southern Africa and Europe.' *Ecology and Society* 12(1): 8.

Kok, M.T., K. Kok, G.D. Peterson, R. Hill, J. Agard, and S.R. Carpenter. 2017. 'Biodiversity and Ecosystem Services Require IPBES to Take Novel Approach to Scenarios.' *Sustainability Science* 12(1): 177–181.

Kosow, H. 2011. 'Consistent Context Scenarios: A New Approach to Story and Simulation.' Paper presented at the *4th International Seville Conference on Future-Oriented Technology Analysis (FTA): FTA and Grand Societal Challenges – Shaping and Driving Structural and Systemic Transformations*, Seville, May 2011.

Kosow, H., and R. Gaßner. 2008. *Methods of Future and Scenario Analysis: Overview, Assessment, and Selection Criteria, Volume 39.* In Studies from German Development Institute/Deutsches Institut für Entwicklungspolitik (DIE). edoc.vifapol.de/opus/volltexte/2013/4381/pdf/Studies_39.2008.pdf.

MA (Millennium Ecosystem Assessment). 2005. *Ecosystems and Human Well-Being: Scenarios*, edited by S.R. Carpenter, P.L. Pingali, E.M. Bennett, and M.B. Zurek. Washington: Island Press.

McGonigal, J. 2011. *Reality is Broken: Why Games Make Us Better and How They Can Change the World.* London: Penguin.

Merrie, A., P. Keys, M. Metian, and H. Österblom. 2018. 'Radical Ocean Futures-scenario Development Using Science Fiction Prototyping.' *Futures* 95: 22–32.

Nelson, G. 2005. 'Drivers of Change in Ecosystem Condition and Services.' Millennium Ecosystem Assessment. *Ecosystems and Human Well-Being: Scenario Assessment.* Washington: Island Press.

Oteros-Rozas, E., B. Martín-López, T. Daw, E.L. Bohensky, J. Butler, R. Hill, J. Martin-Ortega et al. 2015. 'Participatory Scenario Planning in Place-based Social-Ecological Research: Insights and Experiences from 23 Case Studies.' *Ecology and Society* 20(4): 32. doi:10.5751/ES-07985-200432.

Palomo, I., B. Martín-López, C. López-Santiago, and C. Montes. 2011. 'Participatory Scenario Planning for Protected Areas Management under the Ecosystem Services Framework: The Doñana Social-Ecological System in Southwestern Spain.' *Ecology and Society* 16(1): 23.

Pereira, L., T. Hichert, M. Hamann, R. Preiser, and R. Biggs. 2018. 'Using Futures Methods to Create Transformative Spaces: Visions of a Good Anthropocene in Southern Africa.' *Ecology and Society* 23(1): 19.

Peterson, G.D., G.S. Cumming, and S.R. Carpenter. 2003. 'Scenario Planning: A Tool for Conservation in an Uncertain World.' *Conservation Biology* 17(2): 358–366.

Popper, R. 2008. 'Foresight Methodology.' In *The Handbook of Technology Foresight: Concepts and Practices*, edited by L. Georghiou, J. Cassingena Harper, M. Keenan, I. Miles, and R. Popper, 44–88. Cheltenham: Edward Elgar.

Raford, N. 2012. 'Crowdsourced Futures.' In *The Future of Futures*, edited by A. Curry. Houston: Association of Professional Futurists.

Ramirez, R., M. Mukherjee, S. Vezzoli, and A.M. Kramer. 2015. 'Scenarios as a Scholarly Methodology to Produce "Interesting Research".' *Futures* 71: 70–87.

Raskin, P.D. 2005. 'Global Scenarios: Background Review for the Millennium Ecosystem Assessment.' *Ecosystems* 8(2): 133–142.

Raskin, P.D., and E. Kemp-Benedict. 2004. *Global Environment Outlook Scenario Framework.* United Nations Environment Programme.

Raudsepp-Hearne, C., G.D. Peterson, E.M. Bennett, R. Biggs, A.V. Norström, L. Pereira, J. Vervoort et al. 2019. 'Seeds of Good Anthropocenes: Developing Sustainability Scenarios for Northern Europe.' *Sustainability Science.* doi:10.1007/s11625-019-00714-8.

Reid, W.V., and H.A. Mooney. 2016. 'The Millennium Ecosystem Assessment: Testing the Limits of Interdisciplinary and Multi-scale Science.' *Current Opinion in Environmental Sustainability* 19: 40–46.

Rosa, I.M., H.M. Pereira, S. Ferrier, R. Alkemade, L.A. Acosta, H.R. Akcakaya, E. den Belder et al. 2017. 'Multiscale Scenarios for Nature Futures.' *Nature Ecology and Evolution* 1(10): 1416–1419.

Sandker, M., B.M. Campbell, Z. Nzooh, T. Sunderland, V. Amougou, L. Defo, and J. Sayer. 2009. 'Exploring the Effectiveness of Integrated Conservation and Development Interventions in a Central African Forest Landscape.' *Biodiversity and Conservation* 18(11): 2875–2892.

Sardar, Z., and J.A. Sweeney. 2016. 'The Three Tomorrows of Postnormal Times.' *Futures* 75: 1–13.
Schultz, W.L. 2015a. *Crazy Futures: Why Plausibility is Maladaptive.* doi:10.13140/RG.2.1.2897.9921
Schultz, W. 2015b. *Mānoa: The Future is Not Binary.* APF Compass, Methods Anthology Special Edition, 22–26 April 2015.
Schwartz, P. 1991. *The Art of the Long View.* New York: Doubleday.
Sitas N., Z.V. Harmáčková, J.A. Anticamara, A. Arneth, R. Badola, R. Biggs, R. Blanchard et al. 2019. 'Exploring the Usefulness of Scenario Archetypes in Science-Policy Processes: Experience across IPBES Assessments.' *Ecology and Society* 24(3): 35.
Swart, R.J., P. Raskin, and J. Robinson. 2004. 'The Problem of the Future: Sustainability Science and Scenario Analysis.' *Global Environmental Change* 14(2): 137–146.
Tengö, M., E.S. Brondizio, T. Elmqvist, P. Malmer, and M. Spierenburg. 2014. 'Connecting Diverse Knowledge Systems for Enhanced Ecosystem Governance: The Multiple Evidence Base Approach.' *Ambio* 43(5): 579–591.
Vervoort, J.M., A. Palazzo, D. Mason-D'Croz, P.J. Ericksen, P.K. Thornton, P. Kristjanson, W. Förch et al. 2013. 'The Future of Food Security, Environments and Livelihoods in Eastern Africa: Four Socio-Economic Scenarios.' CCAFS Working Paper No. 63. Copenhagen, Denmark: CGIAR Research Program on Climate Change, Agriculture and Food Security (CCAFS). http://hdl.handle.net/10568/34864.
Voros, J. 2017. 'Big History and Anticipation.' In *Handbook of Anticipation: Theoretical and Applied Aspects of the Use of Future in Decision Making*, edited by R. Poli, 1–40. New York: Springer.
Wiebe, K., M. Zurek, S. Lord, N. Brzezina, G. Gabrielyan, J. Libertini, A. Loch, R. Thapa-Parajuli, J. Vervoort, and H. Westhoek. 2018. 'Scenario Development and Foresight Analysis: Exploring Options to Inform Choices.' *Annual Review of Environment and Resources* 43: 545–570.
Wright, G., R. Bradfield, and G. Cairns. 2013. 'Does the Intuitive Logics Method – and its Recent Enhancements – Produce "Effective" Scenarios?' *Technological Forecasting and Social Change* 80: 631–642.

12

Serious games

Olivier Barreteau,[1] *Géraldine Abrami,*[1] *Bruno Bonté,*[1] *François Bousquet*[2] *and Raphaël Mathevet*[3,4]

[1]G-EAU, UNIV MONTPELLIER, AGROPARISTECH, BRGM, CIRAD, IRD, INRAE, INSTITUT AGRO, MONTPELLIER, FRANCE
[2]CIRAD, UMR SENS, F-34398 MONTPELLIER, FRANCE; SENS, UNIVERSITY OF MONTPELLIER, CIRAD, IRD, MONTPELLIER, FRANCE
[3]CEFE, CNRS, UNIVERSITY OF MONTPELLIER, UNIVERSITY PAUL VALÉRY MONTPELLIER III, EPHE, IRD, MONTPELLIER, FRANCE
[4]IFP UMIFRE 21 CNRS/MAEE, PONDICHERRY, INDIA

Key methods discussed in this chapter

Serious games

Connections to other chapters

Serious games are commonly used as part of the toolbox for participatory approaches, including for modelling or planning (Chapter 13). They are designed according to a conceptual model, coming from various types of modelling approaches, including system dynamics (Chapter 26) and agent-based modelling (Chapter 28). They need systems scoping to identify the roles and entities that need to be considered. To describe the dynamics, they may build on methods from state-and-transition modelling (Chapter 27). Serious games are behavioural experiments, but leave participants more freedom of action and exercise less control than controlled behaviour experiments, which are extreme cases of serious games (Chapter 12).

Introduction

Role-playing games as tools to support the understanding and governance of social-ecological systems (SES) emerged by the end of the 1990s, standing on the shoulders of experimental economics (Friedman and Sunder 1994) and policy exercises (Toth 1988; Duke and Geurts 2004). The experimental economics thread is closely related to controlled behavioural experiments (Chapter 21). The policy exercise thread was originally (as far back as ancient China) developed through war games (Mermet 1993), which are strategic simulations of war or crisis situations so that participants can experience virtually the joint outputs of their behavioural patterns. Policy exercises more recently grew as business games and developed as a type of group decision-support system. They have been included in the broader category of serious games.

Although 'serious games' first appeared in 1974 (Abt 1974), they really emerged in the 2000s and mainly after 2010. This category combines role-playing games, policy exercises and business games, among others. It focuses on the fact that these games are used for serious

 DOI: 10.4324/9781003021339-15

SUMMARY TABLE: SERIOUS GAMES	
DISCIPLINARY BACKGROUND	**KNOWLEDGE TYPE**
The methods in this chapter are derived from or have most commonly been used in: Experimental and Behavioural Economics, Social and Cognitive Psychology, Environmental and Cultural Psychology	*The methods in this chapter are primarily used to generate the following types of knowledge:* • Exploratory • Explanatory
RESEARCH APPROACH	**PURPOSE OF METHOD**
The methods in this chapter originate from or most commonly adopt the following research approaches: • Interpretive/subjective • Collaborative/process	*The most common purposes of using the methods in this chapter are:* • System understanding • Stakeholder engagement and co-production • Policy/decision support
TEMPORAL DIMENSION	**SYSTEMIC FEATURES AND PROCESSES**
The methods in this chapter are most commonly applied to the following temporal dimensions: • Present (typically within the last 5–10 years) • Recent past (post-1700s) • Future	*While most methods can do many things, the methods in this chapter are particularly good (i.e. go-to methods) for addressing the following:* • Transformation • Social learning • Collective action and collaborative governance • Evaluating policy options
SPATIAL DIMENSION	
The methods in this chapter are primarily either or both: • Non-spatial *The methods in this chapter are most commonly applied at the following spatial scales:* • Local	

matters, learning and/or decision-making. While serious games include individual games like online awareness-raising games and role-playing games, most policy exercises and business games are based on interactions among players. In this chapter, we refer to collective serious games unless specified otherwise. Although originally mainly based in the disciplines of economics and management sciences (operations research), serious games have now spread to the realm of conservation policies as a tool for land-use planning or ecology. The purpose is to design and/or implement tools to explore, in predefined scenarios, the consequences of interactions among players with diverse behavioural patterns. The main assumptions underpinning the scenarios are collective frames (e.g. existence of collective rules) and external drivers (e.g. weather sequence).

Serious games can complement other methods but are also appropriate alternatives when observation or experimentation is not suitable due to the time scale, harshness of potential outcomes of experiments or disagreement of some subjects with the experimental setting. They are also useful in exploring decisions and interactions in an 'action context', assuming that the interactional context drives the decisions of the actors. Having originated from 'policy exercises' and 'economic experiments', serious games have domains of application far beyond SES, with the military and security being the primary fields. However, application to issues related to environmental and land-use development came quite early. Within the two original threads, some works dealt with SES before the emergence of serious games as a tool to investigate the dynamics of these systems.

Seminal works by Ostrom, Gardner and Walker (1994) used economic experiments to analyse common-pool resources. The International Institute for Applied System Analysis (IIASA) applied policy exercises to foresights and the negotiation of global environmental issues (Mermet 1993). Companion modelling (Bousquet et al. 2002; Etienne 2011) initiated the convergence of both threads with a focus on SES and common-pool resources.

The emergence of games in the late 1990s and early 2000s was facilitated by two additional dynamics: the epistemology of models and the gamification of society. While the use of models as a tool to predict events used to be the norm, modellers in the 1970s started to discuss other potential uses. It was acknowledged that models could fit different uses, including potentially replacing (physical) experience when their implementation is not possible (Legay 1997). Models were then used to explore the consequences of a combination of assumptions to build new knowledge. The next step was to recognise games as models of social dynamic systems by themselves (Meadows 2001) and games became a possible tool to experiment on these systems. The development of computer capacity and the Internet made this easier and standardised the development of an artificial world for fun. Recent works on education theories point out the capacity of fun situations to generate learning, legitimising the use of serious games in these communities (Kapp 2012).

The main assumptions underlying the method of serious games are, first, related to the way in which participants play. In the game, they are supposed to act according to the environment provided, not according to the outcome they would like to see for the sake of achieving their own strategic agenda.

A second assumption is related to the capacity of games to represent an SES with meaningful reduction of the system's complexity. Constraints in game design include operational aspects such as the duration of the game, which has to fit the time available for participants to play the game, and the fun aspect. The game must be fun to incite participants to set their personal strategy aside, i.e. not to act according to their situation outside of the game or to pursue a personal agenda. These constraints mean the number of actions available to players must be reduced, including those related to their interactions with the fake environment simulated in the game.

A third category of assumptions is linked to the relationship of stakeholders to this type of tool: the willingness to play (is it 'serious' enough?) and the validity and suitability of the outcomes in the 'real world'. Hopefully the addition of 'serious' to the word 'game', reports of positive experiences and a deep trend to propose creative environments that enable participants to generate new knowledge and explore scenarios regarding 'serious' stakes will alleviate initial concerns about and prejudice against the futility of 'playing'. The addition of debriefings to game sessions further allows for meaningful outcomes and potentially paves the way for implementation of these outcomes (Ryan 2000; Meadows 2001).

A fourth and final assumption relies on the capacity of games to lead players to accept simulation as an activity echoing real problems without a direct connection to those problems. To be efficient, games should provide a delimited window for collective exploratory behaviour.

SES problems and questions

So far games have mainly been used to (a) disentangle the complexities of SES in order to help participants make sense of these systems, (b) inform participants about the diversity of viewpoints, interests and constraints present in an SES, (c) stimulate the emergence of desirable changes and actions in the real word that the game represents, and to experiment with them (Le Page et al. 2013), and (d) support adaptive governance of SES.

- **Disentangling complexities:** Disentangling the complexity of an SES means raising people's awareness of interdependence and the basic consequences of this interdependence, such as feedback loops and their cascading effects. It is well known that people find dynamic systems with feedback loops difficult to understand and foresee (Sterman 1992). In a game, the concentration of action within a limited time frame and space makes these consequences of complexity more apparent. Pioneering the use of games for dynamic systems, Meadows (2008) led players to acknowledge the existence of dependence among system components. He made them experience (unexpected) feedbacks of actions, which are consequently at best inefficient (Meadows 2008). The relationships emphasised within game settings are either social or social-ecological. In FishBanks Ltd (Meadows and Meadows 1993), players can observe the consequences of choices of some fishermen on the fish population and the cascading effects of these choices on other fishermen and fish populations. Games are powerful tools to demonstrate the existence and consequences of saturation effects (e.g. on land use), of competition for scarce natural resources and money, or the need for coordination to handle all these relationships.
- **Informing participants:** To inform participants about diversity within an SES, the game practice enacts the diversity of situations through role setting and incomplete information delivered to players during the game. Njoobaari Ilnoowo (Barreteau, Bousquet, and Attonaty 2001), for instance, is a game representing the issue of viability of irrigation systems in Northern Senegal as an outcome of multiple constraints on farmers with diverse cropping objectives. This game visualises the diversity of goals and hence behavioural patterns behind the mere activity of irrigation: practices differ significantly between a farmer cultivating to get the maximum out of his field and a farmer

cultivating to keep his access to land, and the coexistence of these practices generates tensions among farmers and collective inefficiencies.

- **Stimulating and experimenting with desirable changes:** Beyond revealing diversity of behavioural patterns, using games is also a way to get players more acquainted with the viewpoints of others. In this case, players endorse a role other than their own and have to handle their constraints and work towards a common objective. SylvoPast is a game representing the capacity of an SES that includes forestry and cattle farming in the same area, with both facing fire risk. In this game, Etienne made forestry agents play herders, and vice versa (Etienne 2003). This process allowed foresters and herders to experience and feel the constraints they have on one another, realise their mutual dependence to achieve their respective objectives, and enter into dialogue for the co-management of forestry with a more open attitude.
- **Supporting adaptive governance:** Games are used in adaptive governance of SES. Even though governance or co-management can be internalised into the game, serious games are also used as exploratory tools to challenge or support governance. Indeed, while designing a game, governance scales are chosen for the processes represented in the game and for the targeted decision processes. Scales of represented and supported management processes can be identical to processes observed in the real world, or they can be simplified and aggregated or embedded to facilitate game play. This choice depends on whether the issue at stake with the game is to explore macroscale management choices or to experiment with framing microscale management choices.

Game sessions can test institutional settings through a game and discuss them on an overarching organisational scale which might facilitate or prevent their occurrence. Mathevet et al. (2007) proposed a role-playing game (ButorStar) based on a multi-agent model that simulates the effects of wetland multi-use on ecosystem and wildlife dynamics. This tool serves as a training support for students to talk about the pros and cons of various negotiation processes and integrated management approaches. Within a ButorStar session, players experiment with co-management meetings as they are usually promoted in local environmental governance. This active experience enables players to understand the critical issue of time management, especially for sharing key information on trends and cause-effect chain understanding. Solutions experienced in games can then be discussed further at different decision-making levels and related spatial scales (i.e. water or land management units, land estates, ecosystem units and local government scale). The ButorStar game was also used with protected area managers and local actors in several Mediterranean wetlands. The results showed that the approach contributes to an increase in the capacity of actors to implement modes of interactions that promote the adaptive management of wetlands (Mathevet et al. 2008).

Brief description of key methods

To deal with the SES issues listed in the previous section, serious games are used to explore the consequences of internal choices or external drivers, raise awareness of diversity, educate people on system complexity, observe behavioural patterns in specific situations and support the governance of SES. A serious game session typically consists of three steps: briefing, playing and debriefing. The first step, briefing, should ensure that all the participants understand the rules so that they can play. The brief should not be too long, otherwise participants get bored. This step should be prepared according to the complexity of the game design, taking into account how much participants know about the technical actions (roles) they may have to endorse. The second step, playing, lasts typically half of the whole session. In this step, the main issue is facilitation to keep the momentum and identifying participants that might get lost or drop off from playing. This may involve asking some participants to explain their behaviours when they seem to act inconsistently with the game. The last step, debriefing, is the most important one because it generates knowledge for facilitators, observers and participants. Table 12.1 provides a summary of key applications for serious games.

Table 12.1 Summary of key applications of serious games

Main applications	*Description*	*References*
Exploring consequences of collective decisions, as a tentative group decision-support system	This category of use is often related to 'management flight simulators', where each player holds a stick to pilot an SES. However, there are also other types of games, more based on an agent-based model structure.	*Applications to SES* Castella, Trung, and Boissau 2005 (land-use change); Martin et al. 2007 (interdependence of local industry and river-basin management); Krolikowska et al. 2009 (land reclamation); Flint 2013 (community development)
Exploring consequences of external drivers	This category of use is close to the previous one in terms of the objective. However, the focus is less on driving the system and more on elaborating consequences of external changes in complex situations.	*Applications to SES* Villamor and Badmos 2015 (adaptation to climate change, e.g. for grazing in Sahelian countries)
Making people learn about others' constraints and building an understanding of system dynamics	Different types of games lead participants either to swap roles or to explain their worldviews.	*Applications to SES* Etienne 2003 (forest/cattle-grazing competition); Mathevet et al. 2007 (land-use competition in wetlands); Richard and Barreteau 2007; Richard-Ferroudji and Barreteau 2012 (simulation of basin management with various worldviews)
Developing a joint representation of an SES and playing with it	This category builds on a set of predefined items that might be used to collectively build a complex representation, as with building bricks, following a 'design-by-playing' approach. The output is a model that might be implemented in a game or in any hybrid form with a computer-based model.	*Applications to SES* Ferrand et al. 2009 (Wat-A-Game); D'Aquino et al. 2017 (TerriStories: terristories.org)

(*Continued*)

Table 12.1 (Continued)

Main applications	*Description*	*References*
Collecting information on collective behaviour	Observation of game simulation with players performing their usual activities in a controlled situation brings knowledge on some tacit behavioural patterns. We refer here to an open frame of action. The objective is to understand how players (re)act in given situations. When players are framed and have a finite set of choices, they are not supposed to perform their usual activities (see Chapter 21). The whole spectrum between these two extremes is possible. Students appear as 'easy-to-grab' players. They need to have more framing, such as information on their roles, but they can still have more possibilities to play than in controlled behavioural experiments.	*Applications to SES* Souchère et al. 2010 (erosion and farming practices); Merrill et al. 2019 (collective investment in security)
Education on the complexity of SES	Games enable visualisation of the hidden complexities of SES. These complexities could stem from physical reasons (underground processes), social reasons (taboos) and social-ecological reasons (time or spatial scales beyond those usually grasped by participants).	*Applications to SES* L'eau en jeu: eauenjeu.org (simplified education games)
Crisis management training	A group of players is placed in a crisis situation that it has to manage collectively. This method is the closest to the original use of military games.	*Applications to SES* Stolk et al. 2001 (fire, flood, terrorist attacks)
Institutional arrangement	Institutional arrangement is a game that supports a group of stakeholders in piloting and adjusting their collective action processes. An initial game evolves with the emergence of new issues or new perspectives on an SES trajectory.	*Applications to SES* Gurung, Bousquet, and Trébuil 2006 (watershed management)

Limitations

Even though serious games are increasingly being promoted, they cannot be considered as panaceas. Several limitations exist, such as a limit on the number of players, limited duration, a lack of social acceptance, and too narrow or too large a representation of the processes. The critical issue of debriefing presents limitations of its own. It is essential to assess the various forms of knowledge that come not only from the game itself but also from debriefing on what happened during the game session. Game design and management must take these objectives into account to keep track of events during the game so that structured discussion can occur during the debriefing. Careful management of a game is essential to transform it into a meaningful learning experience, so paying attention to time and live data-collection management is essential (Daré et al. 2014).

Potential biases could arise due to the limited size of population samples (i.e. the number of players relative to the population they represent) and the difficulties inherent in power games. Social acceptance is also an issue, due to not only the status of the game but also the self-esteem of participants. Playing together means the playing field between participants is levelled and that people of different social status agree to interact directly. This is not always acceptable to those holding the economic, social or political power.

A bottleneck could arise related to the difficulties of upscaling experience from a role-playing game. Groups of people involved in game sessions constitute small societies, but participants and public engagement may not always be appropriate (Reed et al. 2018). Power relationships, social values and epistemologies of participants have to be identified in order to be able to generalise outcomes of game sessions. It should be borne in mind that engagement outcomes are highly scale dependent over time. Spatial scale, decision levels and the legitimate representation of involved stakeholders should all be taken into account (De Vente et al. 2016).

To overcome these biases, positive outputs have been identified and ways to circumvent the limitations proposed. Role-playing game literature shows three main types of social impacts of serious games: (a) the production of socially robust knowledge that fuels a more effective process of public policy construction, (b) social learning to solve practical problems, and (c) empowering actors by putting them in a position where they can participate in a change process and socio-political transformation. These social impacts are more easily reached in small groups with trust. Technology may facilitate a large population of players via remote control. However, it significantly simplifies the richness of environmental information and information gained from the diversity of actions. The Internet or any networking technology is a means to progress in the direction of remote interactions with large groups. However, technical solutions tend to limit face-to-face interactions, which are crucial for trust building among the group of players and for meaningful debriefing. The key characteristic of concentration of time and space is partially lost.

Debriefings often reveal the difficulties local players experience during game sessions. However, role-playing game arenas allow the exploration of various ways to elaborate on the strategic dimensions of upscaling explorations, solutions and rule-change and adapting them to the specific social-ecological problem. Enrolling stakeholders of various decision-making levels in the game could be essential to expand the exploration of the issue at stake and to engage in a real problem-based approach such as ecosystem management. Depending on the simplicity of communication among them, stakeholders acting at different scales may participate in a common arena, or not. Role-playing game design and organisation may thus help to circumvent issues related to upscaling. A few recent experiences have progressed in the direction of dealing with multiple scale issues within a single game. In Uganda, Hassenforder et al. (2016), for instance, have played at local scale but discussed and developed outcomes at regional scale. In Laos, Ornetsmüller, Castella and Verburg (2018) developed a 'metagame' for national experts that summarised findings from a series of local games. On the topic of coastal vulnerability, one game (see Case study 12.1) integrates multiple scales, with players having roles bridging the scales, thanks to a cautious management of space and a rather large number of facilitators (Bonté et al. 2019).

Resource implications

Serious games need skills for crafting the game at suitable levels of complexity to allow participants to play it easily and still be willing to discuss their issues. This means finding the right combination of related items but also providing an environment with suitable pace and willingness to continue. A second set of skills is the facilitation of games. Facilitators have a crucial role to generate a suitable gaming atmosphere for participants to accept the game as an exploration tool. They must keep control of the dynamics and adjust these according to the group. They also have a crucial role to play in the debriefing stage, to lead participants to elaborate on new knowledge derived from the game experience and to set an action plan accordingly.

Case study 12.1: Coastal regional planning under global changes at play in Languedoc, France

The Amenajeu game was developed for and used with a group of 40 stakeholders to support the review process of a regional planning document called the Scheme of Territorial Consistency (*Schéma de Cohérence Territoriale* or SCoT). Among other issues, the game aims to drive the urbanisation process of an area. The SCoT is a mandatory urban planning document reviewed every five years. The elaboration of this document is particularly difficult, due to the long-term and large-scale projection considered and the numerous stakes and sectors of activity involved. In this context, the Amenajeu serious game is designed as a participatory device dedicated to the co-analysis of multi-level and multi-scale adaptations to global change. Actions taken by some to decrease their vulnerability may increase the vulnerability of others. The participatory session is aimed at increasing awareness of potential vulnerability transfers in the group of decision-makers who are in charge of elaborating the SCoT together.

The structure of the Amenajeu game is based on the SES robustness analysis framework proposed by Anderies, Janssen and Ostrom (2004; Anderies, Barreteau, and Brady 2019) that categorises the entities of an SES into four main categories (resources, resource users, public and private infrastructure, and public infrastructure providers). The framework focuses on the interactions between these categories in order to study the impact of exogenous drivers that would affect any of the entities. In the Amenajeu game, we combine the SES robustness analysis framework with the paradigm of a multi-agent system commonly used to discuss natural resource management (Bousquet and Le Page 2004; Le Page et al. 2013) in order to make explicit spatial, multi-scale or multi-sectoral issues.

The stakeholders from the different sectors of activity are viewed as resource users of the SES. They are represented by tokens that are updated by the facilitators during every round. These tokens evolve according to the situation of the area, featuring changes in population, activities and the environment of the SES. Players play the role of infrastructure providers. These infrastructure providers can set up infrastructure on the game boards in order to influence the dynamics of sectors of activities in some locations within the area, or the main attitudes of governance agencies of subregions in the area. Then they can observe the effect of their decisions on the resource users.

The region represented in the game, the SCoT territory, was split into four subregions represented by four game boards placed on four tables (Figure 12.1). At the beginning of the game session, players were given a role of sectoral planner that was close to their functions in the real world (in agriculture, urbanism, nature conservation, tourism or general planning) and then were allocated to one of the four tables. Each player had to write down their objective at the beginning of the game and was asked

The second type of resource required is time. A game session is typically one to two hours long, with 10–15 participants and up to four or five facilitators and observers, depending on the game being played. The preparation stage in the game design might also be time consuming in that it will include test sessions. These test sessions involve colleagues or communities of practice with people involved in serious game facilitation and design (Dionnet et al. 2013).

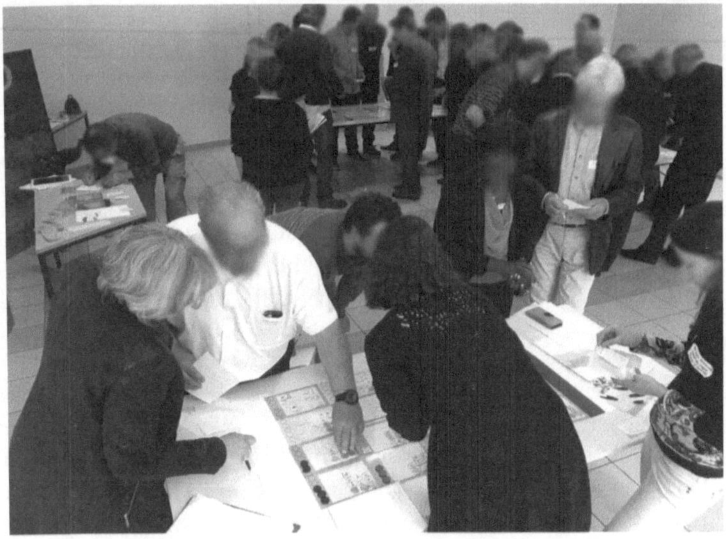

Figure 12.1 Game session of Amenajeu with multiple tables representing various distant but connected places (© Raphaël Mathevet)

to base their strategies on this objective, their assets (infrastructure or money when we represented it) and existing infrastructure. Four rounds of five years each were played. In each round, facilitators came up with various climatic and demographic events to put stress on the players, which they had to adapt to.

Discussion about potential vulnerability transfers first occurred during the game, when players decided how they would set up various infrastructure. Then, during the debriefing part of the session, players discussed how they would reach their objectives. The exercise shows that local adaptations at the subregion level make it possible to temporarily cope with the pressures of global change, by transferring these pressures to other subregions. Players could observe that the good intentions of some are not always followed by positive impacts locally or regionally, sometimes simply owing to a lack of consultation. With the help of this serious game, participants were able to discuss future changes and to experiment with the interplay of social-ecological interdependencies, not only between subregions but also between sectors of activity. This first experience led to the design of a generic and computerised game and method implemented in several places in France and South Africa (Bonté et al. 2021).

Games are increasingly based on the use of computers to improve the representation of ecological dynamics. Even if these dynamics remain quite simple to prevent a black box effect (i.e. participants lose the meaning and relevance of the elements in the game and of game outcomes), there is a need to make laptops or networked terminals and possibly a router available.

According to the context of the game, money might be needed to pay participants (either according to results achieved in the game as in experimental economics, or a flat fee to compensate the players for their time). The game implementation also requires a venue that is acceptable to all participants and easy to reach.

As in any participatory approach (Reed 2008; Etienne 2011), ethics need to be taken into consideration when implementing the game. Contrary to standard workshops, participants are expected to leave aside their own strategies and agendas during the playing phase (step two) of the game session. They might reveal more of themselves than they would have wanted to in the presence of other participants, who could strategically use this in further interaction after the game situation. However, participants may deny the realism of what had taken place in the game during debriefing if they consider that it would harm their position in real life.

New directions

Online games with distributed and highly interactive simulation tools are increasingly being explored and developed to enrol more participants in role-playing game experiments (Becu et al. 2017). Computer simulations are also increasingly incorporating cross-scale and multi-level dimensions, at the risk of losing the easy use and playful or 'fun' characteristic of games. Serious games should remain games, i.e. players should react according to the situation in the game and not take strategic decisions based on their situation outside the game (Kizos et al. 2018). Computer simulations are an emerging area and mostly aim to explore complexity of SES dynamics and especially telecoupling issues. Another area in need of research relates to collecting evidence of the proven impacts (e.g. learning) of role-playing games in real-world settings. Recent progress in this direction should be reinforced by interdisciplinary works involving specialists in psychology and education and social and environmental scientists, with modellers and stakeholders.

Key readings

Bousquet, F., O. Barreteau, P. d'Aquino, M. Etienne, S. Boissau, S. Aubert, C. Le Page, D. Babin, and J-C. Castella 2002. 'Multi-agent Systems and Role Games: An Approach for Ecosystem Co-management', In *Complexity and Ecosystem Management: The Theory and Practice of Multi-agent Approaches*, edited by M. Janssen, 248–285. Cheltenham: Edward Elgar.

Duke, R.D., and J.L.A. Geurts. 2004. *Policy Games for Strategic Management*. Amsterdam: Dutch University Press.

Meadows, D.L. 2001. Tools for Understandings the Limits to Growth: Comparing a Simulation and a Game. *Simulation and Gaming* 32(4): 522–536.

Sterman, J.D. 1992. 'Teaching Takes Off – Flight Simulators for Management Education: "The Beer Game".' http://jsterman.scripts.mit.edu/docs/Sterman-1992-TeachingTakesOff.pdf.

Toth, F.L. 1988. Policy Exercises: Objectives and Design Elements. *Simulation and Games* 19(3): 235–255.

References

Abt, C. 1974. *Serious Games*. New York: Viking.

Anderies, J.M., O. Barreteau, and U. Brady. 2019. 'Refining the Robustness of Social-Ecological Systems Framework for Comparative Analysis of Coastal System Adaptation to Global Change.' *Regional Environmental Change* 19(7): 1891–1908.

Anderies, J.M., M.A. Janssen, and E. Ostrom. 2004. 'A Framework to Analyze the Robustness of Social-EcologicalSystems from an Institutional Perspective.' *Ecology and Society* 9(1): 18. www.ecologyandsociety.org/vol9/iss1/art18.

Barreteau, O., F. Bousquet, and J-M. Attonaty. 2001. 'Role-playing Games for Opening the Black Box of Multi-agent Systems: Method and Teachings of its Application to Senegal River Valley Irrigated Systems.' *Journal of Artificial Societies and Social Simulations* 4(2). http://jasss.soc.surrey.ac.uk/4/2/5.html.

Becu, N., M. Amalric, B. Anselme, E. Beck, X. Bertin, E. Delay, N. Long, N. Marilleau, C. Pignon-Mussaud, and F. Rousseaux. 2017. 'Participatory Simulation to Foster Social Learning on Coastal Flooding Prevention.' *Environmental Modelling and Software* 98: 1–11.

Bonté, B., C. Therville, F. Bousquet, G. Abrami, S. Dhenain, and R. Mathevet. 2019. 'Analyzing Coastal Coupled Infrastructure Systems Through Multi-scale Serious Games in Languedoc, France.' *Regional Environmental Change* 19(7): 1879–1889. doi:10.1007/s10113-019-01523-6.

Bonté, B., C. Therville, F. Bousquet, C. Simi, G. Abrami, C. Guerbois, H. Fritz, O. Barreteau, S. Dhenain, and R. Mathevet. 2021. 'Simulating together Multiscale and Multisectoral Adaptations to Global Change and their Impacts: A Generic Serious Game and its Implementation in Coastal Areas in France and South Africa.' In *Ecosystem and Territorial Resilience*, edited by E. Garbolino and C. Voiron-Canicio, 247–278. Amsterdam: Elsevier. doi:10.1016/B978-0-12-818215-4.00009-2.

Bousquet, F., O. Barreteau, P. d'Aquino, M. Etienne, S. Boissau, S. Aubert, C. Le Page, D. Babin, and J-C. Castella 2002. 'Multi-agent Systems and Role Games: An Approach for Ecosystem Co-management.' In *Complexity and Ecosystem Management: The Theory and Practice of Multi-agent Approaches*, edited by M. Janssen, 248–285. Cheltenham: Edward Elgar.

Bousquet, F., and C. Le Page. 2004. 'Multi-agent Simulations and Ecosystem Management: A Review.' *Ecological Modelling* 176(3–4): 313–332.

Castella, J.C., N.H. Trung, and S. Boissau. 2005. 'Participatory Simulation of Land-use Changes in the Northern Mountains of Vietnam: The Combined Use of an Agent-based Model, a Role-playing Game, and a Geographic Information System.' *Ecology and Society* 10(1): 27.

D'Aquino, P., J. Bourgoin, D. Cefaï, C. Richebourg, S. Hopsort, and T. Pascutto. 2017. 'Du Savoir Local au Pouvoir Central: Un Processus Participatif sur la Reforme Fonciere au Senegal.' *Natures Sciences Sociétés* 25(4): 360–369.

Daré, W.S., A. van Paassen, R. Ducrot, R. Mathevet, J. Queste, G. Trébuil, C. Barnaud, and E. Lagabrielle. 2014. 'Learning about Interdependencies and Dynamics.' In *Companion Modelling*, edited by M. Etienne, 233–262. New York: Springer.

De Vente, J., M. Reed, L. Stringer, S. Valente, and J. Newig. 2016. 'How does the Context and Design of Participatory Decision Making Processes Affect their Outcomes? Evidence from Sustainable Land Management in Global Drylands.' *Ecology and Society* 21(2): 24. doi:10.5751/ES-08053-210224.

Dionnet, M., K.A. Daniell, A. Imache, Y. von Korff, S. Bouarfa, P. Garin, J-Y. Jamin, D. Rollin, and J-E. Rougier. 2013. 'Improving Participatory Processes through Collective Simulation: Use of a Community of Practice.' *Ecology and Society* 18(1): 36. doi:10.5751/ES-05244-180136.

Duke, R.D., and J.L.A. Geurts. 2004. *Policy Games for Strategic Management.* Amsterdam: Dutch University Press.

Etienne, M. 2003. 'SYLVOPAST a Multiple Target Role-playing Game to Assess Negotiation Processes in Sylvopastoral Management Planning.' *Journal of Artificial Societies and Social Simulations* 6(2): http://jasss.soc.surrey.ac.uk/6/2/5.html.

Etienne, M. 2011. *Companion Modelling. A Participatory Approach to Support Sustainable Development.* Versailles: QUAE.

Ferrand, N., S. Farolfi, G. Abrami, and D. du Toit. 2009. 'WAT-A-GAME: Sharing Water and Policies in Your Own Basin.' *40th Annual Conference, International Simulation and Gaming Association.*

Flint, R.W. 2013. 'System's Thinking in Community Development.' In *Practice of Sustainable Community Development*, edited by R.W. Flint, 93–118. New York: Springer.

Friedman, D., and S. Sunder. 1994. *Experimental Methods, A Primer for Economists.* Cambridge: Cambridge University Press.

Gurung, T.R., F. Bousquet, and G. Trébuil. 2006. 'Companion Modeling, Conflict Resolution, and Institution Building: Sharing Irrigation Water in the Lingmuteychu Watershed, Bhutan.' *Ecology and Society* 11(2): 36. www.ecologyandsociety.org/vol11/iss2/art36.

Hassenforder, E., M. Brugnach, B. Cullen, N. Ferrand, O. Barreteau, K.A. Daniell, and J. Pittock. 2016. 'Managing Frame Diversity in Environmental Participatory Processes – Example from the Fogera Woreda in Ethiopia.' *Journal of Environmental Management* 177: 288–297.

Kapp, K. 2012. *The Gamification of Learning and Instruction: Game-based Methods and Strategies for Training and Education.* Hoboken: Johan Wiley & Sons.

Kizos, T., P. Verburg, M. Bürgi, D. Gounaridis, T. Plieninger, C. Bieling, and T. Balatsos. 2018. 'From Concepts to Practice: Combining Different Approaches to Understand Drivers of Landscape Change.' *Ecology and Society* 23(1): 25. doi:10.5751/ES-09910-230125.

Krolikowska, K., A. Dunajski, P. Magnuszewski, and M. Sieczka. 2009. 'Institutional and Environmental Issues en Land Reclamation Systems Maintenance.' *Environmental Science and Policy* 12(8): 1137–1143.

Le Page, C., D. Bazile, N. Becu, P. Bommel, F. Bousquet, M. Etienne, R. Mathevet, V. Souchère, G. Trebuil, and J. Weber. 2013. 'Agent Based Modelling and Simulation Applied to Environmental Management.' In *Simulating Social Complexity*, edited by B. Edmonds and R. Meyer, 499–540. New York: Springer.

Legay, J-M. 1997. *L'expérience et le Modèle. Un Discours sur la Méthode.* INRA éditions.

Martin, L., P. Magnuszewski, J. Sendzimir, F. Rydzak, K. Krolikowska, H. Komorowski, A. Lewandowska et al. 2007. 'Gaming with a Microworld of a Local Product Chain in the Oder River Basin, Lower Silesia, Poland.' *Simulation and Gaming* 38(2): 211–232.

Mathevet, R., C. Le Page, M. Etienne, G. Lefebvre, B. Poulin, G. Gigot, F. Proréol, and A. Mauchamp. 2007. 'BUTORSTAR: A Role-playing Game for Collective Awareness of Wise Reedbed Use.' *Simulation and Gaming* 38(2): 233–262.

Mathevet, R., C. Le Page, M. Etienne, B. Poulin, G. Lefebvre, F. Cazin, and X. Ruffray. 2008. 'Des Roselières et des Hommes: ButorStar un Jeu de Rôles Pour L'aide a la Gestion Collective.' *Revue Internationale de Géomatique* 18(3): 375–395.Meadows, D.L. 2001. 'Tools for Understandings the Limits to Growth: Comparing a Simulation and a Game.' *Simulation and Gaming* 32(4): 522–536.

Meadows, D.H. 2008. *Thinking in Systems: A Primer.* London: Chelsea Green Publishing.

Meadows, D., and D. Meadows 1993. *Fish Banks News.* Fish Banks Limited and Laboratory for Interactive Learning, University of New Hampshire.

Mermet, L. 1993. 'Une Méthode de Prospective: Les Exercices de Simulation de Politiques.' *Nature Sciences Sociétés* 1(1): 34–46.

Merrill, S.C., C.J. Koliba, S.M. Moegenburg, A. Zia, J. Parker, T. Sellnow, S. Wiltshire, G. Bucini, C. Danehy, and J.M. Smith. 2019. 'Decision-making in Livestock Biosecurity Practices Amidst Environmental and Social Uncertainty: Evidence from an Experimental Game.' *PLoS One* 14(4): e0214500.

Ornetsmüller, C., J-C. Castella, and P.H. Verburg. 2018. 'A Multiscale Gaming Approach to Understand Farmer's Decision Making in the Boom of Maize Cultivation in Laos.' *Ecology and Society* 23(2): 35.

Ostrom, E., R. Gardner, and J. Walker. 1994. *Rules, Games and Common-Pool Resources.* Ann Arbor: University of Michigan Press.

Reed, M.S. 2008. 'Stakeholder Participation for Environmental Management: A Literature Review.' *Biological Conservation* 141(10): 2417–2431.

Reed, M.S., S. Vella, E. Challies, J. de Vente, L. Frewer, D. Hohenwallner-Ries, T. Huber, R.K. Neumann, E.A. Oughton, and J. Sidoli del Ceno. 2018. 'A Theory of Participation: What Makes Stakeholder and Public Engagement in Environmental Management Work?' *Restoration Ecology* 26: S7–S17.

Richard, A., and O. Barreteau. 2007. 'Concert'Eau: A Setting to Experiment Difficulties of Pluralisms.' *38th International Simulation and Gaming Association Conference*, Nijmegen, Netherlands.

Richard-Ferroudji, A., and O. Barreteau. 2012. 'Assembling Different Forms of Knowledge for Participative Water Management – Insights from the Concert'eau Game.' In *Environmental Democracy Facing Uncertainty*, edited by C. Claeys-Mekdade and M. Jacque, 97–120. Brussels: Peter Lang.

Ryan, T. 2000. The Role of Simulation Gaming in Policy Making. *Systems Research and Behavioral Science* 17: 359–364.

Souchère, V., L. Millair, J. Echeverria, F. Bousquet, C. Le Page, and M. Etienne. 2010. 'Co-constructing with Stakeholders a Role-playing Game to Initiate Collective Management of Erosive Runoff Risks at the Watershed Scale.' *Environmental Modelling and Software* 25(11): 1359–1370.

Sterman, J.D. 1992. 'Teaching Takes Off – Flight Simulators for Management Education: "The Beer Game".' http://jsterman.scripts.mit.edu/docs/Sterman-1992-TeachingTakesOff.pdf.

Stolk, D., D. Alexandrian, B. Gros, and R. Paggio. 2001. 'Gaming and Multimedia Applications for Environmental Crisis Management Training.' *Computers in Human Behavior* 17(5–6): 627–642.

Toth, F.L. 1988. 'Policy Exercises: Objectives and Design Elements.' *Simulation and Games* 19(3): 235–255.

Villamor, G., and B. Badmos. 2015. 'Grazing Game: A Learning Tool for Adaptive Management in Response to Climate Variability in Semiarid Areas of Ghana.' *Ecology and Society* 21(1): 39. doi:10.5751/ES-08139-210139.

13
Participatory modelling

Géraldine Abrami,[1] William's Daré,[2] Raphaëlle Ducrot,[1] Nicolas Salliou[3] and Pierre Bommel[2]

[1]G-EAU, UNIV MONTPELLIER, AGROPARISTECH, BRGM, CIRAD, IRD, INRAE, INSTITUT AGRO, MONTPELLIER, FRANCE
[2]CIRAD, UR GREEN 47, MUSE UNIVERSITY OF MONTPELLIER, MONTPELLIER, FRANCE
[3]DEPARTMENT OF CIVIL, ENVIRONMENTAL AND GEOMATIC ENGINEERING, INSTITUTE FOR SPATIAL AND LANDSCAPE DEVELOPMENT, PLANNING OF LANDSCAPE AND URBAN SYSTEMS (PLUS), ETH ZÜRICH, ZÜRICH, SWITZERLAND

Key methods discussed in this chapter

Modelling methods: System dynamics (group model building, mediated modelling, shared vision planning), agent-based models (ARDI), role-playing games (Wat-A-Game), expert models (Bayesian networks, fuzzy cognitive maps), state-and-transition models, soft system methodologies (rich pictures, concept maps, decision trees, cognitive maps)
Integrated approaches: Collaborative modelling, companion modelling, participatory system analysis

Connections to other chapters

Methods for generating data and systems scoping (Chapters 5–8), specifically participatory data-collection methods (Chapter 8) or interviews and surveys (Chapter 7), may provide working material or monitoring and evaluation support within participatory modelling processes. Facilitated dialogue methods (Chapter 9) may smooth participatory modelling workshops. Future analysis (Chapter 10), scenario development (Chapter 11) or serious games (Chapter 12) may be articulated with participatory models within broader participatory resilience assessment (Chapter 14) or action research (Chapter 15) projects. Expert modelling (Chapter 16), dynamical systems modelling (Chapter 26), state-and-transition modelling (Chapter 27) and agent-based modelling (Chapter 28) cover the most common types of modelling methods used in participatory modelling, and participatory modelling may use institutional analysis (Chapter 22) conceptual frameworks.

Introduction

Whatever the purpose of a model, e.g. forecasting, prescription, explanation, description, learning and communication or theory building (Kelly et al. 2013; Schlüter, Müller, and Frank 2019), participatory modelling refers to settings where non-scientist stakeholders are involved in any of the stages of the modelling process of their social-ecological systems (SES).

DOI: 10.4324/9781003021339-16

<table>
<tr><th colspan="2">SUMMARY TABLE: PARTICIPATORY MODELLING</th></tr>
<tr><th>DISCIPLINARY BACKGROUND</th><th>KNOWLEDGE TYPE</th></tr>
<tr><td>The methods in this chapter are derived from or have most commonly been used in:
Environmental Management, (soft) Operational Research, Complex Systems Science</td><td>The methods in this chapter are primarily used to generate the following types of knowledge:
• Descriptive
• Exploratory
• Explanatory</td></tr>
<tr><th>RESEARCH APPROACH</th><th>PURPOSE OF METHOD</th></tr>
<tr><td>The methods in this chapter originate from or most commonly adopt the following research approaches:
• Interpretive/subjective
• Collaborative/process</td><td>The most common purposes of using the methods in this chapter are:
• System understanding
• Stakeholder engagement and co-production
• Policy/decision support</td></tr>
<tr><th>TEMPORAL DIMENSION</th><th>SYSTEMIC FEATURES AND PROCESSES</th></tr>
<tr><td>The methods in this chapter are most commonly applied to the following temporal dimensions:
• Present (typically within the last 5–10 years)
• Recent past (post-1700s)
• Future
SPATIAL DIMENSION
The methods in this chapter are primarily either or both:
• Non-spatial
• Explicitly spatial
The methods in this chapter are most commonly applied at the following spatial scales:
• Local
• Multiple places/sites around the world</td><td>While most methods can do many things, the methods in this chapter are particularly good (i.e. go-to methods) for addressing the following:
• Diversity
• Social-ecological dependence and impact
• Social learning
• Collective action and collaborative governance
• Evaluating policy options</td></tr>
</table>

In the 1960s and 1970s, participatory modelling was rooted in the development of system dynamics models (Voinov and Bousquet 2010; Barreteau et al. 2013). At the time, business models were co-developed with business managers at MIT, and citizens were involved by the US Army Corps of Engineers in environmental management by using system dynamics modelling. Various schools of research have since developed consistent principles, practices and methods for modelling and participation, such as participatory simulation (complexity and computer science), collaborative learning (education science), social sciences experiments (experimental economics) or participatory action research. The field of water management is particularly rich in reflections on participation and modelling (Harmonicop Team 2005; Pahl-Wostl et al. 2007; Basco-Carrera et al. 2017).

Stakeholders may be involved in a modelling process to address issues related to the understanding, representation and management of SES, such as water basins, fisheries or forests. Using modelling stimulates knowledge elicitation and creative thinking (Jordan et al. 2018; Van Bruggen, Nikolic, and Kwakkel 2019). Models act as boundary objects by providing explicit and negotiable representations of reality (Star and Griesemer 1989). From a technical perspective, participatory modelling assumes that, within an appropriate interaction setting, non-specialists are able to co-produce models which make sense to them and which generate useful discussions and new knowledge.

SES problems and questions

Questions related to social-ecological systems can be classified according to three rationales for engaging in participatory modelling, as identified by Barreteau et al. (2013).

1. Assuming that the integration of various kinds of knowledge is useful for understanding and managing SES, participatory modelling may be used to increase the intrinsic quality of SES models by addressing questions such as:
   - How does one deal with knowledge gaps and ambiguities? (e.g. using modelling to gather and discuss disparate elements of knowledge from users and managers to reduce uncertainties on complex dynamics (Barreteau et al. 2013))
   - How does one calibrate and validate a model of an SES? (e.g. using a role-playing game as a participatory simulation setting to involve lay stakeholders in the validation of an agent-based model of their system (Barreteau, Bousquet, and Attonaty 2001) or having heterogeneous groups of stakeholders 'suggest and check' a model of their system in workshops (Polhill, Sutherland, and Gotts 2009))
2. Assuming that the involvement of stakeholders, including decision-makers, in the production and spreading of models increases the legitimacy, relevance and impact of these models through the collaborative framing and design of assumptions, elements, outputs and settings (Hare 2011), participatory modelling may improve the adequacy of models in aiding SES management:
   - How does one frame model boundaries, questions and outputs to be relevant to real-world issues? (e.g. organising meetings to discuss and reformulate issues among actors that have divergent stakes and interests in SES (Dewulf, Bouwen, and Taillieu 2006))
   - How does one design and explore management options? (e.g. using participatory simulation sessions with decision-makers, managers and economic actors to examine and virtually test options (Souchère et al. 2010))

- How does one improve appropriation of a model? (e.g. actively involving future users all along the framing, development and deployment of a decision-support system (Jakku and Thorburn 2010))

3. Assuming that the acknowledgement of non-scientific norms, values and interests, and the empowerment of stakeholders are assets for more sustainable transformations and policy pathways of SES (Van Bruggen, Nikolic, and Kwakkel 2019), participatory modelling may be used to support larger processes:
   - How does one engage stakeholders in a shared systemic vision of their SES? (e.g. by engaging them in a full-cycle collaborative modelling process (Langsdale et al. 2013))
   - How does one induce more informed collective decisions? (e.g. through recognising, accepting and exploring uncertainties and ambiguities during the co-design of a model (Salliou et al. 2017))
   - How does one increase the ability of stakeholders to participate in SES governance arenas? (e.g. by using participatory modelling settings to raise the cognitive and deliberative capacities of participants (Daré et al. 2018; Landström et al. 2019))
   - How does one deal with divergent representations? (e.g. by collectively discussing models based on stakeholder heuristics versus science-based models (Smajgl et al. 2015))
   - How does one improve communication among stakeholders? (e.g. by using models resulting from participatory modelling to communicate stakeholders' views among different organisational levels (Daniell et al. 2010))
   - How does one improve deliberation among stakeholders? (e.g. by using models as boundary objects to allow users and managers to negotiate indicators and set a new basis for joint resource management (Barreteau et al. 2012))
   - How does one induce social learning? (e.g. by using group model building to improve the problem-solving capacities of a collective (Vennix 1996))

Brief description of key methods

Participatory modelling is commonly centred on the model information, construction and use stages. However, genuine participation requires one to involve stakeholders early in the modelling process, i.e. during the preparation and organisation stages (including framing and participant selection), all the way to the follow-up stages (including dissemination, monitoring and evaluation). The engagement level varies from observation to co-steering among the different stages and stakeholders. A synthetic description framework described by Bots and Van Daalen (2008) or a participatory engineering method such as the PrePar tool (Ferrand et al. 2017) is useful for planning and reflecting on these processes.

Depending on the characteristics and stage of development of the model, each participatory modelling event has distinct purposes, e.g. to provide advice, clarify the vision, mediate conflicting views or improve the model (e.g. see Bots and Van Daalen 2008; Basco-Carrera et al. 2017). The design of the event should respond to these constraints through organising working groups (size, homogeneity, functioning, rules) for the different activities and choosing participatory settings (e.g. knowledge elicitation with conceptual diagrams or model verification with interactive simulations – see Hare (2011) for an overview). Refer to Voinov et al. (2018) for an extended discussion and guidelines for choosing participatory modelling methods and tools. Hare (2011), Voinov et al. (2016), and Van Bruggen, Nikolic and Kwakkel (2019) also provide lists and descriptions of the

Table 13.1 Summary of key modelling methods used in participatory modelling

Method	*Description*	*References*
System dynamics	System dynamics (SD) represent global state variable dynamics and interdependences and are useful to reflect on global system concepts. Graphical methods of SD (causal loop diagrams and stock and flow diagrams) are good for participatory system thinking and conceptual SD model building. Many available computerised SD tools (e.g. Vensim, Stella) navigate transparently between diagrams, equations and simulation outputs.	*Key introductory text* Costanza and Ruth 1998 *Applications to SES* Sandker et al. 2010; Stave 2010
Agent-based models	Agent-based models (ABMs) represent autonomous entities interacting within spatially explicit levels of organisation. These models are good for heterogeneous knowledge integration because of their close ontological resemblance to the real world. ABMs do not have a specific graphical method. The ARDI (Actors, Resources, Dynamics and Interaction) methodology offers a set of ad hoc diagrams for ABM co-construction. Simplified versions of the unified modelling language (UML) can also be used in participatory settings (see Case study 13.1). ABMs are implemented and simulated with computer platforms or role-playing games (RPGs). Cormas (cormas.cirad.fr) and Netlogo (ccl.northwestern.edu/netlogo) are free ABM platforms that provide features for participatory simulations or computerised RPG.	*Key introductory texts* Barreteau, Bousquet, and Attonaty 2001 (ABMs and RPGs); Bousquet and Le Page 2004; Le Page and Bommel 2005 (UML for ABMs); Etienne, Du Toit, and Pollard 2011 (ARDI) *Applications to SES* Forrester et al. 2014; Hoch et al. 2015; Smajgl et al. 2015
Expert modelling: Bayesian networks and fuzzy cognitive maps	Expert modelling approaches mimic expert thinking. Bayesian networks (BNs) represent the conditional probabilities between variables states. Fuzzy cognitive maps (FCMs) represent causal probabilities propagation between factors. They have simple graphical formalisms that are well suited to participatory settings and are good for coping with variability or divergent opinion and discussing variables or factors and their relationships. Specific methods are available for participative quantification or semi-quantification of probabilities or influences (Das 2004). Netica (norsys.com/netica.html) or Mental Modeler (mentalmodeler.org) are platforms to construct and simulate BNs or FCMs.	*Key introductory texts* Düspohl, Zacharias, and Doell 2012 (BNs); Gray et al. 2015 (FCMs) *Applications to SES* Kok 2009; Celio and Grêt-Regamey 2016; Htun et al. 2016 (FCMs); Salliou et al. 2017 (BNs)
Soft systems methodologies	Soft systems methodologies are modelling approaches developed specifically for human-centred systems in the field of management sciences. They provide graphical methods that can be used alone or in a conceptual stage of most of the models mentioned above, such as concept maps (representation of concepts and semantic relationships within a knowledge domain), decision trees (representation of objectives, actions and system uncertainties and evolution) or cognitive maps (representation of causal or influence relationships).	*Key introductory text* Vidal 2006 *Applications to SES* Mendoza and Prabhu 2006; Hommes et al. 2008

various tools and methods, computerised or not, that may be used for implementing participatory settings.

Because of their expressivity, model paradigms that are most broadly used in participatory modelling of SES are system dynamics models, agent-based models, expert modelling approaches and soft system methodologies (Table 13.1). State-and-transition models are another kind of modelling approach focusing on alternative states, thresholds and transitions (see Chapter 27). Furthermore, any type of model may be developed or used participatively within appropriate processes and settings (see Landström et al. 2019 for a pragmatic approach to participatory hydrological modelling). The relevance of a modelling method depends on the characteristics of the target system and on the available resources in terms of time, skills, money and data. In this regard, criteria and guidelines can be found in Schlüter, Müller and Frank (2019) and Kelly et al. (2013).

Communities of researchers and practitioners developed integrated approaches that provide methodological guidelines and tools for implementing a participatory modelling process (see Van Bruggen, Nikolic and Kwakkel 2019 for a comparative description). Some of the most well known are described in Table 13.2.

Table 13.2 Summary of key integrated approaches used in participatory modelling

Approach	*Description*	*References*
Participatory system analysis	Participatory system analysis combines a broad range of tools and techniques developed in the field of systems thinking with participatory methods. It is presented as an approach that extends participatory rural assessment with system science approaches. It provides guidelines and tools for performing needs and problems analysis, carrying out abstract modelling, exploring decisions or scenario options, and implementing design (maps, spidergrams, Bayesian networks, system dynamics, plans).	*Key introductory text* Lynam 2001 *Applications to SES* Smith, Felderhof, and Bosch 2007; Nguyen and Bosch 2013
Group model building and mediated modelling	Group model building (GMB) and mediated modelling (MM) are historical integrated approaches of participatory system dynamics models building. Group model building focuses on organisational messy problems and strategic decision-making within teams. It provides 'scripts' for standardised protocols (Hovmand et al. 2012) and has been widely used in private and public institutions. Mediated modelling aims at involving broad stakeholder groups in collective learning and consensus building on environmental issues. It provides a structured iterative process for involving the participants between the distinct stages of model development.	*Key introductory texts* Andersen et al. 2007 (GMB); Metcalf et al. 2010 (MM) *Applications to SES* Antunes, Santos, and Videira 2006 (MM); Daniell et al. 2010; Halbe, Pahl-Wostl, and Adamowski 2018 (GMB)

Approach	*Description*	*References*
Shared vision planning and collaborative modelling for decision support	Collaborative modelling for decision support (CMDS) comprises collaborative system dynamics modelling in combination with communication, visualisation and facilitation tools of structured public participation, and mental and cultural models of planning and decision support. It is based on the experience of US government environmental agencies on stakeholders' involvement in water management. It is an evolution of the US Army Corps of Engineers' shared vision planning (SVP), which defines iterative steps for participatory system dynamics modelling. It is supported by principles and best practices established by the joint work of a community of academics and practitioners of modelling, facilitation and water management between 2008 and 2010.	*Key introductory texts* Werick 1994 (SVP); Langsdale et al. 2013 (CMDS); Online sources (principles and guidelines for SVP and CMDS, e.g. labs.wsu.edu/collaborativemodeling/cmds) *Applications to SES* Palmer et al. 1999; Antunes, Santos, and Videira 2006 (MM); Creighton and Langsdale 2009 (SVP); Basco-Carrera et al. 2017 (CMDS)
Companion modelling	Companion modelling (ComMod) promotes the iterative, collaborative and adaptive use of models aimed at sharing and legitimating multiple views of SES and articulating scientific production and collective decision-making. It mostly uses the agent-based model paradigm with alternative or combined use of role-playing games. It emerged in France in the 1990s as an approach for implementing complexity thinking and post-normal theory (Funtowicz and Ravetz 1993) in natural resources management issues. ComMod has an active community of researchers and practitioners hosting regular events and training. Various tools and methods, such as ARDI and Cormas, have been developed (see 'Agent-based models' in Table 13.1). ComMod has also produced spin-off thematic participatory modelling platforms. TerriStories (terristories.org) offers a configurable board game on land issues that combines concepts from participatory modelling, role-playing games and live theatre. Wat-A-Game is a platform for the 'design by playing' of games on water management. It is part of CoOPLAaGE (cooplaage.watagame.info), a larger integrated suite of low-tech tools and protocols based on participatory modelling that supports stakeholder groups in the design of decision procedures.	*Key introductory text* Barreteau et al. 2003 (ComMod charter) *Applications to SES* Souchère et al. 2010; Abrami et al. 2012 (CoOPLAaGE); d'Aquino and Bah 2013; Hassenforder et al. 2015; Bouamrane et al. 2016; Ferrand et al. 2017; Ponta et al. 2019; commod.org (case studies and material)

Limitations

The intrinsic quality of participatory models is highly dependent on how the participation is organised. Biases and interests of participants (including modellers and researchers) may have an impact on the scope and elements of the model, which needs to be carefully considered (Daniell et al. 2010). Skilled facilitation and negotiation are necessary. Special care should be taken with graphical computerised tools because the ease of adding new elements can quickly lead to overly complex models (Kelly et al. 2013).

Participatory modelling may raise false expectations or can be deceptive. Participants may feel that the final model does not represent their views adequately, particularly after some post-processing work (Daniell et al. 2010; Hare 2011). They may perceive it as too narrow or too complex relative to the richness of the modelling process (Sandker et al. 2010). Conceptual models based on stakeholder perspectives may not make sense outside the participation arena, and decision-makers may not need a new integrated model but rather simple, data-rich and trustworthy models that can be used for decision support (Hare 2011). These risks are high when participation happens within several interconnected processes sharing part or all of their teams, participants and events, but each having its own agenda and objectives (Seidl 2015). The risks can be minimised by carefully considering the expectations, purposes and agendas of participants, clients and organising team, and also by designing and communicating as explicitly and transparently as possible about the processes, their outcomes and products, and how they will be shared or used (Sterling et al. 2019). If model uptake is an objective, an answer can be to choose off-the-shelf software that is easy to learn and available to all (Langsdale et al. 2013), or to plan for professional levels of model documentation and maintenance in the project funding (Hare 2011).

Participatory modelling may lead to undesirable effects for the process and for the participants. The framing induced by the model may inhibit deliberation and creativity rather than fostering it (Barnaud and Van Paassen 2013). Hedelin et al. (2017) warn that the modelling may induce a focus on the choice of measures ('How can we get there?') rather than on broader issues of societal development ('Where do we want to go?'). The question of how the modelling approach (modelling method, level of realism, level of integration) is chosen and how it affects the modelling process output and participants is discussed in Langsdale et al. (2013), Le Page and Perrotton (2017), and Schlüter, Müller and Frank (2019). Regarding ethics, inappropriate consideration of participants' motivation or power asymmetries may create biases in the model or harmful impacts in the 'real world' (Barnaud and Van Paassen 2013; Daré and Venot 2017; Hedelin et al. 2017). See for instance criticisms and hints from a socio-political perspective in Tsouvalis and Waterton (2012).

The impact of participatory modelling in larger SES transformation processes may appear limited or difficult to assess. The size of the co-modelling groups and the time frame of research projects are limiting factors. This implies that process extension may be needed to achieve transformative effects (Hare 2011). Joint ownership of an associate decision-making process is an asset in this regard (Van Bruggen, Nikolic, and Kwakkel 2019). Landström et al. (2019) propose a systematic approach to consolidate participants' commitment during and after the project. Extension through widespread adoption of participatory modelling outside research projects is limited not only by low binding participatory obligations but also by the limited human capacity (skills and personnel) of local communities and management institutions (Hare 2011). The monitoring and impact assessment of participatory modelling raises specific questions in data collection and analysis, which requires more research (Jones et al. 2009; Smajgl and Ward 2015; Hassenforder et al. 2016).

Finally, even though participatory modelling case studies have flourished in the past 20 years, the field still lacks common sets of principles and frameworks or roadmaps that would unify existing approaches sharing the same goal. This would support the design of new participatory modelling processes, structure analysis and evaluation of past cases (Jordan et al. 2018; Van Bruggen, Nikolic, and Kwakkel 2019). Generic templates would be useful to systematise approaches to participatory modelling, and thus quality management (Seidl 2015). Many recent review papers demonstrate the current efforts of the community in this direction. Hare (2011), Voinov et al. (2016), Hedelin et al. (2017), Basco-Carrera et al. (2017) and Jordan et al. (2018) focus on the features, limits and challenges of participatory modelling. Langsdale et al. (2013), Voinov et al. (2018) and Sterling et al. (2019) developed principles and best practices. Bots and Van Daalen (2008), Hare (2011), Barreteau et al. (2013), and Van Bruggen, Nikolic and Kwakkel (2019) propose guidelines for basic design choices and the classification of approaches. Hassenforder et al. (2016), Hedelin et al. (2017) and Voinov et al. (2018) propose extended process description frameworks that can be used for assessment and comparison or process design.

Resource implications

As for any participatory process, local social capital is an essential resource for planning and organising participatory modelling (identification, selection and mobilisation of participants, relationships with other arenas). In this regard, established relationships and community co-leads are crucial in navigating the social and political context (Sterling et al. 2019).

Various capacities are needed in the organising team. Some are generic to participatory processes (participatory engineering skills, knowledge of the context, social and facilitation skills, openness to learn, ability to adjust or even dismiss existing objectives and models) or modelling processes (technical skills to implement and explore models). Specific facilitation and modelling skills are needed to ensure knowledge elicitation and make explicit the diversity of views, issues and propositions that have been recognised by the group, even if they cannot be integrated into the model. These are not widespread capacities and make participatory processes sensitive to personnel changes and difficult to sustainably transfer (Langsdale et al. 2013; Sterling et al. 2019).

Participation and modelling both generate important time constraints, and it might prove difficult to coordinate stakeholders' and modellers' time frames (Hedelin et al. 2017). Depending on the type of model, specific material and software are needed. Finally, additional staff may be required for assistance and observation during workshops, and afterwards for dissemination, monitoring and evaluation.

New directions

Participatory modelling needs research and innovation to improve its transfer outside the academic world, its impact, and its uptake by local communities or management institutions. Interesting protocols are based on strategic resource minimalist research projects (Landström et al. 2019) or low-tech tools and adaptive workflows that can be used to engage and autonomise practitioners and stakeholders (Ferrand et al. 2017). Artificial intelligence technologies might be useful by partially automating participatory modelling processes (e.g. transforming discourses into models or assisting groups of participants in a process workflow). Web services and tools that could support online modelling activities in large-scale participatory processes or as alternative settings to face-to-face meetings are reviewed in Voinov et al. (2016).

Case study 13.1: Participatory modelling with cattle breeders in Uruguay

In Uruguay, nicknamed the 'green desert', the pampa offers a natural pasture where the herds graze freely, providing high-quality meat for export. However, climate change is affecting cattle farming – the primary economic source of this small country. From the 1990s onwards, droughts have occurred, killing thousands of animals and causing many bankruptcies. In response to these changes, the Instituto Plan Agropecuario (IPA) of Uruguay launched the 'SequiaBasalto' project, which aims to understand the drought phenomena and develop a participatory methodology to improve the adaptive capacities of livestock farmers using the ComMod approach (Barreteau et al. 2003). The objective of the model was to test several management strategies and facilitate communication between livestock farmers and support services. The study was conducted by Bommel et al. (2014). An agent-based modelling approach was chosen for its capacity to integrate various disciplines and types of expertise.

The first version of the model was collectively designed with livestock and grassland specialists from the project. They used unified modelling language (UML) diagrams to specify and share their expert knowledge on pasture growth processes, herd dynamics and farmers' management strategies into a common vision of the model (Figure 13.1A). The resulting model was then implemented on the Cormas platform. This model refined grass growth, herd behaviour and population dynamics. In contrast, farmers' management strategy options were coarse, corresponding to (a) IPA experts' representations of farmers' traditional practices, guided by profit, or (b) recommended best practices, guided by pasture sustainability.

The second stage of the process was to invite farmers to analyse running simulations and the behaviours of the farmer agents. Not surprisingly, the management strategies were the focus of the farmers' criticisms. What was more surprising was that, after discussions based on interactive simulations, farmers, both men and women, engaged in analysing UML activity diagrams (Figure 13.1C))! In fact, farmers were already accustomed to this formalism because, following the recommendation from an IPA researcher, IPA technicians had been using UML activity diagrams to conduct their interviews. The Cormas specialist in the project considered this to be an opportunity to develop and test an executable activity diagram editor into Cormas (Bommel et al. 2016).

In the third stage of the project, farmers participated in hybrid workshops mixing participatory modelling and interactive simulation to make the assessment livelier and more effective. Using the UML activity diagram editor embedded in Cormas, participants were able to generate new management strategies without programming knowledge and directly observe their impacts in simulations. This increased interactivity with the model revealed two interesting features:

1. By being able to modify the agents' behaviour, participants played with the model and better understood its logic. The immediate response of the model to each change increased their understanding of the underlying mechanisms. This triggered debates on how best to deal with droughts.

Figure 13.1 (A) Design session with experts, (B) first workshop with producers, (C) evaluation by farmers of the UML diagrams of the model, and (D) a farmer explains a simulation (© Pierre Bommel)

2. By testing alternative strategies with the UML editor, participants were able to identify some modelling and implementation biases. A strategy they had defined was not producing the result they had expected and they identified an issue with the way some actions of the agents were scheduled in the agent-based model.

The primary objective of this participatory modelling process was to improve knowledge of the livestock system, and indeed, it brought the IPA experts to acknowledge that their recommendations in terms of management had not always been the best. Outside periods of drought, the traditional strategy seemed in fact more economically profitable.

Although originally designed by the experts, the model highlighted the value of different types of knowledge. Beyond the debates it generated, the agent-based model also helped to identify adaptation strategies that seem to improve producers' resilience. Today, most of the farmers and technicians who participated in the workshops continue to experiment with the model. They use it to seek more effective management strategies in normal and drought periods. The Uruguayan government is now using this project as a methodological example to be followed for other development projects.

A second set of innovations relates to smoothing and tailoring the experience of participatory modelling. Participatory modelling may benefit from a closer collaboration with the fields of design, science communication and human–computer interaction to better design communication elements such as user interfaces, model guides and diagrams (Jordan et al. 2018). Advances in digital and communication technologies provide new forms of visual media for use in participatory modelling (Voinov et al. 2016). There could also be more responsive human–computer interactions to enhance interactions among participants and with the model (Bommel et al. 2018). For a case study where anthropology and arts-based facilitation allowed adjustment to the cultural specificities of participants, see McCarter et al. (2018).

A third set of innovations relates to a better articulation of models among themselves and with real-world processes. Little work has been done on the articulation of different modelling approaches that apply to the same SES (Jordan et al. 2018).

The design science concepts developed by Klabbers (2009) can be used to reflect and design the articulation between a participatory modelling process and the research or decision-making processes with which it is interconnected (Becu 2020). Hedelin et al. (2017) elaborate on organisational integration as a key research question.

Key readings

Barreteau, O., P. Bots, K. Daniell, M. Etienne, P. Perez, C. Barnaud, D. Bazile et al. 2013. 'Participatory Approaches.' In *Simulating Social Complexity: A Handbook* (2nd ed), edited by B. Edmonds and R. Meyer, 197–234. Berlin: Springer. doi:10.1007/978-3-540-93813-2_10.

Hare, M. 2011. 'Forms of Participatory Modelling and its Potential for Widespread Adoption in the Water Sector.' *Environmental Policy and Governance* 21(6): 386–402. doi:10.1002/eet.590.

Van Bruggen, A., I. Nikolic, and J. Kwakkel. 2019. 'Modeling with Stakeholders for Transformative Change.' *Sustainability* 11(3): 825. doi:10.3390/su11030825.

Voinov, A., K. Jenni, S. Gray, N. Kolagani, P.D. Glynn, P. Bommel, C. Prell et al. 2018. 'Tools and Methods in Participatory Modeling: Selecting the Right Tool for the Job.' *Environmental Modelling and Software* 109: 232–255. doi:10.1016/j.envsoft.2018.08.028.

Voinov, A., N. Kolagani, M.K. McCall, P.D. Glynn, M.E. Kragt, F.O. Ostermann, S.A. Pierce, and P. Ramu. 2016. 'Modelling with Stakeholders – Next Generation.' *Environmental Modelling and Software* 77: 196–220. doi:10.1016/j.envsoft.2015.11.016.

References

Abrami, G., F. Nils, M. Sylvie, C. Murgue, A. Popova, H. de Fooij, S. Farolfi, D. du Toit, and W. Aquae-Gaudi. 2012. 'Wat-A-Game, a Toolkit for Building Role-Playing Games about Integrated Water Management.' *International Environmental Modelling and Software Society (IEMSs) – Managing Resources of a Limited Planet*. www.iemss.org/society/index.php/iemss-2012-proceedings.

Andersen, D.F., J.A.M. Vennix, G.P. Richardson, and E. Rouwette. 2007. 'Group Model Building: Problem Structuring, Policy Simulation and Decision Support.' *The Journal of the Operational Research Society* 58(5): 691–694.

Antunes, P., R. Santos, and N. Videira. 2006. 'Participatory Decision Making for Sustainable Development – The Use of Mediated Modelling Techniques.' *Land Use Policy, Resolving Environmental Conflicts: Combining Participation and Multi-Criteria Analysis* 23(1): 44–52. doi:10.1016/j.landusepol.2004.08.014.

Barnaud, C., and A. van Paassen. 2013. 'Equity, Power Games, and Legitimacy: Dilemmas of Participatory Natural Resource Management.' *Ecology and Society* 18(2): 21. doi:10.5751/ES-05459-180221.

Barreteau, O., G. Abrami, W. Daré, D. du Toit, N. Ferrand, P. Garin, V. Souchère, A. Popova, and C. Werey. 2012. 'Collaborative Modelling as a Boundary Institution to Handle Institutional Complexities in Water Management.' In *Restoring Lands – Coordinating Science, Politics and Action: Complexities of Climate and Governance*, 109–127. Amsterdam: Springer.

Barreteau, O., M. Antona, P. d'Aquino, S. Aubert, S. Boissau, F. Bousquet, W. Daré et al. 2003. 'Our Companion Modelling Approach.' *Journal of Artificial Societies and Social Simulation* 6(2). http://jasss.soc.surrey.ac.uk/6/2/1.html.

Barreteau, O., P. Bots, K. Daniell, M. Etienne, P. Perez, C. Barnaud, D. Bazile et al. 2013. 'Participatory Approaches.' In *Simulating Social Complexity: A Handbook* (2nd ed), edited by B. Edmonds and R. Meyer, 197–234. Berlin: Springer. doi:10.1007/978-3-540-93813-2_10.

Barreteau, O., F. Bousquet, and J-M. Attonaty. 2001. 'Role-Playing Games for Opening the Black Box of Multi-Agent Systems: Method and Lessons of Its Application to Senegal River Valley Irrigated Systems.' *Journal of Artificial Societies and Social Simulation* 4(2): 1–5.

Basco-Carrera, L., A. Warren, E. van Beek, A. Jonoski, and A. Giardino. 2017. 'Collaborative Modelling or Participatory Modelling? A Framework for Water Resources Management.' *Environmental Modelling and Software* 91: 95–110. doi:10.1016/j.envsoft.2017.01.014.

Becu, N. 2020. 'Les Courants d'influence et La Pratique de La Simulation Participative: Contours, Design et Contributions Aux Changements Sociétaux et Organisationnels Dans Les Territoires.' Habilitation à diriger des recherches, La Rochelle Université. https://hal.archives-ouvertes.fr/tel-02515352.

Bommel, P., N. Becu, B. Bonte, E. Delay, and C. Le Page. 2018. 'Cormas in 10 Years!' Conference_item. *CoMSES Net's Second Virtual Conference. 2018*. http://agritrop.cirad.fr/589022.

Bommel, P., N. Becu, C. Le Page, and F. Bousquet. 2016. 'Cormas: An Agent-Based Simulation Platform for Coupling Human Decisions with Computerized Dynamics.' In *Simulation and Gaming in the Network Society*. doi:10.1007/978-981-10-0575-6.

Bommel, P., F. Dieguez, D. Bartaburu, E. Duarte, E. Montes, M.P. Machin, J. Corral, C. José, P. de Lucena, and H.M. Grosskopf. 2014. 'A Further Step Towards Participatory Modelling. Fostering Stakeholder Involvement in Designing Models by Using Executable UML.' *Journal of Artificial Societies and Social Simulation* 17(1): 6.

Bots, P.W.G., and C.E. van Daalen. 2008. 'Participatory Model Construction and Model Use in Natural Resource Management: A Framework for Reflection.' *Systemic Practice and Action Research* 21(6): 389. doi:10.1007/s11213-008-9108-6.

Bouamrane, M., M. Spierenburg, A. Agrawal, A. Boureima, M-C. Cormier-Salem, M. Etienne, C. Le Page, H. Levrel, and R. Mathevet. 2016. 'Stakeholder Engagement and Biodiversity Conservation Challenges in Social-Ecological Systems: Some Insights from Biosphere Reserves in Western Africa and France.' *Ecology and Society* 21(4): 25. doi:10.5751/ES-08812-210425.

Bousquet, F., and C. Le Page. 2004. 'Multi-Agent Simulations and Ecosystem Management: A Review.' *Ecological Modelling* 176(3): 313–332. doi:10.1016/j.ecolmodel.2004.01.011.

Celio, E., and A. Grêt-Regamey. 2016. 'Understanding Farmers' Influence on Land-Use Change Using a Participatory Bayesian Network Approach in a Pre-Alpine Region in Switzerland.' *Journal of Environmental Planning and Management* 59(11): 2079–2101. doi:10.1080/09640568.2015.1120713.

Costanza, R., and M. Ruth. 1998. 'Using Dynamic Modeling to Scope Environmental Problems and Build Consensus.' *Environmental Management* 22(2): 183–195. doi:10.1007/s002679900095.

Creighton, J.L., and S. Langsdale. 2009. 'Analysis of Process Issues in Shared Vision Planning Cases.' IWR Report 09-R-05. Institute for Water Resources, US Army Corps of Engineers, Alexandria, VA.

Daniell, K.A., I.M. White, N. Ferrand, I. Ribarova, P. Coad, J.E. Rougier, M. Hare et al. 2010. 'Co-engineering Participatory Water Management Processes: Theory and Insights from Australian and Bulgarian Interventions.' *Ecology and Society* 15(4): 11. www.ecologyandsociety.org/vol15/iss4/art11.

D'Aquino, P., and A. Bah. 2013. 'A Participatory Modeling Process to Capture Indigenous Ways of Adaptability to Uncertainty: Outputs from an Experiment in West African Drylands.' *Ecology and Society* 18(4): 16. doi:10.5751/ES-05876-180416.

Daré, W., and J-P. Venot. 2017. 'Room for Manoeuvre: Users Participation in Water Resources Management in Burkina Faso.' *Development Policy Review* 36. doi:10.1111/dpr.12278.

Daré, W., J-P. Venot, C. Le Page, and A. Aduna. 2018. 'Problemshed or Watershed? Participatory Modeling towards IWRM in North Ghana.' *Water* 10(6): 721. doi:10.3390/w10060721.

Das, B. 2004. *Generating Conditional Probabilities for Bayesian Networks: Easing the Knowledge Acquisition Problem*. Ithaca: Cornell University.

Dewulf, A., R. Bouwen, and T. Taillieu. 2006. 'The Multi-Actor Simulation "Podocarpus National Park" as a Tool for Teaching and Researching Issue Framing.' SSRN Scholarly Paper ID 915943. Rochester: Social Science Research Network. https://papers.ssrn.com/abstract=915943.

Düspohl, M., S. Zacharias, and P. Doell. 2012. 'A Review of Bayesian Networks as a Participatory Modeling Approach in Support of Sustainable Environmental Management.' *International Journal of Sustainable Development* 5: 1–18. doi:10.5539/jsd.v5n12p1.

Etienne, M., D. du Toit, and S. Pollard. 2011. 'ARDI: A Co-construction Method for Participatory Modeling in Natural Resources Management.' *Ecology and Society* 16(1): 44. doi:10.5751/ES-03748-160144.

Ferrand, N., G. Abrami, E. Hassenforder, B. Noury, R. Ducrot, S. Farolfi, P. Garin, B. Bonte, S. Morardet, and D. L'Aot. 2017. 'Coupling for Coping, CoOPLAaGE: An Integrative Strategy and Toolbox Fostering Multi-level Hydrosocial Adaptation.' Conference item. *Proceedings of the ACEWATER2 Scientific Workshop*, Accra, Ghana, 31 October–3 November 2016. http://agritrop.cirad.fr/585578.

Forrester, J., R. Greaves, H. Noble, and R. Taylor. 2014. 'Modeling Social-Ecological Problems in Coastal Ecosystems: A Case Study.' *Complexity* 19(6): 73–82. doi:10.1002/cplx.21524.

Funtowicz, S.O., and J.R. Ravetz. 1993. 'Science for the Post-Normal Age.' *Futures* 25(7): 739–755. doi:10.1016/0016-3287(93)90022-L.

Gray, S., S. Gray, J.L. de Kok, A. Helfgott, B. O'Dwyer, R. Jordan, and A. Nyaki. 2015. 'Using Fuzzy Cognitive Mapping as a Participatory Approach to Analyze Change, Preferred States, and Perceived Resilience of Social-Ecological Systems.' *Ecology and Society* 20(2): 11. doi:10.5751/ES-07396-200211.

Halbe, J., C. Pahl-Wostl, and J. Adamowski. 2018. 'A Methodological Framework to Support the Initiation, Design and Institutionalization of Participatory Modeling Processes in Water Resources Management.' *Journal of Hydrology* 556: 701–716. doi:10.1016/j.jhydrol.2017.09.024.

Hare, M. 2011. 'Forms of Participatory Modelling and its Potential for Widespread Adoption in the Water Sector.' *Environmental Policy and Governance* 21(6): 386–402. doi:10.1002/eet.590.

Harmonicop Team. 2005. 'Learning Together to Manage Together – Improving Participation in Water Management.' EU Research project report. HarmoniCOP. www.harmonicop.uni-osnabrueck.de/HarmoniCOPHandbook.pdf.

Hassenforder, E., N. Ferrand, J. Pittock, K.A. Daniell, and O. Barreteau. 2015. 'A Participatory Planning Process as an Arena for Facilitating Institutional Bricolage: Example from the Rwenzori Region, Uganda.' *Society and Natural Resources* August. doi:10.1080/08941920.2015.1054977.

Hassenforder, E., J. Pittock, O. Barreteau, K.A. Daniell, and N. Ferrand. 2016. 'The MEPPP Framework: A Framework for Monitoring and Evaluating Participatory Planning Processes.' *Environmental Management Journal* 57(1): 79–96. doi:10.1007/s00267-015-0599-5.

Hedelin, B., M. Evers, J. Alkan-Olsson, and A. Jonsson. 2017. 'Participatory Modelling for Sustainable Development: Key Issues Derived from Five Cases of Natural Resource and Disaster Risk Management.' *Environmental Science and Policy* 76: 185–196. doi:10.1016/j.envsci.2017.07.001.

Hoch, C., M. Zellner, D. Milz, J. Radinsky, and L. Lyons. 2015. 'Seeing is not Believing: Cognitive Bias and Modelling in Collaborative Planning.' *Planning Theory and Practice* 16(3): 319–335. doi:10.1080/14649357.2015.1045015.

Hommes, S., J. Vinke-de Kruijf, H.S. Otter, and G. Bouma. 2008. 'Knowledge and Perceptions in Participatory Policy Processes: Lessons from the Delta-Region in the Netherlands.' *Water Resources Management* 23(8): 1641. doi:10.1007/s11269-008-9345-6.

Hovmand, P.S., D.F. Andersen, E. Rouwette, G.P. Richardson, K. Rux, and A. Calhoun. 2012. 'Group Model-Building "Scripts" as a Collaborative Planning Tool.' *Systems Research and Behavioral Science* 29(2): 179–193. doi:10.1002/sres.2105.

Htun, H., S.A. Gray, C.A. Lepczyk, A. Titmus, and K. Adams. 2016. 'Combining Watershed Models and Knowledge-Based Models to Predict Local-scale Impacts of Climate Change on Endangered Wildlife.' *Environmental Modelling & Software* 84(C): 440–457. doi:10.1016/j.envsoft.2016.07.009.

Jakku, E., and P.J. Thorburn. 2010. 'A Conceptual Framework for Guiding the Participatory Development of Agricultural Decision Support Systems.' *Agricultural Systems* 103(9): 675–862. doi:10.1016/j.agsy.2010.08.007.

Jones, N.A., P. Perez, T.G. Measham, G.J. Kelly, P. d'Aquino, K.A. Daniell, A. Dray, and N. Ferrand. 2009. 'Evaluating Participatory Modeling: Developing a Framework for Cross-Case Analysis.' *Environmental Management* 44: 1180–1195. doi:10.1007/s00267-009-9391-8.

Jordan, R., S. Gray, M. Zellner, P.D. Glynn, A. Voinov, B. Hedelin, E.J. Sterling et al. 2018. 'Twelve Questions for the Participatory Modeling Community.' *Earth's Future* 6(8): 1046–1057. doi:10.1029/2018EF000841.

Kelly, R.A., A.J. Jakeman, O. Barreteau, M.E. Borsuk, S. ElSawah, S.H. Hamilton, H.J. Henriksen et al. 2013. 'Selecting among Five Common Modelling Approaches for Integrated Environmental Assessment and Management.' *Environmental Modelling and Software* 47(September): 159–181. doi:10.1016/j.envsoft.2013.05.005.

Klabbers, J.H.G. 2009. *The Magic Circle: Principles of Gaming and Simulation*. Rotterdam: Sens Publishers. gbr.pepperdine.edu/book-corner/the-magic-circle-principles-of-gaming-and-simulation-3rd-and-revised-edition-by-jan-h-g-klabbers.

Kok, K. 2009. 'The Potential of Fuzzy Cognitive Maps for Semi-Quantitative Scenario Development, with an Example from Brazil.' *Global Environmental Change* 19(1): 122–133. doi:10.1016/j.gloenvcha.2008.08.003.

Landström, C., M. Becker, N. Odoni, and S.J. Whatmore. 2019. 'Community Modelling: A Technique for Enhancing Local Capacity to Engage with Flood Risk Management.' *Environmental Science and Policy* 92: 255–261. doi:10.1016/j.envsci.2018.11.009.

Langsdale, S., A. Beall, E. Bourget, E. Hagen, S. Kudlas, R. Palmer, D. Tate, and W. Werick. 2013. 'Collaborative Modeling for Decision Support in Water Resources: Principles and Best Practices.' *JAWRA Journal of the American Water Resources Association* 49(3): 629–638. doi:10.1111/jawr.12065.

Le Page, C., and P. Bommel. 2005. 'A Methodology for Building Agent-Base Simulations of Common-Pool Resources Management: From a Conceptual Model Designed with UML to Its Implementation in CORMAS.' In *Companion Modeling and Multi-Agent Systems for Integrated Natural Resource Management in Asia*, edited by F. Bousquet, G. Trébuil, and B Hardy, 327–350. Los Banos: International Rice Research Institute. http://agritrop.cirad.fr/530538.

Le Page, C., and A. Perrotton. 2017. 'KILT: A Modelling Approach Based on Participatory Agent-Based Simulation of Stylized Socio-Ecosystems to Stimulate Social Learning with Local Stakeholders.' In *Autonomous Agents and Multiagent Systems*, edited by G. Sukthankar and J.A. Rodriguez-Aguilar, 31–44. Lecture Notes in Computer Science. New York: Springer.

Lynam, T. 2001. 'Participatory Systems Analysis – An Introductory Guide.' Special Report No. 22. Institute of Environmental Studies. Harare: University of Zimbabwe.

McCarter, J., E. Sterling, S. Jupiter, G. Cullman, S. Albert, M. Basi, E. Betley et al. 2018. 'Biocultural Approaches to Developing Well-Being Indicators in Solomon Islands.' *Ecology and Society* 23(1): 32. doi:10.5751/ES-09867-230132.

Mendoza, G.A., and R. Prabhu. 2006. 'Participatory Modeling and Analysis for Sustainable Forest Management: Overview of Soft System Dynamics Models and Applications.' *Forest Policy and Economics* 9(2): 179–196. doi:10.1016/j.forpol.2005.06.006.

Metcalf, S.S., E. Wheeler, T.K. BenDor, K.S. Lubinski, and B.M. Hannon. 2010. 'Sharing the Floodplain: Mediated Modeling for Environmental Management.' *Environmental Modelling and Software*, Thematic Issue on Modelling with Stakeholders 25(11): 1282–1290. doi:10.1016/j.envsoft.2008.11.009.

Nguyen, N.C., and O.J.H. Bosch. 2013. 'A Systems Thinking Approach to Identify Leverage Points for Sustainability: A Case Study in the Cat Ba Biosphere Reserve, Vietnam.' *Systems Research and Behavioral Science* 30(2): 104–115. doi:10.1002/sres.2145.

Pahl-Wostl, C., M. Craps, A. Dewulf, E. Mostert, D. Tabara, and T. Taillieu. 2007. 'Social Learning and Water Resources Management.' *Ecology and Society* 12(2): 5. doi:10.5751/ES-02037-120205.

Palmer, R.N., W.J. Werick, A. MacEwan, and A.W. Woods. 1999. 'Modeling Water Resources Opportunities, Challenges and Trade-Offs: The Use of Shared Vision Modeling for Negotiation and Conflict Resolution.' *WRPMD'99'*. doi:10.1061/40430%281999%291.

Polhill, J.G., L-A. Sutherland, and N.M. Gotts. 2009. 'Using Qualitative Evidence to Enhance an Agent-Based Modelling System for Studying Land Use Change.' *Journal of Artificial Societies and Social Simulation* 13(2): 10.

Ponta, N., T. Cornioley, A. Dray, N. van Vliet, P.O. Waeber, and C. Garcia. 2019. 'Hunting in Times of Change: Uncovering Indigenous Strategies in the Colombian Amazon Using a Role-Playing Game.' *Frontiers in Ecology and Evolution*. doi:10.3389/fevo.2019.00034.

Salliou, N., C. Barnaud, A. Vialatte, and C. Monteil. 2017. 'A Participatory Bayesian Belief Network Approach to Explore Ambiguity among Stakeholders about Socio-Ecological Systems.' *Environmental Modelling and Software* 96: 199–209. doi:10.1016/j.envsoft.2017.06.050.

Sandker, M., B. Campbell, M. Ruiz-Pérez, J. Sayer, R. Cowling, H. Kassa, and A. Knight. 2010. 'The Role of Participatory Modeling in Landscape Approaches to Reconcile Conservation and Development.' *Ecology and Society* 15(2): 13. doi:10.5751/ES-03400-150213.

Schlüter, M., B. Müller, and K. Frank. 2019. 'The Potential of Models and Modeling for Social-Ecological Systems Research: The Reference Frame ModSES.' *Ecology and Society* 24(1): 31. doi:10.5751/ES-10716-240131.

Seidl, R. 2015. 'A Functional-dynamic Reflection on Participatory Processes in Modeling Projects.' *Ambio* 44(8): 750–765. doi:10.1007/s13280-015-0670-8.

Smajgl, A., and J. Ward. 2015. 'Evaluating Participatory Research: Framework, Methods and Implementation Results.' *Journal of Environmental Management* 157(July): 311–319. doi:10.1016/j.jenvman.2015.04.014.

Smajgl, A., J. Ward, T. Foran, J. Dore, and S. Larson. 2015. 'Visions, Beliefs, and Transformation: Exploring Cross-Sector and Transboundary Dynamics in the Wider Mekong Region.' *Ecology and Society* 20(2): 15. doi:10.5751/ES-07421-200215.

Smith, C., L. Felderhof, and O.J.H. Bosch. 2007. 'Adaptive Management: Making It Happen through Participatory Systems Analysis.' *Systems Research and Behavioral Science* 24(6): 567–587. doi:10.1002/sres.835.

Souchère, V., L. Millair, J. Echeverria, F. Bousquet, C. Le Page, and M. Etienne. 2010. 'Co-constructing with Stakeholders a Role-playing Game to Initiate Collective Management of Erosive Runoff Risks at the Watershed Scale.' *Environmental Modelling and Software* 25: 1359–1370. doi:10.1016/j.envsoft.2009.03.002.

Star, S.L., and J.R. Griesemer. 1989. 'Institutional Ecology, "Translations" and Boundary Objects: Amateurs and Professionals in Berkeley's Museum of Vertebrate Zoology 1907–1939.' *Social Studies of Science* 19(3). doi:10.1177/030631289019003001.

Stave, K. 2010. 'Participatory System Dynamics Modeling for Sustainable Environmental Management: Observations from Four Cases.' *Sustainability* 2(9): 2762–2784. doi:10.3390/su2092762.

Sterling, E., M. Zellner, K. Jenni, K. Leong, P. Glynn, T. BenDor, P. Bommel et al. 2019. 'Try, Try Again: Lessons Learned from Success and Failure in Participatory Modeling.' *Elementa: Science of the Anthropocene*, February. doi:10.1525/elementa.347.

Tsouvalis, J., and C. Waterton. 2012. 'Building "Participation" upon Critique: The Loweswater Care Project, Cumbria, UK.' *Environmental Modelling and Software*, Thematic issue on Expert Opinion in Environmental Modelling and Management, 36: 111–121. doi:10.1016/j.envsoft.2012.01.018.

Van Bruggen, A., I. Nikolic, and J. Kwakkel. 2019. 'Modeling with Stakeholders for Transformative Change.' *Sustainability* 11(3): 825. doi:10.3390/su11030825.

Vennix, J. 1996. *Group Model Building: Facilitating Team Learning Using System Dynamics*. Chichester: John Wiley and Sons.

Vidal, R.V.V. 2006. 'Operational Research: A Multidisciplinary Field.' *Pesquisa Operacional* 26(1): 69–90. doi:10.1590/S0101-74382006000100004.

Voinov, A., and F. Bousquet. 2010. 'Modelling with Stakeholders.' *Environmental Modelling and Software*, Thematic Issue on Modelling with Stakeholders 25(11): 1268–1281. doi:10.1016/j.envsoft.2010.03.007.

Voinov, A., K. Jenni, S. Gray, N. Kolagani, P.D. Glynn, P. Bommel, C. Prell et al. 2018. 'Tools and Methods in Participatory Modeling: Selecting the Right Tool for the Job.' *Environmental Modelling and Software* 109: 232–255. doi:10.1016/j.envsoft.2018.08.028.

Voinov, A., N. Kolagani, M.K. McCall, P.D. Glynn, M.E. Kragt, F.O. Ostermann, S.A. Pierce, and P. Ramu. 2016. 'Modelling with Stakeholders – Next Generation.' *Environmental Modelling and Software* 77: 196–220. doi:10.1016/j.envsoft.2015.11.016.

Werick, W.J. 1994. 'National Study of Water Management During Drought: Managing Water for Drought.' IWR Report 94-NDS-8. Alexandria, VA: US Army Corps of Engineers, Water Resources Support Centre, Institute for Water Resources. www.iwr.usace.army.mil/docs/iwrreports/94nds8.pdf.

14

Resilience assessment

Allyson Quinlan,[1] My Sellberg[2] and Arthur Perrotton[2]

[1]RESILIENCE ALLIANCE, OTTAWA, CANADA
[2]STOCKHOLM RESILIENCE CENTRE, STOCKHOLM UNIVERSITY, STOCKHOLM, SWEDEN

Key methods discussed in this chapter

Wayfinder, RAPTA, Resilience Assessment Workbook for Practitioners 2.0, STRESS, operationalising systemic resilience

Connections to other chapters

Resilience assessment is an umbrella process within which a wide range of tools and methods can be used. During the first stages of the assessment, systems scoping (Chapter 5) will be performed, commonly also interviews with key informants (Chapter 7) and participatory data collection (Chapter 8), such as timelines and historical profiling. Resilience assessment often benefits from reviewing previous collections of ecological field data (Chapter 6). The more technical part of the assessment includes systems analysis, such as expert modelling (Chapter 16), network analysis (Chapter 23), causal loop diagrams or other dynamical systems modelling (Chapter 26), and state-and-transition modelling (Chapter 27). The entire participatory process will draw on elements from different co-production methods, including facilitated dialogues and change labs (Chapter 9), scenario development (Chapter 11), and participatory modelling and planning (Chapter 13), particularly adaptive environmental management. Resilience assessment could also include action research (Chapter 15), futures analysis (Chapter 10) and qualitative content analysis (Chapter 19).

Introduction

Resilience assessment is a strategic approach grounded in theory that integrates multiple methods relevant to social-ecological systems (SES) research, in order to better understand the dynamics of complex SES and design strategic interventions. Although primarily designed for applying resilience theory in practice, the approach is often used as a research method. While the core technical component focuses on analysing system dynamics,

DOI: 10.4324/9781003021339-17

SUMMARY TABLE: RESILIENCE ASSESSMENT

DISCIPLINARY BACKGROUND	KNOWLEDGE TYPE
The methods in this chapter are derived from or have most commonly been used in: Ecology, Sociology, Environmental Science, Human Geography, Development Studies	*The methods in this chapter are primarily used to generate the following types of knowledge:* • Descriptive • Exploratory
RESEARCH APPROACH	**PURPOSE OF METHOD**
The methods in this chapter originate from or most commonly adopt the following research approaches: • Interpretive/subjective • Collaborative/process	*The most common purposes of using the methods in this chapter are:* • Data collection/generation • System understanding • Stakeholder engagement and co-production • Policy/decision support
TEMPORAL DIMENSION	**SYSTEMIC FEATURES AND PROCESSES**
The methods in this chapter are most commonly applied to the following temporal dimensions: • Present (typically within the last 5–10 years) • Recent past (post-1700s) • Future **SPATIAL DIMENSION** *The methods in this chapter are primarily either or both:* • Non-spatial • Explicitly spatial *The methods in this chapter are most commonly applied at the following spatial scales:* • Local • Regional (provincial/state to continental) • Multiple places/sites around the world	*While most methods can do many things, the methods in this chapter are particularly good (i.e. go-to methods) for addressing the following:* • SES components and linkages • Multiple scales and levels or cross-level interactions • Adaptation and self-organisation • Regime shifts • Transformation • Social learning • Collective action and collaborative governance • Exploring uncertainty

resilience assessment increasingly emphasises the process itself, including strategic engagement, knowledge co-creation and leveraging existing opportunities to help ensure actionable outcomes.

Resilience assessment can be applied in any SES (e.g. rural villages, cities, coastal fisheries, working landscapes) where people depend upon, shape and respond to their environment. Central to resilience assessment is the development of a conceptual model that integrates social and ecological variables and explicitly considers external drivers and system feedbacks (Walker and Salt 2012). The approach assumes an integrated humans-in-nature worldview that encourages multiple types of knowledge and evidence. While a key objective of most assessments is to better understand SES dynamics in order to influence change in the system, it also accepts that this knowledge will always be partial in complex adaptive systems.

The early framework developed by Walker and colleagues (2002) ('A working hypothesis for a participatory approach to applying resilience thinking') introduced a set of methods to help researchers and practitioners view natural resource issues from a systems perspective. These methods included describing the system and historical timelines, mapping external drivers and using future scenarios. While the methods themselves were not new, Walker and colleagues combined them in a framework for the purpose of understanding resilience in SES. Resilience was defined as the amount of change a system can undergo and still retain its structure and function, and its capacity for self-organisation, adaptation and learning (Walker et al. 2002). Building on these core elements, the Resilience Alliance (RA) (2010) developed a practitioner's workbook, integrating concepts such as system thresholds, interactions across scales, social networks and adaptive governance. Most of the resilience assessment approaches available today that are designed to address SES problems and questions can be traced to these original publications.

While early applications of resilience assessment tended to focus on ecological dynamics, over time greater attention was paid to human dimensions and fully integrated SES dynamics (Anderies, Walker, and Kinzig 2006; Walker and Salt 2012). This shift is also reflected in changing definitions of resilience, which more recently has been defined as the capacity of an SES to persist in the face of disturbance and change, while continuing to adapt and develop along a pathway or transform and navigate new pathways in order to sustain human well-being (Biggs, Schlüter, and Schoon 2015; Folke 2016). Increasingly, resilience thinking takes into account and integrates notions of governance systems, ecosystem services and human well-being, adaptive capacity and transformation (Olsson, Folke, and Hughes 2008; Daw et al. 2015; Sellberg et al. 2018b). Engaging with complexity, a core concept of social-ecological resilience thinking (Preiser et al. 2018) is increasingly seen as key to understanding and engaging with SES dynamics. As social-ecological research continues to advance, many new analytical methods, both qualitative and quantitative, are likely to become part of the resilience assessment toolkit, just as the process itself is becoming more of a continuous practice involving reflection and reiteration.

Resilience assessment approaches are now being developed by a range of organisations working in a variety of contexts, including coastal, urban and rural development, conservation and climate change, to name a few. This growing abundance of resilience assessment guides and tools (ODI 2016; Sharifi 2016; Douxchamps et al. 2017) highlights the many different ways in which resilience is conceptualised. However, only a relatively small number of these guides and tools align with an SES framing. In this chapter, we focus on a select few guides that are based on social-ecological perspectives and that engage with complexity.

SES problems and questions

Resilience assessment broadly addresses questions about the capacity of an SES to cope with and respond to change. In the Anthropocene, people and nature increasingly face complex, wicked problems that demand coordinated actions across multiple scales (Steffen et al. 2011). Often there are no simple solutions since actions to improve the conditions and resources for one group might negatively affect another (Enfors-Kautsky et al. 2018). These types of challenges, which span domains and interact across scales, have raised interest in complexity-based approaches to better navigate change while moving towards more desirable futures (Sellberg et al. 2018a). Within the broad framing of just and sustainable futures, resilience assessment is an adaptable approach that employs a variety of social-ecological methods to address multiple issues that suit the context. Each assessment identifies as its focus one or more issues relevant to the particular SES.

The types of questions that resilience assessments generally deal with include:

- Understanding the resilience of an SES, how it has changed over time and what factors build or erode it; resilience assessment usually addresses both resilience to specific changes and potential shifts in a system state, as well as general resilience to unknown change
- Exploring strategies and actions for an SES to continue to deliver important ecosystem services to people in the face of change; these strategies can include buffering change, but also adapting and transforming in response to change
- Exploring how governance and management of an SES can be improved by taking more of the inherent system complexity and dynamics into account; the assumption is that this will better align the governance system with the underlying social-ecological processes and also make it more effective

In practice, these three areas of exploration might be partly overlapping and one resilience assessment could cover all of them. Below are examples of specific questions that resilience assessments have dealt with.

- How has resilience of an SES changed over time? (e.g. understanding how the resilience of a Coastal Pacific herring fishery changed during different management eras (Salomon et al. 2019))
- What factors build or erode resilience? (e.g. comparing cases of resilience and transformation across the Arctic region (Huitric, Peterson, and Rocha 2016))
- How can we increase the resilience of important ecosystem services in the face of future changes? (e.g. exploring strategies for building resilience of food systems in Eskilstuna Municipality in Sweden (Sellberg, Wilkinson, and Peterson 2015))
- How can we shift the system to a trajectory where important ecosystem services are more resilient? (e.g. exploring how the Telecho community in Ethiopia can transition to pathways to a more food-secure system (Maru et al. 2017))
- How can we manage an SES in a way that takes more of its complexity into account? (e.g. co-producing knowledge on multiple ecosystem services in the Helgeå watershed, the synergies and trade-offs between them and potential positive future scenarios (Malmborg 2019))

- How can governance of an SES take more of the social-ecological connections and dynamics of change into account? (e.g. understanding the social and economic dimensions of natural resource management issues and identifying thresholds of potential concern in Australian natural resource management (Sellberg et al. 2018b))
- How can we design development interventions that address the systemic causes of problems, are more effective and have sustained benefits? (Maru et al. 2017)

Brief description of key methods

A small number of resilience assessment guides explicitly engage with the complex nature of SES. These approaches offer ways of exploring social-ecological dynamics and developing strategies to influence how a system might adapt or transform in the face of change. They also offer practical tools grounded in theory that can help researchers and others to assess and influence the resilience of complex adaptive systems (Sellberg et al. 2018a). While there are many other resilience assessment guides available that have been developed for a variety of purposes, the approaches included in Table 14.1 are designed to assess resilience specifically through a social-ecological lens.

Table 14.1 Summary of key approaches used in resilience assessment

Approach	*Description*	*References*
Wayfinder	Wayfinder is a resilience guide for navigating towards sustainable futures. It is a process guide used for resilience assessment, planning and action in SES. It describes a process for engaging stakeholders at multiple levels, co-creating knowledge and exploring system dynamics (e.g. feedbacks, thresholds, cross-scale interactions) and social-ecological dilemmas (e.g. ecosystem service trade-offs). It includes tools for developing strategic actions and deciding when to build resilience and when to adapt or transform. Wayfinder also offers practical guidance and an online toolkit with ready-to-use activity sheets.	*Key introductory text* Enfors-Kautsky et al. 2018 *Applications to SES* Goffner, Sinare, and Gordon 2019; Perrotton, Ka, and Goffner 2019
Resilience, Adaptation Pathways and Transformation Assessment	The Resilience, Adaptation Pathways and Transformation Assessment (RAPTA) framework is a guide to developing and implementing interventions for sustainable development. It includes technical components of system assessment (feedbacks, thresholds, cross-scale interaction) and guidance on filtering options and creating pathways for change. It has been designed to work with project cycles and to enhance or work with existing theory of change methods.	*Key introductory text* O'Connell et al. 2016 *Applications to SES* Maru et al. 2017; Cowie et al. 2019

(Continued)

Table 14.1 (Continued)

Approach	*Description*	*References*
Resilience Assessment Workbook for Practitioners 2.0	The Resilience Assessment Workbook presents a five-phase approach to assessing the resilience of SES. This approach involves defining the system, understanding system dynamics, identifying key relationships, exploring system governance and acting on the assessment.	*Key introductory texts* Walker et al. 2009; Resilience Alliance 2010 *Applications to SES* Haider, Quinlan, and Peterson 2012; Walker and Salt 2012; Wilkinson 2012; Sellberg, Wilkinson, and Peterson 2015; Sellberg et al. 2018b
STRESS	Strategic Resilience Assessment (STRESS) is a learning process for resilience planning that includes a communications plan, a work plan and field-team training. It includes practical guidance on the time and skills required for the assessment, which works towards developing a resilience-focused theory of change. STRESS combines resilience concepts with vulnerability assessment (e.g. developing vulnerability profiles, identifying vulnerable groups).	*Key introductory text* Levine, Vaughan, and Nicholson 2017 *Applications to SES* Mercy Corps 2018
Operationalising systemic resilience	Operationalising systemic resilience is a multi-stakeholder process to build community resilience. The framework is derived from a critical analysis of resilience thinking, systems thinking, community operational research and development studies. Phases in the process include critiquing system boundaries, visioning (negotiating desirable change 'for whom'), setting time frames through asset mapping and back-casting, scenario development to probe uncertainties, locally driven implementation, evaluation learning and re-evaluation.	*Key introductory text* Helfgott 2018 *Applications to SES* No known applications

Limitations

Resilience assessment is a practical, hands-on, transdisciplinary and collaborative approach for exploring critical issues in SES. There are a number of limitations or challenges, many of which also apply to other participatory knowledge co-production processes.

At a conceptual level, a complex adaptive systems mindset is key to resilience assessment, but this takes time to develop and often is in direct contrast to prevailing views. When resilience assessment was used by catchment management authorities in Australia, the practitioners often experienced a clash with existing mainstream approaches to natural resource management that assume linear cause-and-effect relationships (Sellberg et al. 2018b). By contrast, the resilience assessment approach highlights real-world complexity and does not sit neatly within one sector; rather, it acknowledges that outcomes are uncertain, which can sometimes be very challenging. Cross-scale interactions can also appear abstract or too far removed from the system and may be difficult to evaluate, but conceptual models are helpful in this regard (e.g. see Walker et al. 2009).

At a practical level, the approach is process intensive, requiring significant time and resource investments and a commitment to revisiting past steps and challenging assumptions as new knowledge and understanding is gained. As with most transdisciplinary and collaborative approaches, resilience assessment requires time for building relationships and trust and embedding or anchoring the process in an organisation or community. In two Swedish cases, for example, the assessments were mainly side projects to the normal operations, carried forward by engaged key individuals (Sellberg, Wilkinson, and Peterson 2015; Sellberg et al. 2017). As seen in some cases in Australia, where resilience assessment has been used the longest, it takes several years to really embed the approach in an organisation because it requires changes in the organisational culture, structure and processes (Sellberg et al. 2018b). Some have suggested simpler and faster approaches to assessing resilience, but ultimately there are no shortcuts to enabling systemic change, which inevitably involves a long-term commitment (Enfors-Kautsky et al. 2018).

As an ongoing method for understanding system dynamics, resilience assessment is incompatible with short-term project frames that are common in programmes that expect predetermined outcomes according to a set schedule. This points to an ethical consideration of not starting a resilience assessment where there is no possibility of a long-term commitment and follow-through. Resilience assessment has been ongoing in parts of Australia for well over a decade. In Tajikistan, external experts conducted a resilience assessment over a one-year period, but in partnership with an NGO with a long-term engagement in the area (Sellberg et al. 2018a). Capacity building, as part of the Tajikistan project, also ensured that the NGO could continue using and adapting the resilience assessment approach in their operations.

Resource implications

Resilience assessment as described in this chapter is a learning process that requires both a long-term commitment and sufficient resources, including skilled facilitation and people who are trained to guide a participatory process. Resilience assessments also draw on existing sources of data, e.g. regarding different environmental aspects. The quality of and access to these data will determine the depth and quality of the assessment.

The leader or team leading a resilience assessment needs inter- and transdisciplinary skills, since they need to integrate many different types of knowledge and sources of information, e.g. qualitative and quantitative data from natural and social science, as well as practical and experiential knowledge with scientific knowledge. If diverse participants are engaged in the process of assessing resilience and analysing systems, they also need skills that can be translated to ground complex concepts in real-world examples that are relevant to the context. Established networks and relationships with key stakeholders and non-academic partners are not prerequisites but can greatly facilitate the process and decrease the time of preparation.

The assessment leaders or teams also need to be trained in resilience and systems thinking. They also need a complex adaptive systems mindset and the pedagogical skills to teach this mindset to other core people involved in the resilience assessment, if necessary. Additional skills in particular methods and tools, such as scenario development (Chapter 11) or dynamical systems modelling (Chapter 26), will be useful, without getting too attached to any one tool. The case will determine which tool will be useful and it is recommended to have a variety of tools at one's disposal.

Case study 14.1: Collectively redefining a better future in Ranérou-Ferlo, Senegal

Nearly 60 years after independence, rural populations living in the Ferlo, i.e. the northern Sahelian part of Senegal, continue to face immense development challenges. The majority of people living in the region are Fulani pastoralists who rely on extensive livestock herding. Vulnerability persists in the region, despite decades of development initiatives by the Senegalese government and international organisations leading to increased access to health care, education and water. Among the key issues are climate variability (e.g. Herman et al. 2018), malnutrition (e.g. Lazzaroni and Wagner 2016), land degradation (e.g. Hermann, Aziz Diouf, and Sall 2019), and persistent tensions among local actors over the use of pastoral resources. Researchers from the French government-funded Future Sahel Project, in collaboration with the Senegalese Great Green Wall Agency, conducted an 18-month-long participatory process with the goal of co-creating an innovative strategic development plan for the district of Ranérou-Ferlo.

Following the Wayfinder guide (Enfors-Kautsky et al. 2018), the process began by identifying who to engage and doing an initial system exploration. Two coalitions were quickly established: (a) a 'national coalition' (Dakar) involving managers of the national Great Green Wall Agency, and (b) a 'local coalition' (Ranérou) involving the district administrator and the head of the local office of the Directorate of Water, Forestry, Hunting and Soil Conservation. Together with these coalitions, researchers mobilised local citizens to form a multi-stakeholder working group. This group identified a set of local aspirations for development, along with existing constraints that were preventing the realisation of these aspirations. These included, for instance, the general lack of collective action, the lack of accountability of governance actors, the spread of uncontrolled settlements, and prejudices and misconceptions about Fulani herders. The next step used a systems lens to identify key leverage points that were revealed in conceptual models drawn by participants, as well as the networks of interactions between aspirations for the system and constraints.

In the final step of the process, four action strategies were collectively designed. Each strategy revolved around a set of linked aspirations and proposed actions to trigger specific changes in the district of Ranérou-Ferlo in order to reach these aspirations, while bringing the district closer to a more resilient path (future-sahel.blogspot.com).

The resilience assessment process: (a) confirmed the pressing need for improvement of social-ecological conditions in the area, (b) helped researchers and stakeholders to collectively identify development priorities and create strategies that target key leverage points, and (c) along with local actors highlighted the importance of social-oriented development actions in environmental protection. A strategic plan was distributed to all governance and development actors involved in the process. The development planning document included explanations of key concepts in the theoretical framework of the Wayfinder approach, and provided full descriptions of the strategies that had been co-designed with local actors.

The Wayfinder process conducted in Ranérou (Figure 14.1) integrated three groups of actors working at local to national levels. Their respective activities contributed towards developing strategies for change. At a local scale, results of workshops with the multi-stakeholder group were presented by the research team and discussed with the local coalition. To enable the implementation of the strategies, leaders of the Senagalese Great Green Wall Agency were involved through the national coalition.

Many of the challenges encountered with this case study are common to participatory processes in rural areas, including language barriers and the low literacy rates of workshop participants. These were overcome by including Senegalese researchers who could speak Fulani and using drawings and other visual aids during workshops. To address power imbalances among local actors, stakeholder groups met separately first to ensure the inclusion of marginalised voices. Locally relevant metaphors were used to help explain otherwise abstract theoretical concepts that do not always translate well.

The Wayfinder resilience assessment approach was well suited to the objectives and context of research in the Ranérou-Ferlo SES. Beyond the key insights gained regarding local system dynamics, the coalitions helped to maintain a focus on realistic and relevant development strategies, which could be supported and eventually implemented by governance and development actors. Simultaneously, trust among coalition members enabled dialogue about alternative land management options that challenged existing beliefs and habits and opened up new possibilities. Importantly, the involvement of governance actors in the coalitions facilitated uptake of the results in their networks and within the organisations that will implement the development strategies.

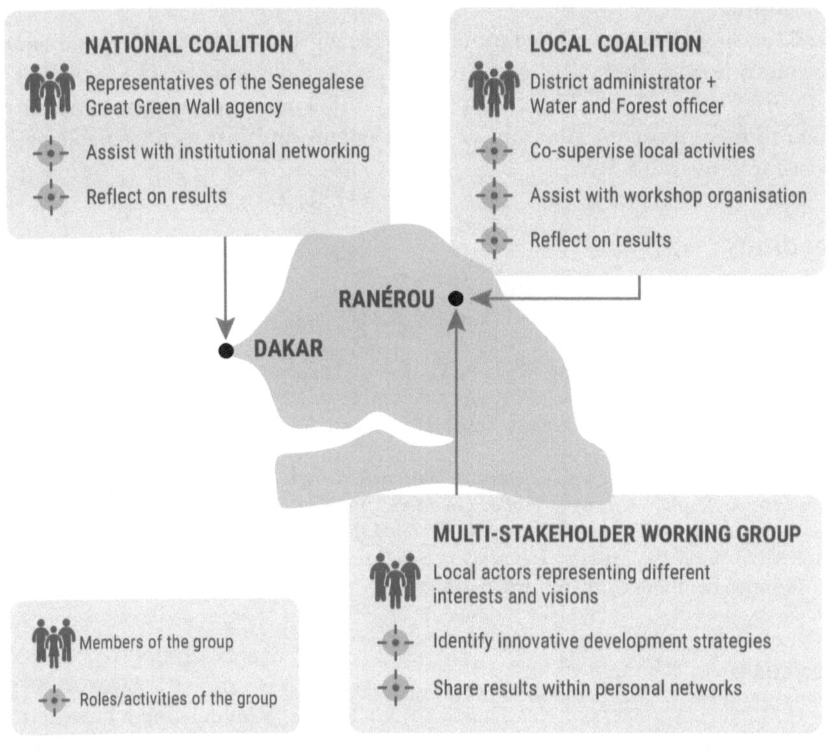

Figure 14.1 A multi-level process framework for resilience assessment in Ranérou, Senegal (© Arthur Perrotton)

New directions

While resilience assessment has traditionally been oriented towards natural resource management and planning processes, it is increasingly being used to inform development programming (Haider, Quinlan, and Peterson 2012; Pollard, Biggs, and Du Toit 2014; Maru et al. 2017; Enfors-Kautsky et al. 2018). A growing number of guides have been designed to streamline the assessment with project cycles and have integrated traditional development methods such as theory of change, capitals approaches and livelihood analysis (OECD 2014; O'Connell et al. 2016; Levine, Vaughn, and Nicholson 2017; UNDP 2017). The intersection of resilience and development practice has the potential to be a source for interdisciplinary innovation by combining and creating new methods. The recently developed Wayfinder approach offers a new framing of change narratives that is informed by theory of change and social innovation, and combines agency, opportunity context and strategic leverage points (Enfors-Kautsky et al. 2018). A number of rapid resilience assessment approaches are also being developed, mostly using quantitative methods, to suit a variety of project objectives (Salomon et al. 2019).

Resilience assessment is increasingly used in urban contexts, as resilience is a key issue for many cities facing extreme weather events, a lack of water and other disruptions (Elmqvist et al. 2019). A recent study assessed the resilience of ecosystem services to climate change and urban growth in southern Stockholm, Sweden, for example (justurbangreen.com/web/en/startpage/enable). This project emphasised spatial aspects of resilience, which are relevant for city planning.

Several recent guides have placed more emphasis on transformation to sustainable and just pathways, in line with the global goals of sustainable development (O'Connell et al. 2016; Enfors-Kautsky et al. 2018). This direction may influence future applications of resilience assessment to focus more on questions of how a system can shift to sustainable pathways, or build transformative capacity.

Key readings

Enfors-Kautsky, E., L. Järnberg, A. Quinlan, and P. Ryan. 2018. 'Wayfinder: A Resilience Guide for Navigating Towards Sustainable Futures.' GRAID programme, Stockholm Resilience Centre. www.wayfinder.earth.

Helfgott, A. 2018. 'Operationalising Systemic Resilience.' *European Journal of Operational Research* 268: 852–864.

Levine, E., E. Vaughan, and D. Nicholson. 2017. *Strategic Resilience Assessment Guidelines.* Portland: Mercy Corps.

O'Connell, D., N. Abel, N. Grigg, Y. Maru, J. Butler, A. Cowie, S. Stone-Jovicich et al. 2016. *Designing Projects in a Rapidly Changing World: Guidelines for Embedding Resilience, Adaptation and Transformation into Sustainable Development Projects. Version 1.0.* Washington: Global Environment Facility.

Resilience Alliance. 2010. *Assessing Resilience in Social-Ecological Systems: Workbook for Practitioners. Version 2.0.* www.resalliance.org/resilience-assessment.

References

Anderies, J.M., B.H. Walker, and A.P. Kinzig. 2006. 'Fifteen Weddings and a Funeral: Case Studies and Resilience-Based Management.' *Ecology and Society* 11(1): 21. www.ecologyandsociety.org/vol11/iss1/art21.

Biggs, R., M. Schlüter, and M.L. Schoon, eds. 2015. *Principles for Building Resilience: Sustaining Ecosystem Services.* Cambridge: Cambridge University Press.

Cowie, A., L.C.M. Waters, F. Garland, S.E. Orgill, A. Baumber, R. Cross, D. O'Connell, and G. Metternicht. 2019. 'Assessing Resilience to Underpin Implementation of Land Degradation Neutrality: A Case Study in the Rangelands of Western New South Wales, Australia.' *Environmental Science and Policy* 100: 37–46.

Daw, T.M., S. Coulthard, W.W.L. Cheung, K. Brown, C. Abunge, D. Galafassi, G.D. Peterson, T.R. McClanahan, J.O. Omukoto, and L. Munyi. 2015. 'Evaluating Taboo Trade-Offs in Ecosystem Services and Human Well-being.' *Proceedings of the National Academy of Sciences* 112(22): 6949–6954.

Douxchamps, S., L. Debevec, M. Giordano, and J. Barron. 2017. 'Monitoring and Evaluation of Climate Resilience for Agricultural Development – A Review of Currently Available Tools.' *World Development Perspectives* 5: 10–23. doi:10.1016/j.wdp.2017.02.001.

Elmqvist, T., E. Andersson, N. Frantzeskaki, T. McPhearson, P. Olsson, O. Gaffney, K. Takeuchi, and C. Folke. 2019. 'Sustainability and resilience for transformation in the urban century.' *Nature Sustainability* 2: 267–273. doi:10.1038/s41893-019-0250-1.

Enfors-Kautsky, E., L. Järnberg, A. Quinlan, and P. Ryan. 2018. Wayfinder: A Resilience Guide for Navigating Towards Sustainable Futures. GRAID Programme, Stockholm Resilience Centre. www.wayfinder.earth.

Folke, C. 2016. 'Resilience.' *Ecology and Society* 21(4): 44. doi:10.5751/ES-09088-210444.

Goffner, D., H. Sinare, and L.J. Gordon. 2019. 'The Great Green Wall for the Sahara and the Sahel Initiative as an Opportunity to Enhance Resilience in Sahelian Landscapes and Livelihoods.' *Regional Environmental Change* 19(5): 1417–1428.

Haider, J.L., A. Quinlan, and G.D. Peterson. 2012. 'Interacting Traps: Resilience Assessment of a Pasture Management System in Northern Afghanistan.' *Planning Theory and Practice* 13(2): 312–319.

Helfgott, A. 2018. 'Operationalising Systemic Resilience.' *European Journal of Operational Research* 268: 852–864. doi:10.1016/j.ejor.2017.11.056.

Herman, R.J., Y. Kushnir, A. Giannini, and M. Biasutti. 2018. 'Understanding Decadal and Interannual Variability in Rainfall over the Sahel.' *AGU Fall Meeting Abstracts* 21. https://ui.adsabs.harvard.edu/abs/2018AGUFM.A21F..07H/abstract.

Hermann, S., A.A. Diouf, and I. Sall. 2019. 'Beyond Productivity: Engaging Local Perspectives in Land Degradation Monitoring and Assessment.' *Journal of Arid Environments* 104002.

Huitric, M., G.D. Peterson, and J.C. Rocha. 2016. 'What Factors Build or Erode Resilience in the Arctic?' In *Arctic Resilience Report*. Stockholm: Stockholm Environment Institute and the Stockholm Resilience Centre. www.arctic-council.org/arr.

Lazzaroni, S., and N. Wagner. 2016. 'Misfortunes Never Come Singly: Structural Change, Multiple Shocks and Child Malnutrition in Rural Senegal.' *Economics and Human Biology* 23: 246–262.

Levine, E., E. Vaughan, and D. Nicholson. 2017. *Strategic Resilience Assessment Guidelines*. Portland: Mercy Corps. www.mercycorps.org/sites/default/files/STRESS-Guidelines-Resilience-Mercy-Corps-2017.pdf.

Malmborg, K. 2019. 'How on Earth: Operationalizing the Ecosystem Service Concept for Local Sustainability.' Licentiate diss., Stockholm University.

Maru, Y., D. O'Connell, N. Grigg, N. Abel, A. Cowie, S. Stone-Jovicich, J. Butler et al. 2017. *Making 'Resilience', 'Adaptation' and 'Transformation' Real for the Design of Sustainable Development Projects: Piloting the Resilience, Adaptation Pathways and Transformation Assessment (RAPTA) Framework in Ethiopia*. CSIRO, Australia.

Mercy Corps. 2018. *Planting Seed of Resilience in Humanitarian Settings: Rapid Strategic Resilience Assessment Report for the Rohingya Crisis, Cox's Bazar, Bangladesh*. www.mercycorps.org/sites/default/files/MercyCorps-IOM_RapidStrategicResilienceAssessment_Report.pdf.

O'Connell, D., N. Abel, N. Grigg, Y. Maru, J. Butler, A. Cowie, S. Stone-Jovicich et al. 2016. *Designing Projects in a Rapidly Changing World: Guidelines for Embedding Resilience, Adaptation and Transformation into Sustainable Development Projects. Version 1.0*. Washington: Global Environment Facility.

ODI (Overseas Development Institute). 2016. 'Analysis of Resilience Measurement Frameworks and Approaches.' *The Resilience Measurement, Evidence and Learning Community of Practice*. www.measuringresilience.org.

OECD. 2014. *Guidelines for resilience systems analysis*. Paris: OECD Publishing.

Olsson, P., C. Folke, and T.P. Hughes. 2008. 'Navigating the Transition to Ecosystem-Based Management of the Great Barrier Reef, Australia.' *Proceedings of the National Academy of Sciences* 105(28): 9489–9494.

Perrotton, A., A. Ka, and D. Goffner. 2019. 'WAYFINDER et Amélioration Collective de la Résilience dans le Département de Ranérou Ferlo: Rapport Stratégique.'

Pollard, S., H. Biggs, and D.R. du Toit. 2014. 'A Systemic Framework for Context-Based Decision Making in Natural Resource Management: Reflections on an Integrative Assessment of Water and Livelihood Security Outcomes Following Policy Reform in South Africa.' *Ecology and Society* 19(2): 63. doi:10.5751/ES-06312-190263.

Preiser, R., R. Biggs, A. de Vos, and C. Folke. 2018. 'Social-Ecological Systems as Complex Adaptive Systems: Organizing Principles for Advancing Research Methods and Approaches.' *Ecology and Society* 23(4): 46. doi:10.5751/ES-10558-230446.

Resilience Alliance (RA). 2010. *Assessing Resilience in Social-Ecological Systems: Workbook for Practitioners. Version 2.0.* www.resalliance.org/resilience-assessment.

Salomon, A.K., A.E. Quinlan, G.H. Pang, D.K. Okamoto, and L. Vazquez-Vera. 2019. 'Measuring Social-Ecological Resilience Reveals Opportunities for Transforming Environmental Governance.' *Ecology and Society* 24(3): 16. doi:10.5751/ES-11044-240316.

Sellberg, M.M., S. Borgstrom, A.V. Norstrom, and G.D. Peterson. 2017. 'Improving Participatory Resilience Assessment by Cross-fertilizing the Resilience Alliance and Transition Movement Approaches.' *Ecology and Society* 22(1): 28. doi:10.5751/ES-09051-220128.

Sellberg, M.M., A. Quinlan, R. Preiser, and G.D. Peterson. 2018a. 'How social-ecological resilience practice engages with complexity.' In *Advancing Resilience Practice: Bridging Social-Ecological Resilience Theory and Sustainable Development Practice* by M.M. Sellberg. PhD diss., Stockholm University.

Sellberg, M.M., P. Ryan, S.T. Borgstrom, A.V. Norstrom, and G.D. Peterson. 2018b. 'From Resilience Thinking to Resilience Planning: Lessons from Practice.' *Journal of Environmental Management* 217: 906–918. doi:10.1016/j.jenvman.2018.04.012.

Sellberg, M.M., C. Wilkinson, and G.D. Peterson. 2015. 'Resilience Assessment: A Useful Approach for Navigating Urban Sustainability Challenges.' *Ecology and Society* 20(1): 43. doi:10.575/ES-07258-200143.

Sharifi, A. 2016. 'A critical Review of Selected Tools for Assessing Community Resilience.' *Ecological Indicators* 69: 629–647. doi:10.1016/j.ecolind.2016.05.023.

Steffen, W., Å. Persson, L. Deutsch, J. Zalasiewicz, M. Williams, K. Richardson, C. Crumley et. al. 2011. 'The Anthropocene: From Global Change to Planetary Stewardship.' *Ambio* 40(7): 739–761. doi:10.1007/s13280-011-0185-x.

UNDP. 2017. *Community Based Resilience Analysis (CoBRA): Implementation Guidelines. Version 2.* New York: UNDP.

Walker, B.H., N. Abel, J.M. Anderies, and P. Ryan. 2009. 'Resilience, Adaptability, and Transformability in the Goulburn-Broken Catchment, Australia.' *Ecology and Society* 14(1): 12. www.ecologyandsociety.org/vol14/iss1/art12.

Walker, B.H., S. Carpenter, J. Anderies, N. Abel, G. Cumming, M. Janssen, L. Lebel et al. 2002. 'Resilience Management in Social-Ecological Systems: A Working Hypothesis for a Participatory Approach.' *Conservation Ecology* 6(1): 14.

Walker, B.H., and D. Salt. 2012. *Resilience Practice: Building Capacity to Absorb Disturbance and Maintain Function.* Washington: Island Press.

Wilkinson, C. 2012. 'Social-Ecological Resilience: Insights and Issues for Planning Theory.' *Planning Theory* 11(2): 148–169. doi:10.1177/1473095211426274.

15
Action research

Jean Boulton[1] *and Rika Preiser*[2]

[1]DEPARTMENT OF SOCIAL AND POLICY SCIENCES, UNIVERSITY OF BATH, BATH, UK
[2]CENTRE FOR SUSTAINABILITY TRANSITIONS, STELLENBOSCH UNIVERSITY, STELLENBOSCH, SOUTH AFRICA

Key methods discussed in this chapter

Narrative inquiry, learning history, cooperative inquiry

Connections to other chapters

This chapter links well with other chapters that explore more qualitative and interpretative ways of dealing with stakeholder engagement as investigated in Chapter 9 (Facilitated dialogues) and Chapter 19 (Qualitative content analysis), and knowledge co-creation as discussed in Chapter 8 (Participatory data collection).

Introduction

The term 'action research' is sometimes co-opted to signify any research that is participative and designed for 'action' and for change to happen as a result. However, action research signifies a coherent and well-established set of approaches, methods and values with a rich history (Torbert 1976; Reason and Rowan 1981; Gustavsen 2003). In this deeper tradition, action research gives a practical and empirical approach to investigating the complex, interconnected and emergent social-ecological world. It is embedded in a view of the world as 'systemic, participative, radically interconnected and evolutionary' (Reason and Bradbury 2001, 12).

Due to this focus on the dynamic and emergent nature of situations, there is no insistence in action research on an initial research hypothesis against which to gain evidence, nor a requirement that methods are defined in their entirety at the beginning. Allowance is made for surprise and shifts in focus through following and illuminating the detailed pathways of what happened and how. As the world does not stand still as we engage with it, project methodologies and interests in part emerge over time as those involved learn more about the issues, try out new ways of doing things, develop relationships and gain confidence in their exploration (Marshall, Coleman, and Reason 2011, 29).

Action research runs counter to the idea that change and agency can be understood through investigating the objective, the average, the universal. It does not give preference

DOI: 10.4324/9781003021339-18

SUMMARY TABLE: ACTION RESEARCH	
DISCIPLINARY BACKGROUND	**KNOWLEDGE TYPE**
The methods in this chapter are derived from or have most commonly been used in: Qualitative Research	*The methods in this chapter are primarily used to generate the following types of knowledge:* • Exploratory • Explanatory
RESEARCH APPROACH	**PURPOSE OF METHOD**
The methods in this chapter originate from or most commonly adopt the following research approaches: • Interpretive/subjective • Collaborative/process	*The most common purposes of using the methods in this chapter are:* • Stakeholder engagement and co-production
TEMPORAL DIMENSION	**SYSTEMIC FEATURES AND PROCESSES**
The methods in this chapter are most commonly applied to the following temporal dimensions: • Present (typically within the last 5–10 years) **SPATIAL DIMENSION** *The methods in this chapter are primarily either or both:* • Non-spatial *The methods in this chapter are most commonly applied at the following spatial scales:* • Local • Regional (provincial/state to continental)	*While most methods can do many things, the methods in this chapter are particularly good (i.e. go-to methods) for addressing the following:* • Diversity • Power relations • Transformation • Social learning • Collective action and collaborative governance • Evaluating policy options • Exploring uncertainty

to objective, 'scientific' evidence but includes evidence that is more subjective. Our subjective experiences – of values, intuitions, relationships – and our perceptions – of what is changing, what is stuck, what is emerging – are valued and included. Action research is an appropriate approach to investigating the complex world precisely because it pays attention to these subjective perspectives, to the particularity of situations and to the way things emerge and change over time. Action research methodologies are designed to surface the complex multi-faceted ways in which people and processes interact and in which change happens and systemic patterns form or dissolve (Boulton, Allen, and Bowman 2015).

Coupled with this view as to 'how the world is constituted', there is a strong emphasis on the ethics of social research: methods must allow for the ownership of the research process and outcomes by those who are involved in it. This is not 'research on' or 'research by' but 'research with'. As part of the approach, issues of power are typically uncovered and made explicit. Indeed, part of the motivation to undertake action research is to 'support people who thought they were powerless to find they have power to do things' (Reason et al. 2009, 10). Action research is 'unashamedly value-laden, asking what is most likely to help us build a freer, better society' (Marshall, Coleman, and Reason 2011) and there is often an intention to create resilience for people and planet.

Action research is also viewed as 'pragmatic' (Greenwood 2007), emphasising the importance of research leading to action and that research outcomes and theories are no use of and in themselves. Whereas there are methodologies that action researchers draw on, as discussed below, 'it is important to understand AR [action research] as an *orientation to inquiry* rather than as a methodology' (Reason and McArdle 2004).

According to Reason (1998), there are five dimensions of action research: participation and democracy, worthwhile purposes, practical challenges, many ways of knowing, and emergent form. Figure 15.1 demonstrates how these five dimensions are related to one another and how the emergent form – what is really emerging in the situation – becomes the centre of the inquiry. Action researchers aim to address practical challenges and bring research into everyday experience and practice. Action research processes aspire to be 'worthwhile' – this is research in order to 'make the world a better place' – and what is deemed to be 'worthwhile' must be addressed as part of the inquiry process (Reason et al. 2009, 9).

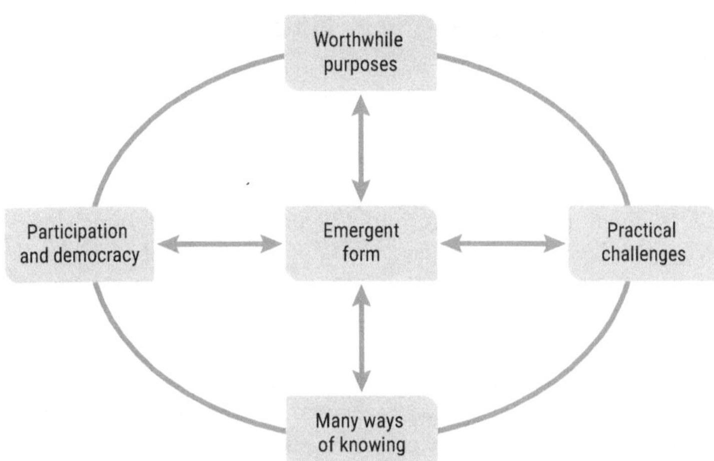

Figure 15.1 Five dimensions of action research (Reason and Bradbury 2001; Reason et al. 2009)

Action research adopts many ways of knowing and favours methods that are experiential and relational and allow for subjectivity and multiple perspectives (Marshall, Coleman, and Reason 2011, 29). Indeed, action research techniques represent key 'ways of knowing' in tune with a perspective from complexity science (Blaikie 2007; Boulton 2011). Action research processes give room for, and do not constrain, the emergence of something unexpected during the research process and enable the tracing of any new factors as they happen. So, for example, in exploring in what ways a project unfolded, the focus is not only on tracing whether and how intended outcomes were achieved; it is also on whether other so-called 'unintended outcomes' occurred and whether outcomes were influenced by other interventions as well as the project, by changes in the wider environment or by factors in combination. Action research needs to take place over time to trace the pathways of how change happens (or not), rather than take a snapshot at a particular point in time.

A key element of action research is to recognise that the mindsets, attitudes and biases of the researchers and participants are pertinent to what is uncovered and valued in the process, and to what action is subsequently taken. Strong emphasis is placed on the need for personal reflective practices to explore the so-called 'inner arc of attention' (Marshall 2016, 336) whereby the researcher seeks 'to notice myself perceiving, making meaning, framing issues'. This is followed by attention to the so-called 'outer arc of attention', which involves

> reaching outside of myself in some way. This might mean actively questioning, raising issues with others or seeking ways to test out my developing ideas … perhaps seeking to change something and learning about situation, self, issues and others in the process.

This so-called 'first-person action research' underpins working with groups, teams or communities (second-person action research) and working with wider systems (third-person action research).

Strong emphasis is placed on following cycles of inquiry. In these cycles, tentative meanings and interpretations are reflected upon with the participants and others. These reflective phases can then lead to reframing understandings and shape further action and further inquiry.

In summary, Reason and Bradbury (2001, xxii) state that action research:

- Responds to practical and pressing issues in the lives of people in organisations and communities
- Engages with people in collaborative relationships
- Is strongly values-oriented, seeking to address issues of significance concerning the flourishing of human persons, their communities and the wider ecology in which we participate
- Is a living, emergent process which cannot be predetermined, but changes and develops as those engaged deepen their understanding of the issues to be addressed

SES problems and questions

Action research techniques are not focused on commonality or general features. It is largely (but not exclusively) a 'bottom-up' approach, focused on capturing the details of situations and changes. It allows the tracking of projects, structures or communities through time and uncovers how patterns emerge through reflexive relationships and how they morph and/or break down. It can allow exploration of which practices lead to adaptation and resilience, or their opposites. This focus on so-called micro-practice allows exploration of the way new qualities or characteristics emerge, and of the impact of participation and the role of power. The inference is that, by understanding situations in detail, the complex ways in which change happens can be uncovered. This knowledge and understanding can lead to learnings which may be obscured in approaches to research which are more statistical (Patton 2011). The focus on the time dimension, on so-called path dependency and history, means that many action research techniques take a narrative approach, which can capture the way things change over time.

Given the strong ethical stance of participation and the collective ownership of the research, action research facilitates exploration and investigation. However, it is emphatically action oriented and can lead to change both of the individuals involved and, potentially, of the systems of which they are a part. Action research approaches provide richness and nuance both in understanding what creates change and in supporting communities, teams or organisations to change things for themselves.

Action research techniques can be used to investigate how people and communities interact with and respond to the environment and are therefore useful to explore the social-ecological or social-technical world. In other words, the way human actions, perceptions and intentions affect and are affected by the wider natural and technological world can be explored systemically (see Boulton, Allen, and Bowman 2015). Attitudes to climate change impact (e.g. the uptake of technology such as solar panels for private dwellings) and political acts (e.g. the removal of incentives to install solar panels) all play their part. These actions and intentions have the potential to shift and be shifted by ecological patterns and norms. What actually emerges is a complex and interdependent weave of the pricing of technology, attitudes to climate change, local peer pressure and other factors. In action research – and this is an important point – there is no presupposition that any particular patterns necessarily exist. The approach therefore allows for the exploration of where various factors (e.g. technology, ecological crises, politics) may have impact, as in the examples above. In some situations, it may also be the case that few stable patterns emerge and that things are chaotic or fast-changing.

Examples of key questions include:

- How do groups of people (in communities, teams, organisations) research the situations in which they find themselves and use these insights to refine their strategies and actions? (Cloete 2017; Lindow, Preiser, and Biggs 2020)
- How can we follow situations over time to inform our understanding of what leads to or mitigates change and share this learning more widely? (Boulton, Allen, and Bowman 2015)
- How can we explore, on a local level, how people, the environment and technology interact reflexively and create norms of behaviour and social-ecological patterns? How can we use insights from these explorations to create resilience and positive change? (Fabre Lewin 2019)
- How can organisations improve the way they address issues of climate change, loss of habitat and pollution, through taking into account human behaviour? (Reason et al. 2009)
- How can we weave shared understandings and intentions about how to address social-ecological issues for our communities and organisations? (Eelderink, Vervoort, and Van Laerhoven 2020)

Brief description of key methods

The methods of action research centre on three highly interdependent levels of inquiry: first-person, second-person and third-person inquiry (Reason and McArdle 2004).

- First-person research practices address the ability of individual researchers to foster an inquiring approach to their own lives, act with awareness and make judicious choices, and assess effects in the outside world while acting.
- Second-person action research practices (e.g. cooperative inquiry) address our ability to inquire face to face with others into issues of mutual concern, usually in small groups.
- Third-person research practices create a wider community of inquiry involving persons who cannot be known to one another face to face. This would include large-scale dialogue and 'whole system' conference designs, the 'learning history' approach, networks of small groups and approaches that are concerned with larger organisations of people.

These different modes of inquiry not only serve to engage the audience in the research but also help that audience to connect their own experience to the narratives and lived experiences and so learn on their own terms (Reason et al. 2009, 12). In the arena of climate change, for example, many projects are complex multi-disciplinary endeavours involving many views and perspectives.

A number of methods fall under the umbrella of action research. The *SAGE Handbook of Action Research*, of which there have been several editions since 2001, provide a key source

Table 15.1 Summary of key methods used in action research

Method	*Description*	*References*
Narrative inquiry	Methods using narrative inquiry follow the stories of the ways in which change happens in a local context over time (Chapter 19). Narratives are accounts that express the character, detail and lived experiences of people and communicate the messiness and complexity of events as they unfold to form a unique situation.	*Key introductory text* Clandinin and Connelly 2000 *Applications to SES* Rogers et al. 2013; Paschen and Ison 2014; Goldstein et al. 2015; Galafasi et al. 2018; Lindow, Preiser, and Biggs 2020
Learning history	Learning history is defined as a shared narrative that reflects on what happened and on what people felt they learnt (Roth and Bradbury 2008). A shared narrative focuses on what happened and on how people felt; it does not seek consensus and is left 'raw'. This kind of research can use many techniques (e.g. drawing, videos, transcripts). The intention is for those involved to reflect and learn together, and for others to engage with the whole, sometimes messy, narrative as a way of learning from the experience. A learning history attempts to stay close to what happened with limited interpretation (or at least, where it occurs, interpretation that is tentative or suggestive) and limited intentional selection. It allows for 'narrative continuity' and for the emergence of patterns and meaning.	*Key introductory texts* Roth and Bradbury 2008; Gearty 2014 *Applications to SES* Fazey, Fazey, and Fazey 2005; Gearty 2009; Gearty et al. 2013

Method	*Description*	*References*
Cooperative inquiry	Cooperative inquiry is a way of supporting a group to consider an issue and own both the questions and the outcomes. The emphasis is on sharing power and on undertaking a number of cycles of inquiry processes, with time in between to reflect. Cooperative inquiry is a form of second-person action research, described by Heron (1996, 1) as follows: '[Cooperative inquiry] involves two or more people researching a topic through their own experience of it, using a series of cycles in which they move between this experience and reflecting on it together. Each person is co-subject in the experience phases and co-researcher in the reflection phases. It is a vision of persons in reciprocal relation using the full range of their sensibilities to inquire together into any aspect of the human condition with which the transparent body-mind can engage'. In a cooperative inquiry, McArdle (2004, 62) clarifies: 'all the active subjects are fully involved as co-researchers in all research decisions – about content and method – taken in the reflection phases'. Cooperative inquiry: • Emphasises inquiring with others (rather than on one's own) • Works reflexively with more than one cycle of inquiry • Moves iteratively between reflection and action • Creates equality between inquirers in developing the process, the content and the interpretation of the inquiry	*Key introductory texts* Heron 1996; McArdle 2004 *Applications to SES* Heron and Reason 2001; Swantz et al. 2008; Lotz-Sisitka et al. 2016

of methods and applications. We highlight narrative inquiry, a learning history approach to narrative inquiry (Gearty 2014), and cooperative inquiry as most relevant for the purposes of exploring social-ecological systems (SES). Table 15.1 provides a summary of key methods used in action research.

Limitations

Active research is an orientation towards inquiry, power sharing, reflexivity, action orientation and the inclusion of the subjective in what is valued and acknowledged. It is a philosophical stance as much as it is a set of methods and in that way it can colour any form of research by reminding researchers to question issues of power and purpose, to reveal hidden assumptions and to ask what is excluded and what is valued.

Having said this, as already discussed, action research is particularly attuned to exploring 'the local', albeit with a view to gaining insight into what creates change and how to act in a complex world. These insights have the potential to inform change practice more generally. Action research techniques can become unwieldy at large scale, when there are attempts to connect together smaller inquiry groups and to engage with larger-scale change. In third-person approaches, attention must be given to hierarchy, power and the impact of the wider context (Gustavsen 2003; Coghlan and Brydon-Miller 2014).

There is no reason why quantitative data cannot be included in action research processes, both in terms of what data are collected and of how these data are investigated in inquiry groups. It is probably fair to say that the local is emphasised over the global and the qualitative over the quantitative, and the practical over the conceptual. However, to use inquiry processes and inquiries that span periods of time can reveal emerging patterns and can suggest

Case study 15.1: Low Carbon Works, UK

Low Carbon Works (Reason et al. 2009) (Figure 15.2) was a long-running action research programme undertaken by Professor Peter Reason, Gill Coleman, David Ballard, Michelle Williams, Margaret Gearty, Carole Bond, Chris Seeley and Esther Maughan McLachlan at the University of Bath. With an increasing concern about climate change, the focus was on 'What is it that encourages and inhibits the adoption of low carbon technologies by business and local authority organisations?' The research assumptions were:

- The barriers to a low carbon economy are not primarily technological.
- Technological, economic and human factors are systemically interlinked.
- Significant human factors in enabling change include awareness of the issues, membership of a community of practice and a sense of agency.
- There are fleeting windows of opportunity for technological transformation.
- The barriers and enablers to significant transformation need to be understood at both micro- and macrolevels.

Six action research engagements were undertaken, including with Ginsters (a food manufacturing company), Holsworthy anaerobic digestion, Thurulie eco-factory (a Sri Lankan manufacturer of lingerie) and Southampton District Energy Scheme.

This process was not a full real-time cooperative inquiry (which can be difficult to set up with busy people in commercial organisations). The process of research consisted of 'engaging intensively with organisation members both in their everyday meetings and through more formal interviews' (Reason et al. 2009, 13). The researchers checked back with participants to ensure accuracy, and

> then worked with the material, crafting an account which used many voices of those involved to present the story back to organisation members so they could engage with it together and draw from it the learning that was most important to them. We worked with them to explore and articulate key learning points, and then developed learning histories.

new lines of inquiry which can then be addressed with more traditional quantitative and wider-reaching methods of both change and research.

There is sometimes a critique of action research orientations that they can entrance individuals into overly focusing on their 'first-person' inquiries, on their own inner worlds and personal actions. This can be transformational but can also, perhaps, take attention away from addressing wider systemic issues. However, not paying enough attention to personal bias and the impact researchers have on the situations with which they engage can be equally problematic.

Although not excluded in action research thinking, there is perhaps not enough emphasis on integrating the small with the large scale, and balancing thinking/conceptualising with the experiential. These tensions are perhaps addressed more explicitly with reference to complexity theory and systems thinking, and the integration of action research with these approaches is of growing interest (e.g. Burns 2007; Birney 2014).

Figure 15.2 A seminal example of action research applied to SES (Reason et al. 2009)

In this way, information gleaned from a variety of sources, including written material, could be combined and then presented back to stakeholders in a way that invited inquiry, discussion and reflection.

Learning histories are narratives, '"jointly told tales" developed in close participation with local actors' (Reason et al. 2009, 16). It is a process that seeks to bring together analysis and story in a way that has value for those originally involved and those seeking to learn from it. In each narrative, key moments and learnings are highlighted

(*Continued*)

(via text boxes of distinctive colours) and quotes from actors and public documents are included. The narratives are presented from multiple perspectives, include pictures, maps and diagrams, and are not designed to reach unequivocal conclusions or show definitive pathways. They allow the reader to explore, compare, dialogue with their own inquiries and reach their own insights.

The research team viewed the narratives through the lens of a range of theories including the social construction of technology (Pinch and Bijker 1984), the socio-technical transition framework (Geels and Schot 2007), theories of power (Lukes 2005) and relational practice (Bouwen and Taillieu 2004). The team then reflected on the overall experience with the Low Carbon Works research programme and drew together key learnings and key messages (Reason et al. 2009, 101). These include:

- The way people talk determines what they can see.
- The factors that lead to either innovation or 'lock in' are systemic and interdependent and include assumptions, worldviews, institutions, narratives, technology and economics.
- Building relationships is important.

A key conclusion was that because the process of transition is complex, 'we need to find ways to help people to step into the messiness and complexity of action' and

Resource implications

Action research techniques can be slow and require not insignificant commitments to time. They also require careful ongoing deep reflection from both those leading the inquiries and the participants. Large quantities of rich data are often collected so as not to preselect what is of value, and to allow for multiple explorations and interpretations of those data. Action research requires a strong commitment to reflect on personal practice as a platform from which to engage in wider inquiries and can thus be arresting, challenging and life changing. It needs to be viewed as much as a change process as it does as an inquiry. In a positive sense this means there is no separation between the research phase and the implementation phase, which suggests a degree of efficiency of effort.

New directions

There are those who are keen to preserve the integrity of action research, with its core of first-person inquiry and the importance of 'holding inquirers to the fire' in terms of honest reflective practice, engagement with issues of power and the ethics of participation (Marshall 2016). This is of vital importance as it is beguiling to step too early into interpretation of collective inquiries and action without paying due attention to personal bias and lacunae and to the way change happens in the minutiae of processes.

For others, however, there is an interest in integrating action research practice more explicitly with theories of systemic change (Burns 2007; Boulton 2011; Birney 2014). In this way, more emphasis is placed on ways of thinking about and engaging with the wider context, the bigger picture, the structures and institutions that shape the wider world – so-called third-person action research. The question becomes: how can we influence and change the wider 'create their own action maps'. Participants who were successful in contributing to transition 'were doing so by being in the thick of it' – reflecting on what they were doing, building relationships, seizing opportunities, questioning their assumptions and recognising the patterns in which they were trapped (Reason et al. 2009, 103).

Key issues for policymakers and research funders included:

- Understanding the systemic nature of change
- Seeking and creating opportunities (when locked-in patterns become unstable)
- Supporting the flourishing of emergent niches
- Actively building coalitions and dialogue
- Spreading accounts of good practice

The work was highly successful in illuminating the complexity of each context and identifying how opportunities and barriers were in general multi-dimensional and synergistic and required the bringing together of the social, the economic and the technical. The immersion in the detailed narrative of each case study inhibited simplistic one-dimensional conclusions and yet still allowed the drawing of broader learnings as to what can support a move to a low carbon future and what gets in the way.

social-economic-environmental systems of which we are a part? There are also links with ideas of deep democracy (Mindell 2002), deep ecology (Næss 1989), participative politics (Bookchin and Colau 2019), new economics (Bronk 2009) and new ways of living (e.g. eco-villages; Dawson 2006), all focused on goals of equality and sustainability, building on deep reflective practice and shared learning and empowerment. Extending action research in this way, better to address these pressing issues and widen the methods and approaches and framing, is an exciting development.

Equally, there is interest in how the overarching theoretical stance of action research, centred on a systemic, emergent and non-deterministic worldview, has been extended by more recent thinkers such as Freya Mathews (2003) and Donna Haraway (2016), and physicists such as Carlo Rovelli (2018), Basarab Nicolescu (2010) and Karen Barad (2007). These philosophers and scientists explore the nature of reality and bring to the fore the essential uncertainty and complexity at the heart of the fabric of the cosmos. Their work supports the need for approaches of inquiry that are subjective, pluralistic, adopt many ways of knowing and allow for uncertainty and emergence.

Key readings

Burns, D. 2007. *Systemic Action Research: A Strategy for Whole Systems Change*. Bristol: The Policy Press.

Marshall, J. 2016. 'Living Life as Inquiry.' In *First Person Action Research*, edited by J. Marshall, 1–2. London: Sage. doi:10.4135/9781473982598.

McArdle, K. 2004. In-powering Spaces: A Co-operative Inquiry with Young Women in Management. www.semanticscholar.org/paper/In-powering-spaces%3A-a-co-operative-inquiry-with-in-Mcardle-McArdle/4473f9f6c8a0f3f91b85b892a94a13e3e1bd05cf.

Pratt, J., P. Gordon, & D. Plamping. 2005. *Working Whole Systems: Putting Theory into Practice in Organisations*. Seattle: Radcliffe.

Reason, P. & H. Bradbury, H., eds. 2001. *Handbook of Action Research: Participative Inquiry and Practice*. London: Sage.

References

Barad, K. 2007. *Meeting the Universe Halfway: Quantum Physics and the Entanglement of Matter and Meaning*. Durham: Duke University Press.

Birney, A. 2014. *Cultivating System Change: A Practitioner's Companion*. Abingdon: Routledge.

Blaikie, N. 2007. *Approaches to Social Enquiry*. Cambridge: Polity.

Bookchin, D., and A. Colau. 2019. *Fearless Cities: A Guide to the Global Municipalist Movement*. Oxford: New Internationalist.

Boulton, J. 2011. 'The Complexity Turn: Science, Narrative and Utility.' MPhil diss., University of Bath.

Boulton, J., P. Allen, and C. Bowman. 2015. *Embracing Complexity*. Oxford: Oxford University Press.

Bouwen, R, and T. Taillieu. 2004. 'Multiparty Collaboration as Social Learning for Interdependence. Developing Relational Knowing for Sustainable Natural Resource Management.' *Journal of Community and Applied Psychology* 14: 137–153.

Bronk, R. 2009. *The Romantic Economist*. New York: Cambridge University Press.

Burns, D. 2007. *Systemic Action Research: A Strategy for Whole Systems Change*. Bristol: The Policy Press.

Clandinin, D.J., and F.M. Connelly. 2000. *Narrative Inquiry Experience and Story in Qualitative Research*. San Francisco: Jossey Bass Publishers.

Cloete, D. 2017. 'Towards Re-imagining the Roles of Change Agents from a Critical Complexity Perspective: An Exploratory Action Research Approach.' PhD diss., Stellenbosch University.

Coghlan, D., and M. Brydon-Miller. 2014. *The SAGE Encyclopedia of Action Research*, Volumes 1–2. London: Sage. doi:10.4135/9781446294406.

Dawson, J. 2006. *Ecovillages. New frontiers for Sustainability*. Totnes: Green Books.

Eelderink, M., J.M. Vervoort, and F. van Laerhoven. 2020. 'Using Participatory Action Research to Operationalize Critical Systems Thinking in Social-Ecological Systems.' *Ecology and Society* 25(1): 16. doi:10.5751/ES-11369-250116.

Fabre Lewin, M. 2019. 'Artful Bodymind: Enlivening Transformative Research Methodologies.' PhD thesis. Centre for Agroecology, Water and Resilience (CAWR) at Coventry University. https://static1.squarespace.com/static/5794c98fe58c6222609cd7bf/t/5e83267e57ad4f6854a6ef7f/1585653381059/Artful+Bodymind+-+Miche+Fabre+Lewin+-+2019+FINAL+%28Online%29.pdf.

Fazey, I., J.A. Fazey, and D.M.A. Fazey. 2005. 'Learning More Effectively from Experience.' *Ecology and Society* 10(2): 4. doi:10.5751/ES-01384--100204.

Galafassi, D., T.M. Daw, M. Thyresson, S. Rosendo, T. Chaigneau, S. Bandeira, L. Munyi, I. Gabrielsson, and K. Brown. 2018. 'Stories in Social-Ecological Knowledge Cocreation.' *Ecology and Society* 23(1): 23. doi:10.5751/ES-09932-230123.

Gearty, M. 2009. 'Achieving Carbon Reduction in the Field of Local Authorities: From Tales to Themes to Transition.' Paper at the *KSI Conference on the Dynamics and Governance of Transitions to Sustainability*, Amsterdam, 4–6 June 2009.

Gearty, M. 2014. 'Learning History.' In *The Sage Encyclopedia of Action Research*, edited by D. Coghlan and M. Miller, 492–496. Los Angeles: Sage.

Gearty, M., M. Williams, P. Pivcevic, and P. Reason. 2013. 'Piloting Digital Storytelling and Action Research as an Approach to Stimulate Pro-Environmental Advocacy and Behaviour Change.' A Report to the Department for Environment, Food and Rural Affairs. University of Bath/DEFRA.

Geels, F., and J. Schot. 2007. 'Typology of Sociotechnical Transition Pathways.' *Research Policy* 36: 399–417.

Goldstein, B.E., A.T. Wessells, R. Lejano, and W. Butler. 2015. 'Narrating Resilience: Transforming Urban Systems Through Collaborative Storytelling.' *Urban Studies* 52: 1285–1303. doi:10.1177/0042098013505653.

Greenwood, D. 2007. 'Pragmatic Action Research.' *International Journal of Action Research* 3(1&2): 131–148.

Gustavsen, B. 2003. 'Action Research and the Problem of the Single Case.' *Concepts and Transformation* 8(1): 93–99.

Haraway, D. 2016. *Staying with the Trouble*. Durham: Duke University Press.

Heron, J. 1996. *Co-operative Inquiry: Research into the Human Condition*. London: Sage.

Heron, J., and P. Reason. 2001. 'The Practice of Cooperative Inquiry.' In *Handbook of Action Research*, edited by P. Reason and H. Bradbury, 144–154. London: Sage.

Lindow, M., R. Preiser, and R. Biggs. 2020. 'Exploring Resilience Capacities with Food Innovators: A Narrative Approach.' *Global Sustainability* 3(e28): 1–12. doi:10.1017/sus.2020.23.

Lotz-Sisitka, H., M. Belay Ali, G. Mphepo, M. Chaves, T. Macintyre, T. Pesanayi, A. Wals et al. 2016. 'Co-Designing Research on Transgressive Learning in Times of Climate Change.' *Current Opinion in Environmental Sustainability* 20: 50–55. doi:10.1016/j.cosust.2016.04.004.

Lukes, S. 2005. *Power: A Radical View. Basingstoke*: Palgrave Macmillan.

Marshall, J. 2016. 'Living Life as Inquiry.' In *First Person Action Research*, edited by J. Marshall, 1–2. London: Sage. doi:10.4135/9781473982598.

Marshall, J., G. Coleman, and P. Reason. 2011. *Leadership for Sustainability: An Action Research Approach*. Sheffield: Greenleaf Publishing.

Mathews, F. 2003. *For Love of Matter: A Contemporary Panpsychism*. Albany: State University of New York Press.

McArdle, K. 2004. 'In-powering Spaces: A Co-operative Inquiry with Young Women in Management.' www.semanticscholar.org/paper/In-powering-spaces%3A-a-co-operative-inquiry-with-in-Mcardle-McArdle/4473f9f6c8a0f3f91b85b892a94a13e3e1bd05cf.

Mindell, A. 2002. *The Deep Democracy of Open Forums: Practical Steps to Conflict Prevention and Resolution for the Family, Workplace, and World*. Charlottesville: Hampton Roads.

Næss, A. 1989. *Ecology, Community and Lifestyle: Outline of an Ecosophy*. Translated by D. Rothenberg. Cambridge: Cambridge University Press.

Nicolescu, B. 2010. 'Methodology of Transdisciplinarity – Levels of Reality, Logic of the Included Middle and Complexity.' *Transdisciplinary Journal of Engineering & Science* 1(January). www.atlas-tjes.org/index.php/tjes/article/view/9.

Paschen, J-A., and R. Ison. 2014. 'Narrative Research in Climate Change Adaptation – Exploring a Complementary Paradigm for Research and Governance.' *Research Policy* 43(6): 1083–1092.

Patton, M. 2011. *Developmental Evaluation: Applying Complexity Concepts to Enhance Innovation and Use*. New York: Guilford Press.

Pinch, T., and W. Bijker. 1984. 'The Social Construction of Facts and Artefacts: Or How the Sociology of Science and the Sociology of Technology Might Benefit Each Other.' *Social Studies of Science* 14(3): 399–444.

Reason, P. 1998. 'Toward a Participatory Worldview.' *Resurgence* 168: 42–44.

Reason, P., and H. Bradbury. 2001. 'Inquiry and Participation in Search of a World Worthy of Human Aspiration.' In *Handbook of Action Research: Participative Inquiry and Practice*, edited by P. Reason and H. Bradbury, 1–14. London: Sage.

Reason, P., G. Coleman, D. Ballard, M. Williams, M. Gearty, C. Bond, C. Seeley, and E.M. McLachlan. 2009. Insider Voices: Human Dimensions of Low Carbon Technology. Bath: Centre for Action Research in Professional Practice, University of Bath. http://people.bath.ac.uk/mnspwr; www.peterreason.net/Papers/lowcarbon_insider_voices.pdf.

Reason, P., and K. McArdle. 2004. 'Brief Notes on the Theory and Practice of Action Research.' In *Understanding Research Methods for Social Policy and Practice*, edited by S. Becker and A. Bryman. London: The Polity Press.

Reason, P., and J. Rowan. 1981. *Human Inquiry: A Sourcebook of New Paradigm Research*. Chichester: John Wiley & Sons.

Rogers, K.H., R. Luton, H. Biggs, R. Biggs, S. Blignaut, A.G. Choles, C.G. Palmer, and P. Tangwe. 2013. 'Fostering Complexity Thinking in Action Research for Change in Social-Ecological Systems.' *Ecology and Society* 18(2): 31.

Roth, G., and H. Bradbury. 2008. 'Learning History: An Action Research Practice in Support of Actionable Learning.' In *Handbook of Action Research: Participative Inquiry and Practice*, edited by P. Reason and H. Bradbury, 350–365. London: Sage.

Rovelli, C. 2018. *The Order of Time*. London: Allen Lane.

Swantz, M. 2008. 'Participatory Action Research as Practice.' In *The SAGE Handbook of Action Research* (2nd ed), edited by P. Reason and H. Bradbury, 31–48. London: Sage.

Torbert, W. 1976. *Creating a Community of Inquiry: Conflict, Collaboration, Transformation*. London: Wiley & Sons.

Methods for analysing systems – system components and linkages

16

Expert modelling

Stuart Kininmonth,[1,2,3] *Steven Gray*[4] *and Kasper Kok*[5]

[1]STOCKHOLM RESILIENCE CENTRE, STOCKHOLM UNIVERSITY, STOCKHOLM, SWEDEN
[2]UNIVERSITY OF OSLO, OSLO, NORWAY
[3]UNIVERSITY OF THE SOUTH PACIFIC, SUVA, FIJI ISLANDS
[4]DEPARTMENT OF COMMUNITY SUSTAINABILITY, MICHIGAN STATE UNIVERSITY, EAST LANSING, MICHIGAN, USA
[5]SOIL GEOGRAPHY AND LANDSCAPE GROUP, WAGENINGEN UNIVERSITY, WAGENINGEN, THE NETHERLANDS

Key methods

Bayesian networks, fuzzy cognitive maps

Connections to other chapters

Fuzzy cognitive maps can be combined with scenario studies (see Chapter 11), and Bayesian networks are based on simple network models (see Chapter 23).

Introduction

The term 'expert model' refers to a computer-based model that can mimic (or outperform) the decisions of a human expert (Jackson 1998). Given the broad nature of this definition, the variety of models that could be classified as 'expert' is substantial. In this chapter, we choose to focus on just two expert methods routinely used in understanding social-ecological systems (SES): Bayesian networks (also referred to as Bayesian belief networks, decision networks) and fuzzy cognitive maps. These two modelling approaches represent two conceptually and practically different ways to develop and apply expert models. At the more applied level, this class of models contributes to the growing toolbox of knowledge co-production methods, which allow the formalisation of stakeholder-generated knowledge as structured representations of complex systems. At the more conceptual level, this class of models belongs to the field of artificial intelligence (AI) and essentially uses inference techniques to process an explicit knowledge base to deduce novel information and increase our understanding of the complex system under study. Note that expert models are a broad category of very flexible tools and methods that are applied in all scientific disciplines. The use of fuzzy cognitive maps, for example, has been reported in close to 20 000 scientific papers ranging from computer science, to medicine, to economics.

The justification for applying expert models in SES analysis is that these models allow researchers to understand the structure and dynamics of complex systems. Overall, the inherent complexity of these systems, which consist of multifaceted interactions between system elements, is difficult to understand without modelling tools. More specifically, diverse SES

DOI: 10.4324/9781003021339-20

SUMMARY TABLE: EXPERT MODELLING

DISCIPLINARY BACKGROUND	KNOWLEDGE TYPE
The methods in this chapter are derived from or have most commonly been used in: Mathematics, Statistics, Ecology, Social Science	*The methods in this chapter are primarily used to generate the following types of knowledge:* • Descriptive • Exploratory • Explanatory
RESEARCH APPROACH	**PURPOSE OF METHOD**
The methods in this chapter originate from or most commonly adopt the following research approaches: • Analytical/objective • Interpretive/subjective	*The most common purposes of using the methods in this chapter are:* • System understanding • Stakeholder engagement and co-production • Policy/decision support
TEMPORAL DIMENSION	**SYSTEMIC FEATURES AND PROCESSES**
The methods in this chapter are most commonly applied to the following temporal dimensions: • Present (typically within the last 5–10 years) • Recent past (post-1700s) • Future **SPATIAL DIMENSION** *The methods in this chapter are primarily either or both:* • Non-spatial • Explicitly spatial *The methods in this chapter are most commonly applied at the following spatial scales:* • Local • Regional (provincial/state to continental) • Multiple places/sites around the world	*While most methods can do many things, the methods in this chapter are particularly good (i.e. go-to methods) for addressing the following:* • SES components and linkages • Social-ecological dependence and impact • Multiple scales and levels or cross-level interactions • Regime shifts • Evaluating policy options • Exploring uncertainty

stakeholders, decision-makers and even scientists have fundamentally different perceptions of how these systems are structured. This disagreement and related miscommunication call for tools that can help our understanding while articulating and mapping these different perceptions. Expert models can help to both understand the complexity of systems and uncover the multiple perceptions of their functionality. Utilising conceptual understanding of the system, the starting point of expert models (Bayesian networks, fuzzy cognitive maps) is commonly cognitive maps, usually represented in the special form of semi-quantitative cognitive maps. These maps, constructed by individual experts or groups of experts (both local and scientific experts), are often used to represent a system of interacting elements visually. In this way both simple and more complex systems can be communicated through visuals with a fair degree of transparency of model assumptions and model structure while allowing for direct input from stakeholders on key relationships of SES.

Although expert models have only recently been applied to understanding SES, the background mathematical concepts that underlie these approaches have been around much longer. In fact, while expert models are now often squarely in the realm of AI and computer science, more basic mechanisms and formulations that allow them to work were developed before computers were commonplace. Bayesian inference, which is the fundamental construct behind Bayesian networks, was for example developed by Thomas Bayes in 1763 and then further developed by Judea Pearl (1982) with graphical models. These developments enabled the mathematical treatment of how one parameter relates to another (conditional inference), particularly when these correlations are connected across a network containing many parameters. The calculated probabilities across this network can be used to predict the probability of one event based on another.

Similarly, including probability and set theory in cognitive maps (Axelrod 1976) was advanced by Bart Kosko (1986) to create fuzzy cognitive maps. In the paper by Kosko, fuzzy cognitive maps are introduced as maps to employ a fuzzy or uncertain description of the causal link between two events. This uncertainty is embedded in a network of influential parameters so that the overall calculation of the causal probability can propagate across the network. Subsequent applications (Van Vliet, Kok and Veldkamp, 2010; Verkerk et al. 2017; Van der Sluis et al. 2018) included a fundamentally different use of fuzzy cognitive maps as influence diagrams, with relationships representing the strength rather than the (un)certainty of a connection. Although this is formally a misinterpretation, this type of application is rapidly gaining importance and acceptance among practitioners (see Helfgott et al. 2015). Likewise, the use of directed and signed digraphs (networks with directional edges that have values of ±1) that are the basis of matrix expert modelling techniques like causal loop diagrams to represent systems of cause-and-effect relationships dates back to Sewall Wright in 1918. Although the use of fuzzy cognitive maps in SES research is relatively new, it is expanding quickly.

The high level of computational requirements for these expert models delayed their common use and restricted the early applications to relatively simple systems. However, modern computing power combined with expanded data libraries and software that emerged over the last decade has enabled the expansion of expert modelling techniques into complex SES (Schlüter et al. 2012).

SES problems and questions

The inherent complexity of SES makes their behaviour difficult to understand or predict. Modelling tools are indispensable to structure and unravel the complex relationships and therefore increase understanding of the system's behaviour. Complexity furthermore gives rise to multiple interpretations and perceptions of how the system works. Expert models are particularly suitable to map out, analyse and compare differences in the perceptions of experts and other stakeholders. It is particularly the combination of understanding system structure and feedback loops and engaging with stakeholders that make expert models an exceptionally strong tool when analysing SES. Key questions include:

- What are different perceptions of key stakeholders? (e.g. understanding the role of land-use intensification in stopping deforestation (Kok 2009), understanding the bushmeat trade based solely on perceived knowledge of stakeholders and combining empirical data with expert knowledge (Htun et al. 2016))
- What are crucial feedbacks in the system? (e.g. understanding how greenhouse gas mitigation and altered hunting practices can increase polar bear persistence (Amstrup et al. 2010), describing the relationship of fish species to financial priorities (Kininmonth et al. 2017))

The particular strength of expert models is the ability to both facilitate fundamental scientific understanding and provide more applied science decision support. These models can be used for initial explorations as well as in-depth analytical assessments, particularly in the realm of using the understanding between system structure and function to support decision-making (Marcot et al. 2006). Unlike many classic statistical approaches, these models can incorporate qualitative and quantitative data types within an interactive framework. This capacity to include empirical and expert-derived data into a comprehensible modelling framework makes the application to SES highly attractive. The transparent nature (being able to see all the components) of the models ensures that stakeholders can gain more confidence in the model outputs while contributing to the model design (Gray et al. 2018). The application to social-ecological models has been hindered only by the access to empirical data that is temporally and spatially relevant to both the ecological and the social domains. However, once sufficient confidence is gained by the analyst in the parameterisation of the models, the generation of projections with estimated levels of certainty is attractive to solve the 'wicked' SES problems (Jentoft and Chuenpagdee 2009).

A particularly useful application of expert models in disentangling SES systems is estimating the impact of scenarios. This allows for extending the understanding of current system perceptions to future configurations and the related impact of external drivers such as climate change or technology development (Jetter and Kok 2014). It also strengthens the link with quantitative models and the comparison with formalised scientific system descriptions in these models (Kok 2009). This capacity to explore the changes in the system can enable the understanding of resilience (Gray et al. 2015) without the constraining simplification of linear models.

Brief description of key methods

Fuzzy cognitive maps are cognitive maps that structure the opinion of individuals or groups of people, allowing the relationships between factors to be fuzzy, thus calculating the degree of certainty stakeholders have. In particular, the application of fuzzy cognitive maps as influence diagrams has recently gained popularity. In this capacity, system descriptions can be combined with scenario studies (see Chapter 11). Bayesian networks are a method to combine the correlation probability between elements in a system using the simplicity of a network model (see Chapter 23). Hence the calculation of how one element relates to another using Bayes theory is restrained to the network connections. Table 16.1 presents a summary of key methods used in expert modelling.

Table 16.1 Summary of key methods used in expert modelling

Method	*Description*	*References*
Bayesian networks	A Bayesian network (BN) (or Bayesian belief network) is the probabilistic graphical model that represents a set of variables and their conditional probability with the use of a directed acyclic graph. In essence, the lines in the network represent the correlations between the system elements and are calculated using Bayes theory. Propagating the correlations across a large number of elements can be unwieldy and the network restrains the number of calculations.	*Key introductory texts* Marcot 2006; Rumpff et al. 2011; Scurati and Denis 2014 *Applications to SES* Stelzenmüller et al. 2010; Kininmonth et al. 2014; Gonzalez-Redin et al. 2016; Kininmonth et al. 2017
Dynamic Bayesian networks	Dynamic Bayesian networks have temporal capacity by linking a time sequence within a probabilistic graphical model. Each time period has a model of the correlations between elements. The models then link to replicates of the model in each time step. The additional complexity of the temporal linkages restricts the individual model complexity simply due to the difficulties in parameterisation.	*Key introductory texts* Dean and Kanazawa 1989; Murphy 2002 *Applications to SES* Pope and Gimblett 2017
Fuzzy cognitive mapping	Fuzzy cognitive mapping (FCM) is a graphical representation of a belief system comprising factors and semi-quantified relationships, with the capacity to examine a variety of scenarios.	*Key introductory texts* Kosko 1986; Özesmi and Özesmi 2004 *Applications to SES* Kok 2009; Penn et al. 2013; Diniz et al. 2015; Gray et al. 2015

Limitations

The expert models described here are based on correlations between variables elicited through stakeholder consultation and/or expert design that are perceptual. Resulting system descriptions, therefore, need to be interpreted with caution as they rely on people's perceptions rather than process-based information. This linkage from correlation to causality is particularly perilous when machine learning of the model structure is used. This is because machine learning is based on data (often observed) without any knowledge of process (such as ecological principles), data-collection bias and expert opinion. An additional issue is the application of logical inference over time (see Chapter 27). If one event has an influence on a second event, then the backward inference has to be handled with care as circular arguments through time are often illogical. This is even more dangerous with fuzzy cognitive maps, as 'time' is inadequately defined. If agricultural yield increases, for example, then the area under that crop will increase (due to perceived benefits of that crop by other farmers). In turn, this will lead to an increase in overall yield within the model, without acknowledging the time steps.

Case study 16.1: Fishers and traders at Lake Nabugabo, Uganda

In the small lake of Nabugabo, situated close to Lake Victoria, Ugandan fishers collect a small variety of fish species to sell to markets or fish traders (Kininmonth et al. 2017). This case study provides us with an opportunity to understand the cross-scale nature of trading and extraction for an open-access resource. Using Bayesian networks, the microeconomic influences on resource use can be described. Collecting the data was a core activity in this modelling exercise and involved being in the community for extended periods, speaking to fishers and traders about their activities and perceptions. To capture this information in a consistent manner, we used structured interview surveys supplemented with fish-catch observations. These data were a diverse mix of categorical, ordinal and count types and formed the basis of the conceptual model of the fishers and the traders.

The models were used to explore different scenarios of fish-trading styles. While the focus was primarily on the harvest of Nile perch (*Lates niloticus*) and Nile tilapia (*Oreochromis niloticus*), the study also collected data on financial, social and ecological dynamics to disentangle the patron–client relationships between traders and fishers. These data ranged from species caught and targeted to the techniques used and the commercial transactions conducted (Figure 16.1). The trading preferences for each fisher and trader were collated and based in a graph theoretical framework where different strategies were observed. Critically, the surveys identified community members who formed the trading relationships.

The results of the study found that if the fishers were incentivised to exchange fish with just one trader ('patron–client') then specific fish species were targeted. If the fisher was able to trade freely ('freelancer') then they were able to create a responsive and flexible extraction practice that matched the market and environmental fluctuations. The use of Bayesian networks enabled a disparate set of data types (from species type, hours spent fishing, to binary yes/no: see Figure 16.1) to be integrated into a quantitative model that could evaluate various scenarios for natural resource extraction.

The model incorporated a diverse and comprehensive parameter set that described the microeconomics of the local fishers at Lake Nabugabo. Scenarios enable the discussion of the effectiveness for management strategies with scope to continue to grow the model with more data (Kleemann et al. 2017).

Figure 16.1 shows the marginal probabilities from the scenario (Kininmonth et al. 2017). Each box in the figure is a system variable that is correlated with a small number of other variables. The arrows highlight the direction of the influence from a Bayesian theory perspective. The rows in the box show the various classes used to group the data while the bars are the frequency of the class. The marginal probability (i.e. no specific case in mind) for the variable 'Gear' shows that 'Gillnet' equipment is used 53.7% of the time, for example.

The success of projects of this nature depends on the quality of the data collected. The range of respondents needs to be comprehensive for the given set of scenarios and the factors being utilised in the model. In this case, although women were under-represented in the surveys, partly as a function of their limited role in the physical act of fishing and trading, the central question about microeconomic influences on fishing practices does involve women in a variety of roles. Another area of omission is the role of allied industries such as boat construction and repair, and hospitality. Seeking data that can illustrate the linkages across the broader community is useful for social-ecological models such as these.

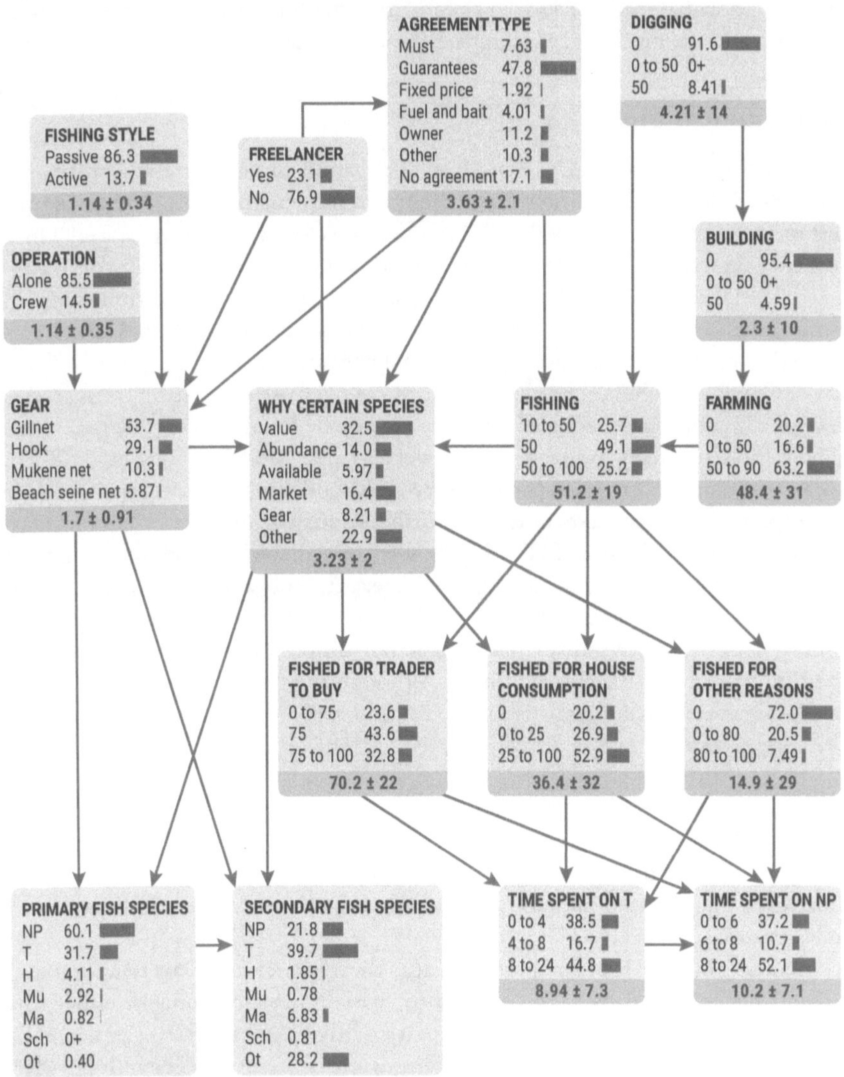

Figure 16.1 Bayesian network of the fisher responses to the questionnaires (Kininmonth et al. 2017)

The methods are also limited by the scope of either the input data or the cognitive mapping derivation. Evaluation and reflective processes are required at all stages to ensure that the models do not deviate from the accepted wisdom of the operators. This constraint of developing an expert model by contemporary thinking yet still being able to deduce emergent new ideas causes continuous tension. Other limitations for specific methods include a lack of feedback loops for Bayesian networks and a lack of explicit time units for fuzzy cognitive maps. In addition, the central limiting constraint for all expert models is based on obtaining data that cover both social and ecological aspects within a logically coherent time frame and spatial scale. Often the temporal and spatial scales of a social system are different to the key components of an ecological system, and the models need to reflect these assumptions embedded in the deductions.

Resource implications

The use of expert models requires a modern application of hardware and software that demands advanced skills in both theory and technical implementation. Due to the popularity of expert models over the last few decades, there is a wealth of software to use. This includes graphical user interface (GUI) programs like Netica (Norsys Software Corp), Bayes Fusion, Banjo, BUGS, Dlib, FBN, JavaBayes, SMILE and UnBBayes. In addition, scripting exists in all major languages but particularly in higher-level languages such as R and Python with packages bnlearn, gRain, abn, catnet and FCMapper.

With new web-enabled technologies, more easy-to-use modelling tools have recently been developed, with some specifically designed for modelling based on diverse expert knowledge. These tools are freely available on the web. Mental Modeler (Gray et al. 2013), for example, has been applied in many participatory SES modelling contexts with both local and scientific experts. These contexts range from fisheries management and agricultural planning to understanding the bushmeat trade. Inputs were based solely on perceived knowledge of stakeholders and combining empirical data with expert knowledge to understand the range of possible futures for endangered wildlife under climate change scenarios (Htun et al. 2016). As 'running' fuzzy cognitive maps relies on rather simple matrix algebra, they can also be developed using a simple Excel spreadsheet. Although this limits flexibility somewhat, it can convince local partners that it really is easy to use and apply.

New directions

Fuzzy cognitive maps are very strong in describing a system of factors and sectors, but rather weak in representing actors. A very promising way forward is the combination of fuzzy cognitive maps and agent-based models (see Chapter 28), which would allow for actor-specific system descriptions.

The weakest point of fuzzy cognitive maps, the representation of temporal and spatial scales, could be improved by linking them to spatially explicit models, either directly or through the use of scenarios. Scenarios, and particularly future narratives or normative strategies, often remain qualitative and based on current systems understanding. Methods to construct fuzzy cognitive maps of future systems perceptions are underdeveloped and could play a role, e.g. in today's discussions of societal transformation to meet the goal of limiting global warming to 1.5 °C.

The exciting development in this field is the accumulation of data that are specifically captured to address strategic questions. These data can be used to develop more powerful expert models since the parameterisation process is significantly more precise. The model design and style are continually evolving with exciting developments in the use of machine learning to disentangle patterns in data. However, the imperative to understand complex behaviour of SES, including predicting tipping points and understanding disturbance events, continues to demand additional modelling and data-collection techniques.

Key readings

Gray, S.A., S. Gray, J.L. de Kok, A.E.R. Helfgott, B. O'Dwyer, R. Jordan, and A. Nyaki. 2015. 'Using Fuzzy Cognitive Mapping as a Participatory Approach to Analyze Change, Preferred States, and Perceived Resilience of Social-Ecological Systems.' *Ecology and Society* 20(2): 11. doi:10.5751/ES-07396-200211.

Marcot, B. G., R.S. Holthausen, M.G. Rowland, and M.J. Wisdom. 2001. 'Using Bayesian Belief Networks to Evaluate Fish and Wildlife Population Viability under Land Management Alternatives from an Environmental Impact Statement.' *Forest Ecology and Management* 153: 29–42. doi:10.1016/S0378-1127(01)00452-2.

Rumpff, L., D.H. Duncan, P.A. Vesk, D.A. Keith, and B.A. Wintle. 2011. 'State-and-transition Modelling for Adaptive Management of Native Woodlands.' *Biological Conservation* 144: 1224–1236.

Acknowledgements

We particularly want to thank Andrew Byekwaso for conducting the questionnaires, and Dennis Twinomugisha for assistance with the field programme outlined in the Kininmonth et al. (2017) case study. This work was partially supported by Mistra through a core grant to the Stockholm Resilience Centre and the Strategic Research Programme EkoKlim.

References

Amstrup, S.C., E.T. DeWeaver, D.C. Douglass, B.G. Marcot, G.M. Durner, C.M. Bitz, and D.A. Bailey. 2010. 'Greenhouse Gas Mitigation Can Reduce Sea-Ice Loss and Increase Polar Bear Persistence.' *Nature* 468: 955–958. doi:10.1038/nature09653.

Axelrod, R. 1976. *Structure of Decision: The Cognitive Maps of Political Elites*. Princeton: Princeton University Press. doi:10.2307/1955121.

Dean, T., and K. Kanazawa. 1989. 'A Model for Reasoning about Persistence and Causation.' *Artificial Intelligence* 93(1–2): 1–27.

Diniz, F.H., K. Kok, M. Hoogstra-Klein, and B. Arts. 2015. 'Mapping Future Changes in Livelihood Security and Environmental Sustainability Based on Perceptions of Small Farmers in the Brazilian Amazon.' *Ecology and Society* 20(2): 26. doi:10.5751/ES-07286-200226.

Gonzalez-Redin, J., S. Luque, L. Poggio, R. Smith, and A. Gimona, A. 2016. 'Spatial Bayesian Belief Networks as a Planning Decision Tool for Mapping Ecosystem Services Trade-Offs on Forested Landscapes.' *Environmental Research*. doi:10.1016/j.envres.2015.11.009.

Gray, S., L. Cox, and S. Henly-Shepard. 2013. 'Mental Modeler: A Fuzzy-logic Cognitive Mapping Modeling Tool for Adaptive Environmental Management.' *Proceedings of the 46th International Conference on Complex Systems*. doi:10.1109/HICSS.2013.399.

Gray, S.A., S. Gray, J.L. de Kok, A.E.R. Helfgott, B. O'Dwyer, R. Jordan, and A. Nyaki. 2015. 'Using Fuzzy Cognitive Mapping as a Participatory Approach to Analyze Change, Preferred States, and Perceived Resilience of Social-Ecological Systems.' *Ecology and Society* 20(2): 11. doi:10.5751/ES-07396-200211.

Gray, S., A. Voinov, M. Paolisso, R. Jordan, T. BenDor, P., Bommel, P. Glynn et al. 2018. 'Purpose, Processes, Partnerships, and Products: Four Ps to Advance Participatory Socio-environmental Modeling.' *Ecological Applications* 28(1): 46–61.

Helfgott, A., S. Lord, N. Bean, M. Wildenberg, S. Gray, S. Gray, J. Vervoort, K. Kok, and J. Ingram. Submitted. 2015. 'Clarifying Fuzziness: Fuzzy Cognitive Maps, Neural Networks and System Dynamics Models in Participatory Social and Environmental Decision-aiding Processes.' Transmango Working Paper 1, Brussels: EC.

Htun, H., S.A. Gray, C.A. Lepczyk, A. Titmus, and K. Adams. 2016. 'Combining watershed models and knowledge-based models to predict local-scale impacts of climate change on endangered wildlife.' *Environmental Modelling and Software* 84: 440–457.

Jackson, P. 1998. *Introduction to Expert Systems* (3rd ed). Boston: Addison-Wesley.

Jentoft, S., and R. Chuenpagdee. 2009. 'Fisheries and Coastal Governance as a Wicked Problem.' *Marine Policy* 33: 553–560.

Jetter, A.J., and L. Kok. 2014. 'Fuzzy Cognitive Maps for Futures Studies. A Methodological Assessment of Concepts and Methods.' *Futures* 61: 45–57. doi:10.1016/j.futures.2014.05.002.

Kininmonth, S., B. Crona, Ö. Bodin, I. Vaccaro, L. Chapman, and C. Chapman. 2017. 'Microeconomic Relationships between and among Fishers and Traders Influence the Ability to Respond to Social-Ecological Changes in a Small-scale Fishery.' *Ecology and Society* 22(2): 26.

Kininmonth, S., S. Lemm, M. Cherie, and T. Hatley. 2014. 'Spatial Vulnerability Assessment of Anchor Damage within The Great Barrier Reef World Heritage Area.' *Ocean and Coastal Management* 100: 20–31.

Kleemann, J., E. Celio, B.K. Nyarko, M. Jiménez-Martínez, and C. Fürst. 2017. 'Assessing the Risk of Seasonal Food Insecurity with an Expert-based Bayesian Belief Network Approach in Northern Ghana, West Africa.' *Ecological Complexity* 32: 53–73.

Kok, K. 2009. 'The Potential of Fuzzy Cognitive Maps for Semi-quantitative Scenario Development, with an Example from Brazil.' *Global Environmental Change* 19: 122–133.

Kosko, B. 1986. 'Fuzzy Cognitive Maps.' *International Journal of Man-Machine Studies* 24: 65–75.

Marcot, B.G., J.D. Steventon, G.D. Sutherland, and R.K. McCann. 2006. 'Guidelines for Developing and Updating Bayesian Belief Networks Applied to Ecological Modeling and Conservation.' *Canadian Journal of Forest Research* 36: 3063–3074.

Murphy, K.P. 2002. 'Dynamic Bayesian Networks: Representation, Inference and Learning.' PhD diss., University of California Berkeley.

Özesmi, U., and S.L. Özesmi. 2004. 'Ecological Models Based on People's Knowledge: A Multi-step Fuzzy Cognitive Mapping Approach.' *Ecological Modelling* 176: 43–64. doi:0.1016/j.ecolmodel.2003.10.027.

Pearl, J. 1982. 'Reverend Bayes on Inference Engines: A Distributed Hierarchical Approach.' *AAAI-82Proceedings*. https://aaai.org/Papers/AAAI/1982/AAAI82-032.pdf.

Penn, A.S., C.J.K. Knight, D.J.B Lloyd, D. Avitabile, K. Kok, F. Schiller, A. Woodward, A. Druckman, and L. Basson. 2013. 'Participatory Development and Analysis of a Fuzzy Cognitive Map of the Establishment of a Bio-Based Economy in the Humber Region.' *PLoS ONE* 8(11): e78319.

Pope, A., and H.R. Gimblett. 2017. 'Linking Bayesian and Agent-based Models to Simulate Complex Social-Ecological Systems in Semi-arid Regions.' In *Hybrid Solutions for the Modelling of Complex Environmental Systems*, edited by C.E. Vincenot, S. Mazzoleni, and L. Parrott. Frontiers e-Books.

Rumpff, L., D.H. Duncan, P.A. Vesk, D.A. Keith, and B.A. Wintle. 2011. 'State-and-transition Modelling for Adaptive Management of Native Woodlands.' *Biological Conservation* 144: 1224–1236.

Schlüter, M., R.R.J. McAllister, R. Arlinghaus, N. Bunnefeld, L. Eisenack, F. Holker, E. Milner-Gulland et al. E. 2012. 'New Horizons for Managing the Environment: A Review of Coupled Social-Ecological Systems Modeling.' *Natural Resource Model* 25: 219–272.

Scurati, M., and J-B. Denis. 2014. *Bayesian Networks: With Examples in R*. Boca Raton: CRC Press.

Stelzenmüller, V., J. Lee, E. Garnacho, and S.I. Rogers. 2010. 'Assessment of a Bayesian Belief Network-GIS Framework as a Practical Tool to Support Marine Planning.' *Marine Pollution Bulletin* 60: 1743–1754.

Van der Sluis, T., M. Bogers, K. Kok, G. Cosor, N. Geamana, E. Crouzat, E. Pavlis et al. 2018. 'Drivers of European Landscape Change: Stakeholders' Perspectives through Fuzzy Cognitive Mapping.' *Landscape Research*. doi:10.1080/01426397.2018.1446074.

Van Vliet, M., K. Kok, and T. Veldkamp. 2010. 'Linking Stakeholders and Modellers in Scenario Studies: The Use of Fuzzy Cognitive Maps as a Communication and Learning Tool.' *Futures* 42(1): 1–14.

Verkerk, P.H., A. Sánchez, S. Libbrecht, A. Broekman, A. Bruggeman, H. Daly-Hassen, E. Giannakis et al. 2017. 'A Participatory Approach for Adapting River Basins to Climate Change.' *Water* 9: 958. doi:10.3390/w9120958.

17

Data mining and pattern recognition

Juan C. Rocha[1,2] *and Stefan Daume*[1,2]

[1]STOCKHOLM RESILIENCE CENTRE, STOCKHOLM UNIVERSITY, STOCKHOLM, SWEDEN
[2]BEIJER INSTITUTE OF ECOLOGICAL ECONOMICS, THE ROYAL SWEDISH ACADEMY OF SCIENCES, STOCKHOLM, SWEDEN

Key methods discussed in this chapter

Data wrangling, clustering analysis, regression trees, neural networks, sentiment analysis, topic models

Connections to other chapters

Classical methods for statistical analysis (Chapter 18), including Bayesian networks (Chapter 16), qualitative data analysis (Chapter 19), network analysis (Chapter 23) and spatial mapping (Chapter 24), complement the methods outlined in this chapter. Data-mining products can be used for mapping ecosystem services (Chapter 31), systems scoping (Chapter 5) and participatory data collection (Chapter 8) in virtual environments.

Introduction

Data science is an interdisciplinary field focused on extracting knowledge and insights from a wide variety of datasets. The general purpose of data science is pattern discovery from unstructured and heterogeneous sources of data (e.g. Twitter, a government census, travel cards, remote-sensing data), through processes of data mining and machine learning. Data mining refers to the process of obtaining unstructured data, which are data collected by a variety of entities (governments, companies, society) without any control of what and how, how often, how complete, or by whom data are generated. As a result, unstructured data are often biased and incomplete, but sometimes the large amount of available data allows us to gain interesting insights into aspects of social-ecological systems (SES) that would otherwise be inaccessible. These insights are gained through a set of methods collectively referred to as machine learning (e.g. regressions, clustering, neural networks). Although these methods are rooted in statistics and carry the same assumptions (see Chapter 18), there are substantial differences in how the methods are applied to unstructured data, compared with their more typical application to structured data (i.e. data that have been collected by a researcher through a designed research process, e.g. ecological field data collection (Chapter 6) and interviews and surveys (Chapter 7)).

DOI: 10.4324/9781003021339-21

SUMMARY TABLE: DATA MINING AND PATTERN RECOGNITION

DISCIPLINARY BACKGROUND

The methods in this chapter are derived from or have most commonly been used in:

Computer Science, Information Science

KNOWLEDGE TYPE

The methods in this chapter are primarily used to generate the following types of knowledge:

- Descriptive
- Exploratory

RESEARCH APPROACH

The methods in this chapter originate from or most commonly adopt the following research approaches:

- Analytical/objective

PURPOSE OF METHOD

The most common purposes of using the methods in this chapter are:

- Data collection/generation
- System understanding
- Policy/decision support

TEMPORAL DIMENSION

The methods in this chapter are most commonly applied to the following temporal dimensions:

- Present (typically within the last 5–10 years)
- Recent past (post-1700s)

SYSTEMIC FEATURES AND PROCESSES

While most methods can do many things, the methods in this chapter are particularly good (i.e. go-to methods) for addressing the following:

- SES components and linkages
- Diversity
- Regime shifts

SPATIAL DIMENSION

The methods in this chapter are primarily either or both:

- Non-spatial
- Explicitly spatial

The methods in this chapter are most commonly applied at the following spatial scales:

- Local
- Regional (provincial/state to continental)
- Global
- Multiple places/sites around the world

The field of data science can be traced back to Alan Turin, who invented the computer and created the first neural network (a machine-learning method) to crack Enigma, the secret code used by the Germans in the Second World War. He was a philosopher (logician) and mathematician (cryptographer) but founded what we know today as computer science, artificial intelligence and machine learning. Data science is agnostic in terms of discipline and data science tools are used by researchers across the sciences and humanities. In fact, the emergence of fields such as digital humanities, computational social science and citizen science highlights the lack of disciplinary borders when it comes to data science. With the advent of new types of data and tools, disciplinary divides become meaningless; what matters is the research problem at hand. The same image-recognition and classification algorithms that are used in medicine to identify and distinguish different cases of cancer, for example, can be used to classify different neighbourhoods and their social dynamics over time (Naik et al. 2017).

In this chapter we focus on how SES researchers have applied data mining and machine-learning methods to date, and provide pointers on how these tools could serve to answer novel research questions in SES science. For more information on the key assumptions of each method, see Chapter 18 on statistical analysis.

SES problems and questions

In contrast to traditional research designs, where one asks a question and then collects data to answer it, data science enables researchers to use available, unstructured data to ask SES research questions. The examples provided here cover a broad range of data types and machine-learning techniques.

- What are the system boundaries? One of the fundamental questions that any systems thinker faces is determining the boundaries of a system (Holland 2012), or distinguishing one type of system from another (see Chapter 2). Machine learning can be used to identify classes of SES through classification tasks. Clustering analysis is a method that enables this classification by detecting how many clusters there are in the data. Social-ecological systems in Ghana and Burkina Faso have been mapped using this approach (Rocha et al. 2019).
- What are the key system relationships and drivers? Machine learning can also be used for prediction in SES. To do so, it is common to split the available data into a training dataset and a validation dataset. The training phase improves the skills of an algorithm at performing a certain task, while the validation phase can be used to predict and assess how good the predictions are. This approach has been used in breakthrough studies that predict global deforestation and its drivers (Curtis et al. 2018), and reduced resilience in tropical forests (Verbesselt et al. 2016) using satellite imagery.
- What are the key social dynamics in SES? Satellite data and social data can be a powerful combination to address questions related to the social dynamics of SES. Jean et al. (2016), for example, used publicly available satellite imagery to predict poverty in data-scarce contexts such as African countries, and validated the findings using national poverty surveys. Other sources of social data, such as publicly available photographs and geotags uploaded to social media platforms (e.g. Twitter, Instagram, Flickr), have been used to quantify nature-based tourism and recreation values of protected areas (Wood et al. 2013) or valuable landscapes (Sonter et al. 2016; Van Zanten et al. 2016). Typically, recreational value data are derived through traditional social science methods such as onsite surveys and interviews. These methods are, however, not always scalable, whereas data mining enables analyses for many more people and places.

- What are the key underlying narratives shaping SES behaviour? Text is an unstructured data type that captures the richness of narratives and meanings, but can also be overwhelming in size and demand for processing time. More papers are published every day than any researcher could possibly read, or more Tweets are broadcast than any user can access. Topic models have been used to identify the impacts of regime shifts on ecosystem services (Rocha and Wikström 2015), and infer citizen-driven reports of biodiversity observations in general and invasive species specifically (Daume, Albert, and Von Gadow 2014b; Daume 2016; Daume and Galaz 2016) (see Case study 17.1).
- How does one detect real-time changes in SES? Cities are landscapes where social-ecological interactions are reshaped all the time, but studying them can be difficult in terms of scalability (i.e. being able to capture social or ecological processes at the city scale account for millions of people's preferences and choices). Data mining and machine learning have been used to assess the presence of biodiversity data in London using sound recordings (Fairbrass et al. 2018) and to quantify the physical and safety improvement of city neighbourhoods in US cities over time using Google Street View (Naik, Raskar, and Hidalgo 2016; Naik et al. 2017). These two examples have also produced online tools so that non-academic parties such as NGOs, municipal governments or citizens may explore their city's biodiversity (londonsounds.org; batslondon.com) or safety scores (streetscore.media.mit.edu).

Brief description of key methods

Data mining and pattern recognition relate to two main aspects of working with large datasets: acquiring the data, and analysing them in a variety of ways to detect patterns. The type of analysis will depend on the nature of the data: qualitative or quantitative analyses, time series versus spatial analyses, or supervised learning (human-guided) versus unsupervised (fully automated) analyses. Table 17.1 provides a summary of commonly used data-mining and pattern-recognition methods in SES research. While an in-depth review of each of the methods falls beyond the scope of this chapter, we mention resources where the reader can learn more.

Many applications of pattern recognition to different SES use large volumes of unstructured data that typically need to be harvested, cleaned and transformed to render them useful. Whereas 'mining' refers to getting the data, 'wrangling' refers to the pre-processing steps that are necessary before analysis. This implies automation and techniques that require the mastery of data-processing tools or programming languages like R or Python, and the interaction with application programming interfaces (APIs).

Cluster analysis is particularly useful to identify patterns that can help define system boundaries. In the context of the global consequences of land-use change, for example, Foley et al. (2005) proposed that landscapes can be classified by the sets of ecosystem services they provide. Other classifications of land use at the global scale include anthromes (Ellis and Ramankutty 2008; Ellis et al. 2010) and general land-use archetypes (Václavík et al. 2013; Ropero, Aguilera, and Rumí 2015; Surendran et al. 2016). Regression trees are helpful for identifying key relationships and drivers. Donovan et al. (2018), for example, used clustering analysis to classify five different regimes in Hawaiian coral reefs as opposed to a previous binary classification, and Jouffray et al. (2019) further investigated the underlying drivers of change for these reef communities using boosted regression trees.

Neural networks are programs (Bayesian regressions) inspired by how the brain is thought to work (Mitchell 2019). They are formed by input nodes, intermediate layers of nodes and an output node. The input nodes receive some data (a photo, video, time series) and the

Table 17.1 Summary of key methods used in data mining and pattern recognition

Method	*Description*	*References*
Data wrangling	Data wrangling is the process of cleaning data that have been obtained and rendering the data in a useful form.	*Key introductory texts* VanderPlas 2016; Wickham and Grolemund 2017; Wilson et al. 2017 *Applications to SES* No known applications
Cluster analysis	Clustering identifies groups of data points with common characteristics. Cluster analysis, in combination with other methods, can help explain why elements cluster in certain ways and not others, and what the underlying causes or drivers of similarity across elements are.	*Key introductory text* Kassambara 2017 *Applications to SES* Raudsepp-Hearne, Peterson, and Bennett 2010; Vackavik et al. 2013; Hamann, Biggs, and Reyers 2015; Meacham et al. 2016
Regression trees	A regression tree is a model that relates a response from a variable to predictors by going through a series of recursive binary splits (the branches of the tree).	*Key introductory texts* Elith, Leathwick, and Hastie 2008; Kuhn and Johnson 2013 *Applications to SES* Jouffray et al. 2015, 2019; Curtis et al. 2018
Neural networks	Neural networks are Bayesian regressions that use a fitness function to assess their performance. Recursively, it learns how to best capture features of the data and become good at predicting them. Application examples include image and sound recognition.	*Key introductory texts* Chollet 2017; Chollet and Allaire 2018 *Applications to SES* Naik, Raskar, and Hidalgo 2016; Naik et al. 2017; Fairbrass et al. 2018
Sentiment analysis	Sentiment analysis is a supervised method – humans perform part of the data classification. It uses dictionaries of words related to emotions and scores large amounts of text, allowing one to detect overall sentiments or how sentiments change over time.	*Key introductory text* Silge and Robinson 2017 *Applications to SES* Dodds et al. 2011, 2015; Spaiser et al. 2014; Vosoughi, Roy, and Aral 2018
Topic models	Topic models allow one to describe large amounts of text and classify the texts by topics that better represent them. It is an unsupervised method – there is no human classification involved, nor is there any bias. One can see how much of a topic is represented in a document, or how it changes over time.	*Key introductory texts* Blei 2012; Silge and Robinson 2017 *Applications to SES* Daume, Albert, and Von Gadow 2014a; Rocha et al. 2015

Case study 17.1: Using Twitter to detect invasive species

Inspired by successful applications of data mining of social media to monitor and predict the spread of epidemic diseases, similar applications have been advocated to obtain early warnings for critical changes in SES (Galaz et al. 2010). Daume, Albert and Von Gadow (2014b), Daume (2016), and Daume and Galaz (2016) focused on the real-time nature of social media and its capacity to deliver early warnings for environmental changes in the form of species observations. This was a global study, but Twitter adoption is biased towards Europe, North America and English-speaking countries. Invasive alien species – species not native to a specific location – were chosen as an example since they are known drivers and indicators of ecosystem change with potentially significant ecological, economic and health impacts (Daume 2016).

The study focused on mentions of three invasive alien species in the northern hemisphere – grey squirrel, emerald ash borer and oak processionary moth – on Twitter. These species represent invasive alien species with different geographies, impacts and invasion stages. Twitter was queried for keywords that indicated potential direct or indirect references to these species (Daume 2016). Data collection was facilitated via the Twitter Search application programming interface (API), a public web interface that returns a sample of Tweets matching the specified search terms. The API only returns Tweets dating back at most nine days. In order to assemble a dataset covering a longer period, a web-based tool was implemented in the Java programming language, which could be run automatically and continuously. The data were then processed and analysed using a combination of bespoke tools programmed in Java and R.

The results confirmed Twitter as a rich source of biodiversity information on various species. Primary observations of the example invasive alien species could be identified in Tweets, but the quality and completeness of this information varied. While images or videos are often included and allow an assessment of the observations, precise geolocation information is rare. At the same time, studies showed that mining Twitter uncovers observations that could not have been gathered in traditional monitoring programmes and can thus complement these. In the case of invasive alien species, even singular observations could be crucial for early detection (Daume 2016).

A key finding was that online conversations frequently emerge around Tweets related to species observation. These conversations led to the formation of ad hoc communities that together determined the name of the observed species (Daume and Galaz 2016). Given that Twitter is both a data source and a communication channel, Tweets can be viewed as an interactive data source, where the data can be actively complemented (e.g. in the case of missing geolocation information). Twitter as a data source is unique in that the gathered Tweets provide not only insights into ecological properties (i.e. species observations) but also social perspectives ranging from an implicit representation of stakeholders and their background to perceptions and sentiments on topics such as invasive species (Daume, Albert, and Von Gadow 2014b). Twitter and similar social media sources can thus be of interest not only in SES research addressing early warnings and long-term biodiversity observations but also for quick thematic assessments and an understanding of public perceptions.

Studies using Twitter data are bound by the Twitter rules and policies. Those forbid the general sharing of the complete data (i.e. all Tweet information collected by a study). Only the identifiers of Tweets may, with certain restrictions, be shared publicly and would then have to be 'rehydrated' using the Twitter API. This presents a potential limitation with regard to the reproducibility of a study: Tweets published by Twitter users who have closed or protected their accounts since the data was originally collected, for example, would no longer be retrievable, thus potentially leading to different results. Figure 17.1 presents a high-level typology of Tweets collected for invasive alien species monitoring.

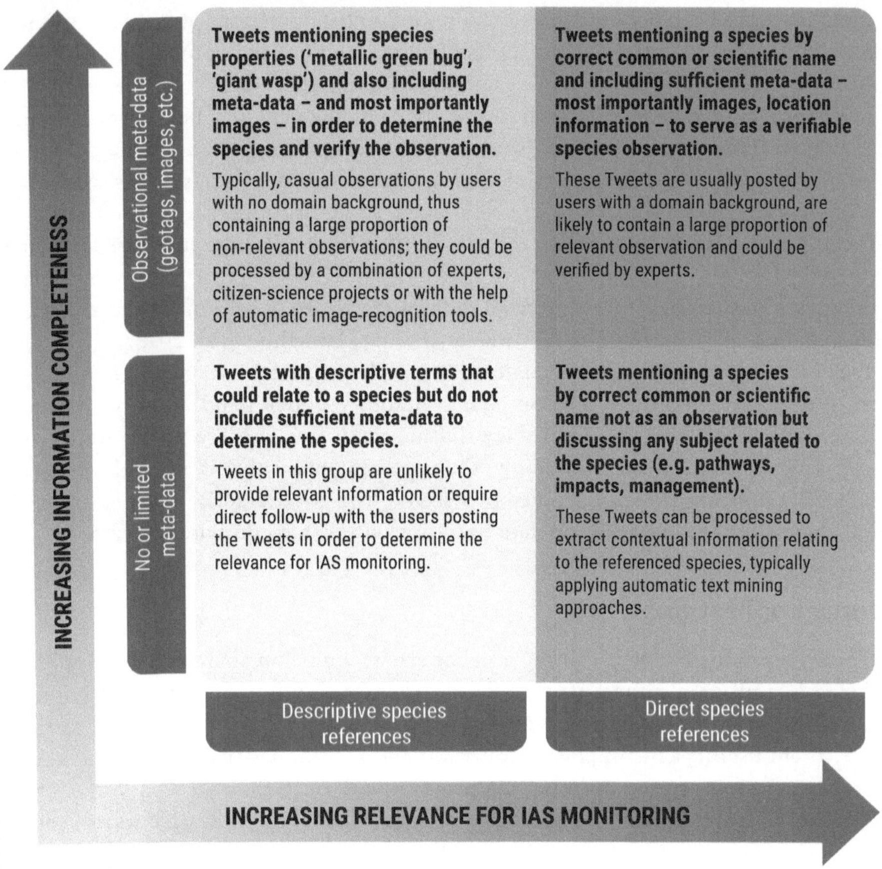

Figure 17.1 High-level typology of Tweets collected for invasive alien species monitoring (adapted from Daume 2016)

output node predicts some aspect of interest. The intermediate layers are weights that are intended to maximise performance of a classification task (e.g. translate text, classify a picture, recognise a sound). Networks with many intermediate layers are known as deep networks. One of the most popular types is convolutional neural networks (CNN) for their performance in classifying images. Natural-language processing (NLP) is another field of artificial intelligence (computer science and linguistics) dealing with the interaction between computers and human (natural) languages. There are many methods that fall within natural-language processing. Among others, Table 17.1 indicates applications of sentiment analysis (a supervised machine-learning technique) to classify the sentiment content of text, and topic modelling (an unsupervised technique) to classify large amounts of text into topics.

Limitations

The fact that data-science approaches are different from classical statistics in terms of the data-generation process implies that the analysis will have to account for potential biases. Platforms such as Twitter, Facebook, Instagram or Flickr, for example, are rich sources of data but their users may not be representative of the underlying populations about which one wants to draw conclusions. Compared with the general population, the composition of Twitter or Facebook users is generally biased with regard to economic status or certain age brackets. However, other sources of data are often lacking, so social media data may be considered a best first approximation. This was the case in a study conducted by Wood et al. (2013), trying to infer the aesthetic value and rates of visitation for natural areas. Interviews or surveys work for small-scale problems but do not scale well, e.g. to national scales or to measure millions of opinions. Being aware of data biases can prevent the misinterpretation of results and may even shed light on how to find ways to verify how biased the dataset is. Knowing that Instagram is biased towards younger populations, for example, a multiplatform assessment improved the assessment of valued landscapes in Europe (Van Zanten et al. 2016; Donahue et al. 2018).

When working with publicly available data such as NASA imagery, Google Earth or Twitter, it is worth keeping an eye on data availability policies and how they change over time. Some data services in the USA were shut down during the Trump administration, for example. Twitter might decide to change the type of content that is available or its accessibility, thereby limiting the replicability of a study. It is good practice to keep up with data availability policies and always keep data back-ups on reliable computer servers.

Resource implications

Scalable and reproducible data-mining and pattern-recognition approaches typically require knowledge of a programming language. Many applications of pattern recognition use large volumes of unstructured data that typically need to be harvested, cleaned and transformed to render them useful. This implies automation and techniques that require the mastery of data-processing tools or programming languages like R or Python.

If the dataset is small (fits on a computer's memory) one might be able to run something simple such as a cluster analysis routine in Microsoft Excel or other click-based programs. However, accessing data and applying state-of-the-art techniques will require an understanding of the code used by developers and modifying it according to one's own application. It is not necessary to be a computer scientist, but a basic understanding of how computers work and the language in which procedures are encoded is a must to apply machine-learning algorithms.

For many applications, researchers need to interact with application programming interfaces. These are services allowing one to access Twitter, Flickr, Instagram or Google Earth data, for example. Even if it is not necessary to query new streams of data, existing data might need to be pre-processed, transformed and organised in a useful form and share code that makes one's research reproducible by colleagues and peers. Popular choices of scientific computing languages include R, Python and Julia, all of which are open access, are well documented and provide many online resources to learn at one's own pace.

Besides the programming skills required to access, clean and analyse data, another key resource implication is the need for sufficient computational resources. Analyses using large datasets such as satellite imagery often cannot be performed on a personal computer and require whole clusters of computers provided through universities or commercial cloud providers. Given the amount of data required in machine-learning applications, parallel processing may be needed. Access to, and familiarity with, the facilities for running computations remotely are therefore key requirements in data-mining and pattern-recognition research.

New directions

Opportunities for applying pattern-detection techniques to better understand SES are substantial and growing. A number of trends and potential emerging applications can be identified. Studies using the clustering of ecosystem services sets to identify and classify SES have been confined to regional and national scales, but a global classification is still lacking. In addition, harmonised datasets such as the Earth Data Cube or World Bank statistics can be used to better understand changes in SES over time (Reichstein et al. 2019). Machine-learning approaches can help researchers uncover development trajectories and anomalies, or develop early-warning systems (Scheffer et al. 2009) to monitor the unfolding of SES regime shifts or transformations.

A key challenge in SES research is to distinguish when sustainable solutions, such as poverty-reduction policies or agricultural innovations, work, and where. Data-mining and machine-learning approaches can help discover types of SES where particular solutions work and where similar SES that may benefit from these approaches are located.

Advances in text mining and language processing can be used to do more rigorous literature reviews by mapping who collaborates with whom, and to scope research opportunities or priorities. These techniques can also empower an emergent field of research in assessing aesthetic and cultural ecosystem service values over time and space at larger scales than were possible before the advent of social media data.

Machine-learning methods are no longer the luxury of computer giants. Today one can run machine-learning experiments on a regular computer or by using cloud services, which makes the methods accessible to students and low-budget research groups.

Key readings

Chollet, F. 2017. *Deep Learning with Python*. New York: Manning.

Chollet, F., and J.J. Allaire. 2018. *Deep Learning with R*. New York: Manning.

Mitchell, M. 2019. *Artificial intelligence: A guide for thinking humans*. London: Pelican Books.

Pearl, J., and D. Mackenzie. 2018. *The Book of Why: The New Science of Cause and Effect*. London: Penguin.

Wickham, H., and G. Grolemund. 2017. *R for Data Science*. Beijing: O'Reilly.

Acknowledgements

This chapter benefited from the careful reading and constructive feedback of Maike Hamann, as well as input ideas from Emma Sundström and Garry Peterson. Juan C. Rocha was supported by FORMAS grant 942-2015-731.

References

Blei, D. 2012. 'Probabilistic Topic Models.' *Communications of the ACM* 55: 77–84.

Chollet, F. 2017. *Deep Learning with Python*. New York: Manning.

Chollet, F., and J.J. Allaire. 2018. *Deep Learning with R*. New York: Manning.

Curtis, P.G., C.M. Slay, N.L. Harris, A. Tyukavina, and M.C. Hansen. 2018. 'Classifying Drivers of Global Forest Loss.' *Science* 361(6407). doi:10.1126/science.aau3445.

Daume, S. 2016. 'Mining Twitter to Monitor Invasive Alien Species – An Analytical Framework and Sample Information Topologies.' *Ecological Informatics* 31: 70–82. doi:10.1016/j.ecoinf.2015.11.014.

Daume, S., M. Albert, and K. von Gadow. 2014a. 'Assessing Citizen Science Opportunities in Forest Monitoring Using Probabilistic Topic Modelling.' *Forest Ecosystems* 1(1). doi:10.1186/s40663-014-0011-6.

Daume, S., M. Albert, and K. von Gadow. 2014b. 'Forest Monitoring and Social Media – Complementary Data Sources for Ecosystem Surveillance?' *Forest Ecology and Management* 316: 9–20. doi:10.1016/j.foreco.2013.09.004.

Daume, S., and V. Galaz. 2016. '"Anyone Know What Species This Is?" – Twitter Conversations as Embryonic Citizen Science Communities.' *PLoS ONE* 11(3). doi:10.1371/journal.pone.0151387.

Dodds, P.S., E.M. Clark, S. Desu, M.R. Frank, A.J. Reagan, J.R. Williams, L. Mitchell et al. 2015. 'Human Language Reveals a Universal Positivity Bias.' *Proceedings of the National Academy of Sciences* 112(8): 2389–2394. doi:10.1073/pnas.1411678112.

Dodds, P.S., K.D. Harris, I.M. Kloumann, C.A. Bliss, and C.M. Danforth. 2011. 'Temporal Patterns of Happiness and Information in a Global Social Network: Hedonometrics and Twitter.' *PLoS ONE* 6(12): e26752. doi:10.1371/journal.pone.0026752.

Donahue, M.L., B.L. Keeler, S.A. Wood, D.M. Fisher, Z.A. Hamstead, and T. McPhearson. 2018. 'Using Social Media to Understand Drivers of Urban Park Visitation in the Twin Cities, MN.' *Landscape and Urban Planning* 175: 1–10. doi:10.1016/j.landurbplan.2018.02.006.

Donovan, M.K., A.M. Friedlander, J. Lecky, J-B. Jouffray, G.J. Williams, L.M. Wedding, L.B. Crowder et al. 2018. 'Combining Fish and Benthic Communities into Multiple Regimes Reveals Complex Reef Dynamics.' *Scientific Reports* 8(1). doi:10.1038/s41598-018-35057-4.

Elith, J., J.R. Leathwick, and T. Hastie. 2008. 'A Working Guide to Boosted Regression Trees.' *Journal of Animal Ecology* 77: 802–813.

Ellis, E.C., K.K. Goldewijk, S. Siebert, D. Lightman, and N. Ramankutty. 2010. 'Anthropogenic Transformation of the Biomes, 1700 to 2000.' *Global Ecology and Biogeography* 19(5): 589–606. doi:10.1111/j.1466-8238.2010.00540.x.

Ellis, E.C., and N. Ramankutty. 2008. 'Putting People in the Map: Anthropogenic Biomes of the World.' *Frontiers in Ecology and the Environment* 6(10). doi:10.1890/070062.

Fairbrass, A.J., M. Firman, C. Williams, G.J. Brostow, H. Titheridge, and K.E. Jones. 2018. 'CityNet – Deep Learning Tools for Urban Ecoacoustic Assessment.' *Methods in Ecology and Evolution* 4(2): 206–197. doi:10.1111/2041-210X.13114.

Foley, J.A., R. DeFries, G.P. Asner, C. Barford, G. Bonan, S.R. Carpenter, F.S. Chapin et al. 2005. 'Global Consequences of Land Use.' *Science* 309(5734): 570–574. doi:10.1126/science.1111772.

Galaz, V., B. Crona, T. Daw, Ö. Bodin, M. Nyström, and P. Olsson. 2010. 'Can Web Crawlers Revolutionize Ecological Monitoring?' *Frontiers in Ecology and the Environment* 8(2): 99–104. doi:10.1890/070204.

Hamann, M., R. Biggs, and B. Reyers. 2015. 'Mapping Social-Ecological Systems: Identifying "Green-loop" and "Red-loop" Dynamics Based on Characteristic Bundles of Ecosystem Service Use.' *Global Environmental Change* 34: 218–226. doi:10.1016/j.gloenvcha.2015.07.008.

Holland, J.H. 2012. *Signals and Boundaries*. Cambridge: MIT Press.

Jean, N., M. Burke, M. Xie, W.M. Davis, D.B. Lobell, and S. Ermon. 2016. 'Combining Satellite Imagery and Machine Learning to Predict Poverty.' *Science* 353(6301): 790–794. doi:10.1126/science.aaf7894.

Jouffray, J-B., M. Nyström, A.V. Norström, I.D. Williams, L.M. Wedding, J.N. Kittinger, and G.J. Williams. 2015. 'Identifying Multiple Coral Reef Regimes and Their Drivers Across the Hawaiian Archipelago.' *Biological Sciences* 370(1659). doi:10.1098/rstb.2013.0268.

Jouffray, J-B., L.M. Wedding, A.V. Norström, M.K. Donovan, G.J. Williams, L.B. Crowder, A.L. Erickson et al. 2019. 'Parsing Human and Biophysical Drivers of Coral Reef Regimes.' *Proceedings of the Royal Society B* 286(1896). doi:10.1098/rspb.2018.2544.

Kassambara, A. 2017. *Practical Guide to Cluster Analysis in R: Unsupervised Machine Learning.* STHDA.

Kuhn, M., and K. Johnson. 2013. *Applied Predictive Modeling.* New York: Springer.Meacham, M., C. Queiroz, A. Norström, and G. Peterson. 2016. 'Social-Ecological Drivers of Multiple Ecosystem Services: What Variables Explain Patterns of Ecosystem Services Across the Norrström Drainage Basin?' *Ecology and Society* 21(1): 14. doi:10.5751/ES-08077-210114.

Mitchell, M. 2019. *Artificial Intelligence: A Guide for Thinking Humans.* London: Pelican Books. Naik, N., S.D. Kominers, R. Raskar, E.L. Glaeser, and C.A. Hidalgo. 2017. 'Computer Vision Uncovers Predictors of Physical Urban Change.' *Proceedings of the National Academy of Sciences* 197(29). doi:10.1073/pnas.1619003114.

Naik, N., R. Raskar, and C.A. Hidalgo. 2016. 'Cities Are Physical Too: Using Computer Vision to Measure the Quality and Impact of Urban Appearance.' *American Economic Review* 106(5): 128–132. doi:10.1257/aer.p20161030.

Raudsepp-Hearne, C., G.D. Peterson, and E.M. Bennett. 2010. 'Ecosystem Service Bundles for Analyzing Tradeoffs in Diverse Landscapes.' *Proceedings of the National Academy of Sciences* 107(11): 5242–5247. doi:10.1073/pnas.0907284107.

Reichstein, M., G. Camps-Valls, B. Stevens, M. Jung, J. Denzler, and N. Carvalhais. 2019. 'Deep Learning and Process Understanding for Data-Driven Earth System Science.' *Nature* 566(7743): 195–204. doi:10.1038/s41586-019-0912-1.

Rocha, J.C., K. Malmborg, L.J. Gordon, K.A. Brauman, and F. DeClerck. 2019. 'Mapping Social Ecological Systems Archetypes.' *Environmental Research Letters.* doi:10.1088/1748-9326/ab666e.

Rocha, J.C., and R. Wikström. 2015. 'Detecting Potential Impacts on Ecosystem Services Related to Ecological Regime Shifts – A Matter of Wording.' *Conference paper.*

Ropero, R.F., P.A. Aguilera, and R. Rumí. 2015. 'Analysis of the Socioecological Structure and Dynamics of the Territory Using a Hybrid Bayesian Network Classifier.' *Ecological Modelling* 311: 73–87. doi:10.1016/j.ecolmodel.2015.05.008.

Scheffer, M., J. Bascompte, W.A. Brock, V. Brovkin, S.R. Carpenter, V. Dakos, H. Held, E.H. van Nes, M. Rietkerk, and G. Sugihara. 2009. 'Early-Warning Signals for Critical Transitions.' *Nature* 461(7260): 53–59. doi:10.1038/nature08227.

Silge, J., and D. Robinson. 2017. *Text Mining with R.* Beijing: O'Reilly.

Sonter, L.J., K.B. Watson, S.A. Wood, and T.H. Ricketts. 2016. 'Spatial and Temporal Dynamics and Value of Nature-Based Recreation, Estimated via Social Media.' *PLoS ONE* 11(9): e0162372. doi:10.1371/journal.pone.0162372.

Spaiser, V., T. Chadefaux, K. Donnay, F. Russmann, and D. Helbing. 2014. 'Social Media and Regime Change: The Strategic Use of Twitter in the 2011–12 Russian Protests.' *SSRN Electronic Journal* November. doi:10.2139/ssrn.2528102.

Surendran, N.S., B.L. Preston, A.W. King, and R. Mei. 2016. 'Using Landscape Typologies to Model Socioecological Systems: Application to Agriculture of the United States Gulf Coast.' *Environmental Modelling & Software* 79: 85–95.

Václavík, T., S. Lautenbach, T. Kuemmerle, and R. Seppelt. 2013. 'Mapping Global Land System Archetypes.' *Global Environmental Change – Human and Policy Dimensions* 23(6): 1637–1647. doi:10.1016/j.gloenvcha.2013.09.004.

VanderPlas, J. 2016. *Python Data Science Handbook.* Beijing: O'Reilly.

Van Zanten, B.T., D.B. van Berkel, R.K. Meentemeyer, J.W. Smith, K.F. Tieskens, and P.H. Verburg. 2016. 'Continental-scale Quantification of Landscape Values Using Social Media Data.' *Proceedings of the National Academy of Sciences*, October. doi:10.1073/pnas.1614158113.

Verbesselt, J., N. Umlauf, M. Hirota, M. Holmgren, E.H. van Nes, M. Herold, A. Zeileis, and M. Scheffer. 2016. 'Remotely Sensed Resilience of Tropical Forests.' *Nature Climate Change* 6(11): 1028–1031. doi:10.1038/nclimate3108.

Vosoughi, S., D. Roy, and S. Aral. 2018. 'The Spread of True and False News Online.' *Science* 359(6380): 1146–1151. doi:10.1126/science.aap9559.

Wickham, H., and G. Grolemund. 2017. *R for Data Science.* Beijing: O'Reilly.

Wilson, G., J. Bryan, K. Cranston, J. Kitzes, L. Nederbragt, and T.K. Teal. 2017. 'Good Enough Practices in Scientific Computing.' *PLoS Computational Biology* 13(6): e1005510. doi:10.1371/journal.pcbi.1005510.

Wood, S.A., A.D. Guerry, J.M. Silver, and M. Lacayo. 2013. 'Using Social Media to Quantify Nature-based Tourism and Recreation.' *Scientific Reports* 3(1): 17. doi:10.1038/srep02976.

18

Statistical analysis

Ingo Fetzer,[1,2] *Luigi Piemontese,*[1]
Juan C. Rocha[1,4] *and Berta Martín-López*[3]

[1]STOCKHOLM RESILIENCE CENTRE, STOCKHOLM UNIVERSITY, STOCKHOLM, SWEDEN
[2]BOLIN CENTRE FOR CLIMATE RESEARCH, STOCKHOLM, SWEDEN
[3]LEUPHANA UNIVERSITY, LÜNEBURG, GERMANY
[4]BEIJER INSTITUTE OF ECOLOGICAL ECONOMICS, THE ROYAL SWEDISH ACADEMY OF SCIENCES, STOCKHOLM, SWEDEN

Key methods discussed in this chapter

Descriptive statistics, group comparison, regression models (linear, generalised linear), multivariate analysis (including clustering, non-metric multidimensional scaling (n-MDS), principal component analysis (PCA), redundancy analysis (RDA), canonical correspondence analysis (CCA), factor analysis (FA) and multiple correspondence analysis), time series analysis.

Connections to other chapters

Statistical methods are used to analyse data from a wide variety of sources, such as ecological field data (Chapter 6) for understanding drivers and dynamics, data from participatory data collection (Chapter 8) or interview data from surveys (Chapter 7). Statistical methods can give an overview of spatial datasets (Chapter 24) but also help to discover patterns in time series results of agent-based modelling (Chapter 28). Statistical approaches are, among others, also used to estimate how well dynamic systems models (Chapter 26), expert models (Chapter 16) and ecosystem service models (Chapter 31) represent empirical data. Finally, statistics provide methods for historical assessments (Chapter 25) and quantitative pattern recognition (Chapter 17).

Introduction

Statistical methods are mathematical tools that can help to aggregate, present and explore complex datasets from various sources. These include field data, which are often used in social-ecological systems (SES) research. These methods can be very useful for understanding interactions, dependencies and relationships between social, environmental and ecological variables. Most statistical approaches and tests allow researchers to identify and determine the main significant controlling (environmental or social) parameters (Dalgaard 2008; Crawley 2015).

Although the use of statistical methods dates back to at least the 5th century BCE, the origin of statistics had little in common with what it is used for today, as the original use of statistics was limited to governance, mainly by providing demographic data. It was only during

 DOI: 10.4324/9781003021339-22

SUMMARY TABLE: STATISTICAL ANALYSIS

DISCIPLINARY BACKGROUND	KNOWLEDGE TYPE
The methods in this chapter are derived from or have most commonly been used in: Ecology, Social Science, Political Science, Economics, Demography, Psychology, Earth Sciences	*The methods in this chapter are primarily used to generate the following types of knowledge:* • Descriptive • Exploratory • Explanatory
RESEARCH APPROACH	**PURPOSE OF METHOD**
The methods in this chapter originate from or most commonly adopt the following research approaches: • Analytical/objective	*The most common purposes of using the methods in this chapter are:* • System understanding
TEMPORAL DIMENSION	**SYSTEMIC FEATURES AND PROCESSES**
The methods in this chapter are most commonly applied to the following temporal dimensions: • Present (typically within the last 5–10 years) • Recent past (post-1700s) • Pre-industrial revolution (pre-1700s) • Future **SPATIAL DIMENSION** *The methods in this chapter are primarily either or both:* • Non-spatial • Explicitly spatial *The methods in this chapter are most commonly applied at the following spatial scales:* • Local • Regional (provincial/state to continental) • Global • Multiple places/sites around the world	*While most methods can do many things, the methods in this chapter are particularly good (i.e. go-to methods) for addressing the following:* • SES components and linkages • Diversity • Social-ecological dependence and impact • Regime shifts • Exploring uncertainty

the 18th century that the approach was extended to various scientific fields, when the newly developed probability theory (or 'law of great numbers') was integrated and recognised as a new branch of mathematics. Jakob Bernoulli, Abraham de Moivre and others introduced the idea of representing complete certainty by the number one, and probability (as a measurement for uncertainty) as a number between zero and one (Varberg 1963; Fienberg 1992).

Modern statistical methods consist of a vast toolbox of different tests and analytical procedures. The choice of the most appropriate method requires: (a) a good understanding of the type of data (categorical or continuous data), (b) knowledge of the distribution pattern and shape of the data (e.g. normally distributed or skewed) in order to apply the appropriate statistical tests, and (c) an understanding of statistical methods to correctly interpret the outcomes. This is especially crucial because most statistical methods, as already mentioned above, do not provide an explicit outcome of 'yes' or 'no', but rather give the probability of an event happening or the likelihood of an item belonging to a certain group. So, researchers have to be clear about what is meant by assigning statements such as 'unlikely' or 'likely' to the outcome of an analysis, expressed by a 'probability value' (p-value) (Hon 2010; Crawley 2015).

Statistical tools generally test a hypothesis, usually derived from a specific scientific question, and investigate the likelihood for the hypothesis to be either accepted or refuted. Statisticians have agreed that an event is unlikely if it occurs less than 5% of the time ($p < 0.05$). Sometimes, for stricter tests (e.g. in medical studies), thresholds of 1% ($p < 0.01$) or lower are used. Thus, a probability value of less than 5% suggests that this event is unlikely to occur by chance. However, there is still a small probability that two unrelated events could take place simply by chance (leading to a 'false positive' or type I error), or that a relationship between two related events is not detected by the test (leading to a 'false negative' or type II error). Thus, poor insight into basic statistical principles combined with a lack of knowledge about data quality, biased data or too few data may lead to wrong conclusions (Dalgaard 2008; Hon 2010; Crawley 2015).

Generally, statistical methods do not directly test the likelihood that the question being asked is true, but rather the likelihood of a 'null hypothesis' being true. A null hypothesis is a general statement or default position that there is 'no relationship' between two measured phenomena, or 'no association' among groups. Thus, dynamics between them are uncoupled and random. The null hypothesis often starts in the opposite position of the question being asked! In other words, the null hypothesis is the logical opposite of the scientific question that one is interested in. The logic is that if it turns out that the null hypothesis is very unlikely to be true (e.g. with a probability of $p < 0.05$ (Dalgaard 2008; Crawley 2015)), then its logical opposite (the alternative hypothesis reflecting the research question) must be valid. The real art is to pose a question that can lead to a research hypothesis of which its logical opponent, the null hypothesis, can be formulated. Finally, the null hypothesis has to be tested with the appropriate test (Quinn and Keough 2002; Hon 2010).

Example
Research question: Is there a difference between measurements of Group A and B?
Research hypothesis: Measurements of Group A and B differ
Null hypothesis: Measurements of Group A and B do not differ
Only if a null hypothesis needs to be refuted by a statistical test then the research hypothesis, as the only logical alternative, is accepted to be true.

Social-ecological systems research uses many popular statistical methods to analyse quantitative and qualitative data, including statistical visualisation, analyses and modelling.

- **Statistical visualisation:** Visualisation of statistical results, mainly through plots and diagrams, is used to aggregate data and understand patterns (e.g. social-ecological interactions) and trends (forecasting and development), and to help build conceptual models (e.g. causal loop diagrams).
- **Statistical analyses:** Statistical analyses include traditional aggregation of datasets (e.g. by estimating mean, variance, standard error) that help to summarise data, understand patterns, estimate dynamics, dependencies and correlations between or within elements of data, and calculate differences between groups. Different statistical techniques (e.g. confidence intervals, errors) are often used to review the quality and certainty of data collections and inquiries, and to understand relationships between different components of complex SES and their contexts (e.g. estimating covariance, correlation strengths). These analyses can be used for data-based scientific and political decision support, for instance by creating statistical models or calculating metrics as means and modes from data points by comparing deviation from similar studies done elsewhere or earlier (e.g. the number of disease outbreaks in a community before and after the introduction of a waste-water treatment plant).
- **Statistical modelling:** Statistical modelling includes modelling and forecasting future trends or filling in gaps for missing data. Statistical models are often used to improve strategic and political planning, e.g. fitting empirical data to a function to allow one to make general statements, even for situations where no data exist. Clustering multifactorial data into an easy-to-read visual plot to understand which factors are relevant is another common use of statistical model outputs.

SES problems and questions

A number of common SES features and questions can be explored using statistical analyses of empirical SES data, such as ecological field data (Chapter 6), interview and survey data (Chapter 7), or data obtained from participatory data collection (Chapter 8) or via data-mining techniques (Chapter 17). Statistical methods allow an understanding of a huge range of phenomena, such as social behaviour, ecological or environmental changes, and social-ecological interactions. Typical questions include:

- What are the typical attributes of key SES features in a particular region? (e.g. what is the average amount of fuelwood harvested or used by communities in a particular region, and how much does it vary between households? (Paumgarten and Shackleton 2011; Keane et al. 2019)). Statistical analysis can be used to determine the mean or central value of a dataset containing multiple data points, as well as the range and variability in the dataset, and allows comparing the values.
- How do SES features in a particular area compare with those of different areas or with other studies from literature? (e.g. how does the average income of rural and urban communities differ (Keane et al. 2019; Robinson, Zheng, and Peng 2019); how does mean summer precipitation values differ between regions?). Statistical methods can be used to determine whether there are significant differences between different regions or groups.
- How are different SES features related? (e.g. regional farm assessments investigate how environmental factors such as irrigation water availability, soil type, mean summer temperatures and growing periods are related to aspects such as farm size, income, distance to markets and the number of family members (Keane et al. 2019; Robinson, Zheng, and Peng 2019)). Regression analyses can be applied to determine relationships, and the

strength of relationships, between system components (e.g. how does the application of a pesticide relate to its effect on crops while increasing income but also affecting human health or an ecosystem (Carvalho 2017), or how do different levels of land cover change the impact on biodiversity, ecosystem services provision and human well-being? (Iniesta-Arandia et al. 2014; Meacham et al. 2016; Martín-López et al. 2017)). Once these relationships are established, regression models can be used to forecast effects on the impact of future changes, or to help fill gaps in empirical data (Pollnac et al. 2010; Pinsky and Fogarty 2012).

- Which places have similar SES characteristics? (e.g. which farms are more similar in their socio-economic and environmental characteristics? (Martín-López et al. 2012; Meacham et al. 2016; see also Case study 18.1)). Multivariate analysis can be used to bring all the SES characteristics (with all their different units) together, allowing researchers to identify clusters of similar SES sites. Similar sites often either share similar problems (Meacham et al. 2016) or can be used as archetypes to identify other places sharing similar SES conditions where a particular intervention could also be successfully applied (Piemontese et al. 2020).
- How do key SES features change across years? (e.g. time series analyses allow researchers to estimate how early identification of changing rainfall conditions can be a critical indicator to ensure national food security; these analyses can also allow conducting risk assessments to determine the likelihood and potential impacts of droughts and flooding for small-scale farmers (Shongwe et al. 2011; Husak et al. 2013)).

Brief description of key methods

Prior to any analysis, a researcher must be aware of the characteristics of the investigated data (Hon 2010), as the selection of appropriate statistical methods depends on the data types. This includes both the type and the distribution of data. A first step in any analysis is usually to summarise and visualise the data to gain an understanding of the dataset and detect obvious data entry errors. Following this, the data may be analysed using a variety of statistical methods.

Data types

Two main characteristic types of data exist: discrete and continuous. It is essential to make this distinction because each data type is analysed using different statistical methods and tests (so-called 'parametric' and 'non-parametric' statistical approaches – see details below).

Discrete (categorical) data are descriptive data that can be counted according to group attributes (or their 'quality'). Group attributes can be categorical factors which are either non-rankable (e.g. sex, colour) or rankable (e.g. age classes, size groupings). The number of male and female participants in surveys can be counted, for example. The total number of male and female participants cannot be ranked, whereas the numbers of females and males between several survey batches can be ranked. Discrete data that cannot be ranked are referred to as 'nominal data', whereas data that can be ranked or ordered are referred to as 'ordinal data' (see 'Statistical notation' in Table 18.1).

Continuous data are data that cannot be counted, e.g. temperature can be measured using a thermometer but cannot be counted. Unlike ordinal data, continuous data include the size or a magnitude (the 'quantity' of a parameter that can be put on a standardised equidistant (= interval)

Table 18.1 Overview of data types, characteristics and terms used for the statistical categorisation of data

Data type	*Characteristics (and scale)*	*Statistical notation*	*Category*	*Distribution*	*Example*
Count data (*n*)	Discrete (= countable)	Nominal data	Qualitative	Non-parametric	15 students
Count with 2 groups (dichotomous data)	Discrete categorical Non-rankable	Nominal data	Qualitative	Non-parametric	yes–no, male–female
Count data with > 2 groups	Discrete categorical Non-rankable	Nominal data	Qualitative	Non-parametric	7 black, 4 white, 6 red
Count data with rankable groups	Discrete categorical Rankable	Ordinal data	Qualitative	Non-parametric	2 small, 4 medium, 5 large
Measurements (but no true zero)	Continuous	Interval data	Quantitative	Parametric	2.0 °C, 2.1 °C, 3.2 °C, etc.
Relative measurements (with true zero)	Relative continuous	Interval ratio data	Quantitative	Parametric	23 kg 34 kg 45 kg, etc.
Time series	Continuous data linked by a time vector	Interval data along a time axis	Quantitative	Parametric	Day 1: 2.3; Day 2: 3.1; Day 3: 4.3; etc.

scale). Continuous data therefore always take numerical values: the distance between two points, when put on a scale, is placed at an equal representative distance from each other. Continuous data are therefore statistically categorised confusingly as 'interval data'. With no zero reference point, values below zero are possible (e.g. for temperature, altitude).

For continuous data with a zero reference point present, all data are placed relative (or as a 'ratio') to this zero reference point. Values below zero are thus not possible (e.g. percentage, height or weight). These data are termed 'ratio data' (see 'Statistical notation' in Table 18.1). When working with ratio data, but not interval data, the ratio of two measurements has fundamentally different consequences: for the ratio variable 'weight', a weight of 8 grams is always exactly twice as heavy as a weight of 4 grams. However, for the interval variable 'temperature', a temperature of 10 °C need not necessarily be considered to be twice as hot as 5 °C.

Continuous data always contain a higher 'information quantity' than rankable, discrete (ordinal) data, which in turn contain a higher information quantity than discrete (nominal) data (Table 18.1). Data with a higher information quantity can be easily converted into data with a lower information quantity (e.g. continuous temperature measurements can be placed into groups of 'cold' and 'warm' and events can be counted), but the reverse is rarely possible. However, data of high information quantity usually require more effort and resources to gather and, depending on the system under investigation, may often be even impossible to attain. Moreover, comparison of data is often only possible between datasets of similar types. In their research planning, researchers have to balance the data with high 'information quantity' with the amount of 'data quantity' actually needed to answer the research question (Dalgaard 2008; Hon 2010; Crawley 2015).

Data distribution

In addition to knowing the type of data, it is critical to understand the distribution of the data in order to select an appropriate statistical method. The data distribution can be described by a continuous distribution function, which is called the 'probability density function'. Continuous data in statistical data analysis are often assumed to be symmetrically or 'normally' distributed around a central value (the 'mean') with some variability. However, the assumption that data are normally distributed needs to be tested (e.g. by running a 'normality' test) before any further analysis, as the results may be invalid if this assumption does not hold. As discrete data are usually not normally distributed, most standard statistical tests cannot be applied. 'Parametric' data refer to data that follow a normal or other specified distribution, allowing the more statistically reliable parametric tests to be applied, while data that do not follow a specified distribution are termed 'non-parametric' (or distribution free). Here only the less robust non-parametric class of tests is applicable.

Table 18.1 provides an overview of data types and technical statistical terms used to categorise data.

Summarising and visualising data

The first step in statistical analysis is typically to summarise and visualise the data. Parameters such as mean, variance and standard deviation are typical overall descriptors for quantitative, normally distributed parametric data. One may, for example, summarise mean (average) crop production across different study sites or conditions, and the variability in crop production using other characteristic metrics such a range, variance or standard deviation. For non-parametric data and discrete data, descriptive parameters such as median, quartiles or mode (relative maximum) are typical descriptive indices (Crewson 2006; Brown and Saunders 2007).

It is also useful to explore the data by visualising them in order to get a 'feel' for the data and spot any obvious data-entry errors. Discrete data are usually presented as numbers in a contingency table, or graphically as bar or pie plots. If the discrete (countable) data are several repeated measurements (e.g. number of participants in several survey batches), then overall descriptions around a median or mode can be done. Variability of this type of data is described using quantiles or percentiles and is visualised in 'box-and-whisker' plots (Dalgaard 2008; Hon 2010; Crawley 2015).

Continuous ungrouped data are typically visually represented as ranked or non-ranked scatter plots, single-line plots or simple bar plots. With grouped continuous data, stacked bar plots, sequential means with error bars or overlapping line plots are often applied. For the representation of size distributions, bar graphs with binned data or frequency histograms are useful. These are often also used for visual detection of the data's distribution for normality, skewness and biases (often with a standard normal plot overlay). Alternatively, normal probability plots can be applied directly to visualise how far all or single extremes of the measurement values deviate from a normal distribution (Dalgaard 2008; Hon 2010; Crawley 2015).

Analysing data

Many SES datasets consist of multiple variables. Each of these variables can be analysed separately using descriptive statistics to understand their central tendency and variability (Table 18.2). Where the same variables (e.g. height, species abundance) have been measured across different sites or groups, they can be compared using various methods for group comparison techniques (e.g. to test whether bird abundance differs between different sites). If the prerequisites for parametric data are not fulfilled, then less sensitive non-parametric alternatives for the parametric test have to be used instead.

Alternatively, relationships between different variables can be explored. The simplest relationship is an explanatory–response relationship, where a change in an explanatory variable 'determines' a change in a response variable. Regression models can test the relationship between a response and single or multiple explanatory variables. Most regression models assume that the response and explanatory variables are normally distributed, but there are also options for variables with other distributions. In SES research, regression models are commonly used to: (a) determine the strength of predictors, (b) forecast an effect, or (c) find trends. Linear regression models are also used to fill in missing values in time series, and other types of datasets estimate gaps in measurements using values anticipated by the model. With a well-fitted model, back- and forecasting estimates, across the minimum and maximum limits where own measurements have been made, are possible.

In contrast to 'univariate' analyses that focus on predicting a single outcome, 'multivariate' analyses investigate how multiple outcome variables co-occur. The aim of these approaches usually centres on comparing how similar two or more objects (e.g. sites, households, ecological state) are, allowing the combination of relevant characteristics with different units and data classes. Multivariate techniques include, among others, hierarchical clustering, principal component analysis (PCA), factor analysis (FA), multiple correspondence analysis (MCA) or non-metric multidimensional scaling (n-MDS) (Hon 2010).

In an SES context, variables in multivariate analyses can contain combinations of social and/or environmental aspects simultaneously, but they should all be on quasi-identical scale; otherwise, the variable with the largest range will dominate the outcomes. To avoid this, normalisation of the data (by converting them into relative values of a range from 0 to 1) is usually necessary. Multivariate analyses typically entail a mathematical procedure of 'dimensionality reduction'. This allows the dimensionality of the data to be reduced to a two-dimensional table matrix or graph that can be plotted and is easier to understand. Thus, multivariate analysis allows interpretation of complex interactions of multiple variables where no direct explanatory–response relationship is possible.

Many events in SES research are ordered in time and can be analysed using time series analysis (TSA) models. These models can have different purposes. Some aim to interpret plausible descriptions of sampled data; others to test hypotheses about the drivers behind the observed variability or to quantify response similarity (e.g. synchrony of two or more events). Time series analysis models are sometimes also intended to forecast and simulate scenarios, e.g. conducting risk assessment or projecting trajectories into the future for different management policies. Table 18.2 provides a summary of the key methods used in statistical analysis.

Table 18.2 Summary of key methods used in statistical analysis

Method	*Description*	*References*
Descriptive statistics	Descriptive statistics are indices that summarise a given dataset so that it can be easily understood. The coefficients can be divided into measures of: • Central tendency, including mean and median mode • Variability (spread), including range, quartiles, variance and standard deviation • Correlation	*Key introductory texts* Dalgaard 2008; Hon 2010 *Applications to SES* Scheffer et al. 2012; Dearing et al. 2014; Jaramillo and Destouni 2014; Oteros-Rozas et al. 2014
Group comparison	Several statistical tests are available to compare whether the means of two or more groups differ. To compare two groups, the most common tests are the T-test (normal data), the Mann–Whitney *U* test (non-parametric ordinal data) and Chi-Square (non-parametric nominal data). To compare more than two groups, the ANOVA (normal data), Kruskal–Wallis (non-parametric ordinal) or Chi-Square (non-parametric nominal) test can be used. To determine which of more than three groups differ, a 'post hoc' test, e.g. a Tukey or Scheffé test, needs to be applied.	*Key introductory texts* Dalgaard 2008; Crawley 2015 *Applications to SES* Cummings et al. 2010; Iniesta-Arandia et al. 2014; Martín-López et al. 2017
Regression models	Regression models (simple linear, multiple linear, non-linear, generalised linear) are statistical methods allowing one to summarise and study the relationships between two or more variables. Simple regression models are used to study explanatory–response relationships between two continuous variables. Linear models are the most common regression models, used for normally distributed continuous data. Multiple regression models estimate the response of a single variable from a combination of several predictor variables. Using least-square fitting to non-linear distribution models allows one to understand the characteristics of non-linear data. Data transformation (e.g. log, square root, double square root) of non-linear data may make it possible to linearise data and allow the application of linear models. Generalised linear models can be used for non-normally distributed, binary (logistic regression) and count (Poisson regression) data. Model optimisation (e.g. finding the correct set of variables that best explains the behaviour of the variable of interest) is often supported by using an Aikaike (AIC) or a Bayesian (BIC) information criterion and avoids critical over-fitting.	*Key introductory texts* Dalgaard 2008; Hon 2010; Crawley 2015 *Applications to SES* Pollnac et al. 2010; Martín-López et al. 2012; Pinsky and Fogarty 2012; Meacham et al. 2016; Kim and Kim 2018

Method	*Description*	*References*
Multivariate analysis	Multivariate clustering methods originate from ecology. These techniques include clustering, non-metric multidimensional scaling (n-MDS), principal component analysis (PCA), redundancy analysis (RDA), canonical correspondence analysis (CCA), factor analysis (FA) and multiple correspondence analysis (MCA). They are used for the classification (grouping) of multivariate data objects (i.e. datasets with multiple response or outcome variables) into sets with similar characteristics ('clusters') according to the similarity of their overall variable values. These techniques are often used with large datasets with many variables and help to transform high-dimensional data into more easily understandable two-dimensional graphs or tables. Similarity of objects is determined based on how similar objects are with regard to their variable composition and values. While clustering and n-MDS are based on a similarity index (typically Euclidean distance), PCA, RDA, CCA, FA and MCA define the dimension reduction based on the correlation or covariance between objects.	*Key introductory texts* Wilmink and Uytterschaut 1984; Quinn and Keough 2002; Oksanen 2015 *Applications to SES* Cinner et al. 2013; González-Orozco et al. 2014; Iniesta-Arandia et al. 2014; Kotzee and Reyers 2016; Maione, Nelson, and Barbosa 2018
Time series analysis	Time series analysis (TSA) is applied to time series data. The most common statistical models of time series are multivariate autoregressive state-space models (MARSS), also known as vector autoregression in econometrics. These models are typically composed by a drift term that represents a deterministic dynamic (e.g. a growth rate) and a process error term that is stochastic (e.g. some random variable that represents environmental variation). The latter is important because even in a model with a positive growth rate, a population can collapse due to the environmental noise. TSA is often a necessary condition to identify causality, although not all time series data will allow for the detection of causal effects.	*Key introductory texts* Angrist and Pischke 2009; Holmes, Ward, and Wills 2012; Pearl and Mackenzie 2018 *Applications to SES* Sabo et al. 2017

Limitations

Social-ecological systems are intrinsically difficult to understand due to their complex relationships, interactions between internal and external parts of the system, and the effect of random fluctuations and chance events. A key objective in SES research is to understand how different elements of an SES are related and how relationships change over time. However, it is often difficult to distinguish between the main influencing factors in a system and so-called 'noise'. Noise can derive from either insignificant influencing elements or random processes (Crewson 2006; Dalgaard 2008). Random processes are patterns that are measured but lack structure or predictability. Random noise can often be a large component in data, and it can be difficult to determine its magnitude. It decreases the quality of data by obscuring the main signal, i.e. the effect of the main influencing factors.

Data quality can often be already increased during the data-gathering process, e.g. through excluding avoidable random processes (e.g. taking unequal numbers of samples at different points during a field study) and biases (e.g. accidentally avoiding certain groups during surveys and favouring others) by a good consistent sampling protocol, and well-structured experimental design (Dalgaard 2008; Hon 2010).

Statistical methods strongly rely on the concept that the observed dataset that is sampled is a valid representative of a larger (unknown) 'population'. A good sampling design (also called 'casing' in social sciences) helps to avoid known biases. Biased data are data with a disproportionate weight in favour of or against one thing, person or group, so that the data are no longer a perfect representative of the larger population about which conclusions are being drawn. Sampling biases often result from uneven or patchy sampling, or from an underlying preconceived idea of the structure of the unknown population that results in the wrong type of sampling strategy. In time series analysis, bias can also arise when the variance of errors (random noise) changes over time, or when the data distribution is not normal, potentially invalidating results when the wrong analysis method is applied. Reducing bias helps to enable better discovery of underlying patterns and detection of relationships in an SES. Systematic biases may be reduced through increased sample sizes, increased randomisation of sampling (locations) or better selection of the sampling strategy (e.g. avoidance of explicit inclusion or exclusion of certain groups or system constellations) that could bias the data (see Crewson 2006; Dalgaard 2008; Hon 2010; Crawley 2015).

Thus, the overall aim should be, generally, to reduce the so-called 'sampling error', which denotes the gap between the sampled data as a representative and the current real status of the system. Because statistical methods are very sensitive to biased, patchy and noisy data, the outcome of most statistical approaches depends mainly on the overall quality of the investigated data.

Data collection therefore aims to collect unbiased data with as little noise as possible (Brown and Saunders 2007). However, high-quality data are often difficult to achieve in appropriate (temporal and spatial) resolution with fully excluding potential (often unknown) biases. Empirical data obtained from fieldwork often barely cover the minimum required representative data needed, as data are often either difficult to obtain given personnel capacity and time constraints, are simply not available, or are available in limited quantities.

In some situations, accurately estimating the effect or strength of specific explanatory variables requires experimental comparison (in the form of 'before–after' situations or as intentional experimental manipulations). However, in many real-world situations, and at the scale that most SES operate, experimental exclusion of a given factor is realistically impossible (e.g. removing the effect of global warming in order to see its influence, or excluding all insects from a large region to estimate the contribution of pollinators).

Moreover, results of most statistical methods focus on average outcomes, patterns or relationships in a dataset and cannot be used to draw conclusions about individual characteristics or cases. Another problem with statistical analysis is the tendency to jump to unjustified conclusions about causal relationships. One can often find evidence that two variables are highly correlated, but that does not prove that one variable actually causes another. A strong statistical relationship in the past can also not be assumed to hold in the future if conditions change. Statistical analysis is not capable of proving a causal relationship between two variables but requires a good understanding of the system investigated by the investigator (Pearl 2009).

Lastly, even with perfect data, the incorrect application of a method may result in drawing incorrect conclusions, as every method can only be applied to certain data types and under certain assumptions. Thus, the analysing researcher must have knowledge of the data type and structure to ensure that the method or test is appropriate and valid.

Resource implications

Complex statistical methods require computers and specialised software to conduct calculations or create graphical visualisations of the data and results. In the past, accessibility and computational power were the main constraints for rigorous and complex analyses of large datasets or the application of advanced statistical methods. However, given the technical capabilities of current computers, the application of statistical methods on even medium- to large-size datasets (up to several gigabytes) generally does not require specialised hardware or computational power any more. Also, given the multitude of free statistical software packages, specialised commercial software is rarely needed, as most free-of-charge statistical software packages now cover a wide range of statistical methods typically applied in SES research. Freely available open-source software includes, for example, R, Octave and Pandas, while many universities also provide access to commercially available products such as Statistica, SPSS, SAS, JMP and Strata. More specialised software packages such as JmulTi and Microfit (econometric analysis), OpenEpi (epidemiology), Simfit (simulations) or OpenNN, Torch or Weka (machine learning) are also freely available (see en.wikipedia.org/wiki/List_of_statistical_packages for a full list).

Many universities also offer statistical resource centres to assist students and researchers, given the specialised and highly technical aspects of applying many statistical methods. These include: (a) posing a correctly formulated research question (that can create a null hypothesis, that then can be falsified), (b) choosing the appropriate sampling design and type of data needed to answer the question, (c) having expertise in data acquisition, manipulation and interpretation, (d) selecting the appropriate statistical method for the type of data, and (e) being able to correctly interpret statistical results, which often requires significant expertise and understanding of statistical methods.

There are also a multitude of information courses (e.g. at DataCamp, Udemy and other online learning platforms) and tutorials available online (e.g. KhanAcademy.org, R-project.org).

New directions

Increasing computational power and accessibility of data from large online databases enables the analysis of 'big data' to test SES relationships and novel hypotheses (see also Chapter 17). 'Big data' are created by increasing numbers of electronic sensors, the automatic collection of data available through the Internet, and more long-term studies and national statistics data (e.g. the World Bank, the UN and other databases) being made available online (Bhadani and Jothimani 2016). The increasing number of people with mobile devices, active participation in social media and Internet use means data on human social interactions have become much more accessible. Analysis of social network data has recently radically transformed the way applied social science is done (Foster et al. 2016), allowing for better predictions of social dynamics, human decision-making and the potential consequences (Thai, Wu, and Xiong 2017).

A completely different field is the increasing accessibility of geophysical data by monitoring satellites and satellite-based sensor techniques (provided by e.g. NASA, ESA or platforms such

Case study 18.1: Drivers of ecosystem vulnerability in Andalusia, Spain

Different statistical methods can be used to unravel information about social-ecological dynamics and associations between social and ecological aspects. This is well illustrated by SES research conducted in the semi-arid watersheds of Adra and Nacimiento in Andalusia, Spain. The aim of this research was to find patterns of social-ecological interactions between stakeholders' perceptions of ecosystem services, stakeholder well-being and drivers of change (Iniesta-Arandia et al. 2014).

Altogether 381 face-to-face questionnaires were administered from May 2009 to February 2010. Based on these surveys, different groups of stakeholders who used and managed ecosystem services in each of the watersheds were identified by conducting hierarchical cluster analysis (HCA) and principal component analysis (PCA) (Figure 18.1A). Five stakeholder groups were identified: (a) local actors who depend on provisioning ecosystem services, (b) local actors not directly dependent on provisioning ecosystem services, (c) environmental and local development professionals, (d) rural tourists, and (e) nature tourists.

Next, different perceptions of vulnerable ecosystem services according to the five groups of stakeholders were explored by conducting the non-parametric Kruskal–Wallis test. This test showed that there were statistically significant differences in the perceptions about vulnerable ecosystem services among stakeholders. Local actors perceived a higher degree of vulnerability of provisioning ecosystem services, whereas environmental professionals perceived a higher degree of vulnerability of cultural ecosystem services than other stakeholders (Figure 18.1B).

Finally, to determine whether there were associations between vulnerable ecosystem services, human well-being and the effect of drivers of change according to the perceptions of different stakeholders, a principal component analysis (multivariate ordination analysis) was conducted. This statistical analysis showed that whereas local actors perceived the importance and vulnerability of provisioning ecosystem services for their well-being, environmental professionals perceived cultural ecosystem services, specifically the aesthetic values of landscape and local ecological knowledge, and soil fertility as vulnerable services. Locals also perceived the effects of drivers of change on ecosystem services. The principal component analysis showed that nature tourists perceived regulating services as vulnerable ecosystem services (Figure 18.1C).

Results from this study suggest that perceptions of the interlinkages between ecosystem services, human well-being and the effect of drivers of change on ecosystem services differ among stakeholder groups in the Adra and Nacimiento watersheds.

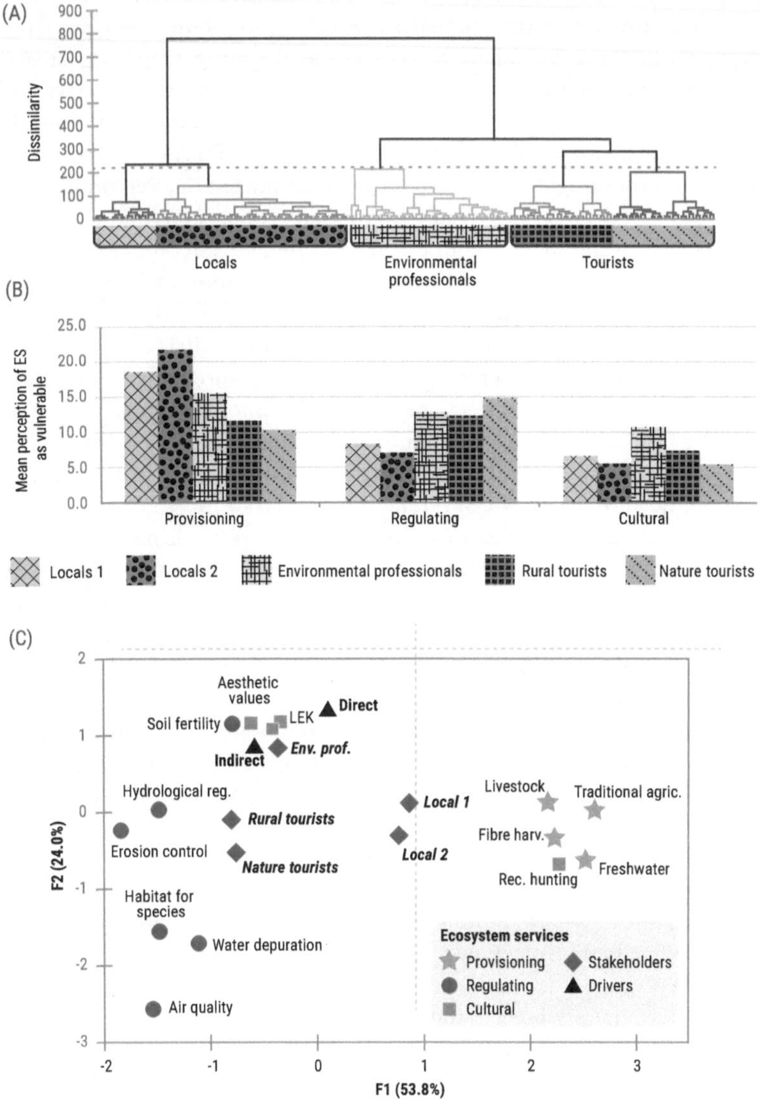

Figure 18.1 (A) A dendrogram showing the different stakeholder groups based on a hierarchical cluster analysis, (B) a bar diagram illustrating the perceptions of vulnerable ecosystem services among stakeholder groups, and (C) a biplot of the principal component analysis to find patterns between the associations (between stakeholders and their perceptions of vulnerable ecosystem services) and the effect of drivers of change (Iniesta-Arandia et al. 2014)

as Google's Earth Engine or Earth System Data Cube – see Chapter 24). These developments have helped to see SES interactions and problems in a whole new light (Rejcek 2017). Hampton et al. (2013) showed how the statistical analysis of large ecosystem dynamics data could help understand SES interlinkages. The analysis is done by aggregating, searching, cross-referencing and mining large volumes of data. The resulting outcomes finally generate new understanding that can inform decision-making about emergent properties of complex systems. Dubey et al. (2019) showed how big datasets and predictive analytics could improve social and environmental sustainability by combining data from 205 studies conducted in a region in India.

The analysis of large datasets may lead to erroneous conclusions and unexpected failures if blind trust is placed in the sheer amount of data available. Theoretical understanding and knowledge of the system remain fundamental to good, rigorous analysis. Furthermore, 'big data' are often strongly biased towards developed countries (Reis, Braatz, and Chiang 2016). Satellite or environmental sensor data are, thus, often only available for risk areas in developed Western countries at usable temporal and spatial resolution for conducting profound analyses. In contrast, these types of data are often lacking for particular SES risk regions of interest, especially in many undeveloped countries. Similarly, high-quality social media and trading data are only available in industrialised and rich countries that rely heavily on Internet-mediated communication for social interactions and trading.

Another important emerging statistical field is the use of machine learning. In the wake of big data, artificial intelligence methods are opening new approaches to investigating complex problems. Machine-learning approaches allow for statistical pattern recognition in datasets of very high complexity (see Chapter 17). However, these methods are typically not based on predefined verifiable statistical models and hypotheses, and often act as 'black boxes'. Typical neural network approaches use dynamically adaptive statistical models and therefore do not necessarily provide outcomes that can contribute to the logical understanding of interactions (Kleinberg, Ludwig, and Mullainathan 2016; Knight 2017; Pearl and Mackenzie 2018). However, novel analysis methods based on machine learning enabled by better computational grids allow for much faster analysis of data and near-instantaneous and automated decision-making. This is habitually done in economic trading when it comes to the buying and selling of commodities and stocks. These analysis methods may open up new possibilities for political and other national and international stakeholders to become better informed.

Some studies even suggest that artificial intelligence approaches and machine learning are ushering us into a new era in which abundant data and mathematics will replace classical statistical analysis methods (Anderson 2008). While scientists in the past had to rely on sample testing and statistical analysis to understand a process, computer scientists today have access to the 'entire population' (not only a 'representative sample') and therefore would not need classical statistical analysis methods any more. Thus, some authors speculate that traditional statistical testing approaches could become obsolete (Anderson 2008).

Key readings

Bors, D.A. 2018. *Data Analysis for the Social Sciences: Integrating Theory and Practice.* Thousand Oaks: Sage.

Brown, R.B., and M.P. Saunders. 2007. *Dealing with Statistics: What You Need to Know.* Berkshire: McGraw-Hill.

Crawley, M.J. 2015. *Statistics: An Introduction Using R.* Chichester: John Wiley & Sons.

Pearl, J. 2003. *Causality: Models, Reasoning, and Inference.* Cambridge: Cambridge University Press.

Quinn, G.P., and M.J. Keough. 2002. *Experimental Design and Data Analysis for Biologists.* Cambridge: Cambridge University Press.

Acknowledgements

Ingo Fetzer acknowledges funding from the European Research Council under EU H2020 (grant no. ERC-2016-ADG 743080). Luigi Piemontese was funded by the Swedish Research Council for Environment, Agricultural Sciences and Spatial Planning FORMAS (942-2015-740). Juan C. Rocha also received support from the FORMAS grant (942-2015-731). Ingo Fetzer, Luigi Piemontese and Juan C. Rocha especially thank the Stockholm Resilience Centre for support that enabled the writing of this chapter. The authors also especially want to thank Dr Goeff Wells for his fantastic feedback, general help and useful references to this chapter.

References

Anderson, C. 2008. 'The End of Theory: The Data Deluge Makes the Scientific Method Obsolete.' *Wired*. www.wired.com/2008/06/pb-theory.

Angrist, J.D., and J-S. Pischke. 2009. *Mostly Harmless Econometrics: An Empiricist's Companion*. Princeton: Princeton University Press.

Bhadani, A., and D. Jothimani. 2016. 'Big Data: Challenges, Opportunities and Realities.' In *Effective Big Data Management and Opportunities for Implementation*, edited by M.K. Singh and D.G. Kumar, 1–24. Pennsylvania: IGI Global. http://arxiv.org/abs/1705.04928.

Brown, R.B., and M.P. Saunders. 2007. *Dealing with Statistics: What You Need to Know*. Berkshire: McGraw-Hill.

Carvalho, F.P. 2017. 'Pesticides, Environment, and Food Safety.' *Food and Energy Security* 6: 48–60. doi:10.1002/fes3.108.

Cinner, J.E., C. Huchery, E.S. Darling, A.T. Humphries, N.A.J. Graham, C.C. Hicks, N. Marshall, and T.R. McClanahan. 2013. 'Evaluating Social and Ecological Vulnerability of Coral Reef Fisheries to Climate Change.' *PLoS ONE* 8(9): e74321. doi:10.1371/journal.pone.0074321.

Crawley, M.J. 2015. *Statistics: An Introduction Using R* (2nd ed). Chichester: John Wiley & Sons. www.wiley.com/en-se/Statistics%3A+An+Introduction+Using+R%2C+2nd+Edition-p-9781118941096.

Crewson, P. 2006. 'Applied Statistics Handbook.' AcaStat Software. www.acastat.com.

Cummings, E.M., C.E. Merrilees, A.C. Schermerhorn, M.C. Goeke-Morey, P. Shirlow, and E. Cairns. 2010. 'Testing a Social Ecological Model for Relations between Political Violence and Child Adjustment in Northern Ireland.' *Development and Psychopathology* 22(2): 405–418. doi:10.1017/S0954579410000143.

Dalgaard, P. 2008. *Introductory Statistics with R* (2nd ed). New York: Springer. doi:10.1007/978-0-387-79054-1.

Dearing, J.A., R. Wang, K. Zhang, J.G. Dyke, H. Haberl, M.S. Hossain, P.G. Langdon et al. 2014. 'Safe and Just Operating Spaces for Regional Social-Ecological Systems.' *Global Environmental Change* 28: 227–238. doi:10.1016/j.gloenvcha.2014.06.012.

Dubey, R., A. Gunasekaran, S.J. Childe, T. Papadopoulos, Z. Luo, S.F. Wamba, and D. Roubaud. 2019. 'Can Big Data and Predictive Analytics Improve Social and Environmental Sustainability?' *Technological Forecasting and Social Change* 144: 534–545. doi:10.1016/j.techfore.2017.06.020.

Fienberg, S.E. 1992. 'A Brief History of Statistics in Three and One-Half Chapters: A Review Essay.' *Statistical Science* 7(2): 208–225. doi:10.1214/ss/1177011360.

Foster, I., R. Ghani, R.S. Jarmin, F. Kreuter, and J. Lane. 2016. *Big Data and Social Science: A Practical Guide to Methods and Tools*. Boca Raton: CRC Press.

González-Orozco, C.E., A.H. Thornhill, N. Knerr, S. Laffan, and J.T. Miller. 2014. 'Biogeographical Regions and Phytogeography of the Eucalypts.' *Diversity and Distributions* 20(1): 46–58. doi:10.1111/ddi.12129.

Hampton, S.E., C.A. Strasser, J.J. Tewksbury, W.K. Gram, A.E. Budden, A.L. Batcheller, C.S. Duke, and J.H. Porter. 2013. 'Big Data and the Future of Ecology.' *Frontiers in Ecology and the Environment* 11(3): 156–162. doi:10.1890/120103.

Holmes, E.E., E.J. Ward, and K. Wills. 2012. 'MARSS: Multivariate Autoregressive State-space Models for Analyzing Time-series Data.' *The R Journal* 4(1): 30.

Hon, K. 2010. *An Introduction to Statistics*. Scotts Valley: CreateSpace.

Husak, G.J., C.C. Funk, J. Michaelsen, T. Magadzire, and K.P. Goldsberry. 2013. 'Developing Seasonal Rainfall Scenarios for Food Security Early Warning.' *Theoretical and Applied Climatology* 114(1–2): 291–302. doi:10.1007/s00704-013-0838-8.

Iniesta-Arandia, I., M. García-Llorente, P.A. Aguilera, C. Montes, and B. Martín-López. 2014. 'Socio-cultural Valuation of Ecosystem Services: Uncovering the Links between Values, Drivers of Change, and Human Well-Being.' *Ecological Economics* 108: 36–48. doi:10.1016/j.ecolecon.2014.09.028.

Jaramillo, F., and G. Destouni. 2014. 'Developing Water Change Spectra and Distinguishing Change Drivers Worldwide: Worldwide Water Change Spectra.' *Geophysical Research Letters* 41(23): 8377–8386. doi:10.1002/2014GL061848.

Keane, A., J.F. Lund, J. Bluwstein, N.D. Burgess, M.R. Nielsen, and K. Homewood. 2019. 'Impact of Tanzania's Wildlife Management Areas on Household Wealth.' *Nature Sustainability*. doi:10.1038/s41893-019-0458-0.

Kim, J.I., and G. Kim. 2018. 'Effects on Inequality in Life Expectancy from a Social Ecology Perspective.' *BMC Public Health* 18(1): 243. doi:10.1186/s12889-018-5134-1.

Kleinberg, J., J. Ludwig, and S. Mullainathan. 2016. 'A Guide to Solving Social Problems with Machine Learning.' *Harvard Business Review*. https://hbr.org/2016/12/a-guide-to-solving-social-problems-with-machine-learning.

Knight, W. 2017. 'There's a Big Problem with AI: Even Its Creators Can't Explain How It Works.' *MIT Technology Review*. www.technologyreview.com/s/604087/the-dark-secret-at-the-heart-of-ai.

Kotzee, I., and B. Reyers. 2016. 'Piloting a Social-Ecological Index for Measuring Flood Resilience: A Composite Index Approach.' *Ecological Indicators* 60: 45–53. doi:10.1016/j.ecolind.2015.06.018.

Maione, C., D.R. Nelson, and R.M. Barbosa. 2018. 'Research on Social Data by Means of Cluster Analysis.' *Applied Computing and Informatics*. doi:10.1016/j.aci.2018.02.003.

Martín-López, B., I. Iniesta-Arandia, M. García-Llorente, I. Palomo, I. Casado-Arzuaga, D.G. del Amo, E. Gómez-Baggethun et al. 2012. 'Uncovering Ecosystem Service Bundles through Social Preferences.' *PLoS ONE* 7(6): e38970. doi:10.1371/journal.pone.0038970.

Martín-López, B., I. Palomo, M. García-Llorente, I. Iniesta-Arandia, A.J. Castro, D.G. del Amo, E. Gómez-Baggethun, and C. Montes. 2017. 'Delineating Boundaries of Social-Ecological Systems for Landscape Planning: A Comprehensive Spatial Approach.' *Land Use Policy* 66: 90–104. doi:10.1016/j.landusepol.2017.04.040.

Meacham, M., C. Queiroz, A. Norström, and G. Peterson. 2016. 'Social-Ecological Drivers of Multiple Ecosystem Services: What Variables Explain Patterns of Ecosystem Services across the Norrström Drainage Basin?' *Ecology and Society* 21(1): 14. doi:10.5751/ES-08077-210114.

Oksanen, J. 2015. 'Multivariate Analysis of Ecological Communities in R: Vegan Tutorial.' http://cc.oulu.fi/~jarioksa/opetus/metodi/vegantutor.pdf.

Oteros-Rozas, E., B. Martín-López, J. González, T. Plieninger, C. López-Santiago, and C. Montes. 2014. 'Socio-cultural Valuation of Ecosystem Services in a Transhumance Social-Ecological Network.' *Regional Environmental Change* 14: 1269–1289. doi:10.1007/s10113-013-0571-y.

Paumgarten, F., and C. Shackleton. 2011. 'The Role of Non-Timber Forest Products in Household Coping Strategies in South Africa: The Influence of Household Wealth and Gender.' *Population and Environment* 33(1): 108–131. www.jstor.org/stable/41487565.

Pearl, J. 2009. *Causality: Models, Reasoning and Inference* (2nd ed). Cambridge: Cambridge University Press.

Pearl, J., and D. Mackenzie. 2018. *The Book of Why: The New Science of Cause and Effect*. New York: Basic Books. http://search.ebscohost.com/login.aspx?direct=true&scope=site&db=nlebk&db=nlabk&AN=1592572.

Piemontese, L., G. Castelli, I. Fetzer, J. Barron, H. Liniger, N. Harari, E. Bresci, and F. Jaramillo. 2020. 'Estimating the Global Potential of Water Harvesting from Successful Case Studies.' *Global Environmental Change* 63. doi:10.1016/j.gloenvcha.2020.102121.

Pinsky, M.L., and M.l. Fogarty. 2012. 'Lagged Social-Ecological Responses to Climate and Range Shifts in Fisheries.' *Climatic Change* 115(3): 883–891. doi:10.1007/s10584-012-0599-x.

Pollnac, R., P. Christie, J.E. Cinner, T. Dalton, T.M. Daw, G.E. Forrester, N.A.J. Graham, and T.R. McClanahan. 2010. 'Marine Reserves as Linked Social-Ecological Systems.' *Proceedings of the National Academy of Sciences* 107(43): 18262–18265. doi:10.1073/pnas.0908266107.

Quinn, G.P., and M.J. Keough. 2002. *Experimental Design and Data Analysis for Biologists*. Cambridge: Cambridge University Press.

Reis, M.S., R.D. Braatz, and L.H. Chiang. 2016. 'Big Data: Challenges and Future Research Directions.' www.aiche.org/resources/publications/cep/2016/march/big-data-challenges-and-future-research-directions.

Rejcek, P. 2017. 'The Farms of the Future Will Be Automated from Seed to Harvest.' *Singularity Hub* (blog). 30 October. https://singularityhub.com/2017/10/30/the-farms-of-the-future-will-run-on-ai-and-robots.

Robinson, B.E., H. Zheng, and W. Peng. 2019. 'Disaggregating Livelihood Dependence on Ecosystem Services to Inform Land Management.' *Ecosystem Services* 36(100902). doi:10.1016/j.ecoser.2019.100902.

Sabo, J., A. Ruhi, G.W. Holtgrieve, V. Elliott, M.E. Arias, P.B. Ngor, T.A. Räsänen, and S. Nam. 2017. 'Designing River Flows to Improve Food Security Futures in the Lower Mekong Basin.' *Science* 358(6368): eaao1053. doi:10.1126/science.aao1053.

Scheffer, M., S.R. Carpenter, T.M. Lenton, J. Bascompte, W. Brock, V. Dakos, J. van de Koppel et al. 2012. 'Anticipating Critical Transitions.' *Science* 338(6105): 344–348. doi:10.1126/science.1225244.

Shongwe, M.E., G.J. van Oldenborgh, B. van den Hurk, and M. van Aalst. 2011. 'Projected Changes in Mean and Extreme Precipitation in Africa under Global Warming. Part II: East Africa.' *Journal of Climate* 24(14): 3718–3733.

Thai, M.T., W. Wu, and H. Xiong, eds. 2017. *Big Data in Complex and Social Networks*. Boca Raton: CRC Press.

Varberg, D.E. 1963. 'The Development of Modern Statistics.' *The Mathematics Teacher* 56(4): 252–257.

Wilmink, F.W., and H.T. Uytterschaut. 1984. 'Cluster Analysis, History, Theory and Applications.' In *Multivariate Statistical Methods in Physical Anthropology: A Review of Recent Advances and Current Developments*, edited by G.N. van Vark and W.W. Howells, 135–175. Dordrecht: Springer. doi:10.1007/978-94-009-6357-3_11.

19

Qualitative content analysis

Rika Preiser,[1] *María Mancilla García,*[2] *Lloyd Hill*[3] *and Louis Klein*[4]

[1]CENTRE FOR SUSTAINABILITY TRANSITIONS, STELLENBOSCH UNIVERSITY, STELLENBOSCH, SOUTH AFRICA
[2]STOCKHOLM RESILIENCE CENTRE, STOCKHOLM UNIVERSITY, STOCKHOLM, SWEDEN
[3]DEPARTMENT OF SOCIOLOGY AND SOCIAL ANTHROPOLOGY, STELLENBOSCH UNIVERSITY, STELLENBOSCH, SOUTH AFRICA
[4]EUROPEAN SCHOOL OF GOVERNANCE, BERLIN, GERMANY; INTERNATIONAL FEDERATION FOR SYSTEMS RESEARCH, AUSTRIA

Key methods discussed in this chapter

Discourse analysis, critical discourse analysis, thematic analysis, narrative analysis, critical narrative analysis, interpretative phenomenological analysis

Connections to other chapters

Methods used for qualitative analysis treat language as a resource that has agency to shape societal practices and institutions. By using certain words or phrases, language and discourse convey certain power relations and influence worldviews, societal forms and actions. By analysing how language and other forms of non-textual representation (such as descriptions, accounts, opinions, feelings) are being used in various contexts, this chapter links well to Chapter 5 (Systems scoping), Chapter 7 (Interviews and surveys), Chapter 8 (Participatory data collection), Chapter 9 (Facilitated dialogues), Chapter 10 (Futures analysis), Chapter 11 (Scenario development) and Chapter 20 (Comparative case study analysis). The set of methods proposed in this chapter are all well equipped to discover trends in how content is developed and knowledge generated.

Introduction

Qualitative methods of content analysis seek to find and examine patterns of sense-making and meaning creation in the communicative characteristics of language, by focusing on the content and underlying themes and meaning that emerge in a text (in either written or spoken form). The word 'text' here points to a wide range of phenomena such as descriptions, accounts, opinions and feelings that are conveyed in a variety of representations, not only as letters on a white page or screen. Qualitative data can be represented in words, pictures and even sounds.

Nowadays the word 'text' can be applied to landscapes, heritage sites, technologies, urban spaces or institutional practices – all phenomena that can be 'read' in a certain way. Drawing on the implications of the 'linguistic turn' that was introduced in the humanities by French linguist Ferdinand de Saussure (1974), researchers from other disciplines have developed

DOI: 10.4324/9781003021339-23

SUMMARY TABLE: QUALITATIVE CONTENT ANALYSIS

DISCIPLINARY BACKGROUND	KNOWLEDGE TYPE
The methods in this chapter are derived from or have most commonly been used in: Semiotics, Psycholinguistics, Psychology, Sociolinguistics, Pragmatism, Sociology, Management and Organisational Studies, Media and Cultural Studies	*The methods in this chapter are primarily used to generate the following types of knowledge:* • Descriptive • Exploratory • Explanatory
RESEARCH APPROACH	**PURPOSE OF METHOD**
The methods in this chapter originate from or most commonly adopt the following research approaches: • Interpretive/subjective	*The most common purposes of using the methods in this chapter are:* • System understanding • Policy/decision support
TEMPORAL DIMENSION	**SYSTEMIC FEATURES AND PROCESSES**
The methods in this chapter are most commonly applied to the following temporal dimensions: • Present (typically within the last 5–10 years) • Recent past (post-1700s) • Future **SPATIAL DIMENSION** *The methods in this chapter are primarily either or both:* • Non-spatial *The methods in this chapter are most commonly applied at the following spatial scales:* • Local • Regional (provincial/state to continental) • Global • Multiple places/sites around the world	*While most methods can do many things, the methods in this chapter are particularly good (i.e. go-to methods) for addressing the following:* • SES components and linkages • Power relations • Transformation • Evaluating policy options

various analytical methods to break away from theories and methods that assume that language is an autonomous system in which the meaning of words and sentences can always be constructed in an objective and internally fixed manner. In a sense, language is treated as being significantly more than just the medium through which we conduct research or communicate with one another.

Language and texts are seen as phenomena that have agency. Linked to this interpretation of the representational power of language, the notion of 'discourse' is then understood to be an 'interrelated set of texts, and the practices of their production, dissemination, and reception, that brings an object into being' (Parker 1992). The stories and narratives that are linked to organisations, traditions, practices and communities have the power to shape how these social forms are structured. From this view we can derive that qualitative methods aim to demonstrate how language and discourse produce social realities that shape certain practices and social-ecological interactions. As researchers, the use of qualitative methods of content analysis thus allows us to explore the relationship between constructed discourses and epistemic realities.

Qualitative methods for analysing linguistic content are used broadly to analyse textual and narrative-based content such as documents, interviews, observation notes and stories. There is a variety of qualitative analysis methods – some focusing on the explicit themes presented in the text, and others trying to disentangle the implicit or hidden patterns of meaning that are present in the text. Many researchers who use thematic analysis, for example, assume that the explicit content can be coded and analysed to unveil its inherent meaning. By contrast, an inherent and subtler or hidden pattern of meaning might be shaping certain perceptions, norms and societal orientations that only become explicit when highlighted through more reflective methods of textual and semiotic analysis like critical narrative inquiry. Some interpretive approaches to qualitative content analysis also take into consideration that meaning emerges as a result of researchers' relationship with the textual content in which their subjectivity comes to interact with the text.

The general purpose of qualitative content analysis methods is to make the various forms of 'meaning' that emerge through different methods of analysis, more explicitly noticeable. In some disciplines, this generative quality of meaning is called 'intertextuality', which highlights the fact that meaning also emerges through the way in which language and discourses are embedded in certain historical and social contexts and that meaning might change when the contexts change. These temporally linked changes can then also be tracked by methods such as narrative analysis or critical discourse analysis.

Interpretive and qualitative methods of content analysis explore and consider the different possible meanings people might attach to situated events or phenomena, or their connection to a certain place and the values that certain practices or objects hold for them. These methods aim to discover and explore the dominant discourses and how they are sustained or include and exclude people and practices in social-ecological systems (SES) relations. These methods are well suited to studying notions of identity construction, sense of place, sense-making, power relations and historical memory. They can also be used to discern what value judgements people might have towards certain events or objects.

Qualitative modes of research have a strong contemporary association with the humanities and social sciences, and different disciplines have developed diverse approaches for analysing textual or narrative-based phenomena. Practices and philosophies of interpretive content analysis also vary between academic disciplines. They all engage in a systematic reading or observation of fragmented texts or artefacts, which are then clustered or assigned labels or codes to extrapolate or infer meaning from them. The labels used in this analysis are not necessarily used consistently across disciplines.

Qualitative methods of content analysis allow researchers to expose the diversity of perspectives different actors or groups might hold in relation to a specific problem, the nuances in meaning attributed to phenomena, and the dominant views that are present in a given context. They also provide insights as to why some views might be more dominant than others. As such, interpretive and qualitative methods of content analysis are likely to draw attention to the multiplicity of meanings and interpretations, and their consequences for how governance and stewardship interventions are implemented and enacted. These methods can also be used to understand how individuals position their own and others' personal and social identities in relation to some shared values or common resource-use problem, for example. In addition, the methods can highlight which forms of knowledge are considered valid by whom and what the consequences of certain positions are regarding access to resources and governance arrangements.

SES problems and questions

In seeking to understand social-ecological linkages, qualitative data can help researchers to address questions related to human–nature interactions (e.g. values, stewardship, resource use). To understand SES, research is needed on the institutional, social-relational, contextual and individual human dimensions of the system. Within each of these dimensions, a variety of themes and concepts can be explored and understood through qualitative data analysis methods. These methods are also useful for exploring the diverse understandings of SES held by various stakeholders and how the blind spots in their different ways of meaning creation shape people's experiences of using, governing and valuing certain resources, for example.

Typical questions that qualitative content analysis methods can help frame to address SES challenges include the following:

- What diverse understandings or interpretations of a problem exist in a given context?
- How do people group themselves in relation to the different understandings?
- What are the dominant views of a problem or challenge? How did these views become dominant?
- Whose interests are given advantage under the current management system? What are the underlying causes for that?
- Why do people have a specific relationship with their environment?
- What meanings do they attribute to that relationship?
- How do different meanings and interpretations affect the way SES are governed?
- How are meanings created and put into practice?
- Have perceptions of a place or definitions of a problem changed over time? How? Why?

Typical problems these methods seek to understand include:

- Why do conflicts of interest exist in the management of resources?
- How do different groups regard different approaches to environmental management as being successful (or not)?
- What are the power dynamics that govern the use of certain resources?

Brief description of key methods

In general, qualitative methods that focus on analysing narratives aim to understand how people make sense of, and reconstruct, their experiences from a particular standpoint in

Table 19.1 Summary of key methods used in qualitative content analysis

Method	*Description*	*References*
Discourse analysis	Discourse analysis originated in linguistics, where it is commonly defined as the analysis of a unit of language (utterance or written text) larger than the sentence. Language at this level cannot usefully be analysed with methods traditionally associated with the decomposition of sentences or verbal utterances/words. Discourse analysis migrated into other social science domains, where it has tended to be associated with the material culture or 'texts' (e.g. novels, various forms of print/digital media and formal publications more generally) rather than more or less 'naturally' occurring (and transcribed) speech. Discourse analysis is traditionally distinguished from the qualitative analysis of interview transcripts or conversation transcripts (using conversation analysis, which originated in sociology). However, the post-1960 digitalisation of culture has blurred this distinction.	*Key introductory texts* Jupp and Norris 1993; Potter 1996 *Applications to SES* Dryzek 1997; Feindt and Oels 2005; Hajer and Versteeg 2005
Critical discourse analysis	Critical discourse analysis (CDA) takes account of factors beyond the text itself and emphasises the role of language as a powerful resource that is related to ideology and socio-cultural change. As a context-sensitive approach, CDA tends to uncover the representational properties of language as a vehicle for the exercise of power associated with the production and circulation of texts. CDA is often combined with ethnography to explore situated practices in a manner that can also differentiate it from interviews and focus groups.	*Key introductory texts* Foucault 1974; Weiss and Wodak 2003; Blommaert 2005; Hajer, Van den Brink, and Metz 2006 *Applications to SES* Huitema 2002
Thematic analysis (text)	The process of thematic analysis aims to identify themes – also called patterns – in the dataset. These themes can be explicitly stated in the data or can be implicitly present and identified by the researcher. Thematic analysis can be developed deductively, when themes are developed after theory or a research question, or inductively, when the researcher searches for themes in the data. Researchers can choose themes based on their ability to provide an answer to the research question and adequately synthesise the studied phenomenon. How frequently a theme occurs does not necessarily determine its importance, but rather indicates what constitutes specific patterns. There are fundamentally three stages in the identification of themes: (a) the researcher begins with an initial and holistic reading of a text (e.g. an interview), (b) the researcher analyses or fragments the text by identifying meaningful subsections ('meaning' is defined in terms of prior reading, research questions, etc.) and by coding or categorising the subsections, and (c) the researcher elaborates and refines the emerging code system as more texts (e.g. additional interviews) are put through this process.	*Key introductory texts* Jupp and Norris 1993; Boyatzis 1998; Braun and Clarke 2006 *Applications to SES* Stojanovic et al. 2016; Sitas et al. 2019

Method	Description	References
Narrative analysis	Narrative analysis provides researchers with a set of qualitative approaches whereby stories and the events that shape the temporal unfolding of events can be interpreted. These stories can be collected through participatory data collection methods (see Chapter 8) or shared through the everyday lived experiences of people. The researcher interprets the stories in terms of how the story is structured, what functions the story has in the context of the storyteller, what the core themes of the story are, and how the story is performed or communicated. Narrative analysis is effective in exploring how individuals confer meaning onto objects. It provides a means of sense-making and meaning creation of experiences within the individual's social environment.	*Key introductory texts* Bruner 1987; Cortazzi 1993; Boje 2008 *Applications to SES* Paschen and Ison 2014
Critical narrative inquiry	As a form of 'sceptical reading', critical narrative inquiry (CNI) exposes established and dominant narratives on the one hand and informal narrative speculations and ante-narratives on the other. By making this distinction apparent, the researcher can look for the fragmented, non-linear, incoherent, collective, unplotted stories in a specific context that do not or do not yet form part of the main or dominant narratives, but bear the potential to change these. CNI reveals the unquestioned and taken-for-granted assumptions, their internal tensions, contradictory forces and paradoxes lurking behind the words in the text that have the power to shape dominant narratives and support the power structures that are propagated in this way. By exposing the nature and construction of these narratives, people can reframe their worldviews and sense-making and meaning-creation practices and are prompted to consider alternative options for sense-making and acting in their specific context.	*Key introductory texts* Boje 2011; Edson and Klein 2016 *Applications to SES* Jørgensen and Largacha-Martinez 2014; Klein and Weiland 2014
Interpretative phenomenological analysis	Having developed from the field of experiential and psychological research, interpretative phenomenological analysis aims to explore and interpret the particularities of people's lived experiences (treated as 'text') in a given context and make sense of a given phenomenon. Rooted in the theoretical origins of phenomenology and hermeneutics, the methodology employs a 'double hermeneutic' in which the researcher uses qualitative data gathered through interviews, diaries or focus groups. Engaging in flexible and open-ended inquiry, the researcher adopts an exploratory mode of facilitation while trying to make sense of the participants' stories as they make sense of how they assign meaning to their lived experiences in relation to certain phenomena. After data collection, the researcher analyses the data for recurring themes that form patterns of meaning (ideas, thoughts, feelings) throughout the text. The themes are used to identify what issues matter to the participants (i.e. an object of concern, topic of some import) and also how participants ascribe meaning to certain phenomena and events.	*Key introductory texts* Reid, Flowers, and Larkin 2005; Smith, Flowers, and Larkin 2009; Gill 2014; Kurtz 2014 *Applications to SES* Lejano, Ingram, and Ingram 2013; Lindow 2017

Case study 19.1: The technical discourse in water governance: who shapes SES in Peru and Brazil?

Participatory stakeholder processes and collaborative governance have been recommended as key means for ensuring the sustainable management of common-use resources. The underlying hypothesis is that if actors with different interests come together, they will need to agree on the management of the resource, and thus come up with initiatives that allow them to sustain their own uses and conserve the resource. Yet, whether collaborative governance effectively manages to ensure more sustainable governance is strongly debated. In particular, the literature questions whether participatory forms of governance effectively manage to incorporate interests (e.g. the environmental protection/conservation interest) and perspectives (e.g. those of indigenous peoples) that have been historically excluded from management. While actors representing historically excluded interests might be physically present in participatory forums for environmental governance, the question that remains to be answered is whether they can actively participate in the discussions that lead to making decisions over the use of the resource.

In this case study, researchers analysed the discourses at play in four water-basin councils in Latin America, two in Peru and two in Brazil (Figure 19.1). The purpose of this study was to understand whether water-basin councils (i.e. participatory organisations set up to ensure the collegiate management of rivers) effectively include

Figure 19.1 The degradation of the Paraíba do Sul River in Caçapava, south-eastern Brazil (pt.wikipedia.org/wiki/Rio_Para%C3%ADba_do_Sul#/media/Ficheiro:Rio_para%C3%ADba_do_sul.jpg)

actors who had historically been excluded from water governance (e.g. small NGOs, peasant communities) and what the conditions are that help or hinder their inclusion.

For data collection, the researchers used interviews, surveys and observation notes (Mancilla García and Bodin 2019). A thematic analysis was then applied to the data, i.e. the researchers identified the themes most frequently occurring in the interviews and observation notes. A deductive and an inductive approach were combined to determine these themes. This means they drew a first list of themes from the literature on participatory governance and inclusion. As they analysed the data to account for the themes or subthemes (specifications of more general themes) that appeared in the empirical material, the themes were revised.

The researchers complemented thematic analysis with critical discourse analysis to analyse which discourses or perspectives were considered valuable and appropriate by different participants in the council discussions. They examined what was considered valid and what not, and which types of actors were perceived as representative of valid, legitimate discourses. They also investigated what could not be said or was considered inappropriate and by whom. Besides identifying topics and linking those to discourses, the researchers explored which stakeholders perceived which discourses as either positive or negative.

The method allowed the researchers to identify the different themes present in the data and the perspective from which the themes had been presented. The theme 'environmental protection' was identified, for example, and linked to diverse discourses such as 'indigenous understanding', 'traditional management' or 'technical management'. This type of analysis allowed an organisation of the data that then helped to identify which discourses were considered valid and by whom. It provided insights into what power positions are associated with which discourses, and how power dynamics are embedded in discursive framings.

Since the purpose of the study was to assess each forum's capacity for inclusion of different actors, critical discourse analysis was found to be a most appropriate method. Indeed, the literature focusing on issues related to power distribution frequently has recourse to critical discourse analysis. The method is, however, very time consuming, since the data need to be read and classified multiple times – first to complete the list of themes, then to identify diverse perspectives on themes and finally to assess and organise the values attributed to the different discourses. While this fine analysis makes the method challenging to use, it provides nuanced data that allow one to understand the context specificities that explain why certain discourses are closer to power than others.

The method helped to expose the different discourses at play in the same forum, i.e. the different voices present and the relations between those voices in terms of power. Indeed, some discourses occupied more 'talking space' than others and were more likely to be guiding action than others. Certain voices were excluded: some of the participants said the forums did not provide them with the opportunity to express their vision of the environment in their own terms, which the researchers could observe through this method. In general terms – and despite the differences between the four forums studied – it can be said that technical and scientific language dominated in the forums. In some of the forums where indigenous peoples participated, they felt

(Continued)

the forum was not a space in which their vision of management could be expressed. However, some of the previously excluded actors appropriated dominant discourses (technical, scientific) and used these to their advantage. Some of the environmentalists participating in the forums, for example, used technical knowledge to defend their position on the maximum volume of water that could be extracted from a river, and their concerns – expressed in technical terms – were taken into consideration by the rest of the participants.

Through discourse analysis, the researchers showed that discourses are actively transformed and performed as they are used to put forward different interests. However, interests that cannot be expressed in terms of scientific, technical or expert discourses, such as indigenous understandings, were difficult to integrate into the forums. The adoption of a critical perspective on discourse analysis made it possible to distinguish how discourses played out differently in each of the cases studied and how these discourses carried the weight of history in each of the countries studied. The researchers observed the dominance of technical discourses on irrigation in Peru

time. Through narratives, people make sense of events and experiences in order to orientate themselves and respond to events in the world. Through narrative we create coherence and unity from many different forces present in the context in which we live and act. Qualitative content analysis methods in an SES context focus on how meaning is attributed to certain situations and experiences and how that has material consequences in terms of access to and the governance of resources.

The different methods listed in Table 19.1 seek to explore how meaning emerges from texts and narratives and what can be inferred from these emergent patterns of meaning. As mentioned in the 'Introduction', the notion of 'meaning' varies from one approach to the next, depending on whether the method assumes that meaning is explicit and objectively present in the data (e.g. in the form of responses to interview questions) or whether meaning needs to be interpreted. Meaning can also be embedded in discursive practices that distribute power and define dominant or more marginal positions, i.e. the identity and social positioning of different actors are tightly linked to what they consider as the appropriate meaning of a resource or phenomenon.

Limitations

Interpretive and qualitative methods of content analysis are mostly open-ended inquiries where participants have more control over the content of the data collected. The nature of the research output changes, as do the challenges associated with generating these outputs. In a sense the researcher deals with 'warm data', i.e. the subjective perceptions of participants and researchers, relational interdependencies between different actors and human–nature interdependencies, and the contextual experiences of the participants. This makes the researcher's task of analysing the content challenging as it is difficult to verify the results objectively against the scenarios stated by the respondents. As a result, the reliability and validity of the research will not be verified in terms of its reproducibility, but in terms of whether or not the findings generated by the researcher provide deeper insights to synergise general themes.

where the forums existed for a short period and still struggled to establish themselves as a permanent institution. In Brazil, where the forums had existed for longer, they seemed better able to accommodate different perspectives.

Critical discourse analysis was particularly useful to investigate the power dynamics and the distribution of roles in natural resource management. It situated issues of access to resources and their management in a historical perspective as the dominant discourses could be identified through time. It therefore made it possible to pinpoint changes in terms of who used dominant discourses and which discourses became valid. In addition, using critical discourse analysis in association with thematic analysis allowed the researchers to identify which topics are considered interesting within a given discourse.

The main challenge remained the time to analyse large amounts of data collected through interviews and observation notes. Only with the support of software was it possible to identify the co-occurrence and patterns of association of themes related to sentiments, and of discourses related to topics and actors.

Through this process, the researcher can uncover counter-discourses that inform mainstream conceptions of the phenomena under investigation. The transparency of the methods used allows the researcher to go beyond data collection and to analyse content by questioning how themes were identified, how discourses are distinguished from one another and how the researcher's position influenced particular results.

No perfect measure can be developed to remove the subjectivity of human experience. The researcher should therefore adopt a critical and reflexive attitude to identify and deal with the intersubjective biases and blind spots that inevitably arise when interpreting the modes and methods participants use to make sense of and create meaning from their lived experiences. The ethics of interpreting another person's lived experience can sometimes be challenging and could confront the researcher with uncomfortable situations. Allowing participants to read through the data and analyses and provide feedback on a researcher's interpretation of their responses will allow researchers to check for inconsistencies and reflect on their own assumptions. It will also indicate whether researchers should re-analyse their findings.

Key qualities required of a researcher are open-mindedness, patience, empathy, insight into human nature, emotional maturity and the willingness to enter into, and respond to, the participant's world.

Resource implications

Contrary to quantitative methods of data analysis, such as methods using objective coding or data-mining tools, interpretive and qualitative methods of content analysis do not rely on well-established software programs to run data analyses. Deriving the meaning of texts and narratives calls for good judgement, and careful reading with and against the dominant use of language and structural forms of meaning creation.

It is also likely that researchers will collect a much broader range of data than they can actually use in any specific research article. Because interactions need to be examined in fine-grained detail, the analysis will probably involve 'deep dives' into the data to identify critical incidents or interactions that are particularly revealing of the processes examined.

Researchers need to plan for having enough time to go through the data and find patterns of meaning. The process of analysing which words are associated with which, which sentiment is associated with certain words or ideas, and uncovering who defends which positions and why, could be very time consuming.

Good data collection is essential for sound data analysis. Training in computer-assisted qualitative analysis software (CAQDAS) such as NVivo, Atlas.ti, MaxQDA or Python might be essential. CAQDAS is often used to help manage and code very large amounts of data in well-organised ways as a first step in the process of doing narrative or discourse analyses. These software packages also provide automatic tools to search for or count words, and to see patterns in coding (e.g. which pieces of text were coded under several codes). This helps the researcher to identify associations that can support or reject different interpretations. Learning how to use these software programs requires a time investment and running them requires a financial one.

New directions

Responding to the limitations of discourse analysis and critical discourse analysis, the method of discourse practice analysis is on the rise. It acknowledges the difficulty of appropriating truth claims in qualitative interviews and focus groups by drawing on the assumption that people believe their own lies and build their sense-making and meaning-creation practices on paradigmatic references by hearsay. This assumption is based on the work of French literature scholar and author Pierre Bayard (2010). He related this phenomenon to the manner in which people create 'idiosyncratic' discourses about events and experiences they did not have in the same manner that they speak about books they have never read. The analysis of qualitative data in discourse practice analysis explores foci of attention, semantics and important distinctions to understand how the relevant stories are built and how they feed into the narratives in use. The discourse practice analysis reveals the so-called realms of possibilities for social systems which facilitate their ability to change.

Key readings

Bryman, A. 2016. 'Language in Qualitative Research.' In *Social Research Methods* (5th ed), 525–544. Oxford: Oxford University Press.

Dryzek, J. 1997. *The Politics of the Earth: Environmental Discourses.* Oxford: Oxford University Press.

Fairclough, N. 2003. *Analysing Discourse: Textual Analysis for Social Research.* London: Routledge.

Hajer, M.A., M. van den Brink, M., and T. Metz. 2006. 'Doing Discourse Analysis: Coalitions, Practices, Meaning.' In *Words Matter on Policy and Planning: Discourse Theory and Method in the Social Sciences*, edited by M. Brink and T. Metze, 65–74. Utrecht: Koninklijk Nederlands Aardrijkskundig Genootschap. http://hdl.handle.net/11245/1.289572.

Weiss G., and R. Wodak. 2003. 'Introduction: Theory, Interdisciplinarity and Critical Discourse Analysis.' In *Critical Discourse Analysis*, edited by G. Weiss and R. Wodak, 1–32. London: Palgrave Macmillan. doi:10.1057/9780230514560_1.

References

Bayard, P. 2010. *How to Talk about Books You Haven't Read.* New York: Bloomsbury.

Blommaert, J. 2005. *Discourse – A Critical Introduction.* Cambridge: Cambridge University Press.

Boje, D.M. 2008. *Storytelling Organizations.* Thousand Oaks: Sage.

Boje, D.M. 2011. 'Introduction to Agential Antenarratives that Shape the Future of Organizations.' In *Storytelling and the Future of Organizations: An Antenarrative Handbook*, edited by D.M. Boje, 1–19. New York: Routledge.

Boyatzis, R. 1998. *Transforming Qualitative Information: Thematic Analysis and Code Development*. Thousand Oaks: Sage.

Braun, V., and V. Clarke. 2006. 'Using Thematic Analysis in Psychology.' *Qualitative Research in Psychology* 3(2): 77–101.

Bruner, J. 1987. 'Life as Narrative.' *Social Research* 54: 12–32.

Cortazzi, M. 1993. *Narrative Analysis*. London: Falmer Press.

De Saussure, F. 1974. *Course in General Linguistics*. London: Fontana.

Dryzek, J. 1997. *The Politics of the Earth: Environmental Discourses*. Oxford: Oxford University Press.

Edson, M.C., and L. Klein. 2016. 'Problem Structuring and Research Design in Systemic Inquiry.' In *A Guide to Systems Research: Philosophy, Processes and Practice, Volume 10*, edited by M.C. Edson, P. Buckle Henning, and S. Sankaran, 59–80. Dordrecht: Springer.

Feindt, P.H., and A. Oels. 2005. 'Does Discourse Matter? Discourse Analysis in Environmental Policy Making.' *Journal of Environmental Policy & Planning* 7(3): 161–173. 10.1080/15239080500339638.

Foucault, M. 1974. *The Order of Things: An Archaeology of the Human Sciences*. New York: Vintage.

Gill, M.J. 2014. 'The Possibilities of Phenomenology for Organizational Research.' *Organizational Research Methods* 17(2): 118–137.

Hajer, M.A., M. van den Brink, M., and T. Metz. 2006. 'Doing Discourse Analysis: Coalitions, Practices, Meaning.' In *Words Matter on Policy and Planning: Discourse Theory and Method in the Social Sciences*, edited by M. Brink and T. Metze, 65–74. Utrecht: Koninklijk Nederlands Aardrijkskundig Genootschap. http://hdl.handle.net/11245/1.289572.

Hajer, M., and W. Versteeg. 2005. 'A Decade of Discourse Analysis of Environmental Politics: Achievements, Challenges, Perspectives.' *Journal of Environmental Policy & Planning* 7(3): 175–184.

Huitema, D. 2002. *Hazardous Decisions: Hazardous Waste Siting in The UK, the Netherlands, and Canada*. Dordrecht: Kluwer Academic Publishers.

Jørgensen, K., and C. Largacha-Martinez. 2014. *Critical Narrative Inquiry: Ethics, Sustainability and Action to Critical Narrative Inquiry Storytelling, Sustainability and Power*. New York: Nova Science Publishers.

Jupp, V., and C. Norris. 1993. 'Traditions in Documentary Analysis.' In *Social Research – Philosophy, Politics and Practice*, edited by M. Hammersley, 37–51. London: Sage.

Klein, L., and C.A.P. Weiland. 2014. 'Critical Systemic Inquiry: Ethics, Sustainability and Action.' In *Critical Narrative Inquiry: Storytelling, Sustainability and Power*, edited by K.M. Jørgensen and C. Largacha-Martinez, 145–158. New York: Nova Science Publishers.

Kurtz, C. 2014. *Working with Stories* (3rd ed). Creative Commons, Kindle edition.

Lejano, R., M. Ingram, and H. Ingram. 2013. *The Power of Narrative in Environmental Networks*. Cambridge: MIT Press.

Lindow, M. 2017. 'Exploring Resilience Capacities through the Art of Storymaking: The Case of Food Innovators in the Western Cape.' MPhil diss., Stellenbosch University. http://hdl.handle.net/10019.1/102910.

Mancilla García, M., and Ö. Bodin. 2019. 'Participatory Water Basin Councils in Peru and Brazil: Expert Discourses as Means and Barriers to Inclusion.' *Global Environmental Change* 55: 139–148. doi:10.1016/j.gloenvcha.2019.02.005.

Parker, I. 1992. *Discourse Dynamics*. London: Routledge.

Paschen, J-A., and R. Ison. 2014. 'Narrative Research in Climate Change Adaptation—Exploring a Complementary Paradigm for Research and Governance.' *Research Policy* 43(6): 1083–1092.

Potter, J. 1996. *Representing Reality: Discourse, Rhetoric and Social Construction*. London: Sage.

Reid, K., P. Flowers, and M. Larkin. 2005. 'Exploring Lived Experience: An Introduction to Interpretative Phenomenological Analysis.' *The Psychologist* 18(1): 20–23.

Sitas, N., Z.V. Harmáčková, J.A. Anticamara, A. Arneth, R. Badola, R. Biggs, R. Blanchard et al. 2019. 'Exploring the Usefulness of Scenario Archetypes in Science-Policy Processes: Experience Across IPBES Assessments.' *Ecology and Society* 24(3): 35. doi:10.5751/ES-11039-240335.

Smith, J.A., P. Flowers, and M. Larkin. 2009. *Interpretative Phenomenological Analysis*. Thousand Oaks: Sage.

Stojanovic, T., H. McNae, P. Tett, T.W. Potts, J. Reis, H.D. Smith, and I. Dillingham. 2016. 'The "Social" Aspect of Social-Ecological Systems: A Critique of Analytical Frameworks and Findings from a Multisite Study of Coastal Sustainability.' *Ecology and Society* 21(3): 15. doi:10.5751/ES-08633-210315.

Weiss, G., and R. Wodak. 2003. 'Introduction: Theory, Interdisciplinarity and Critical Discourse Analysis.' In *Critical Discourse Analysis*, edited by G. Weiss and R. Wodak, 1–32. London: Palgrave Macmillan. doi:10.1057/9780230514560_1.

20

Comparative case study analysis

Claudia Pahl-Wostl,[1] Xavier Basurto[2] and Sergio Villamayor-Tomas[3]

[1]INSTITUTE OF GEOGRAPHY, UNIVERSITY OF OSNABRÜCK, OSNABRÜCK, GERMANY
[2]NICHOLAS SCHOOL OF THE ENVIRONMENT, DUKE UNIVERSITY, DURHAM, NORTH CAROLINA, USA
[3]INSTITUTE OF ENVIRONMENTAL SCIENCE AND TECHNOLOGY, AUTONOMOUS UNIVERSITY OF BARCELONA, BARCELONA, SPAIN

Key methods discussed in this chapter

Variable-oriented analysis, archetype analysis (formal concept analysis, qualitative comparative analysis)

Connections to other chapters

This chapter refers to direct comparisons guided by social-ecological systems frameworks. A prominent SES framework was developed by Elinor Ostrom and contributors of the institutional analysis tradition (Chapter 22). The vulnerability and sustainability livelihood framework (see Chapter 32) has often been used for comparative work. This chapter also connects with data mining and pattern recognition (Chapter 17).

Introduction

Individual, in-depth case study analyses are quite common in SES research. These analyses provide a deeper understanding of the complexity of social-ecological systems (SES). However, they often deliver case-specific insights and are thus limited in their potential for theory development. Structured case study comparisons are a way forward to leverage theoretical lessons from particular cases and elicit general insights from a population of phenomena that share certain characteristics (Pahl-Wostl 2015).

This chapter focuses on comparative analyses comprising a small or medium number of cases (i.e. 2 to about 30) and qualitative data. It first does so by distinguishing between variable-oriented and case-oriented analyses (Ragin 2004). The aim of variable-oriented analysis is to establish potentially generalisable relationships between features of cases (i.e. variables). These structured comparisons of variables can be guided by SES frameworks (Binder et al. 2013) that facilitate a comparable representation of different case studies. Social-ecological systems frameworks can host comparisons among small numbers of cases and also guide 'qualitative meta-analyses' of a larger number of cases (Rudel 2008).

 DOI: 10.4324/9781003021339-24

SUMMARY TABLE: COMPARATIVE CASE STUDY ANALYSIS

DISCIPLINARY BACKGROUND	KNOWLEDGE TYPE
The methods in this chapter are derived from or have most commonly been used in: Comparative Sociology, Comparative Politics	*The methods in this chapter are primarily used to generate the following types of knowledge:* • Descriptive • Exploratory • Explanatory
RESEARCH APPROACH	**PURPOSE OF METHOD**
The methods in this chapter originate from or most commonly adopt the following research approaches: • Analytical/objective • Interpretive/subjective	*The most common purposes of using the methods in this chapter are:* • System understanding
TEMPORAL DIMENSION	**SYSTEMIC FEATURES AND PROCESSES**
The methods in this chapter are most commonly applied to the following temporal dimensions: • Present (typically within the last 5–10 years) • Recent past (post-1700s)	*While most methods can do many things, the methods in this chapter are particularly good (i.e. go-to methods) for addressing the following:* • Transformation • Collective action and collaborative governance
SPATIAL DIMENSION	
The methods in this chapter are primarily either or both: • Explicitly spatial *The methods in this chapter are most commonly applied at the following spatial scales:* • Local • Multiple places/sites around the world	

Case-oriented research seeks to understand complex interdependencies rather than simple cause-and-effect relationships. To explain a certain characteristic of an SES (e.g. adaptive capacity), one searches for typical configurations of variables. Here qualitative comparative analysis (QCA) and other qualitative archetype analysis techniques such as formal concept analysis (FCA) have gained increasing popularity in recent years (Ragin and Sonnett 2004; Oberlack et al. 2019). Archetype analysis encompasses a variety of mostly quantitative techniques to identify patterns of complex causation (archetypes of models) and/or state conditions (archetypes of traits).

Qualitative comparative analysis was developed by Charles Ragin, a comparative sociologist, and originates from comparative sociology and comparative politics. Ragin became frustrated by the lack of an analytical technique that dealt with limited diversity (there tends to be less diversity among sets of cases than theoretically possible) and equifinality (i.e. multiple paths leading to the same outcome).

Formal concept analysis (originally *Formale Begriffsanalyse* in German) is a method used for knowledge representation, information management and data analysis. It was devised in the early 1980s by Rudolf Wille, a German mathematician. He defined 'formal concept' as a unit of analysis comprising a set of objects and their shared attributes (i.e. sets of attributes as they co-occur across groups of cases).

Both qualitative comparative analysis and formal concept analysis allow researchers to group cases and identify configurations of variables as they explain outcomes. Qualitative comparative analysis is a distinct method in the archetype family. It relies on set theory and Boolean algebra, which allow synthesising the configurations to their minimum expression. Formal concept analysis is based on assessing the co-occurrence of variables across cases, which in turn makes it possible to identify subsets of closely related variables within the configurations (Oberlack et al. 2016; Villamayor-Tomas et al. 2020). Conditions that influence the choice of the variable or the case-oriented approaches to comparative analysis range from practical considerations such as stages in a research cycle and the data and methodological skills available to epistemological preferences.

As hinted above, comparative case study research has great potential to build and test middle-range SES theories, i.e. explanations that are generalisable in specific SES contexts. This is accomplished in variable-oriented research by distinguishing between explanatory variables and scope variables (also called 'state variables' or 'parameters'). Explanatory variables feature explanations, whereas scope variables set the empirical boundaries of the explanations (i.e. of generalisations). In case-oriented research the contextualisation of generalisations is less explicit and usually embedded in the description of the phenomenon being studied (e.g. deforestation due to agricultural expansion in small tropical forests) and/or in the scaling of measurements (e.g. small size defined according to the standards for tropical forests).

SES problems and questions

Variable-oriented comparisons (i.e. both small-*n* and meta-analyses) have tended to address a wide variety of questions about variable-to-outcome effects. These comparisons can be grouped depending on whether they inquire about influences of social variables on ecological outcomes, the influence of ecological variables on social outcomes, or the reciprocal impact of social and ecological variables (Binder et al. 2013). The questions and social-ecological ambition of comparative analyses are reflected in the SES frameworks used. A fair number of works on ecosystem services have addressed the impact of social variables on ecological variables (i.e. services) in comparative perspective. Typical research questions focus on the effects of different land-use strategies on specific services or a variety of services (Schwenk et al. 2012; Ray et al. 2015), and across contexts or scenarios (Fontana et al. 2013; Matthews et al. 2014).

The vulnerability and sustainability livelihood framework (see Chapter 32) has often been used for comparative work addressing the impact of ecological variables on social variables. Comparative studies have mostly taken the form of assessments focusing on social vulnerability to different drivers of change (e.g. climate change and variability, globalisation, migration) and/or in different contexts (e.g. rural areas in a region, cities, coastal zones). There has also been an interest in the influence that different sensitivity variables (e.g. resource dependence, livelihood strategies) and adaptive capacity variables (e.g. education, income, livelihood assets) have on this vulnerability (Turner et al. 2003; Eakin 2005; Simelton et al. 2009). In addition, studies have compared different livelihood strategies in or across local communities and explained those strategies by looking at the influence of biophysical assets (Ferrol-Schulte et al. 2013; Córdova, Hogarth, and Kanninen 2018), among others.

Various comparative works assess reciprocal relations between social and ecological variables. In the water management context, for example, the management and transition framework has inspired studies of environmental change, adaptive management and social learning across basins and countries (Knieper et al. 2010; Kranz, Menniken, and Hinkel 2010). The SES framework was created in response to the comparability issue mentioned in the introduction to this chapter and has since been used to articulate a fair number of studies on the fit between ecological and institutional diversity and cooperation dynamics at local and larger scales (Ostrom 2009; Leslie et al. 2015) (see Chapter 22 for more details).

All the examples above refer mostly to cross-case comparisons, but there are also cross-time comparisons, which can be embedded in cross-case comparisons or just in a single case (Yin 2014). Not many analyses make cross-case comparisons explicit. A number of studies relying on the SES framework have started to do so to assess natural resource management decentralisation processes (Baldwin et al. 2015; Chavez et al. 2019) or the emergence of international environmental regimes (Fleischman et al. 2014; Villamayor-Tomas et al. 2016).

Case-oriented comparisons like those carried out with qualitative comparative analysis and formal concept analysis are particularly well suited to analysing the emergence of phenomena (i.e. how conditions lead to or are associated with outcomes). Neither qualitative comparative analysis nor formal concept analysis looks into processes explicitly, but requires that the researcher thinks and draws hypotheses about them. Rudel (2008), for instance, identified sets of conditions for deforestation around the world and concluded, among other things, that deforestation has shifted from a state-initiated to an enterprise-driven process in the last decades; Basurto's (2013) studies of community-based biodiversity conservation in Costa Rica showed that pathways available for the emergence of collective action for conservation are more limited than those available to maintain it; and Oberlack et al. (2016) identified seven typical processes through which large-scale land acquisitions affect local livelihood vulnerability around the world.

Case-oriented comparisons can also be applied to address integrated system design or institutional fit questions because they make explicit reference to the appropriate combination of different measures to achieve a certain outcome or the conditions under which a certain intervention might achieve a certain outcome (Lam and Ostrom 2010; Roggero 2015; Baggio et al. 2016). Lam and Ostrom (2010), for example, assessed the performance of an innovative governmental programme for irrigation infrastructure improvement in Nepal and found that the effectiveness of the programme was contingent on the existence of strong community-based irrigation associations and local entrepreneurs. Baggio et al. (2016) tested Elinor Ostrom's eight design principles theory for community-based natural resource management and concluded that the relevance of some of the principles depended to a great extent on the natural and hard human-made infrastructure available (see also Chapter 22). Roggero (2015) explored the feasibility of coordination among local administrations for the implementation of adaptation plans and found that this coordination can be accomplished if administrators share worldviews and values and have sufficient discretion to make decisions that take one another's interest into account.

Finally, case-based comparisons can support the synthesis of SES features, processes and/or outcomes into types for descriptive and/or explanatory purposes. Pahl-Wostl and Knieper (2014), for example, used qualitative comparative analysis to explore variation of water-governance regimes along the two dimensions of decentralisation and coordination. They used the results to confirm the existence of four ideal-typical configurations (i.e. one per combination of features) that are more or less prone to facilitate adaptive capacity (see also Oberlack et al. 2016 for an example).

On a more general note, it is important to note the potential of comparative case studies to develop theory. Although comparative case studies can be used to test theory, the research process usually does not end there. The relative depth of knowledge gained from cases usually provides details that allow researchers to qualify theories (e.g. about the conditions that are fulfilled), which translates into new hypotheses for testing. This is clear in the Basurto (2013) and Baggio et al. (2016) studies mentioned above, for example.

Brief description of key methods

Variable-oriented comparisons strongly rely on counterfactual analysis, i.e. Mill's comparative methods of most similar and most different systems design (Toshkov 2016). In a most similar systems design, the researcher assesses whether differences between otherwise very similar cases correlate with variation in outcomes. In the most different systems design, the researcher looks for similarities across otherwise very different cases that nevertheless have similar outcomes. Whether in the form of small-*n* or 'qualitative meta-analyses', the replicability of these direct comparisons can benefit from some tools, including variable books, two-way tables, and rule- and case-based reasoning (Table 20.1).

Table 20.1 Summary of key methods used in comparative case study analysis

Method	*Description/tools*	*References*
Direct comparisons guided by an SES framework	Direct comparisons (Mill's comparative methods; most similar and most different systems design) require variable books that specify definitions and operationalisations of attributes, and ensure the comparability of cases. They can also include explanations of theoretical importance and examples. Two-way tables are useful as they facilitate visualisation and the identification of differences across cases in terms of the SES framework variables. When using case-based reasoning, the relevance of these differences with regard to whether outcomes are also different is important.	*Key introductory texts* Yin 2014; Toshkov 2016 *Applications to SES* Basurto, Gelcich, and Ostrom 2013; Epstein et al. 2013; Fleischman et al. 2014

Method	Description/tools	References
	Rule-based reasoning uses the variable differences to predict or explain outcomes on the basis of one or more ecological or social mechanisms (rules) that link those facts to an outcome (Cox 2011).	
Qualitative comparative analysis	Input data for qualitative comparative analysis are represented in matrix form so that each row represents a case and each column represents one of the defined attributes and the outcome. Each case can therefore be seen as a configuration of present or absent attributes. Truth tables synthesise raw data matrices by collapsing all cases with the same configuration and then adding a column with the count of cases per row (i.e. configurations). Then Boolean algebra is used to synthesise the configurations into their minimal expression (i.e. solutions). The impact of attributes on the outcome is assessed in terms of necessity and sufficiency. The attributes contained in the solutions are INUS (insufficient but necessary part of an unnecessary but sufficient configuration). The explanatory capacity of solutions can be assessed through measures of consistency (the percentage of cases with the outcome out of those matching a solution) and coverage (the percentage of cases that match a solution out of all cases with the outcome).	*Key introductory texts* Ragin 2008; Rihoux and Ragin 2009; Schneider and Wagemann 2010 *Applications to SES* Sutton and Rudd 2015; Villamayor-Tomas, Iniesta-Arandia, and Roggero 2020
Formal concept analysis	Formal concept analysis is similar to truth tables in its purpose. As in qualitative comparative analysis, input data are represented by a matrix of cases and attributes (here called a context). A formal concept is defined as the co-occurrence of a set of attributes in a set of cases. Set theory algebra allows one to reason about the nesting of concepts into one another. Less detailed (i.e. in number of attributes involved) concepts are supersets of more detailed concepts. This enables the construction of concept lattices. A concept lattice is a hierarchical relationship of all the concepts of a context in the form of a line diagram. The attributes and groups of cases of each concept are displayed in the nodes. The nodes are in turn linked to one another depending on whether they are part of the same formal concepts or not. 'Higher' attributes in the network are supersets of lower attributes; by the same token, the size of the groups of cases decreases from the 'top down'.	*Key introductory texts* Ganter and Wille 2012; Škopljanac-Mačina and Blaškovic 2014 *Applications to SES* Oberlack et al. 2016; Oberlack and Eisenack 2018

Case-oriented analyses, in turn, aim to identify recurrent sets of variables (i.e. configurations of attributes) that are associated with outcomes. Qualitative comparative analysis and formal concept analysis both rely on tabular input data (i.e. two-way tables of cases and attributes) and set theory (Table 20.1). Qualitative comparative analysis provides a systematic approach to identifying how the combination of certain conditions leads to an outcome, or how different configurations of conditions lead to the presence of an outcome (resilience, breakdown of a system, cooperation, regime shift, etc.) or its absence. In contrast to regression analyses, qualitative comparative analysis works much better if the scholar has a certain familiarity with cases since one deviant case may lead to the re-evaluation of the conditions associated with the outcome and the potential causal pathways. This iterative process encourages explicit conversation between empirical evidence and theory, and constitutes a particular strength of the method.

In formal concept analysis, a data matrix can contain many concepts. Formal concept analysis allows one to find, visualise and order concepts hierarchically (less detailed concepts encompass more detailed ones). Formal concept analysis has been applied in various fields such as mathematics, medicine, biology, sociology, psychology and economics, mostly for descriptive purposes (Škopljanac-Mačina and Blaškovic 2014). The advantage of formal concept analysis compared to other data representation methods is that it provides information about attribute interdependencies (e.g. dendrograms) as well as a hierarchy of those attributes according to their relevance (e.g. tree diagrams).

Finally, it is worth mentioning process tracing. This is not a technique that allows comparative analysis per se but it can complement both direct comparisons and configurational analyses. Process tracing consists of the identification and questioning of the sequence of events that are supposed to connect cases and outcomes (Collier 2011). In direct comparisons, process tracing is frequently used to confirm that covariance among variables actually reflects causality. In qualitative comparative analysis, process tracing enhances the identification of causal mechanisms and allows addressing the implications that arise from deviant cases (Schneider and Rohlfing 2013). In formal concept analysis, process tracing can be used to examine the relationships that are behind the co-occurrence of variables in explanations (e.g. whether they reflect interaction effects, multiple chains of causality).

Limitations

Small-*n* comparative analysis has some limitations that apply to both variable- and case-oriented analyses. An important limitation has to do with the 'too many variables, too few cases', roughly akin to a 'degrees of freedom' issue (which is particularly acute in SES research). A high ratio of variables to cases translates into difficulties to find counterfactuals and make inferences in variable-oriented analyses. In case-oriented analyses, the number of possible configurations increases exponentially with the number of variables. This poses problems to meaningfully synthesise the configurations that are actually observed in the cases at hand. The challenge can be overcome by carefully selecting the variables being studied. In variable-oriented analyses, the selection can benefit from clearly distinguishing scope and explanatory variables, i.e. a good conceptualisation of estate parameters or distant causes, and proximate causes. In case-oriented analyses, it is usually useful to select variables based on expectations about how the interactions of these variables affect outcomes.

Then, there is the issue of data standardisation. Comparisons require the aggregation of deep case study knowledge, which involves missing information and issues of replicability. In variable-oriented analysis, variable books can include some of the missing information as well as information to ensure replicability, but only to some extent. In case-oriented analysis, standardisation is more challenging because the scale of variables is set relative to the cases at hand, i.e. they are calibrated. A few works have recently emerged to guide calibration of this nature and set transparency standards (Basurto and Speer 2012).

Neither variable-oriented nor case-oriented analysis methods have been developed to study dynamics. There is thus a lack of guidelines about how to carry out comparisons over time (e.g. how to select time periods and units) and about the difference between over-time comparisons (e.g. comparing well-delimited time periods versus historical analyses, such as characterising a development over time) (Bartolini 1993).

Variable-oriented and case-oriented comparisons also have their own limitations. Variable-oriented comparisons are not particularly suited to assessing interaction effects (at least not as suited as case-oriented comparisons). This is because most similar and most different systems designs are based on isolating the effects of potentially relevant attributes rather than exploring their joint effects. Case-oriented analysis (e.g. qualitative comparative analysis), in turn, has not been generally used to test hypotheses but to build them. Testing hypotheses would require thinking in terms of conditional effects (e.g. variable A has an effect on variable B depending on the levels of variable C), which is a promising but so far rather unexplored inroad (Yamakasi 2003; Hellström 2011). In case-oriented analysis, a related limitation is the fact that it is challenging to make sense of variable configurations because of the general lack of theories that inform interaction effects, particularly if configurations are very large or complex.

Resource implications

Case comparisons can be resource demanding if data collection requires fieldwork at different field sites. In variable-oriented analysis, case selection can be particularly resource consuming. Both the most similar and the most different systems designs require prior knowledge and a good understanding of which variables will operate as scope or context conditions and which will be more relevant for explanation. Also, the development of variable books in variable-oriented analyses can be time consuming and may require overcoming coordination problems if the data from the cases are collected by different scholars (Cox et al. 2020).

In qualitative comparative analysis, data calibration requires a very good understanding of how the cases at hand vary and the resources needed to revise cases as the analysis unfolds (e.g. to understand deviant cases and/or refine calibrations). Also, both qualitative comparative analysis and formal concept analysis can benefit from the use of software (see fsQCA, R, Concept Explorer) but these need to be learnt. By the same token, training in basic Boolean algebra is essential for a meaningful application of both qualitative comparative analysis and formal concept analysis methods, and the use of software. As with statistics, the methods should not be used as a 'black box', but should be based on a good understanding of the underlying concepts, benefits and trade-offs.

New directions

Progress in comparative analysis is being made on several fronts. In variable-oriented analysis, an increasing number of case study scholars are becoming aware of the importance of guaranteeing the comparability of their findings and are progressively improving the transparency and interoperability of their data. Platforms like the SES Library at ASU (seslibrary.asu.edu), the Social-Ecological Systems Meta-Analysis Database (SESMAD) at Dartmouth

Case study 20.1: The influence of water governance on capacity to adapt to climate change

Failure at multiple levels of governance lies at the core of many water crises. Despite increasing scholarly research on water governance and efforts towards policy reform, the overall situation has not substantially improved and major transformation in water governance is yet to be implemented. Numerous recommendations, often relying on simplistic 'standard' panaceas, have been put forward for water governance reform without testing their appropriateness in diverse socio-economic and environmental contexts. A diagnostic approach and more systematic comparative case study analyses are urgently needed to improve this situation.

Pahl-Wostl et al. (2012) conducted the first comprehensive comparative analysis of complex water governance and management systems in national river basins, compiling insights from 29 basins in developed and developing/emerging countries. The researchers used qualitative approaches and statistical analyses to analyse the interdependencies between the water-governance regime, regime performance, and the environmental and socio-economic context. The example presented here takes this analysis one step further using fsQCA to analyse the importance of polycentricity for performance with respect to climate change adaptation. The analysis built on the hypotheses that ideal types, as depicted in Figure 20.1 (from Pahl-Wostl and Knieper 2014), have a strong explanatory power with respect to understanding the influence of regime configurations on regime performance. The figure depicts the categorisation of governance regimes in a two-dimensional grid of distribution of power and degree of coordination/cooperation. The shaded boxes in the corners denote the ideal-typical configurations.

Case study data were provided by regional experts by means of a questionnaire comprising 81 indicators. For the application of fsQCA, the original score-based data-set had to be converted to fuzzy-set membership values (calibration). These indicators were then aggregated to obtain values for the conditions to be included in the analyses. The conditions included distribution of power (DIS), vertical coordination (VCOR), horizontal coordination (HCOR) and adaptive capacity (ADAP).

College (sesmad.dartmouth.edu) and the Illuminating Hidden Harvests project by Duke University, the Food and Agriculture Organization (FAO) and WorldFish (fao.org/voluntary-guidelines-small-scale-fisheries/ihh/en) are references in that regard. Another promising front in variable-oriented analysis is the systematic study of interactions and processes (Cumming et al. 2020; Villamayor-Tomas et al. 2020). Case study methods are particularly suitable for studying interactions and processes; however, despite recent improvements, these methods tend to be obscured by deeply descriptive and relatively unstructured narratives. Social-ecological systems frameworks (e.g. Ostrom's SES framework) and theory on types of interactions and processes can assist in structuring findings and accumulating knowledge (Villamayor-Tomas et al. 2020).

Case-oriented comparisons in the form of configurational analyses are also blooming. Recent works on syndromes of sustainability (Manuel-Navarrete, Gómez, and Gallopín 2007) and archetypes (Oberlack et al. 2016; Eisenack et al. 2019) are good examples of that. There is

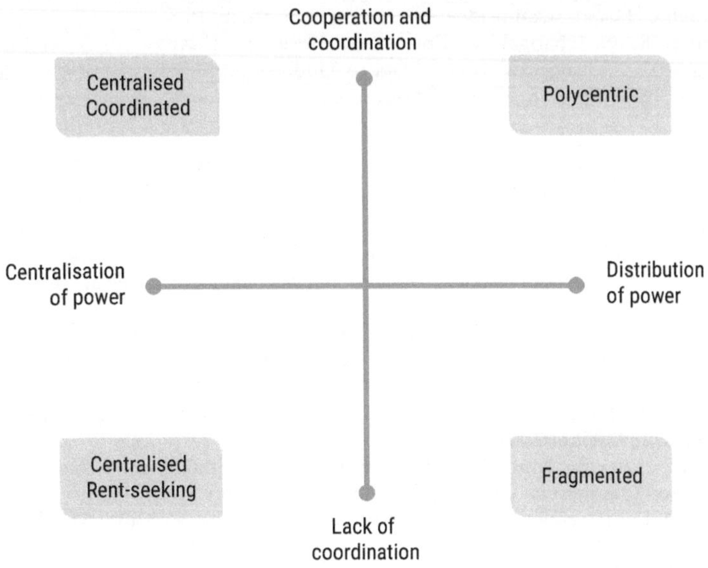

Figure 20.1 Categorisation of governance regimes (Pahl-Wostl and Knieper 2014)

fsQCA was then used to test models for high (ADAP) or low (adap) adaptive capacity, respectively: ADAP = f(VCOR, HCOR, DIS) and adap = f(VCOR, HCOR, DIS). The results confirmed the hypotheses regarding the influence of different regime configurations. They showed that a set of factors associated with polycentricity has the highest explanatory power for high performance regarding climate change adaptation. Factors associated with fragmented and centralised regimes can be identified for paths leading to low performance. Furthermore, the analysis identified the effectiveness of formal institutions as an important condition, in particular for paths leading towards low performance.

remarkable potential to develop innovative approaches in mixed-methods research for comparative analysis by combining variable- and case-oriented research, or small- and large-*n* analyses, for example. Although there have been some advances on this front (Heikkila 2004; Villamayor-Tomas, Iniesta-Arandia, and Roggero 2020), there is still progress to be made, particularly to standardise the methods most suitable to combining, and for specific research purposes.

Key readings

Basurto, X., S. Gelcich, and E. Ostrom. 2013. 'The Social-Ecological System Framework as a Knowledge Classificatory System for Benthic Small-Scale Fisheries.' *Global Environmental Change* 23(6): 1366–1380.

Binder, C., J. Hinkel, P.W.G. Bots, and C. Pahl-Wostl. 2013. 'Comparison of Frameworks for Analyzing Social-Ecological Systems.' *Ecology and Society* 18(4): 26.

Ragin, C. 2000. *Fuzzy-Set Social Science*. Chicago: University of Chicago Press.
Schneider C.Q., and C. Wagemann. 2012. *Set-theoretic Methods for the Social Sciences. A Guide to Qualitative Comparative Analysis*. Cambridge: Cambridge University Press.
Villamayor-Tomas, S., C. Oberlack, G. Epstein, S. Partelow, M. Roggero, E. Kellner, M. Tschopp, and M. Cox. 2020. 'Using Case Study Data to Understand SES Interactions: A Model-centered Meta-analysis of SES Framework Applications.' *Current Opinion in Environmental Sustainability*.

References

Baggio, J., A. Barnett, I. Perez-Ibarra, U. Brady, E. Ratajczyk, N. Rollins, C. Rubiños, H. Shin, D. Yu, and R. Aggarwal. 2016. 'Explaining Success and Failure in the Commons: The Configural Nature of Ostrom's Institutional Design Principles.' *International Journal of the Commons* 10(2): 417–439. doi:10.18352/ijc.634.
Baldwin, E., C. Washington-Ottombre, J. Dell'Angelo, D. Cole, and T. Evans. 2015. 'Polycentric Governance and Irrigation Reform in Kenya.' *Governance* 29(2): 207–225. doi:10.1111/gove.12160.
Bartolini, S. 1993. 'On Time and Comparative Research.' *Journal of Theoretical Politics* 5(2): 131–167. doi:10.1177/0951692893005002001.
Basurto, X. 2013. 'Linking Multi-level Governance to Local Common-pool Resource Theory Using Fuzzy-set Qualitative Comparative Analysis: Insights from Twenty Years of Biodiversity Conservation in Costa Rica.' *Global Environmental Change* 23(3): 573–587. doi:10.1016/j.gloenvcha.2013.02.011.
Basurto, X., and E. Ostrom. 2009. 'Beyond the Tragedy of the Commons.' *Economia Delle Fonti Di Energia e Dell'Ambiente* LII(1): 35–60.
Basurto, X., and J. Speer. 2012. 'Structuring the Calibration of Qualitative Data as Sets for Qualitative Comparative Analysis (QCA).' *Field Methods* 24(2): 155–174. doi:10.1177/1525822x11433998.
Binder, C.R., J. Hinkel, P.W.G. Bots, and C. Pahl-Wostl. 2013. 'Comparison of Frameworks for Analyzing Social-Ecological Systems.' *Ecology and Society* 18(4): 26.
Chavez, C., I. Ivania, S. Partelow, R. Madrigal-Ballestero, A. Schlüter, and I. Gutierrez-Montes. 2019. 'Do Responsible Fishing Areas Work? Comparing Collective Action Challenges in Three Small-Scale Fisheries in Costa Rica.' *International Journal of the Commons* 13(1): 705. doi:10.18352/ijc.923.
Collier, D. 2011. 'Understanding Process Tracing.' *Political Science & Politics* 44(4): 823–830. doi:10.1017/S1049096511001429.
Córdova, R., N. Hogarth, and M. Kanninen. 2018. 'Sustainability of Smallholder Livelihoods in the Ecuadorian Highlands: A Comparison of Agroforestry and Conventional Agriculture Systems in the Indigenous Territory of Kayambi People.' *Land* 7(2): 45. doi:10.3390/land7020045.
Cox, M. 2011. 'Advancing the Diagnostic Analysis of Environmental Problems.' 2011. *International Journal of the Commons* 5(2): 346–363. www.thecommonsjournal.org/index.php/ijc/article/view/273.
Cox, M., S. Villamayor-Tomas, N.C. Ban, G. Epstein, L. Evans, F. Fleischman, M. Nenadovic et al. 2020. 'From Concepts to Comparisons: A Resource for Diagnosis and Measurement in Social-Ecological Systems.' *Environmental Science & Policy* 107: 211–216. doi:10.1016/j.envsci.2020.02.009.
Cumming, G.S., G. Epstein, J.M. Anderies, C.I. Apetrei, J. Baggio, Ö. Bodin, S. Chawla et al. 2020. 'Advancing Understanding of Natural Resource Governance: A Post-Ostrom Research Agenda.' *Current Opinion in Environmental Sustainability* 44: 26–34. doi:10.1016/j.cosust.2020.02.005.
Eakin, H. 2005. 'Institutional Change, Climate Risk, and Rural Vulnerability: Cases from Central Mexico.' *World Development* 33(11): 1923–1938. doi:10.1016/j.worlddev.2005.06.005.
Eisenack, K., S. Villamayor-Tomas, G. Epstein, C. Kimmich, N. Magliocca, D. Manuel-Navarrete, C. Oberlack, M. Roggero, and D. Sietz. 2019. 'Design and Quality Criteria for Archetype Analysis.' *Ecology and Society* 24(3): 6. doi:10.5751/es-10855-240306.
Epstein, G., J.M. Vogt, S.K. Mincey, M. Cox, and B. Fischer. 2013. 'Missing Ecology: Integrating Ecological Perspectives with the Social-Ecological System Framework.' *International Journal of the Commons* 7(2): 432–453. doi:10.18352/ijc.371.
Ferrol-Schulte, D., M. Wolff, S. Ferse, and M. Glaser. 2013. 'Sustainable Livelihoods Approach in Tropical Coastal and Marine Social-Ecological Systems: A Review.' *Marine Policy* 42: 253–258. doi:10.1016/j.marpol.2013.03.007.

Fleischman, F.D., B. Loken, G.A. Garcia-Lopez, and S. Villamayor-Tomas. 2014. 'Evaluating the Utility of Common-Pool Resource Theory for Understanding Forest Governance and Outcomes in Indonesia between 1965 and 2012.' *International Journal of the Commons* 8(2): 304–336. www.thecommonsjournal.org/index.php/ijc/article/view/409.

Fontana, V., A. Radtke, V.B. Fedrigotti, U. Tappeiner, E. Tasser, S. Zerbe, and T. Buchholz. 2013. 'Comparing Land-use Alternatives: Using the Ecosystem Services Concept to Define a Multi-Criteria Decision Analysis.' *Ecological Economics* 93: 128–136. doi:10.1016/j.ecolecon.2013.05.007.

Ganter, B., and R. Wille. 2012. *Formal Concept Analysis: Mathematical Foundations*. New York: Springer.

Heikkila, T. 2004. 'Institutional Boundaries and Common-pool Resource Management: A Comparative Analysis of Water Management Programs in California.' *Journal of Policy Analysis and Management* 23(1): 97–117. doi:10.1002/pam.10181.

Hellström, J. 2011. 'Conditional Hypotheses in Comparative Social Science: Mixed-method Approaches to Middle-sized Data Analysis.' *Methodological Innovations Online* 6(2): 71–102. doi:10.4256/mio.2010.0036.

Knieper, C., G. Holtz, B. Kastens, and C. Pahl-Wostl. 2010. 'Analysing Water Governance in Heterogeneous Case Studies – Experiences with a Database Approach.' *Environmental Science & Policy* 13(7): 592–603.

Kranz, N., T. Menniken, and J. Hinkel. 2010. 'Climate Change Adaptation Strategies in the Mekong and Orange-Senqu Basins: What Determines the State-of-Play?' *Environmental Science and Policy* 13(7): 648–659. doi:10.1016/j.envsci.2010.09.003.

Lam, W., and E. Ostrom. 2010. 'Analyzing the Dynamic Complexity of Development Interventions: Lessons from an Irrigation Experiment in Nepal.' *Policy Sciences* 43(1): 1–25. doi:10.1007/s11077-009-9082-6.

Leslie, H., X. Basurto, M. Nenadovic, L. Sievanen, K.C. Cavanaugh, J.J. Cota-Nieto, B.E. Erisman et al. 2015. 'Operationalizing the Social-Ecological Systems Framework to Assess Sustainability.' *Proceedings of the National Academy of Sciences (PNAS)* 112(19): 5979–5984. doi/10.1073/pnas.1414640112.

Manuel-Navarrete, D., J.J. Gómez, and G. Gallopín. 2007. 'Syndromes of Sustainability of Development for Assessing the Vulnerability of Coupled Human-Environmental Systems. The Case of Hydrometeorological Disasters in Central America and the Caribbean.' *Global Environmental Change* 17(2): 207–217. doi:10.1016/j.gloenvcha.2006.07.002.

Matthews, S.N., L.R. Iverson, M.P. Peters, A.M. Prasad, and S. Subburayalu. 2014. 'Assessing and Comparing Risk to Climate Changes among Forested Locations: Implications for Ecosystem Services.' *Landscape Ecology* 29(2): 213–228. doi:10.1007/s10980-013-9965-y.

Oberlack, C., and K. Eisenack. 2018. 'Archetypical Barriers to Adapting Water Governance in River Basins to Climate Change.' *Journal of Institutional Economics* 14(3): 527–555. doi:10.1017/S1744137417000509.

Oberlack, C., D. Sietz, E.B. Bonanomi, A. de Bremond, J. Dell'Angelo, K. Eisenack, E.C. Ellis et al. 2019. 'Archetype Analysis in Sustainability Research: Meanings, Motivations, and Evidence-Based Policy Making.' *Ecology and Society* 24(2): 26. doi:10.5751/ES-10747-240226.

Oberlack, C., L. Tejada, P. Messerli, S. Rist, and M. Giger. 2016. 'Sustainable Livelihoods in the Global Land Rush? Archetypes of Livelihood Vulnerability and Sustainability Potentials.' *Global Environmental Change* 41: 153–171.

Pahl-Wostl, C. 2015. 'A Methodological Framework for Empirical Analyses.' In *Water Governance in the Face of Global Change – From Understanding to Transformation* by C. Pahl-Wostl, 181–201. New York: Springer.

Pahl-Wostl, C., and C. Knieper. 2014. 'The Capacity of Water Governance to Deal with the Climate Change Adaptation Challenge: Using Fuzzy Set Qualitative Comparative Analysis to Distinguish between Polycentric, Fragmented and Centralized Regimes.' *Global Environmental Change* 29: 139–154.

Pahl-Wostl, C., L. Lebel, C. Knieper, and E. Nikitina. 2012. 'From Applying Panaceas to Mastering Complexity: Toward Adaptive Water Governance in River Basins.' *Environmental Science & Policy* 23: 24–34. doi:10.1016/j.envsci.2012.07.014.

Ragin, C. 2004. 'Turning the Tables: How Case-Oriented Research Challenges Variable-oriented Research.' In *Rethinking Social Inquiry: Diverse Tools, Shared Standards*, edited by H.E. Brady and D. Collier, 123. Oxford: Rowman & Littlefield.

Ragin, C. 2008. *Redesigning Social Inquiry: Fuzzy Sets and Beyond.* Chicago: University of Chicago Press.

Ragin, C., and J. Sonnett. 2004. 'Between Complexity and Parsimony: Limited Diversity, Counterfactual Cases, and Comparative Analysis.' In *Vergleichen in Der Politikwissenschaft*, edited by S. Kropp and M. Minkenberg, 180–197. Wiesbaden: VS Verlag für Sozialwissenschaften.

Ray, D., S. Bathgate, D. Moseley, P. Taylor, B. Nicoll, S. Pizzirani, and B. Gardiner. 2015. 'Comparing the Provision of Ecosystem Services in Plantation Forests under Alternative Climate Change Adaptation Management Options in Wales.' *Regional Environmental Change* 15(8): 1501–1513. doi:10.1007/s10113-014-0644-6.

Rihoux, B., and C. Ragin. 2009. *Configurational Comparative Methods. Applied Social Research Methods Series.* Thousand Oaks: Sage.

Roggero, M. 2015. 'Adapting Institutions: Exploring Climate Adaptation through Institutional Economics and Set Relations.' *Ecological Economics* 118: 114–122. doi:10.1016/j.ecolecon.2015.07.022.

Rudel, T.K. 2008. 'Meta-Analyses of Case Studies: A Method for Studying Regional and Global Environmental Change.' *Global Environmental Change* 18(1): 18–25. doi:10.1016/j.gloenvcha.2007.06.001.

Schneider, C.Q., and I. Rohlfing. 2013. 'Combining QCA and Process Tracing in Set-theoretic Multimethod Research.' *Sociological Methods & Research* 42(4): 559–597. doi:10.1177/0049124113481341.

Schneider, C.Q., and C. Wagemann. 2010. 'Standards of Good Practice in Qualitative Comparative Analysis (QCA) and Fuzzy-sets.' *Comparative Sociology* 9(3): 397–418. doi:10.1163/1569132 10x12493538729793.

Schwenk, W.S., T.M. Donovan, W.S. Keeton, and J.S. Nunery. 2012. 'Carbon Storage, Timber Production, and Biodiversity: Comparing Ecosystem Services with Multi-criteria Decision Analysis.' *Ecological Applications* 22(5): 1612–1627. doi:10.1890/11-0864.1.

Simelton, E., E.D.G. Fraser, M. Termansen, P.M. Forster, and A.J. Dougill. 2009. 'Typologies of Crop-drought Vulnerability: An Empirical Analysis of the Socio-economic Factors that Influence the Sensitivity and Resilience to Drought of Three Major Food Crops in China (1961–2001).' *Environmental Science & Policy* 12(4): 438–452.

Škopljanac-Mačina, F., and B. Blašković. 2014. 'Formal Concept Analysis – Overview and Applications.' In *Procedia Engineering* 69: 1258–1267. Amsterdam: Elsevier. doi:10.1016/j.proeng.2014.03.117.

Sutton, A.M., and M.A. Rudd. 2015. 'The Effect of Leadership and Other Contextual Conditions on the Ecological and Socio-economic Success of Small-scale Fisheries in Southeast Asia.' *Ocean and Coastal Management* 114: 102–115. doi:10.1016/j.ocecoaman.2015.06.009.

Toshkov, D. 2016. 'Comparative Designs.' In *Research Design in Political Science*, edited by D. Toshkov, 258–284. New York: Macmillan.

Turner, B.L., P.A. Matson, J.J. McCarthy, R.W. Corell, L. Christensen, N. Eckley, G.K. Hovelsrud-Broda et al. 2003. 'Illustrating the Coupled Human–Environment System for Vulnerability Analysis: Three Case Studies.' *Proceedings of the National Academy of Sciences* 100(14). doi:10.1073/pnas.1231334100.

Villamayor-Tomas, S., M. Avagyan, M. Firlus, G. Helbing, and M. Kabakova. 2016. 'Hydropower vs. Fisheries Conservation: A Test of Institutional Design Principles for Common-pool Resource Management in the Lower Mekong Basin Social-Ecological System.' *Ecology and Society* 21(1): 3. doi:10.5751/ES-08105-210103.

Villamayor-Tomas, S., I. Iniesta-Arandia, and M. Roggero. 2020. 'Are Generic and Specific Adaptation Institutions Always Relevant? An Archetype Analysis of Drought Adaptation in Spanish Irrigation Systems.' *Ecology and Society* 25(1): 32. doi:10.5751/ES-11329-250132.

Villamayor-Tomas, S., C. Oberlack, G. Epstein, S. Partelow, M. Roggero, E. Kellner, M. Tschopp, and M. Cox. 2020. 'Using Case Study Data to Understand SES Interactions: A Model-centered Meta-analysis of SES Framework Applications.' *Current Opinion in Environmental Sustainability* 44(June): 48–57. doi:10.1016/j.cosust.2020.05.002.

Yamakasi, S. 2003. 'Testing Hypotheses with QCA: Application to the Nuclear Phase-out Policy in 9 OECD Countries.' *ECPR General Conference.* Germany: Marburg.

Yin, R.K. 2014. *Case Study Research: Design and Methods.* Thousand Oaks: Sage.

21

Controlled behavioural experiments

Therese Lindahl,[1,2] *Marco A. Janssen*[3] *and Caroline Schill*[1,2]

[1]BEIJER INSTITUTE OF ECOLOGICAL ECONOMICS, THE ROYAL SWEDISH ACADEMY OF SCIENCES, STOCKHOLM, SWEDEN
[2]STOCKHOLM RESILIENCE CENTRE, STOCKHOLM UNIVERSITY, STOCKHOLM, SWEDEN
[3]SCHOOL OF SUSTAINABILITY, AND SCHOOL OF COMPLEX ADAPTIVE SYSTEMS, ARIZONA STATE UNIVERSITY, TEMPE, ARIZONA, USA

Key methods discussed in this chapter

Controlled behavioural experiments

Connections to other chapters

Some sort of statistical analysis (Chapter 18) is needed and used to analyse the experimental data obtained from controlled behavioural experiments. Interviews or surveys (Chapter 7) are often conducted in combination with the experiment to collect relevant data on potential drivers and motivations for the observed behaviour. More recently, these experiments have also been used in combination with agent-based modelling (Chapter 28) to develop and test different mechanisms that can explain observed outcomes. Finally, the game designs of these experiments can also be set up as role-playing serious games (Chapter 12) and some researchers also use them in the classroom as a pedagogical tool.

Introduction

Many sustainability challenges confronting us are associated with some sort of collective action problem: while a group benefits from joint action, no one has the incentive to take that action individually. We would all enjoy the benefits from investments made to reduce greenhouse gas emissions, for example, but progress is remarkably slow because it is not in the individual countries' interest to halt emissions on their own. Likewise, a group of fishers with access to the same fishing ground would benefit from individual efforts to reduce overfishing, but each fisher would be better off by catching as many fish as possible. If everybody acts only in their self-interest, the aggregate outcome can be disastrous for both people and ecosystems. Solving these collective action problems is challenging and has preoccupied scholars from various disciplines for decades (see Ostrom 1990; Bromley 1992; Barrett 2016).

DOI: 10.4324/9781003021339-25

SUMMARY TABLE: CONTROLLED BEHAVIOURAL EXPERIMENTS

DISCIPLINARY BACKGROUND	KNOWLEDGE TYPE
The methods in this chapter are derived from or have most commonly been used in: Experimental and Behavioural Economics, Social and Cognitive Psychology, Environmental and Cultural Psychology	*The methods in this chapter are primarily used to generate the following types of knowledge:* • Explanatory
RESEARCH APPROACH	**PURPOSE OF METHOD**
The methods in this chapter originate from or most commonly adopt the following research approaches: • Analytical/objective	*The most common purposes of using the methods in this chapter are:* • Data collection/generation • System understanding
TEMPORAL DIMENSION	**SYSTEMIC FEATURES AND PROCESSES**
The methods in this chapter are most commonly applied to the following temporal dimensions: • Present (typically within the last 5–10 years) **SPATIAL DIMENSION** *The methods in this chapter are primarily either or both:* • Non-spatial *The methods in this chapter are most commonly applied at the following spatial scales:* • Local • Multiple places/sites around the world	*While most methods can do many things, the methods in this chapter are particularly good (i.e. go-to methods) for addressing the following:* • Social-ecological interactions over time • Adaptation and self-organisation • Collective action and collaborative governance

This chapter focuses on two types of controlled behavioural experiments designed to study collective action in social-ecological systems (SES): so-called public good and common-pool resource game experiments (Janssen, Lindahl, and Murphy 2015). In both types of games, each individual can either take an action that would benefit the entire group or choose an action that would maximise individual returns. In a public good game, the individual decides whether or not to invest in a public good, e.g. an irrigation system or pollution mitigation. In a common-pool resource game, the individual decides how much to extract from a common-pool resource, e.g. how much to fish from a shared fishing ground or how much timber to harvest from a forest.

Controlled experiments are grounded in a positivist perspective, which emphasises empirical evidence with the intent to uncover 'objective' patterns and regularities. These experiments have been, and still are, a widely used methodology in the natural sciences. Over the past few decades, the experimental methodology has been increasingly adopted by social scientists, especially in economics and psychology, as it enables the researcher to explore and test hypotheses about human behaviour when people face different situations (Falk and Heckman 2009). Participants are randomly assigned to different groups (called 'treatments') so that the only difference between these groups is the variable of interest. In this sense, the researcher 'controls' these different situations. By virtue of this control, experiments allow the researcher to establish a causal link between the observed behaviour and the variable of interest.

The experimental methods discussed in this chapter mainly follow the experimental economics tradition (Smith 1976). This tradition emphasises the importance of providing participants with sufficient incentives (often, but not necessarily, monetary) in order to make participants act as they would in a real situation. Moreover, this compensation should be directly linked to decisions taken to avoid a potential hypothetical bias, e.g. in the form of behaving more (or less) cooperatively because decisions have no monetary consequence. Also, unlike psychological experiments, economic experiments do not allow for the deception of participants. Whereas economists design experiments mainly to analyse market outcomes, psychologists (and later behavioural economists) are predominantly interested in individual behaviour. This chapter aligns more with the general interest in human behaviour that guides psychologists (and later behavioural economists) than with experiments designed by traditional experimental economists to analyse market outcomes.

The controlled behavioural experiments discussed in this chapter are particularly suitable for studying collective action in SES because they allow the experimenter to mimic key social-ecological interactions and interdependencies over time. Each action an individual takes in the experiment affects the shared resource (public good or common-pool resource), which in turn affects the individuals' returns (i.e. their livelihoods in reality).

SES problems and questions

The overall purpose of using controlled experiments in the form of public good and common-pool resource games is to test under what conditions we can expect collective action to emerge, i.e. resource users to cooperate (or not) around shared resources, and by extension to attain sustainable resource use. Since the first public good experiments (Dawes, McTavish, and Shaklee 1977) and the first common-pool resource experiments (Jorgenson and Papciak 1981; Ostrom, Walker, and Gardner 1992) were undertaken, a huge number of variants and extensions of both these types of experiments have been performed. Experiments have been designed to answer research questions such as:

- What is the role of communication and costly sanctioning for achieving successful cooperation around shared resources? (e.g. see Ostrom, Walker, and Gardner 1992; Cárdenas 2000; Ostrom 2006 for a review)
- Are larger groups less likely to form cooperative agreements around shared resources than smaller groups? (e.g. see Kerr 1989; Pereda, Capraro, and Sanchez 2019)
- What motives are important for sustaining cooperation around shared resources? (e.g. see Ostrom 2000; Fehr and Fischbacher 2002; Kopelman, Weber, and Messick 2002; Cárdenas and Ostrom 2004; Velez, Strandlund, and Murphy 2009 for reviews)
- What is the effect of different institutions (such as an externally imposed regulation) on cooperation around shared resources? (e.g. see Cárdenas, Strandlund, and Willis 2000; Vollan 2008; Moreno-Sánchez and Maldonado 2010; Lopez et al. 2012)

Most public good and common-pool resource experiments to date (including the listed examples above) focus on social interactions and dynamics (the different social conditions are represented by different treatments, e.g. comparing outcomes where groups are allowed to communicate to outcomes where groups are not allowed to communicate). These experiments do not include some of the important challenges experienced by resource users associated with natural resource dynamics (notable exceptions include Walker and Gardner 1992; Herr, Gardner, and Walker 1997).

In recent years, a 'new generation' of public good and common-pool resource experiments has emerged (Cárdenas, Janssen, and Bousquet 2013). In this new generation of experiments, which are especially relevant for SES research, there is specific emphasis on including relevant ecological features and dynamics, such as thresholds and regime shifts, spatial heterogeneity, asymmetrical resource access and past ecological conditions. The overall purpose of using the 'new generation' of public good and common-pool resource experiments is still to test under what conditions we can expect relative changes in the level of cooperation, or when cooperation emerges (or not). What separates these new experiments from previous experiments is primarily the specific experimental context, which may also be represented in the treatments themselves. In a nutshell, the different treatments can represent different social conditions in a specific ecological context, or different ecological conditions in a specific social context. These experiments have been designed to answer questions such as:

- What is the role of communication and punishment in a common-pool resource setting with spatial and temporal stochastic resource dynamics? (e.g. see Janssen et al. 2010)
- How will asymmetrical resource access influence users' willingness to invest in shared infrastructure for shared resource provision? (e.g. see Janssen, Anderies, and Cárdenas 2011; Anderies et al. 2013)
- What is the role of potential ecological regime shifts (driven by resource users' own actions) for cooperation and resource use? (e.g. see Schill, Lindahl, and Crépin 2015; Lindahl, Crépin, and Schill 2016; Lindahl and Jarungrattanapong 2018; Schill and Rocha 2019; also see Case study 21.1 for more details)
- How will global, externally driven uncertainties about future resource flows influence behavioural strategies of local resource users? (e.g. see Cárdenas et al. 2017; Finkbeiner et al. 2018)

- What is the impact of culture and past ecological constraints on cooperation concerning shared resources? (e.g. see Prediger, Vollan, and Frölich 2011; Gneezy, Leibbrandt, and List 2015)

Brief description of key methods

This chapter discusses different types of game designs (public good and common-pool resource) (Table 21.1). These game designs can in turn be implemented with different types of experiments. A laboratory (lab) experiment is, for example, performed with students as participants and often with neutral instructions, speaking about the other participants, costs and benefits, choice options A or B, among other things. A framed lab experiment also uses students as participants, but here the researcher uses instructions containing context-specific elements (e.g. letting participants know that they represent fictive resource users such as fishers and that they, together with other fictive fishers, have access to a common fishing ground). Framed instructions are common in public good and common-pool resource game experiments as it can be hard for a researcher to inform participants of certain resource

Table 21.1 Summary of key applications of controlled behavioural experiments

Main applications	*Description*	*References*
Public good experiment	Each participant decides in each round how much to contribute to a public good that is shared by the group. The socially preferred outcome is that everyone contributes. But from the individual's perspective, the rational egoistic choice is to contribute zero (in finite games).	*Applications to SES in the lab* Barrett and Dannenberg 2012 *Applications to SES in the field* Cárdenas et al. 2017
Common-pool resource investment experiment	Each participant decides in each round how much to invest in two types of goods (or markets), where one of the goods is a common-pool resource and the other a private good. Investment in the common-pool resource (such as allocating time to harvesting from the common-pool resource) means more exploitive behaviour. The socially preferred outcome is associated with more moderate investments in the common-pool resource compared to the individually preferred choice based on a rational egoistic decision-maker (in finite games).	*Applications to SES in the lab* Ostrom 2006 *Applications to SES in the field* Cárdenas 2000
Common-pool resource extraction experiment	Each participant decides in each round how much of the common-pool resource to extract. The socially preferred outcome is associated with less extraction of the common-pool resource compared to the individually preferred choice based on a rational egoistic decision-maker (in finite games).	*Applications to SES in the lab* Hine and Gifford 1996 *Applications to SES in the field* Gelcich et al. 2013

dynamics or other ecological conditions without mentioning natural resources or ecosystem dynamics. The researcher can also take the lab design to the field and analyse the behaviour of non-students (e.g. resource users), in which case we talk about field experiments. Again, if the instructions contain specific contextual elements, we talk about a framed field experiment (see Harrison and List 2004 for an overview of different types of experiments). Most common-pool resource and public good experiments relevant for social-ecological research are either framed lab experiments or framed field experiments.

The choice of the type of game and experiment is at the discretion of the researcher and will depend on the research question. However, it is not uncommon for the researcher to use several types of games and experiments, e.g. evaluating a design in the lab before taking it to the field. Regardless of type, the experimental implementation of these games typically involves four to eight participants. Over a number of rounds, which can be one or several (known or unknown to the participants), each participant makes a private and anonymous decision about how much to contribute to the provision of a public good or to appropriate/extract from a common-pool resource. Table 21.1 provides some examples of different types of experiments (lab and field) applied to the different games (public good and common-pool resource).

Public good and common-pool resource experiments are often complemented by one or several other standardised controlled behavioural experiments. Eliciting preferences towards risk or uncertainty, for example, often involves a design where participants are asked to make a choice between different lotteries that are more or less risky (see e.g. Cárdenas and Carpenter 2009). Altruistic tendencies are typically measured through the dictator game (see Engel 2011 for a review). The willingness to invest in trust and the willingness to reciprocate trust are often measured through the trust game (see Johnson and Mislin 2011 for a review).

Besides analysing collective action problems and eliciting risk preferences or social preferences, controlled behavioural experiments are also valuable when evaluating different types of interventions (e.g. policies), where the different interventions represent the different treatments. Experiments of this nature that use citizens or consumers (non-students) as participants without them knowing they are participating in an experiment are called natural field experiments or randomised control trials (see Harrison and List 2004 for more details).

Limitations

With a controlled experiment it is only possible to test the effect of one variable at a time, which can be a limitation if the researcher wants to study different aspects of a complex system. Experimental programmes can be set up to test several variables and their interactions, but that requires substantial resources. Performing quantitative statistical analysis requires large enough sample sizes and collecting experimental data is costly and time consuming. Also, in the field it can be difficult to get enough participants, especially because of potential 'contamination' effects in smaller communities. An experimenter cannot stay too long in the same community because community members will talk to one another and share experiences about the experiment, which can affect outcomes.

The external validity of controlled behavioural experiments, i.e. to what extent results can be generalised beyond the experiment, is a question frequently asked (Levitt and List 2007; Falk and Heckman 2009; Gelcich et al. 2013). As a first step, researchers may want to reflect on the instructions and framing used – do they capture the situation and conditions the researcher wants to capture? Researchers could evaluate the representativeness of their sample. Have the results been obtained with the kind of people the researcher wants to say something about? Researchers could also ask themselves if they need to generalise the results beyond the experimental context, and if this is the case, what pieces of the overall puzzle are missing. As with all research, it is important to keep in mind that one experimental study is not a proof of anything. The positive side of using controlled experiments is that the designs can be easily replicated, especially since practically all scientific peer-review journals require experimental instructions and protocols to be reported.

Resource implications

The type of experiments presented here can be more or less costly, depending on the size of (monetary) incentives used, the number of treatments and consequently the number of participants to be paid. Other potential costs involve facilities and whether the experiment involves fieldwork or not. Moreover, some type of ethical clearance will be required but the exact requirements depend on the rules and regulations of the researcher's home institution and where the experiment will be conducted. Besides ethical clearance, experiments can be administratively burdensome for other reasons, e.g. the regulations of the researcher's home institution regarding paying the participants. Experiments need to be thoroughly planned and experimenters and assistants well trained because tight control is essential for the method (one minor mistake can ruin a whole dataset). Experiments can be pen-and-paper based or computer based, in which case some programming skills are required (Janssen, Lee, and Waring 2014). To analyse the results, statistical skills and skills for using adequate statistical software are necessary.

New directions

Most public good and common-pool resource experiments illustrate (more or less explicitly) that the social dynamics of groups are crucial for determining overall outcomes. Although we often equate social dynamics with communication, making agreements or cooperation, social dynamics also include many other aspects, e.g. how group members perceive their fellow group members, or biophysical conditions, which in turn influence communication and cooperation. Emergent social dynamics depend not only on individual and group attributes and the design and framing of the experiment but also on broader contextual factors such as the social groups the participants belong to and the broader socio-cultural and biophysical (i.e. social-ecological) contexts in which they live (Schill et al. 2019). Thus, if we want to further our understanding of the social-ecological conditions under which cooperation and sustainable resource use can be attained, controlled behavioural experiments should continue to focus on cross-cultural studies (such as those conducted by Cárdenas et al. 2017), especially systematic ones.

The controlled behavioural experiments discussed in this chapter are good for revealing behavioural outcomes in different experimental conditions, but it is more challenging for these experiments to unravel the specific motivations, drivers and mechanisms underlying behavioural outcomes. We therefore see a research frontier in combining controlled behavioural experiments with complementary methods and approaches beyond the usual post-experimental surveys. To advance understanding about the emotional drivers of decision-making, for example, insights and applications from neuroscience could be used, building on the work done in neuroeconomics (Rilling and Sanfey 2011; Glimcher and

Case study 21.1: Framed lab experiments to explore effects of potential ecological regime shifts on cooperative behaviour and sustainable resource use

While our understanding of the drivers and impacts of regime shifts has advanced significantly, empirical research on how human behaviour relates to regime shifts and their associated uncertainties has received hardly any attention – specifically, how resource users deal with the possibility that their actions might induce these shifts (endogenously driven). This led Lindahl, Crépin and Schill (2016) to ask the following question: what is the effect of an endogenously driven ecological regime shift on human behaviour, particularly in relation to the emergence of collective action and sustainable resource use?

To answer this research question, the authors for several reasons made use of controlled behavioural experiments in the form of a framed common-pool resource game. First, pre- and post-shift social-ecological data are seldom available, which limited possibilities for case study analyses. The experimental method allowed the authors to gather observational data on individual and group behaviour and to test causal relationships. By using the experimental method, they could isolate and causally link behavioural responses to specific resource dynamics. Furthermore, they observed revealed behaviour rather than stated behaviour (which can be subjected to hypothetical biases). Finally, they could compare and contrast their insights with the vast amount of previous experimental research in the commons literature.

The authors conducted a series of framed lab experiments between 2014 and 2015 in Stockholm (Schill, Lindahl, and Crépin 2015; Lindahl, Crépin, and Schill 2016). Groups of four participants each represented fictive resource users who had common access to a renewable resource. Over a number of rounds (unknown to the participants), they made individual and anonymous decisions about how many units of the shared resource they would like to harvest, where each harvested unit was worth money (real). Participants belonging to the same group were allowed to communicate with one another throughout the game. The groups faced different resource dynamics (treatments). Some groups played a game where there was no risk of a latent regime shift (no threshold treatment). Other groups faced resource dynamics with a latent regime shift below a certain resource stock size (threshold treatment) and some groups were told that they faced a risk of a latent regime shift with a certain percentage (90%, 50% and 10%). The authors' main challenge with this project was to create an experimental design that was able to capture the ecological complexity while being understandable to the participants. They solved this by keeping the institutional aspects of the game as simple as possible.

In short, the results showed that cooperation was endogenous to the treatment, i.e. it depended on the resource dynamics each group faced. A latent regime shift (or when

Fehr 2013). Other interesting directions are to combine controlled behavioural experiments with interpretive approaches that seek to understand diverse contextual meanings (potential methods include in-depth interviews or focus groups, see Chapter 7) or with agent-based models (see e.g. Schill et al. 2016 for an example of this; also see Chapter 28) to unpack micro- and meso-level mechanisms.

Figure 21.1 Experimental set-up for study with Thai fishers (© Therese Lindahl)

the risk of a regime shift was high) led to more communication and cooperative agreements emerging than when there was no shift or when the risk of a latent shift was low. The authors also found that communication led to knowledge sharing and more efficient management. This implies that a latent regime shift was also associated with less overexploitation compared to a case without such a shift. The authors also found that behaviour was affected by how risk was communicated, where familiar examples (like flipping a coin) triggered more cooperative behaviour.

Our main conclusion from these experiments relates to the importance of communicating about potential regime shifts, and how this is done. The results highlight the importance of enabling arenas for knowledge sharing and communication. Field experiments with Colombian and Thai fishers using a similar (although somewhat simplified) version have also been conducted. Figure 21.1 shows the table set-up for the Thai experimental participants (fishers) at a local school in a Thai fishing village. Fish bait (shaped as fish) was used to mimic fish and fish dynamics. The dynamics were illustrated visually in a table-like format on a paper board. Preliminary results (Lindahl and Jarungrattanapong 2018; Schill and Rocha 2019) show that socio-economic conditions (community effects) and individual background variables (linked to resource dependency) influence behaviour in these games. This highlights the need for more systematic explorations of the role of contextual factors and how they interact with ecological conditions for cooperation and sustainable resource use.

Finally, an interesting development in the application of public good and common-pool resource experiments is to use them to improve understanding of key social-ecological feedbacks to motivate behavioural change or facilitate local self-governance beyond the experiment or 'field lab', into the participants' everyday life (see Meinzen-Dick et al. 2018 for an example on how 'playing games' can save water).

Key readings

Anderies, J.M., M.A Janssen, F. Bousquet, J.C. Cárdenas, D. Castillo, M-C. Lopez, R. Robias, B. Vollan, and A. Wutich. 2011. 'The Challenge of Understanding Decisions in Experimental Studies of Common-pool Resource Governance.' *Ecological Economics* 70: 1571–1579.

Friedman D., and S. Sunder. 1994. *Experimental Methods: A Primer for Economists*. Cambridge: Cambridge University Press.

Kopelman, S., J.M. Weber, and D.M. Messick. 2002. 'Factors Influencing Cooperation in Commons Dilemmas: A Review of Experimental Psychological Research.' In *The Drama of the Commons*, edited by E. Ostrom, T. Dietz, N. Dolšak, P.C. Stern, S. Stonich, E.U. Weber, and The Committee on the Human Dimensions of Global Change, Division of Behavioral and Social Sciences and Education. Washington: National Academy Press.

Ostrom, E. 2006. 'The Value-added of Laboratory Experiments for the Study of Institutions and Common-pool Resources.' *Journal of Economic Behavior and Organization* 61(2): 149–163.

Poteete, A., M.A. Janssen, and E. Ostrom. 2009. 'Experiments in the Laboratory and the Field.' In *Working Together: Collective Action, The Commons, and Multiple Methods in Practice*, edited by A. Poteete, M.A. Janssen, and E. Ostrom, 229–265. Princeton: Princeton University Press.

Acknowledgements

Therese Lindahl and Caroline Schill thank the Swedish Research Council (No. 2017–05641) and the Kjell and Märta Beijer Foundation for support. Marco Janssen thanks the US National Science Foundation (no. 1658608) for support.

References

Anderies, J.M., M.A. Janssen, A. Lee, and H. Wasserman. 2013. 'Environmental Variability and Collective Action: Experimental Insights from an Irrigation Game.' *Ecological Economics* 93: 166–176.

Barrett, S. 2016. 'Collective Action to Avoid Catastrophe: When Countries Succeed, When They Fail, and Why.' *Global Policy* 7(Supplement 1). doi:10.1111/1758-5899.12324.

Barrett, S., and A. Dannenberg. 2012. 'Climate Negotiations under Scientific Uncertainty.' *Proceedings of the National Academy of Sciences* 109(43): 17372–17376.

Bromley, D.W. 1992. 'The Commons, Common Property, and Environmental Policy.' *Environmental and Resource Economics* 2(1): 1–17.

Cárdenas, J.C. 2000. 'How do Groups Solve Local Commons Dilemmas? Lessons from Experimental Economics in the Field.' *Environment, Development and Sustainability* 2: 305–322.

Cárdenas, J.C., and S. Carpenter. 2009. 'Risk Attitudes and Well-being in Latin America.' *Journal of Development Economics* 103: 52–61.

Cárdenas, J.C., M.A. Janssen, M. Ale, R. Bastakoti, A. Bernal, J. Chalermphol, Y. Gong, H. Shin, G. Shivakoti, Y. Wang, and J.M. Anderies. 2017. 'Fragility of the Provision of Local Public Goods to Private and Collective Risks.' *Proceedings of the National Academy of Sciences* 114(5): 921–925.

Cárdenas, J.C., M.A. Janssen, and F. Bousquet. 2013. 'Dynamics of Rules and Resources: Three New Field Experiments on Water, Forests and Fisheries.' In *Handbook on Experimental Economics and the Environment*, edited by J.A. List and M.K. Price. Cheltenham: Edward Elgar.

Cárdenas, J.C., and E. Ostrom. 2004. 'What do People Bring into the Game? Experiments in the Field about Cooperation in the Commons.' *Agricultural Systems* 82: 307–326.

Cárdenas, J.C., J. Strandlund, and C. Willis. 2000. 'Local Environmental Control and Institutional Crowding-out.' *World Development* 28(10): 1719–1733.

Dawes, R.M., J. McTavish, and H. Shaklee. 1977. 'Behavior, Communication and Assumptions about Other People's Behavior in a Commons Dilemma Situation.' *Journal of Personality and Social Psychology* 35: 1–11.

Engel, C. 2011. 'Dictator Games: A Meta Study.' *Experimental Economics* 14(4): 583–610.

Falk, A., and J.J. Heckman. 2009. 'Lab Experiments are a Major Source of Knowledge in the Social Sciences.' *Science* 326(5952): 535–538.

Fehr, E., and U. Fischbacher. 2002. 'Why Social Preferences Matter: The Impact of Non-selfish Motives on Competition, Cooperation and Incentives.' *The Economic Journal* 112(478): doi:10.1111/1468-0297.00027.

Finkbeiner E.M., F. Micheli, F., A. Saenz-Arroyo, L. Vazquez-Vera, C.A. Perafan, and J.C. Cárdenas. 2018. 'Local Response to Global Uncertainty: Insights from Experimental Economics in Small-scale Fisheries.' *Global Environmental Change* 48: 151–157.

Gelcich, S., R. Guzman, C. Rodriguez-Sickert, J.C. Castilla, and J.C. Cárdenas. 2013. 'Exploring External Validity of Common Pool Resource Experiments: Insights from Artisanal Benthic Fisheries in Chile.' *Ecology and Society* 18(3): 2.

Glimcher, P.W., and E. Fehr. 2013. 'Introduction: A Brief History of Neuroeconomics.' In *Neuroeconomics: Decision Making and the Brain* (2nd ed), edited by P.W. Glimcher and E. Fehr, xvii–xxviii. Cambridge: Academic Press.

Gneezy, U., A. Leibbrandt, and J.A. List. 2015. 'Ode to the Sea: Workplace Organizations and Norms of Cooperation.' *The Economic Journal* 126(595): 1856–1883.

Harrison, G.W., and J.A. List. 2004. 'Field Experiments.' *Journal of Economic Literature* 42(4): 1009–1055.

Herr, A., R. Gardner, and J.M. Walker. 1997. 'An Experimental Study of Time-independent and Time-dependent Externalities in the Commons.' *Games and Economic Behavior* 19(1): 77–96.

Hine, R., and D.W. Gifford. 1996. 'Individual Restraint and Group Efficiency in Commons Dilemmas: The Effects of Two Types of Environmental Uncertainty.' *Journal of Applied Social Psychology* 26(11): 993–1009.

Janssen, M.A., J.M. Anderies, and J.C. Cárdenas. 2011. 'Head-enders as Stationary Bandits in Asymmetric Commons: Comparing Irrigation Experiments in the Laboratory and the Field.' *Ecological Economics* 70: 1590–1598. 10.1016/j.ecolecon.2011.01.006.

Janssen, M.A., R. Holahan, A. Lee, and E. Ostrom. 2010. 'Lab Experiments for the Study of Social-Ecological Systems.' *Science* 328: 613–617.

Janssen, M.A., A. Lee, and T.M. Waring. 2014. 'Experimental Platforms for Behavioral Experiments on Social-Ecological Systems.' *Ecology and Society* 19(4): 20.

Janssen, M.A., T. Lindahl, and J. Murphy. 2015. 'Advancing the Understanding of Behavior in Social-Ecological Systems: Results from Lab and Field Experiments.' *Ecology and Society* 20(4): 34.

Johnson, N.D., and A.A. Mislin. 2011. 'Trust Game: A Meta-analysis.' *Journal of Economic Psychology* 32(5): 865–889. doi:10.1016/j.joep.2011.05.007.

Jorgenson, D.O., and A.S. Papciak. 1981. 'The Effects of Communication, Resource Feedback, and Identifiability on Behavior in a Simulated Commons.' *Journal of Experimental Social Psychology* 17(4): 373–385.

Kerr, N.L. 1989. 'Illusions of Efficacy: The Effects of Group Size on Perceived Efficacy in Social Dilemmas.' *Journal of Experimental Social Psychology* 25(4): 287–313.

Kopelman, S., J.M. Weber, and D.M. Messick. 2002. 'Factors Influencing Cooperation in Commons Dilemmas: A Review of Experimental Psychological Research.' In *The Drama of the Commons*, edited by E. Ostrom, T. Dietz, N. Dolšak, P.C. Stern, S. Stonich, E.U. Weber, and The Committee on the Human Dimensions of Global Change, Division of Behavioral and Social Sciences and Education. Washington: National Academy Press.

Levitt, S.D., and J.A. List. 2007. 'What Do Laboratory Experiments Measuring Social Preferences Reveal about the Real World?' *Journal of Economic Perspectives* 21(2): 153–174.

Lindahl, T., A-S. Crépin, and C. Schill. 2016. 'Potential Disasters Can Turn the Tragedy into Success.' *Environmental and Resource Economics* 65(3): 657–676.

Lindahl, T., and R. Jarungrattanapong. 2018. 'Avoiding Catastrophic Collapse in Small Scale Fisheries through Inefficient Cooperation: Evidence from a Framed Field Experiment.' *Beijer Discussion Papers* 263. Stockholm: Beijer Institute of Ecological Economics.

Lopez, M.C., J.J. Murphy, J.M. Spraggon, and J.K. Strandlund. 2012. 'Comparing the Effectiveness of Regulation and Pro-social Emotions to Enhance Cooperation: Experimental Evidence from Fishing Communities in Colombia.' *Economic Inquiry* 50(1): 131–142.

Meinzen-Dick, R., M.A. Janssen, S. Kandikuppa, R. Chaturvedi, K. Rao, and S. Theis. 2018. 'Playing Games to Save Water: Collective Action Games for Groundwater Management in Andhra Pradesh, India.' *World Development* 107: 40–53. 10.1016/j.worlddev.2018.02.006.

Moreno-Sánchez, R., and J. Maldonado. 2010. 'Evaluating the Role of Co-management in Improving Governance of Marine Protected Areas: An Experimental Approach in the Colombian Caribbean.' *Ecological Economics* 69: 2557–2567. 10.1016/j.ecolecon.2010.07.032.

Ostrom, E. 1990. *Governing the Commons: The Evolution of Institutions for Collective Action*. Cambridge: Cambridge University Press.
Ostrom, E. 2000. 'Collective Action and the Evolution of Social Norms.' *Journal of Economic Perspectives* 14(3): 137–158.
Ostrom, E. 2006. 'The Value-added of Laboratory Experiments for the Study of Institutions and Common-pool Resources.' *Journal of Economic Behavior and Organization* 61: 149–163.
Ostrom, E., J. Walker, and R. Gardner. 1992. 'Covenants with and without a Sword: Self-governance is Possible.' *American Political Science Review* 86(2): 404–417.
Pereda, M., V. Capraro, and A. Sanchez. 2019. 'Group Size Effects and Critical Mass in Public Goods Games.' *Scientific Reports* 9: 5503.
Prediger, S., B. Vollan, and M. Frölich. 2011. 'The Impact of Culture and Ecology on Cooperation in a Common-pool Resource Experiment.' *Ecological Economics* 70(9): 1599–1608.
Rilling, J.K., and A.G. Sanfey. 2011. 'The Neuroscience of Social Decision-making.' *Annual Review of Psychology* 62: 23–48. doi:10.1146/annurev.psych.121208.131647.
Schill, C., J.M. Anderies, T. Lindahl, C. Folke, S. Polasky, J.C. Cárdenas, A-S. Crépin, M.J. Janssen, J. Norberg, and M. Schlüter. 2019. 'A More Dynamic Understanding of Human Behaviour for the Anthropocene.' *Nature Sustainability* 2: 1075–1082.
Schill, C., T. Lindahl, and A-S. Crépin. 2015. 'Collective Action and the Risk of Ecosystem Regime Shifts: Insights from a Laboratory Experiment.' *Ecology and Society* 20(1): 48.
Schill, C., and J.C. Rocha. 2019 'Uncertainty Can Help Protect Local Commons in the Face of Climate Change.' *Beijer Discussion Papers* 270. Stockholm: Beijer Institute of Ecological Economics.
Schill, C., N. Wijermans, M. Schlüter, and T. Lindahl. 2016. 'Cooperation Is Not Enough – Exploring Social-Ecological Micro-foundations for Sustainable Common-pool Resource Use.' *PLoS ONE* 11: e0157796. doi:10.1371/journal.pone.0157796.
Smith, V.L. 1976 'Experimental Economics: Induced Value Theory.' *The American Economic Review* 66(2): 274–279.
Velez, M.A., J.K. Strandlund, and J.J. Murphy. 2009. 'What Motivates Common Pool Resource Users? Experimental Evidence from the Field.' *Journal of Economic Behavior and Organization* 70(3): 485–497.
Vollan, B. 2008. 'Socio-ecological Explanations for Crowding-out Effects from Economic Field Experiments in Southern Africa.' *Ecological Economics* 67(4): 560–573.
Walker, J.M., and R. Gardner. 1992. 'Probabilistic Destruction of Common-pool Resources: Experimental Evidence.' *The Economic Journal* 102: 1149–1161.

22

Institutional analysis

Graham Epstein,[1] Sergio Villamayor-Tomas[2] and Michael Schoon[3]

[1]SCHOOL OF POLITICS, SECURITY, AND INTERNATIONAL AFFAIRS, UNIVERSITY OF CENTRAL FLORIDA, FLORIDA, USA
[2]INSTITUTE OF ENVIRONMENTAL SCIENCE AND TECHNOLOGY, AUTONOMOUS UNIVERSITY OF BARCELONA, BARCELONA, SPAIN
[3]SCHOOL OF SUSTAINABILITY, ARIZONA STATE UNIVERSITY, TEMPE, ARIZONA, USA

Key methods discussed in this chapter

Institutional analysis and development framework, SES framework, action situations, networks of action situations, institutional grammar tool, rule typology

Connections to other chapters

Institutional analysis relies on other methods to assess the institutions being studied. Frequently, this entails interviews and participant observation (Chapter 7) and comparative case study analysis (Chapter 20). While many of the methodological approaches in this book could be used in conjunction with institutional analysis, a few other commonly featured approaches include behavioural experiments (Chapter 21), network analysis (Chapter 23) and agent-based modelling (Chapter 28).

Introduction

Institutional analysis is a term that is shared by a number of different intellectual traditions, including the Bloomington or Ostrom School of Institutional Analysis, which is the focus of this chapter. Readers interested in situating the Bloomington School within the broader discourse on institutions should refer to Hall and Taylor (1996) and Mitchell (1988). For other overviews on institutions and institutions and the environment, see also Hodgson (1998) and Vatn (2005).

Institutional analysis has rapidly gained traction as a leading interdisciplinary approach for analysing the structure of social-ecological problems and developing institutional solutions to address them (Van Laerhoven and Ostrom 2007). It emerged from the field of public administration where administrative consolidation and centralisation were promoted as a panacea for the problem of delivering public services in metropolitan areas based on a general assumption that all public services exhibited economies of scale. Institutional analysis was used to demonstrate the folly of these assumptions, highlighting a range of factors that may mediate the impacts of scale on the costs and benefits associated with different public goods and services (Ostrom, Tiebout, and Warren 1961; Ostrom, Parks, and Gordon 1973).

DOI: 10.4324/9781003021339-26

SUMMARY TABLE: INSTITUTIONAL ANALYSIS

DISCIPLINARY BACKGROUND	KNOWLEDGE TYPE
The methods in this chapter are derived from or have most commonly been used in: Political Science, Human Geography, Interdisciplinary	*The methods in this chapter are primarily used to generate the following types of knowledge:* • Descriptive • Exploratory • Explanatory
RESEARCH APPROACH	**PURPOSE OF METHOD**
The methods in this chapter originate from or most commonly adopt the following research approaches: • Analytical/objective	*The most common purposes of using the methods in this chapter are:* • System understanding
TEMPORAL DIMENSION	**SYSTEMIC FEATURES AND PROCESSES**
The methods in this chapter are most commonly applied to the following temporal dimensions: • Present (typically within the last 5–10 years) • Recent past (post-1700s) **SPATIAL DIMENSION** *The methods in this chapter are primarily either or both:* • Non-spatial *The methods in this chapter are most commonly applied at the following spatial scales:* • Local • Multiple places/sites around the world	*While most methods can do many things, the methods in this chapter are particularly good (i.e. go-to methods) for addressing the following:* • SES components and linkages • Collective action and collaborative governance

A similar logic, meanwhile, prevailed in the context of natural resource management, where Hardin's (1968) tragedy of the commons narrative proclaimed that society was destined for ruin in the absence of enforceable private property rights or strong central government command and control. Elinor Ostrom's (1990) *Governing the Commons* convincingly refuted this narrative and identified the institutional foundations of sustainable community-based natural resource management, which continue to enjoy considerable empirical support (Cox, Arnold, and Villamayor-Tomas 2010; Baggio et al. 2016). This chapter therefore provides an overview of the institutional analysis and development and social-ecological systems (SES) frameworks that have been used to develop, organise and undertake highly influential and impactful research on SES over the past 30 years.

Institutional analysis is oriented around the role of institutions in shaping the incentives, opportunities and constraints that actors face as they interact with the environment and one another. However, unlike many branches of institutional analysis (see Shepsle 2006), Ostrom's frameworks tend to reject the notion of the all-knowing and calculative rational optimiser in favour of a model of bounded rationality in which actors are presumed to be goal-seeking but rely upon heuristics, such as satisficing, in order to make complex and time-sensitive decisions (Ostrom 1998). Institutions, which are defined broadly to include formal and informal rules, norms and shared strategies (Crawford and Ostrom 1995), play a particularly important role in decision-making and the subsequent social and ecological outcomes that are realised. Boundary rules that define the eligibility of actors to harvest resources, for instance, can provide powerful incentives to invest in the management and sustainable exploitation of resources by internalising the costs and benefits of resource use (Ostrom 1990).

Although institutions are, unsurprisingly, central to institutional analysis, attributes of communities (e.g. group size, cultural homogeneity, economic status) and resources (e.g. clarity of boundaries, spatial distribution, mobility) also play an important role in influencing the decisions that actors make and the outcomes that are observed (Agrawal 2003; Ostrom 2005, 2007). Indeed, institutional analysis is defined to a great extent by its attentiveness to a large number of attributes of actors, institutions and the environment that potentially influence the sustainability of SES, as shown in the institutional analysis and development framework illustrated in Figure 22.1.

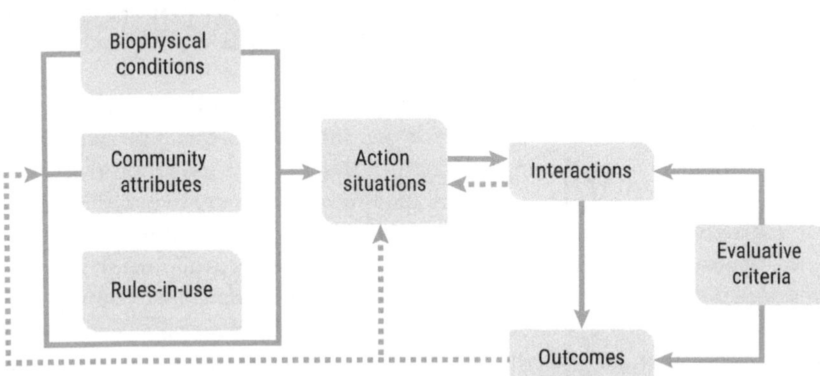

Figure 22.1 The institutional analysis and development framework (Ostrom, Gardner and Walker 1994, 37)

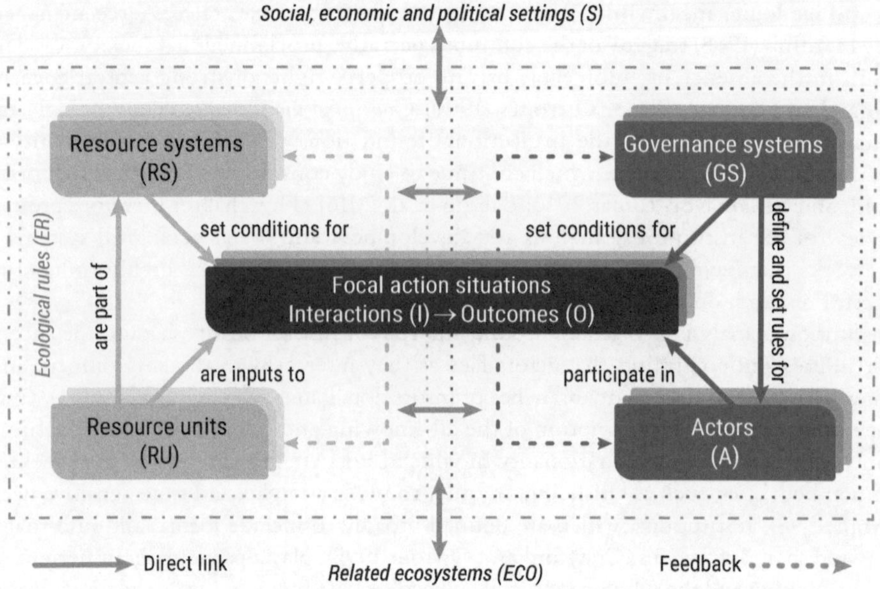

Figure 22.2 The SES framework (Monroy-Sais et al. 2016)

The institutional analysis and development (IAD) and SES frameworks are the primary tools with which institutional analysts approach and undertake research on decision-making and the sustainability of SES. The IAD framework was developed as a general tool for conducting research on the development and impacts of institutions in a range of empirical settings (Kiser and Ostrom 1982; Ostrom 2005); while the SES framework, shown in Figure 22.2, was developed to provide a better understanding of how different configurations of social, ecological and institutional factors affect the sustainability of SES, and to facilitate comparisons (Ostrom 2005, 2007; McGinnis and Ostrom 2014). Although the frameworks differ in terms of their respective levels of detail and the ways in which they tend to be applied in research, they are both organised around the study of one or more action situations in which actors make decisions and generate outcomes. Action situations can be broadly understood as spaces of interdependent decision-making (individuals make decisions that generate positive and negative externalities) (Ostrom, Gardner, and Walker 1994).

Environmental governance invariably involves a large number of potentially salient action situations pertaining to tasks such as resource use, infrastructure maintenance, rulemaking, social and environmental monitoring, sanctioning and conflict resolution (McGinnis 2011). Whereas applications of the IAD framework tend to be more attentive to the analysis of dynamics across networks of interrelated action situations (McGinnis 2011; Villamayor-Tomas et al. 2015), applications of the SES framework tend to neglect the underlying processes to focus on associations between variables and social-ecological outcomes (Gutierrez, Hilborn, and Defeo 2011). Nonetheless, there is nothing that would preclude an analyst from analysing networks of action situations with the SES framework, or from using the IAD framework to develop insights concerning the social, ecological and institutional correlates of environmental sustainability (Cole, Epstein, and McGinnis 2019). Indeed, both frameworks operate under the general hypothesis that decisions in action situations are influenced by the incentives, opportunities and constraints that actors face, which are in turn determined by the institutions that apply to each action situation as well as other attributes of the broader social, ecological and institutional context in which decisions are made.

Other longstanding developments within the 'family of IAD tools' (McGinnis 2011) include the institutional grammar tool, which has been used to carry formal analyses of institutions and the networks of rights and responsibilities they generate (Crawford and Ostrom 1995); and the 'levels of action' distinction, which facilitates the study of institutions operating at different levels of individual and collective decision-making (Kiser and Ostrom 1982). Others, meanwhile, have extended these frameworks to address gaps or specific types of research questions. These include the robustness framework, which provides a heuristic to study interactions among SES components (Anderies, Janssen, and Ostrom 2004); the nature-related transactions framework, which draws attention to the characteristics of the interdependencies generated around the joint use of natural resources (Hagedorn 2008); and the politicised IAD framework, which focuses on the role of discourses and the broader political-economic context in shaping interactions and outcomes in action situations (Clement 2010).

SES problems and questions

Institutional analysis is generally promoted as an interdisciplinary approach for developing, organising and undertaking research on collective action and environmental sustainability in heterogeneous social and ecological contexts (Ostrom 2005; Poteete, Janssen, and Ostrom 2010). This has resulted in the development of several tools that facilitate research on institutions and their role in SES, as briefly outlined in Table 22.1. The utility of institutional analysis is clearly demonstrated by applications across a wide range of environmental issues, including forests, fisheries, irrigation and rangelands (see Partelow 2018); methods, including case studies, experiments and statistical analysis (see Poteete, Janssen, and Ostrom 2010); and levels of analysis, ranging from the study of individuals to the study of global environmental regimes (Ostrom, Gardner, and Walker 1994; Fleischman et al. 2014). Nonetheless, its most significant contributions to SES theory and practice relate to questions concerning (a) the institutional foundations of sustainable environmental governance regimes, (b) the micro-foundations of cooperative behaviour in resource-dependent communities, and (c) the implications of different social and ecological contexts for institutional design (see also Chapter 10 in Poteete, Janssen and Ostrom 2010 for an overview on the theory generated around these three questions).

Expanding on the points above, first, institutional analysis of SES has been defined to a great extent by Elinor Ostrom's (1990) institutional design principles, which emerged from an empirical analysis of several cases of community-based natural resource management. The primacy of knowledge from empirical research is, in fact, a defining characteristic of institutional analysis, stemming from concerns about the inappropriate use of abstracted theories and models in environmental policy and planning (Schlager 1999). Noteworthy contributions from applications of the design principles to SES theory include highlighting the general importance of participation in rulemaking, social monitoring and adjusting institutional arrangements to fit the contexts in which they are used (Ostrom 1990; Agrawal 2003; Cox, Arnold, Villamayor-Tomas 2010; Baggio et al. 2016).

Second, research on the micro-foundation of cooperative behaviour, meanwhile, seeks to better understand the conditions in which actors are more (or less) likely to cooperate, relying on case studies and ethnographic research, common-pool resource experiments and other methodological approaches highlighted below and discussed in other chapters on behavioural experiments (Chapter 21), interviews (Chapter 7) and case study analysis (Chapter 20), among others. This research has offered an important counterpoint to rational choice theory by clearly demonstrating that institutional arrangements that support endogenous rule choice, such as communication and voting, and/or allow participants to sanction one another, can yield high levels of cooperation in both laboratory and field settings (Ostrom, Gardner, and Walker 1994; Cárdenas, Stranlund, and Willis 2000; Janssen et al. 2010; DeCaro, Janssen, and Lee 2015). Although insights from laboratory (lab) experiments, as discussed in the chapter on behavioural experiments (Chapter 21), enjoy strong internal validity, important questions remain concerning the extent to which they reflect essential features of the resource use and management problems experienced by communities as identified in the case studies. Researchers have begun to test these questions through field experiments as well. As a result, scholars are increasingly shifting their attention towards understanding the implications of different social and ecological characteristics for cooperation and institutional design, including resource dynamics (Janssen et al. 2010), thresholds and regime shifts (Schill, Lindahl, and Crépin 2015), interlinked resources (Lindahl, Bodin, and Tengö 2015), uncertainty (Janssen 2013) and past experience with successful self-organisation (Gelcich et al. 2013).

Lastly, with reference to environmental governance, institutional analysis is frequently used to address questions concerning the fit between institutions and the broader social, ecological and institutional context in which they are found (Young 2002; Galaz et al. 2008; Epstein et al. 2015). Indeed, the concept of contingency is a recurrent theme in the scholarship of Vincent and Elinor Ostrom (Dietz 2005), urging scholars and decision-makers to attend carefully to the structure of problems when developing institutional arrangements to address them. Participatory rulemaking is often recommended as a general strategy to improve the fit of institutions by providing mechanisms for incorporating knowledge of the local context (Chhatre and Agrawal 2009). Others, meanwhile, have examined the fit between institutions and local contexts by considering how variability in the attributes of communities such as group size (Agrawal and Goyal 2001) and inequality (Andersson and Agrawal 2011), and resource attributes such as mobility and storage (Schlager, Blomquist, and Tang 1994) influence the efficacy of alternative institutional arrangements. More recently, social-ecological networks have been used to highlight the importance of consolidating governance functions or establishing governance networks to coordinate the management of interlinked resources (Bodin 2017).

Table 22.1 Summary of key approaches used in institutional analysis

Main approach	*Description*	*References*
Institutional analysis and development framework	An institutional analysis and development (IAD) framework is a structured approach used to investigate the process by which attributes of a community, rules-in-use and biophysical conditions structure interactions in action situations to generate outcomes. *Methods* Single case studies, comparative case studies, process tracing, statistical analysis, meta-analysis	*Key introductory text* Ostrom 2005 *Applications to SES* Ostrom 1990; Gibson, McKean, and Ostrom 2000

Main approach	*Description*	*References*
SES framework	An SES framework is a diagnostic approach used to investigate the combinations of social, ecological and institutional factors that contribute to social and environmental outcomes. *Methods* Single case studies, comparative case studies, process tracing, statistical analysis, meta-analysis	*Key introductory text* Ostrom 2007 *Applications to SES* Basurto and Ostrom 2009; Gutierrez, Hilborn, and Defeo 2011
Action situations	Action situations are an approach used to develop and test models of decision-making through manipulation of rules (i.e. communication, sanctioning) that structure interactions. *Methods* Laboratory experiments, field experiments, agent-based models	*Key introductory text* Ostrom 2005 *Applications to SES* Ostrom, Gardner, and Walker 1994; Anderies 2000; Cárdenas and Ostrom 2004
Networks of action situations	Networks of action situations are an approach used to analyse decision-making processes across systems of linked action situations that jointly influence outcomes. *Methods* Single case studies, comparative case studies, process tracing	*Key introductory text* McGinnis 2011 *Applications to SES* Villamayor-Tomas et al. 2015; Jones, Rigg, and Pinkerton 2017; McCord et al. 2017
Institutional grammar tool	The institutional grammar tool is a systematic approach used to characterise the design of institutional arrangements to facilitate analysis of formal institutions, their change and their interlinkages. *Methods* Single case studies, comparative case studies, process tracing, statistical analysis, meta-analysis	*Key introductory texts* Crawford and Ostrom 1995; Siddiki et al. 2011 *Applications to SES* Siddiki, Basurto, and Weible 2012; Heikkila and Weible 2018; Lien, Schlager, and Lona 2018
Rule typology	Rule typology is a systematic approach used to feature institutional configurations as they affect behaviour in action situations and change over time.	*Key introductory text* Crawford and Ostrom 2005 *Applications to SES* Ostrom and Basurto 2011; Villamayor-Tomas et al. 2015

Table 22.1 highlights several approaches, tools and frameworks commonly used in institutional analysis and the methods used in conjunction with them. Key references and application to the study of SES are also given.

Limitations

Despite being promoted as a flexible and widely applicable tool for empirical research on environmental sustainability, institutional analysis is often applied to a relatively limited range of cases, problems and outcomes. The vast majority of cases that inform institutional theory, for instance, are framed in terms of local communities exploiting a single common-pool resource, relatively isolated from the broader social, ecological and institutional context in which they are found (Agrawal 2003). Although this greatly enhances the tractability of research, it also neglects critical interactions among resources and across scales that may drive resource outcomes at local, regional or even global scales. The sustainability of a resource, for instance, often depends on the management of other related resources, such as predator and prey species as well as inputs of water and energy for food production (Bodin et al. 2014). Similarly, efforts to manage local resources are often challenged by connections to global markets that can rapidly overwhelm the capacity of local communities to respond effectively (Berkes et al. 2006).

A further challenge stems from the early empirical foundations of institutional research which focused on understanding institutional robustness, or more generally the attributes of institutions that allow them to persist, promote long-term cooperation and avoid overexploitation of resources (Ostrom 1990; Anderies, Janssen, and Ostrom 2004). This has allowed scholars to rapidly develop a general understanding of the institutional ingredients for long-term sustainable community-based management, but with little guidance about the underlying processes through and conditions in which those institutions emerged and changed over time. Although there are growing efforts to address these gaps (see Section 'New directions'), institutional analysis generally appears to facilitate relatively static empirical research on the impacts of variables and create challenges for analysing the coevolution of institutions and social-ecological processes.

Institutional analysis has also been criticised for its emphasis on analysing cooperation at the expense of other social processes and mechanisms premised on values, conflict or power dynamics

Case study 22.1: Environmental governance of Mexican community fisheries

Basurto and Ostrom (2009) provide an illustrative example of how institutional analysis can be used to support research on environmental governance. Their paper begins by arguing that fishers will elect to invest in developing rules when presented with favourable incentives, and then draws on previous research to highlight several attributes of the resource and fishing community that are likely to influence the nature of these incentives. Three benthic fisheries in Mexico were used to evaluate the conditions in which groups are successful in self-organising to govern the use of local resources. As predicted, the two cases in which groups had strong local leadership, high levels of social capital and high levels of dependence on the resource in addition to several favourable attributes of the resource system had successfully self-organised to develop new institutions, whereas the one case that lacked many of these conditions failed to self-organise.

The authors extended their analysis to examine the robustness of the governance systems developed by the successful communities using Ostrom's (1990) design principles. They found that, although both communities had successfully self-organised to develop rules, only one of these systems proved robust to an external

(Agrawal 2003; Clement 2010). In fact, several decentralisation initiatives meant to promote cooperation at the local level have failed to achieve their intended objectives precisely because powerful actors exploit their knowledge, resources and authority to retain control over the governance of natural resources (Blaikie 2006; Ribot, Agrawal, and Larson 2006). Similarly, important challenges remain with respect to integrating institutional and ecological theory to foster a truly interdisciplinary understanding of the sustainability of SES (Epstein et al. 2013). Finally, institutional analysis has struggled immensely in its attempt to manage trade-offs between flexibility to allow scholars to adjust their approach to different contexts and methods and consistency to support cross-case comparison and empirical synthesis. In particular, it has thus far failed to provide clear guidelines for measuring core concepts and variables (Thiel, Adamseged, and Baake 2015; Partelow 2018; Schlager and Cox 2018), resulting in a patchwork of empirical findings that are difficult to integrate and compare.

Resource implications

Institutional analysis does not in and of itself require significant resources in terms of materials, technology or financing to undertake meaningful and impactful research on sustainable environmental governance. In practice, knowledge of the language of institutional analysis is important to understand its core theories, therefore training (see Section 'Key readings') is often a critical precondition for effectively engaging with institutional research. Furthermore, certain methods used for institutional analysis, such as agent-based models (Schoon et al. 2014) and dynamic multi-player experimental environments (Janssen et al. 2010), require access to relatively advanced hardware and software packages and skills in computer programming. In contrast, an ethnographic study may require a significant investment of

shock. Whereas the governance system employed by the Seri community was characterised by all eight design principles, the Peñasco community lacked external recognition of their rights to self-organise and a system of nested governance. As a result, when the Peñasco community was faced with a rapid influx of fishers from other communities, they lacked the legal authority and adequate resources to exclude them, resulting in the rapid depletion of the resource.

Basurto and Ostrom (2009) exemplify a number of defining characteristics of institutional analysis and its approach to the study of the sustainability of SES, although it is important to note that these characteristics can vary widely across different methods or studies. Nonetheless, the paper begins by clearly specifying the action situation(s) that it aims to investigate (rule change) and considers how attributes of the resource system and actors are likely to influence the incentives that groups face in that situation. The empirical analysis, meanwhile, allows them to test their hypotheses and specify the combinations of attributes that appear to give rise to outcomes.

Finally, the design principles analysis aggregates several action situations to explore how the presence or absence of a suite of principles affects robustness of the system to shocks. In fact, aggregation of multiple action situations is implicit to many institutional studies that seek to understand the relationships between variables and a range of social and ecological outcomes.

time and financial resources to build trust with resource-dependent communities to reveal the important, but often unwritten, rules that structure human interactions with the environment (Acheson 2003).

New directions

The field of institutional analysis continues to evolve swiftly to address many of the gaps discussed above and respond to current needs for research on the sustainability of SES. These efforts include broadening the diversity of cases that inform institutional theory; characterising and analysing relationships among actors, the environment and decision-making arenas at multiple scales; and developing tools to support empirical research and synthesis.

Several research programmes have been developed in recent years in response to critiques of the relatively limited range of cases and outcomes that inform institutional analysis. Empirical studies, for instance, are increasingly analysing relationships between institutions and multiple social and ecological outcomes to identify opportunities for often elusive 'win-win' outcomes (Persha, Agrawal, and Chhatre 2011; Cinner et al. 2012). Others, meanwhile, have explored the extent to which principles derived from the study of the small-scale community-based governance of natural resources might apply to cases involving communities facing significant external socio-economic and ecological disturbances (Brondizio, Ostrom, and Young 2009; Villamayor-Tomas and García-López 2017), large-scale resource systems (Fleischman et al. 2014), pollution (Epstein et al. 2014b) and energy systems (Bauwens, Gotchev, and Holstenkamp 2016). As efforts continue to expand the diversity of cases and outcomes, there is a great opportunity to gain traction on important questions concerning contextually explicit institutional design.

Institutional analysis has also been influenced by developments in network theory to analyse the network structures and processes that underlie sustainable environmental governance. First, social-ecological networks are increasingly being used as a tool to clarify the structure of environmental problems to develop insights for addressing them. This line of research generally highlights the importance of developing mechanisms to coordinate the governance of interlinked resources (Bodin 2017). Second, networks among multiple arenas of decision-making are used to analyse the development and implementation of environmental governance systems. The ecology of games (Lubell 2013) and the networks of action situation approaches (McGinnis 2011) facilitate efforts to develop insights about how institutions and patterns of behaviour emerge from networks of interdependent decisions.

Finally, several tools have been developed that aim to contribute to the development of knowledge by formalising our understanding of concepts to facilitate the comparison and integration of research findings. The Social-Ecological Systems Meta-Analysis Database (SESMAD) project, for instance, has operationalised many of the variables used in previous studies and used them to formally express a number of environmental governance theories (Cox 2014; Cox et al. 2016). The grammar of institutions has also been adapted in recent years to provide a tool to systematically characterise institutional arrangements and study institutional change over time (Ostrom and Basurto 2011). Further initiatives, meanwhile, have offered conceptual clarifications on the concept of institutional fit (Epstein et al. 2015) and begun to formalise the study of power in institutional research (Epstein et al. 2014a; Morrison et al. 2017).

Key readings

Crawford, S.E., and E. Ostrom. 1995. 'A Grammar of Institutions.' *American Political Science Review* 89(3): 582–600.

McGinnis, M.D. 2011. 'An Introduction to IAD and the Language of the Ostrom Workshop: A Simple Guide to a Complex Framework.' *Policy Studies Journal* 39(1): 169–183.

Ostrom, E. 2005. *Understanding Institutional Diversity*. Princeton: Princeton University Press.

Ostrom, E. 2007. 'A Diagnostic Approach for Going Beyond Panaceas.' *Proceedings of the National Academy of Sciences* 104(39): 15181–15187.

Thiel, A., M.E. Adamseged, and C. Baake. 2015. 'Evaluating an Instrument for Institutional Crafting: How Ostrom's Social-Ecological Systems Framework is Applied.' *Environmental Science and Policy* 53: 152–164.

Acknowledgements

The authors would like to thank the Ostrom Workshop at Indiana University where they received training in understanding institutions. Michael Schoon would also like to thank the Center for Behavior, Institutions and the Environment at Arizona State University and the support and insight of his colleagues there.

References

Acheson, J.M. 2003. *Capturing the Commons: Devising Institutions to Manage the Maine Lobster Industry*. Lebanon: University Press of New England.

Agrawal, A. 2003. 'Sustainable Governance of Common Pool Resources: Context, Methods, and Politics.' *Annual Review of Anthropology* 32(1): 243–262.

Agrawal, A., and S. Goyal. 2001. 'Group Size and Collective Action.' *Comparative Political Studies* 34: 63–93.

Anderies, J.M. 2000. 'On Modeling Human Behavior and Institutions in Simple Ecological Economic Systems.' *Ecological Economics* 35(3): 393–412.

Anderies, J.M., M.A. Janssen, and E. Ostrom. 2004. 'A Framework to Analyze the Robustness of Social-Ecological Systems from an Institutional Perspective.' *Ecology and Society* 9(1): 18.

Andersson, K., and A. Agrawal. 2011. 'Inequalities, Institutions, and Forest Commons.' *Global Environmental Change* 21(3): 866–875.

Baggio, J., A. Barnett, I. Perez-Ibarra, U. Brady, E. Ratajczyk, N. Rollins, C. Rubiños, H. Shin, D. Yu, and R. Aggarwal. 2016. 'Explaining Success and Failure in the Commons: The Configural Nature of Ostrom's Institutional Design Principles.' *International Journal of the Commons* 10(2): 417–439.

Basurto, X., and E. Ostrom. 2009. 'Beyond the Tragedy of the Commons.' *Economia delle Fonti di Energia e dell'Ambiente* LII(1): 35–60.

Bauwens, T., B. Gotchev, and L. Holstenkamp. 2016. 'What Drives the Development of Community Energy in Europe? The Case of Wind Power Cooperatives.' *Energy Research and Social Science* 13: 136–147.

Berkes, F., T.P. Hughes, R.S. Steneck, J.A. Wilson, D.R. Bellwood, B. Crona, C. Folke et al. 2006. 'Globalization, Roving Bandits, and Marine Resources.' *Science* 311: 1557–1558.

Blaikie, P. 2006. 'Is Small Really Beautiful? Community-based Natural Resource Management in Malawi and Botswana.' *World Development* 34: 1942–1957.

Bodin, Ö. 2017. 'Collaborative Environmental Governance: Achieving Collective Action in Social-Ecological Systems.' *Science* 357(6352): eaan1114.

Bodin, Ö., B. Crona, M. Thyresson, A.L. Golz, and M. Tengö. 2014. 'Conservation Success as a Function of Good Alignment of Social and Ecological Structures and Processes.' *Conservation Biology* 28(5): 1371–1379.

Brondizio, E.S., E. Ostrom, and O.R. Young. 2009. 'Connectivity and the Governance of Multilevel Social-Ecological Systems: The Role of Social Capital.' *Annual Review of Environment and Resources* 34: 253–278.

Cárdenas, J-C., and E. Ostrom. 2004. 'What Do People Bring into the Game? Experiments in the Field about Cooperation in the Commons.' *Agricultural Systems* 82(3): 307–326.

Cárdenas, J-C., J. Stranlund, and C. Willis. 2000. 'Local Environmental Control and Institutional Crowding-Out.' *World Development* 28(10): 1719–1733.

Chhatre, A., and A. Agrawal. 2009. 'Trade-offs and Synergies Between Carbon Storage and Livelihood Benefits from Forest Commons.' *Proceedings of the National Academy of Sciences* 106(42): 17667–17670.

Cinner, J.E., T.R. McClanahan, M.A. MacNeil, N.A.J. Graham, T.M. Daw, A. Mukminin, D.A. Feary et al. 2012. 'Comanagement of Coral Reef Social-Ecological Systems.' *Proceedings of the National Academy of Sciences* 109: 5219–5222.

Clement, F. 2010. 'Analysing Decentralised Natural Resource Governance: Proposition for a "Politicised" Institutional Analysis and Development Framework.' *Policy Sciences* 43: 129–156.

Cole, D.H., G. Epstein, and M.D. McGinnis. 2019. 'Combining the IAD and SES Frameworks.' *International Journal of the Commons* 13(1): 244–275.

Cox, M. 2014. 'Understanding Large Social-Ecological Systems: Introducing the SESMAD Project.' *International Journal of the Commons* 8(2): 265–276.

Cox, M., G. Arnold, and S. Villamayor-Tomas. 2010. 'A Review of Design Principles for Community-based Natural Resource Management.' *Ecology and Society* 15(4): 38.

Cox, M., S. Villamayor-Tomas, G. Epstein, L. Evans, N.C. Ban, F. Fleischman, M. Nenadovic, and G. Garcia-Lopez. 2016. 'Synthesizing Theories of Natural Resource Management and Governance.' *Global Environmental Change* 39: 45–56.

Crawford, S.E.S., and E. Ostrom. 1995. 'A Grammar of Institutions.' *The American Political Science Review* 89(3): 582–600.

DeCaro, D.A., M.A. Janssen, and A. Lee. 2015. 'Synergistic Effects of Voting and Enforcement on Internalized Motivation to Cooperate in a Resource Dilemma.' *Judgment and Decision Making* 10: 511–537.

Dietz, T. 2005. 'The Darwinian Trope in the Drama of the Commons: Variations on Some Themes by the Ostroms.' *Journal of Economic Behavior and Organization* 57: 205–225.

Epstein, G., A. Bennett, R. Gruby, L. Acton, and M. Nenadovic. 2014a. 'Understanding Power with the Social-Ecological Systems Framework.' In *Understanding Society and Natural Resources: Forging New Strands of Integration Across the Social Sciences*, edited by M. Manfredo, J. Vaske, A. Rechkemmer, and E. Duke, 111–135. New York: Springer.

Epstein, G., I. Pérez, M. Schoon, and C.L. Meek. 2014b. 'Governing the Invisible Commons: Ozone Regulation and the Montreal Protocol.' *International Journal of the Commons* 8(2): 337–360.

Epstein, G., J. Pittman, S.M. Alexander, S. Berdej, T. Dyck, U. Kreitmair, K.J. Raithwell, S. Villamayor-Tomas, J. Vogt, and D. Armitage. 2015. 'Institutional Fit and the Sustainability of Social-Ecological Systems.' *Current Opinion in Environmental Sustainability* 14(June): 34–40.

Epstein, G., J.M. Vogt, S.K. Mincey, M. Cox, and B. Fischer. 2013. 'Missing Ecology: Integrating Ecological Perspectives with the Social-Ecological System Framework.' *International Journal of the Commons* 7(2): 432–453.

Fleischman, F.D., N.C. Ban, L.S. Evans, G. Epstein, G. Garcia-Lopez, and S. Villamayor-Tomas. 2014. 'Governing Large-scale Social-Ecological Systems: Lessons from Five Cases.' *International Journal of the Commons* 8(2): 428–456.

Galaz, V., P. Olsson, T. Hahn, C. Folke, and U. Svedin. 2008. 'The Problem of Fit among Biophysical Systems, Environmental and Resource Regimes, and Broader Governance Systems: Insights and Emerging Challenges.' In *Institutions and Environmental Change – Principal Findings, Applications, and Research Frontiers*, edited by O.R. Young, L.A. King, and H. Schröder, 147–182. Cambridge: MIT Press. www.stockholmresilience.org/publications/artiklar/2009-12-22-the-problem-of-fit-between-governance-systems-and-environmental-regimes.html.

Gelcich, S., R. Guzman, C. Rodríguez-Sickert, J.C. Castilla, and J.C. Cárdenas. 2013. 'Exploring External Validity of Common Pool Resource Experiments: Insights from Artisanal Benthic Fisheries in Chile.' *Ecology and Society* 18(3): 2.

Gibson, C.C., M.A. McKean, and E. Ostrom, eds. 2000. *People and Forests: Communities, Institutions, and Governance*. Cambridge: MIT Press.

Gutiérrez, N.L., R. Hilborn, and O. Defeo. 2011. 'Leadership, Social Capital and Incentives Promote Successful Fisheries.' *Nature* 470(7334): 386.

Hagedorn, K. 2008. 'Particular Requirements for Institutional Analysis in Nature-related Sectors.' *European Review of Agricultural Economics* 35(3): 357–384.

Hall, P.A., and R.C.R. Taylor. 1996. 'Political Science and the Three New Institutionalisms.' *Political Studies* 44: 936–957.

Hardin, G. 1968. 'The Tragedy of the Commons.' *Science* 162: 1243–1248.

Heikkila, T., and C.M. Weible. 2018. 'A Semiautomated Approach to Analyzing Polycentricity.' *Environmental Policy and Governance* 28(4): 308–318.

Hodgson, G.M. 1998. 'The Approach of Institutional Economics.' *Journal of Economic Literature* 36(1): 166–192.

Janssen, M.A. 2013. 'The Role of Information in Governing the Commons: Experimental Results.' *Ecology and Society* 18(4): 4.

Janssen, M.A., R. Holahan, A. Lee, and E. Ostrom. 2010. 'Lab Experiments for the Study of Social-Ecological Systems.' *Science* 328(5978): 613–617.

Jones, R., C. Rigg, and E. Pinkerton. 2017. 'Strategies for Assertion of Conservation and Local Management Rights: A Haida Gwaii Herring Story.' *Marine Policy* 80: 154–167.

Kiser, LL., and E. Ostrom. 1982. 'The Three Worlds of Action: A Metatheoretical Synthesis of Institutional Approaches.' In *Polycentric Games and Institutions: Readings from the Workshop in Political Theory and Policy Analysis*, edited by M.D. McGinnis, 56–89. Ann Arbor: University of Michigan Press.

Lien, A.M., E. Schlager, and A. Lona. 2018. 'Using Institutional Grammar to Improve Understanding of the Form and Function of Payment for Ecosystem Services Programs.' *Ecosystem Services* 31: 21–31.

Lindahl, T., O. Bodin, and M. Tengö. 2015. 'Governing Complex Commons – The Role of Communication for Experimental Learning and Coordinated Management.' *Ecological Economics* 111: 111–120.

Lubell, M. 2013. 'Governing Institutional Complexity: The Ecology of Games Framework.' *Policy Studies Journal* 41: 537–559.

McGinnis, M.D. 2011. 'An Introduction to IAD and the Language of the Ostrom Workshop: A Simple Guide to a Complex Framework.' *Policy Studies Journal* 39(1): 169–183.

McGinnis, M.D., and E. Ostrom. 2014. 'Social-Ecological System Framework: Initial Changes and Continuing Challenges.' *Ecology and Society* 19(2): 30.

Mitchell, W.C. 1988. 'Virginia, Rochester, and Bloomington: Twenty-Five Years of Public Choice and Political Science.' *Public Choice* 56: 101–119.

Monroy-Sais, S., A. Castillo, E. García-Frapolli, and G. Ibarra-Manríquez. 2016. 'Ecological Variability and Rule-Making Processes for Forest Management Institutions: A Social-Ecological Case Study in the Jalisco Coast, Mexico.' *International Journal of the Commons* 10(2): 1144–1171.

Morrison, T.H., W.N. Adger, K. Brown, M.C. Lemos, D. Huitema, and T.P. Hughes. 2017. 'Mitigation and Adaptation in Polycentric Systems: Sources of Power in the Pursuit of Collective Goals.' *Wiley Interdisciplinary Reviews: Climate Change* 8(5): e479.

Ostrom, E. 1990. *Governing the Commons*. New York: Cambridge University Press.

Ostrom, E. 1998. 'A Behavioral Approach to the Rational Choice Theory of Collective Action: Presidential Address, American Political Science Association 1997.' *The American Political Science Review* 92: 1–22.

Ostrom, E. 2005. *Understanding Institutional Diversity*. Princeton: Princeton University Press.

Ostrom, E. 2007. 'A Diagnostic Approach for Going Beyond Panaceas.' *Proceedings of the National Academy of Sciences* 104(39): 15181–15187.

Ostrom, E., and X. Basurto. 2011. 'Crafting Analytical Tools to Study Institutional Change.' *Journal of Institutional Economics* 7(Special Issue 3): 317–343.

Ostrom, E., R. Gardner, and J. Walker. 1994. *Rules, Games and Common Pool Resources*. Ann Arbor: Michigan University Press.

Ostrom, E., R.B. Parks, and P.W. Gordon. 1973. 'Do We Really Want to Consolidate Urban Police Forces? A Reappraisal of Some Old Assertions.' *Public Administration Review* 33(5): 423–432.

Ostrom, V., C.M. Tiebout, and R. Warren. 1961. 'The Organization of Government in Metropolitan Areas: A Theoretical Inquiry.' *The American Political Science Review* 55: 831–842.

Partelow, S. 2018. 'A Review of the Social-Ecological Systems Framework: Applications, Methods, Modifications, and Challenges.' *Ecology and Society* 23(4): 36.

Persha, L., A. Agrawal, and A. Chhatre. 2011. 'Social and Ecological Synergy: Local Rulemaking, Forest Livelihoods, and Biodiversity Conservation.' *Science* 331: 1606–1608.

Poteete, A.R., M.A. Janssen, and E. Ostrom. 2010. *Working Together: Collective Action, the Commons, and Multiple Methods in Practice*. Princeton: Princeton University Press.

Ribot, J.C., A. Agrawal, and A.M. Larson. 2006. 'Recentralizing While Decentralizing: How National Governments Reappropriate Forest Resources.' *World Development* 34: 1864–1886.

Schlager, E. 1999. 'A Comparison of Frameworks, Theories, and Models of Policy Processes.' *Theories of the Policy Process* 1: 233–260.
Schlager, E., W. Blomquist, and S.Y. Tang. 1994. 'Mobile Flows, Storage, and Self-Organized Institutions for Governing Common-pool Resources.' *Land Economics* 70: 294–317.
Schlager, E., and M. Cox. 2018. 'The IAD Framework and the SES Framework: An Introduction and Assessment of the Ostrom Workshop Frameworks.' In *Theories of the Policy Process*, edited by C.M. Weible and P.A. Sabatier, 225–262. Abingdon: Routledge.
Schill, C., T. Lindahl, and A-S. Crépin. 2015. 'Collective Action and the Risk of Ecosystem Regime Shifts: Insights from a Laboratory Experiment.' *Ecology and Society* 20(1): 48.
Schoon, M., J.A. Baggio, K.R. Salau, and M. Janssen. 2014. 'Insights for Managers from Modeling Species Interactions across Multiple Scales in an Idealized Landscape.' *Environmental Modelling and Software* 54: 53–59.
Shepsle, K.A. 2006. 'Rational Choice Institutionalism.' In *The Oxford Handbook of Political Institutions*, edited by R.A. Rhodes, S.A. Binder, and B.A. Rockman. Oxford: Oxford University Press.
Siddiki, S, X. Basurto, and C.M. Weible. 2012. 'Using the Institutional Grammar Tool to Understand Regulatory Compliance: The Case of Colorado Aquaculture.' *Regulation & Governance* 6(2): 167–188.
Siddiki, S., C.M. Weible, X. Basurto, and J. Calanni. 2011. 'Dissecting Policy Designs: An Application of the Institutional Grammar Tool.' *Policy Studies Journal* 39(1): 79–103.
Thiel, A., M.E. Adamseged, and C. Baake. 2015. 'Evaluating an Instrument for Institutional Crafting: How Ostrom's Social-Ecological Systems Framework is Applied.' *Environmental Science and Policy* 53: 152–164.
Van Laerhoven, F., and E. Ostrom. 2007. 'Traditions and Trends in the Study of the Commons.' *International Journal of the Commons* 1: 3–28.
Vatn, A. 2005. 'Rationality, Institutions and Environmental Policy.' *Ecological Economics* 55(2): 203–217.
Villamayor-Tomas, S., and G. García-López. 2017. 'The Influence of Community-based Resource Management Institutions on Adaptation Capacity: A Large-N Study of Farmer Responses to Climate and Global Market Disturbances.' *Global Environmental Change* 47: 153–166.
Villamayor-Tomas, S., P. Grundmann, G. Epstein, T. Evans, and C. Kimmich. 2015. 'The Water-Energy-Food Security Nexus Through the Lenses of the Value Chain and IAD Frameworks.' *Water Alternatives* 8(1): 735–755.
Young, O.R. 2002. *The Institutional Dimensions of Environmental Change: Fit, Interplay, and Scale.* Cambridge: MIT Press.

23
Network analysis

Kristine Maciejewski[1] *and Jacopo Baggio*[2,3]

[1]CENTRE FOR SUSTAINABILITY TRANSITIONS, STELLENBOSCH UNIVERSITY, STELLENBOSCH, SOUTH AFRICA
[2]SCHOOL OF POLITICS, SECURITY, AND INTERNATIONAL AFFAIRS, UNIVERSITY OF CENTRAL FLORIDA, FLORIDA, USA
[3]SUSTAINABLE COASTAL SYSTEMS CLUSTER, NATIONAL CENTER FOR INTEGRATED COASTAL RESEARCH, UNIVERSITY OF CENTRAL FLORIDA, FLORIDA, USA

Key methods discussed in this chapter

Network analysis

Connections to other chapters

Network analysis connects to various other methods, particularly as it relies on the collection of data which is covered in ecological field data collection (Chapter 6), interviews and surveys (Chapter 7) and participatory data collection (Chapter 8). The metrics generated through network analysis can be statistically analysed and therefore linked to statistical analysis (Chapter 18). As a tool, network analysis can also be linked to ecosystem service modelling (Chapter 31) to study the linkages between social and ecological components in a system.

Introduction

Network analysis, based on graph theory and statistics, provides a rigorous, systematic approach to studying how relationships and their structuring influence social-ecological systems (SES). A network perspective allows a researcher to analyse how landscape, species, individuals and organisations, among others, are connected and how these structures enable specific processes such as information sharing, opinions, policy adoption, species migration, the spread of epidemics and ecosystem flows (Dakos et al. 2015). Network analysis is widely used to analyse the structural properties of complex systems and gives insight into how a system works by understanding the role that individual parts play in the system through their connections to other parts. This method frees us from the typical assumptions that individuals act independently. Instead, it embraces the importance of relationships and provides a potential bridge between different disciplines.

Generally speaking, a network is defined as a set of nodes connected via edges. That is, a network consists of two types of components: (a) nodes (also called 'vertices' or 'actors'), which can represent people, places, organisations, species and so on, and (b) edges (ties,

DOI: 10.4324/9781003021339-27

SUMMARY TABLE: NETWORK ANALYSIS	
DISCIPLINARY BACKGROUND	**KNOWLEDGE TYPE**
The methods in this chapter are derived from or have most commonly been used in: Ecology, Resource Economics, Computational Geography, Systems Dynamics, Computer Science, Information Science	*The methods in this chapter are primarily used to generate the following types of knowledge:* • Descriptive • Exploratory • Explanatory
RESEARCH APPROACH	**PURPOSE OF METHOD**
The methods in this chapter originate from or most commonly adopt the following research approaches: • Analytical/objective	*The most common purposes of using the methods in this chapter are:* • System understanding
TEMPORAL DIMENSION	**SYSTEMIC FEATURES AND PROCESSES**
The methods in this chapter are most commonly applied to the following temporal dimensions: • Present (typically within the last 5–10 years) • Recent past (post-1700s)	*While most methods can do many things, the methods in this chapter are particularly good (i.e. go-to methods) for addressing the following:* • SES components and linkages • Power relations • Collective action and collaborative governance
SPATIAL DIMENSION	
The methods in this chapter are primarily either or both: • Non-spatial *The methods in this chapter are most commonly applied at the following spatial scales:* • Local • Regional (provincial/state to continental)	

relationships, links), which represent a 'connection' existing between two nodes. Edges can represent friendships, legal authorities, collaborations, conflicts and also species migration, ecosystem flows, resource exchanges (including predation) and extraction. Networks can either be directed or undirected, and weighted or unweighted. Directed networks often represent a flow of information, e.g. from A to B but not vice versa, whereas undirected networks represent an interaction between two nodes (A and B) but with no direction. Edges can be unweighted, representing only instances in which A and B are connected or not; or weighted, representing also the 'strength' of the connection between A and B. Weights can be based on distance, frequency of contact, amount of resources exchanged, or 'value judgement of the relationship'.

The first study of networks can be ascribed to Euler in 1739 when he abstracted a geographical space and for the first time described entities as nodes and edges, representing land and bridges connecting different parts of Königsberg in Germany (Euler and Euler 1736). Two centuries later, modern network analysis originated from the social sciences. Jacob L. Moreno, an American psychiatrist, was interested in the dynamics of social interactions and used network analysis to map a social network of Hudson School for Girls, following an epidemic of runaway children in 1932 (Borgatti et al. 2009). He graphically mapped individuals' feelings towards one another to identify the channels of social influence and ideas among the girls. In the 1940s and 1950s, work in social network mapping or analyses advanced towards the use of matrix algebra and graph theory as researchers began to study the effects of different communication network structures to solve problems (Moreno 1934).

About 20 years later, another step towards modern network analysis was undertaken by two Hungarian mathematicians, Paul Erdös and Alfréd Rényi, who mathematically described the properties of random graphs (Erdös and Rényi 1959, 1960). However, it was not until the end of the 1990s and the beginning of the 2000s, with the advancement of computing power to analyse data, that network analysis really took off as a method to assess the structural properties of complex systems (Watts and Strogatz 1998; Albert and Barabási 2001). The use of network analysis enabled researchers to start disentangling the relationships between human and organisational interactions and how structure facilitates specific outcomes, such as the ranking of economists and understanding psychological well-being. It was also used in social and policy analysis. Today, network analysis is used in a wide variety of different fields including neuroscience, economics, political science, genetics, ecology, biology, sociology, psychology, engineering, computer science and physics (Caldarelli 2007; Borgatti et al. 2009; Barthélemy 2011; Costa et al. 2011).

Although network analysis has a long interdisciplinary history, its application to SES (involving the construction specifically of interlinked social-ecological networks) has started only recently (Bodin and Tengö 2012; Baggio et al. 2016; Bodin et al. 2016; Barnes et al. 2017; Sayles and Baggio 2017a; Baggio and Hillis 2018; Sayles et al. 2019). Network analysis can be used to analyse and assess how social and ecological processes are influenced by the underlying connectivity structure.

This chapter focuses on applications of network analysis to SES. Network analysis can be used in combination with other methods in this handbook, such as interviews and surveys (Chapter 7) and ecological field data collection (Chapter 6) to better understand SES.

SES problems and questions

In the context of SES, network analysis can be used to understand how social and ecological system components are connected and how the structure or pattern of connections affects the function of the SES, thus taking account of how systems are constituted relationally.

Ecological and social processes 'propagate' along specific networks and connectivity. Network structure plays a central role in constraining and/or facilitating these different processes. Examples of key social processes (flows) in SES include sharing ideas, management strategies, information, knowledge, economic/financial flows, and conflict and cooperation. Examples of key ecological processes (flows) in SES include nutrient flows, the transfer of energy from one species to another (food webs), water flowing within rivers and species migration. These social and ecological processes are connected via specific processes related to management (e.g. altering specific social and/or ecological processes such as creating forums for sharing ideas on natural resource management, damming a river, building fences), resource extraction (e.g. water, food, hunting) and resource production (e.g. pollution). These processes are often interdependent and interact with one another.

Both local and global network characteristics affect how specific social-ecological processes unfold. In fact, specific 'structures' (values of metrics, distribution of nodal metrics) can facilitate or hinder species migration, the spread of pests, biological invasion, knowledge sharing, management strategies, innovations, learning, financial flows and food sharing (Garlaschelli 2004; Bodin and Norberg 2007; Baggio et al. 2011; Barthélemy 2011; Costa et al. 2011; Granell, Gomez, and Arenas 2014; De Domenico et al. 2016). More so, network analysis can show how structural characteristics of SES can affect their ability of a system to withstand specific disturbances, and can give insight into how processes and flows may change in response to those disturbances (Albert, Jeong, and Barabási 2000; Nicosia et al. 2012; De Domenico et al. 2014; Brummitt, Barnett, and D'Souza 2015; Poledna et al. 2015; Baggio et al. 2016).

Typical questions that network analysis can be used for to understand SES include:

- Who are the key stakeholders, potential leaders or agents of change in the system? (e.g. a study by Bodin and Crona (2008) used network analysis to identify key individuals in a fishing community to explain the lack of common initiative to deal with the overexploitation of fisheries)
- Which specific human–environmental dependencies are key for the functioning of the overall system? (e.g. network analysis was used to link processes of change in SES to decision-making across multiple layers of rules underpinning societal organisation (Barnes et al. 2017))
- Are there spatial-scale mismatches between social and ecological systems within an SES? (e.g. are there issues of fit between ecological processes to be managed and the socio-political unit tasks involved in managing these systems? (Sayles and Baggio 2017b))
- How do individuals affect ecological processes? (e.g. a wildfire transmission network, developed through simulation of wildfires, was compared to a governance network to determine 'risk interdependence archetypes' based on the spatial configurations by which one actor is exposed to risk through the actions of another actor (Hamilton, Fischer, and Ager 2019))
- How can social perceptions of an SES be assessed? (e.g. using data collected from participatory workshops, network analysis was used to map out how communities perceive the performance of a community-based natural resource management system (Delgado-Serrano et al. 2015))
- How do systems respond to exogenous effects? (e.g. Frank and Fahrbach (1999) studied the interactions between actors and their sentiments (values, attitudes, beliefs, opinions) to understand how natural resource use is influenced by others)
- How do collaborations emerge and function within the context of natural resource management? (e.g. Bodin et al. (2017) used network analysis to show that different network characteristics can give rise to similar ecosystem-based management outcomes)

- How is collaboration influenced by the origin of the relationship? (e.g. Sayles and Baggio (2017b) looked at the difference in perceived collaboration productivity between relationships that were mandated, born out of shared interest, funded, or a mix of the three categories)

Network analysis may also be used to facilitate collaborative processes for scoping the nature of the relationships that shape SES and interactions (e.g. using NET-MAP, an interview-based mapping tool that helps people to understand, visualise, discuss and improve situations in which many different actors influence outcomes (netmap.wordpress.com/about) in order to map stakeholder networks in facilitated processes aimed at the co-production of knowledge related to a specific SES).

Brief description of key methods

Network analysis can be used to understand how entities (social and/or ecological structures) are connected and how they relate to one another. In network analysis, these properties can be quantified using network metrics that have been developed to answer specific questions.

By calculating the centrality value of a network, one can determine the most central node in the network, which can be seen as the most important node that connects all the nodes. An outbreak or spread of a virus in a population, for example, can be illustrated using network analysis, where the nodes represent the infected individuals and the interaction between individuals, and the transmission or spread of the virus is depicted as the edges of the network. The individuals' position in the population can be described by calculating the centrality value of each node in the network using centrality analysis (Table 23.1). The most central node represents the individual with the highest number of contacts. This information can then be used to identify the most contagious individual in the network, or where the virus originally started from.

As shown in Table 23.1, network analysis can be used to understand species interactions in food webs by simulating species loss. For each food web, species can be sequentially removed, focusing on the most connected species, randomly chosen species and the least connected species. The number of prey and predator links can then be counted to determine total trophic connections, or network degree. This can also be used to determine the robustness of food webs, as the fraction of species that had to be removed to result in a total loss of less than 50% of the species (Dunne, Williams, and Martinez 2002). Network analysis can also be used to track species movement patterns across habitat patches to quantify habitat connectivity. This is particularly relevant for managing conservation landscapes as it helps to identify critical 'stepping stone' patches that, when removed, may cause changes in habitat connectivity (Keitt, Urban, and Milne 1997).

Table 23.1 summarises a few of the most common uses of network analysis metrics in SES research.

Limitations

As SES are characterised by dynamic interactions between many social and/or ecological components, it is challenging to capture all the relevant actors and social and ecological relationships. Social-ecological systems are also 'open' systems that have interactions with other systems, and are therefore not clearly bounded. The structure of the network depends on the context of the system, which is usually clarified by setting system boundaries. The choice of boundary therefore has direct implications.

Table 23.1 Summary of key metrics used in network analysis

Main application	*Description*	*References*
Centrality analysis	Centrality analysis is used to identify the role of key individuals to enable natural resource governance, and/or species/ landscape patches that are key to the stability of an ecosystem.	Bodin and Crona 2008
Centrality/ participation coefficient analysis mixing network and 'reason for existing relationship'	Centrality/participation coefficient analysis is used to identify how different types of relationships may affect productivity and/or outcomes in SES.	Sayles and Baggio 2017b
Mono or multiplex network analysis of 'sharing' relationship	Understanding the perceptions of local communities and sharing relationships may assist in building resilience in community-based natural resource management.	Delgado-Serrano et al. 2015; Baggio et al. 2016
Modelling motifs of social-ecological networks	Modelling motifs of social-ecological networks help researchers to understand how natural resource users make decisions and how these decisions are influenced by different types of relationships.	Frank and Fahrbach 1999; Bodin et al. 2016; Barnes et al. 2017
Food web analysis	Food web analysis is used to understand species interactions and identify keystone species.	Dunne, Williams, and Martinez 2002; Garlaschelli 2004
Map out species movement patterns	Animal movements between habitat patches are used to map out the movement patterns of species to identify improved habitat connectivity.	Keitt, Urban, and Milne 1997; Urban and Keitt 2001; Minor and Urban 2007; Baggio et al. 2011
Map out and analyse social-ecological interactions	By mapping out and analysing social-ecological interactions, researchers can understand how actors and their relationship with other actors, and with different interconnected ecosystem components, contribute to different management/governance outcomes.	Bodin and Tengö 2012; Sayles and Baggio 2017a
Model processes in networks	By modelling the processes in networks, researchers can assess how structural properties joined with specific nodal characteristics influence ecological and social flows.	Baggio and Hillis 2018
Plot multi-level networks	Plotting multi-level networks enable researchers to compare two or more interconnected networks.	Bodin and Tengö 2012; Barnes et al. 2017

Since many network analysis methods demand complete network data (with all relevant links and nodes within the system boundaries being defined and measured), they are often limited to a particular and contextual scale, and therefore not always useful for capturing the radical openness of systems. Collecting the empirical data that underlie a network analysis often requires extensive fieldwork, which typically involves interacting with or surveying as many actors as possible. Network analysis therefore requires extensive effort in terms of both time and money for a network to be correctly inferred. In fact, given the specificity of network data, one cannot rely on random sampling but needs to rely on different sampling techniques (e.g. snowball, census).

Network analysis often represents merely a snapshot of how the system is connected. However, this limitation can be overcome by a dynamic network analysis, which is particularly useful for assessing causalities and answering questions such as what factors contribute to the formation of certain networks. Still, given time and cost constraints, as well as the potential fatigue of individuals in the case of social networks, longitudinal network studies are rare.

Resource implications

Network analysis software consists of either packages based on graphical user interfaces (GUIs) or packages built for scripting and coding. The GUI packages are generally easier to learn and are widely used. There are many open-source examples such as Gephi, NodeXL, EgoNet, MPnet and UCINet.

Scripting tools used for network analysis include NetMiner with a Python scripting engine; the statnet suite of packages for the R statistical programming language, igraph; the NetworkX library for Python; and the SNAP packages for network analysis in C++ and Python. Scripting tools can also be based in R via the sna, igraph and statnet packages. For advanced network analysis that relates to the use of ERGM, statnet in R and the stand-alone Pnet can be used. For multi-layer/multiplex network analysis in R, one can use the multinet package or MuxViz, which has its own interface. Multiplex networks can also be analysed in Python via the Pymnet package.

New directions

Although social-ecological network analysis (SENA) is still in its infancy, it has shown promise in advancing difficult SES problems such as identifying potential social-ecological scale mismatches (Sayles and Baggio 2017a) and assessing the robustness of social-ecological networks to social or ecological perturbations (Markowetz 2010). Looking towards the future, better integration of qualitative data-collection methods and protocols, with a strong and rigorous mathematical framework to analyse the complex web of interactions and interdependencies that exist in SES, is still needed.

Promising tools from a mathematical or statistical perspective are the implementation of multi-level exponential random graphs (Wang et al. 2009) and multi-layer/multiplex networks (De Domenico et al. 2014; Kivelä et al. 2014). Exponential random graphs (and by extension multi-level exponential random graphs) enable the analysis of SES via the presence of specific microlevel configurations called 'motifs' (Bodin and Tengö 2012; Bodin et al. 2016; Barnes et al. 2017; Guerrero et al. 2018). Exponential random graph motifs allow a researcher to understand how macrolevel structures are probabilistically related to specific network motifs, and how those motifs may affect outcomes of interest. Multi-layer networks allow researchers to analyse the overall macrolevel network properties and local nodal properties in multiple interdependent networks (Bodin et al. 2019).

Case study 23.1: Protected area networks in South Africa

Protected area networks respectively situated in the Western and Eastern Cape provinces of South Africa illustrate how network analysis can be used to measure the resilience of SES and to identify individual actors that are important for overall network connectivity (Maciejewski and Cumming 2015).

Interviews were conducted with managers from various protected areas in the two protected area networks to understand how those managers were interacting with managers from surrounding protected areas. Interaction was defined as exchanging ideas; sharing equipment; trading in wildlife; engaging in discussions with regard to management, education, tourism and research; and forming collaborations, among others. This information was used to generate a graph network, where nodes represented the protected areas where interviewed managers were based (source) as well as the protected areas (targets) to which they were connected through various interactions. The edges (links) of the protected area network concerned consisted of directed linkages between source and target protected areas.

Network analysis indicated that the role of protected areas differed in management strategies between the two protected area networks, as illustrated by the positioning and size of the nodes and the density of connections (Figure 23.1). Nodes were sized according to each node's eigenvector (characteristic vector) centrality, which is a measure of the influence of a node in a network. A high eigenvector centrality score means a node is connected to many nodes which themselves have high scores, and indicate that they are in a position to receive and control flows (Borgatti 2005). Both protected area networks had shorter diameters than expected. This suggests frequent interactions between the managers, which may be beneficial to the spread of information (Janssen et al. 2006). A small network diameter also indicates high reachability, which increases the ability of the system to respond to change.

Both networks exhibited socio-economic interactions that were more intense between protected areas situated near one another in geographic space than between protected areas belonging to the same organisation. In other words, as would be expected for ecological connectivity, geographic proximity mattered more than organisational membership in the formation of socio-economic interactions.

The networks in Figure 23.1 were generated using network analysis. Nodes are coloured or shaded according to organisation (salmon/black: national parks; yellow/

Key readings

Barabási, A-L. n.d. *Network Science*. www.networksciencebook.com/chapter/1 (Online and free).

Bodin, Ö., and B. Crona. 2008. 'Community-Based Management of Natural Resources – Exploring the Role of Social Capital and Leadership in a Rural Fishing Community.' *World Development* 36(12): 2763–2779.

Bodin, Ö., and M. Tengö. 2012. 'Disentangling Intangible Social-Ecological Systems.' *Global Environmental Change* 22(2): 430–439. doi:10.1016/j.gloenvcha.2012.01.005.

Gonzalez, A., P. Thompson, and M. Loreau. 2017. 'Spatial Ecological Networks: Planning for Sustainability on the Long-term.' *Current Opinion in Environmental Sustainability* 29: 187–197.

Wasserman, S., and K. Faust. 1994. *Social Network Analysis: Methods and Applications*. Cambridge: Cambridge University Press.

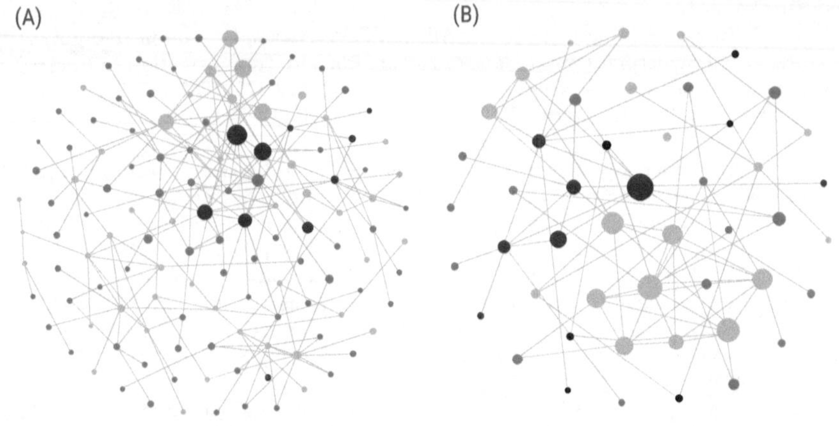

Figure 23.1 Protected area network of (A) the Western Cape, and (B) the Eastern Cape province of South Africa (Maciejewski and Cumming 2015)

dark grey: provincial parks; green/grey: private protected areas) and were sized and positioned according to centrality.

This example illustrates how network analysis provides a platform to spatially present the relational arrangement between SES components and how this can be used to understand the most important role-players in these networks, as well as those acting as stepping stones. Understanding the arrangement of different socio-economic interactions and the role that individual protected areas play in the network is important when it comes to making decisions about the network as a whole. In the Western Cape, for example, national parks should be consulted when designing the protected area expansion strategy or when making national overarching decisions about the future development of protected areas. However, when dealing with the management of private parks, it would be more strategic to consult with the surrounding provincial parks. This is clearly illustrated in the Biodiversity Stewardship Programme initiated by CapeNature in 2003, which facilitates conservation on privately owned land by setting up agreements between landowners and the provincial parks.

References

Albert, R., and A-L. Barabási. 2001. 'Statistical Mechanics of Complex Networks.' doi:10.1103/RevModPhys.74.47.

Albert, R., H. Jeong, and A-L. Barabási. 2000. 'Error and Attack Tolerance of Complex Networks.' *Nature* 406(6794): 378–382. doi:10.1038/35019019.

Baggio, J.A., S.B. BurnSilver, A. Arenas, J.S. Magdanz, G.P. Kofinas, and M. de Domenico. 2016. 'Multiplex Social Ecological Network Analysis Reveals How Social Changes Affect Community Robustness More Than Resource Depletion.' *Proceedings of the National Academy of Sciences* 113(48): 13708–13713. doi:10.1073/pnas.1604401113.

Baggio, J.A., and V. Hillis. 2018. 'Managing Ecological Disturbances: Learning and the Structure of Social-Ecological Networks.' *Environmental Modelling and Software* 109(August): 32–40. doi:10.1016/j.envsoft.2018.08.002.

Baggio, J.A., K. Salau, M.A. Janssen, M.L. Schoon, and Ö. Bodin. 2011. 'Landscape Connectivity and Predator-Prey Population Dynamics.' *Landscape Ecology* 26(1): 33–45. doi:10.1007/s10980-010-9493-y.

Barnes, M.L., Ö. Bodin, A. Guerrero, R. McAllister, S.M. Alexander, and G. Robins. 2017. 'Theorizing the Social Structural Foundations of Adaptation and Transformation in Social-Ecological Systems.' *Social Science Research Network* (January): 1–19. doi:10.5751/ES-09769-220416.

Barthélemy, M. 2011. 'Spatial Networks.' *Physics Reports* 499(1–3): 1–101. doi:10.1016/j.physrep.2010.11.002.

Bodin, Ö., S.M. Alexander, J.A. Baggio, M.L. Barnes, R. Berardo, G.S. Cumming, L.E. Dee et al. 2019. 'Improving Network Approaches to the Study of Complex Social–Ecological Interdependencies.' *Nature Sustainability* 2(June): 551–559. www.nature.com/articles/s41893-019-0308-0.

Bodin, Ö., M.L. Barnes, R.R.J. McAllister, J.C. Rocha, and A.M. Guerrero. 2017. 'Social–Ecological Network Approaches in Interdisciplinary Research: A Response to Bohan et al. and Dee et al.' *Trends in Ecology & Evolution* 32(8): 547–549. doi:10.1016/j.tree.2017.06.003.

Bodin, Ö., and B. Crona. 2008. 'Community-based Management of Natural Resources – Exploring the Role of Social Capital and Leadership in a Rural Fishing Community.' *World Development* 36(12): 2763–2779.

Bodin, Ö., and J. Norberg. 2007. 'A Network Approach for Analyzing Spatially Structured Populations in Fragmented Landscape.' *Landscape Ecology* 22(1): 31–44. doi:10.1007/s10980-006-9015-0.

Bodin, Ö., G. Robins, R.R. McAllister, A.M. Guerrero, B. Crona, M. Tengö, and M. Lubell. 2016. 'Theorizing Benefits and Constraints in Collaborative Environmental Governance: A Transdisciplinary Social-Ecological Network Approach for Empirical Investigations.' *Ecology and Society* 21(1): 40. doi:10.5751/ES-08368-210140.

Bodin, Ö., and M. Tengö. 2012. 'Disentangling Intangible Social-Ecological Systems.' *Global Environmental Change* 22(2): 430–439. doi:10.1016/j.gloenvcha.2012.01.005.

Borgatti, S.P. 2005. 'Centrality and Network Flow.' *Social Networks* 27(1): 55–71.

Borgatti, S.P., A. Mehra, D.J. Brass, and G. Labianca. 2009. 'Network Analysis in the Social Sciences.' *Science* 323(5916): 892–895. doi:10.1126/science.1165821.

Brummitt, C.D., G. Barnett, and R.M. d'Souza. 2015. 'Coupled Catastrophes: Sudden Shifts Cascade and Hop Among Interdependent Systems.' *Journal of the Royal Society Interface* 12(112). doi:10.1098/rsif.2015.0712.

Caldarelli, G. 2007. *Scale-Free Networks: Complex Webs in Nature and Technology.* Oxford: Oxford University Press.

Costa, L. da F., O.N. Oliveira Jr, G. Travieso, F.A. Rodrigues, P.R. Villas Boas, L. Antiqueira, M. Palhares Viana, and L.E. Correa Rocha. 2011. 'Analyzing and Modeling Real-World Phenomena with Complex Networks: A Survey of Applications.' *Advances in Physics* 60(3): 329–412. doi:10.1080/00018732.2011.572452.

Dakos, V., A. Quinlan, J.A. Baggio, E. Bennett, Ö. Bodin, and S. BurnSilver. 2015. 'Principle 2 – Manage Connectivity.' In *Principles for Building Resilience*, edited by R. Biggs, M. Schlüter, and M.L. Schoon, 80–104. Cambridge: Cambridge University Press. doi:10.1017/CBO9781316014240.005.

De Domenico, M., C. Granell, M.A. Porter, and A. Arenas. 2016. 'The Physics of Spreading Processes in Multilayer Networks.' *Nature Physics* 12(10): 901–906. doi:10.1038/nphys3865.

De Domenico, M., A. Solé-Ribalta, E. Cozzo, M. Kivelä, Y. Moreno, M.A. Porter, S. Gómez, and A. Arenas. 2014. 'Mathematical Formulation of Multilayer Networks.' *Physical Review X* 3(4): 1–15. doi:10.1103/PhysRevX.3.041022.

Delgado-Serrano, M., E. Oteros-Rozas, P. Vanwildemeersch, C. Ortíz Guerrero, S. London, and R. Escalante. 2015. 'Local Perceptions on Social-Ecological Dynamics in Latin America in Three Community-based Natural Resource Management Systems.' *Ecology and Society* 20(4): 24. doi:10.5751/ES-07965-200424.

Dunne, J.A., R.J. Williams, and N.D. Martinez. 2002. 'Network Structure and Biodiversity Loss in Food Webs: Robustness Increases with Connectance.' *Ecology Letters* 5(4): 558–567.

Erdös, P., and A. Rényi. 1959. 'On Random Graphs.' *Publicationes Mathematicae* 6: 290–297. doi:10.2307/1999405.

Erdös, P., and A. Rényi. 1960. 'On the Evolution of Random Graphs.' *Publications of the Mathematical Institute of the Hungarian Academy of Sciences* 5: 17–61.

Euler, L., and L. Euler. 1736. 'Solutio Problematis ad Geometrian Situs Pertinentis.' *Comentarii Academiae Scientarum Petropolitanae*. doi:002433.d/232323.

Frank, K.A., and K. Fahrbach. 1999. 'Organisational Culture as a Complex System: Balance and Information in Models of Influence and Selection.' *Organization Science* 10(3): 253–277.

Garlaschelli, D. 2004. Universality in Food Webs. *European Physical Journal B* 38(2): 277–285. doi:10.1140/epjb/e2004-00120-3.

Granell, C., S. Gomez, and A. Arenas. 2014. 'Competing Spreading Processes on Multiplex Networks: Awareness and Epidemics.' *Physical Review E – Statistical, Nonlinear, and Soft Matter Physics* 90(1): 1–7. doi:10.1103/PhysRevE.90.012808.

Guerrero, A.M., N.J. Bennett, K.A. Wilson, N. Carter, D. Gill, M. Mills, C.D. Ives et al. 2018. 'Achieving the Promise of Integration in Social-Ecological Research: A Review and Prospectus.' *Ecology and Society* 23(3): 38. doi:10.5751/ES-10232-230338.

Hamilton, M., A.P. Fischer, and A. Ager. 2019. 'A Social-Ecological Network Approach for Understanding Wildfire Risk Governance.' *Global Environmental Change* 54: 113–123.

Janssen, M.A., Ö. Bodin, J.M. Anderies, T. Elmqvist, H. Ernstson, R.R.J. McAllister, P. Olsson, and P. Ryan. 2006. 'Toward a Network Perspective of the Study of Resilience in Social-Ecological Systems.' *Ecology and Society* 11(1): 15.

Keitt, T., D. Urban, and B. Milne 1997. 'Detecting Critical Scales in Fragmented Landscapes.' *Conservation Ecology* 1(1): 4.

Kivelä, M., A. Arenas, M. Barthélemy, J.P. Gleeson, Y. Moreno, and M.A. Porter. 2014. 'Multilayer Networks.' *ArXiv* 2: 203–271. doi:arXiv:1309.7233.

Maciejewski, K., and G. Cumming. 2015. 'The Relevance of Socio-economic Interactions for the Resilience of Protected Area Networks.' *Ecosphere* 6(9): 1–14.

Markowetz, F. 2010. 'How to Understand the Cell by Breaking It: Network Analysis of Gene Perturbation Screens.' *PLoS Computational Biology* 6(2): e1000655.

Minor, E.S., and D.L. Urban. 2007. 'Graph Theory as a Proxy for Spatially Explicit Population Models in Conservation Planning.' *Ecological Applications* 17(6): 1771–1782. doi:10.1890/06-1073.1.

Moreno, J.L. 1934. 'Who Shall Survive: A New Approach to the Problem of Human Interrelations.' *Nervous and Mental Diseases Monographs* 58. https://archive.org/details/whoshallsurviven00jlmo

Nicosia, V., R. Criado, M. Romance, G. Russo, and V. Latora. 2012. 'Controlling Centrality in Complex Networks.' *Scientific Reports* 2: 1–7. doi:10.1038/srep00218.

Poledna, S., J.L. Molina-Borboa, S. Martinez-Jaramillo, M. van der Leij, and S. Thurner. 2015. 'The Multi-layer Network Nature of Systemic Risk and its Implications for the Costs of Financial Crises.' *Journal of Financial Stability* 20: 70–81. doi:10.2307/2597748.

Sayles, J.S., and J.A. Baggio. 2017a. 'Social-Ecological Network Analysis of Scale Mismatches in Estuary Watershed Restoration.' *Proceedings of the National Academy of Sciences* 114(10): 201604405. doi:10.1073/pnas.1604405114.

Sayles, J.S., and J.A. Baggio. 2017b. 'Who Collaborates and Why: Assessment and Diagnostic of Governance Network Integration for Salmon Restoration in Puget Sound, USA.' *Journal of Environmental Management* 186: 64–78. doi:10.1016/j.jenvman.2016.09.085.

Sayles, J., M.M. Garcia, M. Hamilton, S. Alexander, J. Baggio, A.P. Fischer, and J. Pittman. 2019. 'Social-Ecological Network Analysis for Sustainability Sciences: A Systematic Review and Innovative Research Agenda for the Future.' *Environmental Research Letters*, 1–5. doi:10.1088/1748-9326/ab2619.

Urban, D., and T. Keitt. 2001. 'Landscape Connectivity: A Graph-theoretic Perspective.' *Ecology* 82(5): 1205–1218.

Wang, P., K. Sharpe, G.L. Robins, and P.E. Pattison. 2009. 'Exponential Random Graph (P*) Models for Affiliation Networks.' *Social Networks* 31(1): 12–25. doi:10.1016/j.socnet.2008.08.002.

Watts, D.J., and S.H. Strogatz. 1998. 'Collective Dynamics of "Small-World" Networks.' *Nature* 393(6684): 440–442. doi:10.1038/30918.

24

Spatial mapping and analysis

Graeme S. Cumming[1] *and Ralf Seppelt*[2,3]

[1]ARC CENTRE OF EXCELLENCE FOR CORAL REEF STUDIES, JAMES COOK UNIVERSITY, TOWNSVILLE, AUSTRALIA
[2]UFZ – HELMHOLTZ CENTRE FOR ENVIRONMENTAL RESEARCH, DEPARTMENT OF COMPUTATIONAL LANDSCAPE ECOLOGY, LEIPZIG, GERMANY
[3]INSTITUTE OF GEOSCIENCE AND GEOGRAPHY, MARTIN LUTHER UNIVERSITY HALLE-WITTENBERG, HALLE, GERMANY

Key methods discussed in this chapter

Spatial mapping and analysis, including geography, landscape ecology, remote sensing, statistics, land surveying, brief overview of relevant mapping and analytical approaches

Connections to other chapters

This chapter connects to many others through the fundamental properties of location, connectivity and context. Spatial mapping is used to bound study systems, and any analysis for which the location, context or connectivity of an element or phenomenon is relevant requires familiarity with these methods. Chapters on statistics (Chapter 18), network analysis (Chapter 23), agent-based modelling (Chapter 28), participatory data collection (Chapter 8), participatory modelling (Chapter 13), ecosystem service modelling (Chapter 31), historical assessment (Chapter 25) and ecological field data collection (Chapter 6) are particularly relevant.

Introduction

Space is part of the fabric of our existence. We live in four dimensions: time, and three dimensions of space. We can exist in only one location at any point in time. Space is the matrix in which we live. Any analysis can be conducted spatially and every problem has spatial elements. The key questions are (a) whether there are occasions on which one can safely ignore the spatial elements of a problem, and (b) what one loses by ignoring them.

Spatial mapping and analysis are among the oldest scientific techniques in social-ecological systems (SES) research, dating back to early biogeographic analyses of plant kingdoms and the relevance of ecological variation for human agriculture (Von Humboldt and Bonpland 1807; reprint 2010). The disciplines of geography, biogeography and landscape ecology focus on the relevance of spatial variation, spatial context and spatial location as influences on abiotic, biotic and anthropogenic patterns and processes (Turner, Gardner, and O'Neill 2001). A comparison of maps from two different points in time is a long-standing approach to exploring temporal change, dynamics and social-ecological feedbacks.

DOI: 10.4324/9781003021339-28

SUMMARY TABLE: SPATIAL MAPPING AND ANALYSIS

DISCIPLINARY BACKGROUND

The methods in this chapter are derived from or have most commonly been used in:

Land Surveying, Geography, Landscape Ecology, Remote Sensing, Statistics

KNOWLEDGE TYPE

The methods in this chapter are primarily used to generate the following types of knowledge:

- Descriptive
- Exploratory
- Explanatory
- Prescriptive

RESEARCH APPROACH

The methods in this chapter originate from or most commonly adopt the following research approaches:

- Analytical/objective

PURPOSE OF METHOD

The most common purposes of using the methods in this chapter are:

- Data collection/generation
- System understanding
- Stakeholder engagement and co-production
- Policy/decision support

TEMPORAL DIMENSION

The methods in this chapter are most commonly applied to the following temporal dimensions:

- Present (typically within the last 5–10 years)
- Recent past (post-1700s)
- Pre-industrial revolution (pre-1700s)
- Future

SPATIAL DIMENSION

The methods in this chapter are primarily either or both:

- Explicitly spatial

The methods in this chapter are most commonly applied at the following spatial scales:

- Local
- Regional (provincial/state to continental)
- Global
- Multiple places/sites around the world

SYSTEMIC FEATURES AND PROCESSES

While most methods can do many things, the methods in this chapter are particularly good (i.e. go-to methods) for addressing the following:

- SES components and linkages
- Diversity
- Social-ecological interactions over time
- Regime shifts

Recent technological advances in collecting and analysing spatially explicit data (i.e. data associated with spatial coordinates), coupled with improved processing capabilities, have led to a rapid and potentially overwhelming explosion of both data and methods for spatial analysis and mapping. High-resolution, broad-extent spatial and temporal datasets for biophysical variables are now available for periods of approximately 50 years. Mapping of human elements of SES has also improved through using the Internet and mobile devices to map spatial and temporal patterns in human demography, preferences, resource use and movements. Census data in many countries are now linked to zip codes or survey districts through a geographic information system (GIS), for example. By virtue of their shared location in space, these datasets can be rapidly linked to satellite-derived maps of the biophysical system to explore relationships and social-ecological feedbacks (Cord et al. 2017).

For the social-ecological researcher, the primary challenge in spatial mapping and analysis is less one of documenting patterns, which can be done in a wide variety of ways, and more one of inferring mechanisms. Spatial patterns arise from many different processes and it is easy to jump to false assumptions about cause and effect. People often assume, for example, that agglomeration, or clustering, in the distribution of organisms or businesses reflects the presence of clustered resources. But clustering may equally be produced by simple mechanisms, such as limited dispersal capacity and selective mortality, that have nothing to do with underlying resource availability (Skellam 1951). Ecologists and geographers use a variety of techniques, such as neutral landscape models, autocorrelation analysis and matching methods, to explore counterfactuals and to avoid going astray when testing hypotheses that have spatial elements (e.g. Gardner and Urban 2007; Geldmann et al. 2013). One of the core principles for good spatial analysis and mapping is that assumptions about spatial causality must always be treated as hypotheses and compared against other alternatives.

SES problems and questions

Spatial mapping is a common starting point for SES studies. Maps are used in nearly every SES study, whether explicitly or implicitly, to bound study areas, select and move between sampling locations, and identify important social and ecological heterogeneity within the study site. Biophysical variation in elevation, climate, water and soils drives patterns in human and ecological systems. The basic building blocks of ecological analysis are estimates of species diversity and the abundances of organisms; the units for these quantities are spatial and are estimated using spatially explicit techniques such as quadrats, line transects or camera traps. Similarly, most socio-economic studies occur in particular locations such as urban areas or villages; respondents for interviews and surveys are often selected based on their proximity to resources or their membership of a particular community; and in economic analyses, both market size and market access are heavily contingent on location, membership in spatial networks of supply and demand, and spatial context.

Spatial mapping and analysis can relate social and ecological elements of systems of interest. The use of tools for mapping and analysis depends to a very large extent on the goals of the analyst. Spatial relations may define or bound an analytical context (e.g. analysing water management approaches across different catchments), provide or inform mechanisms that explain outcomes of interest (e.g. understanding how household location influences ecosystem service preferences by people), or confound attempts to relate non-spatial variables in analyses of cause and effect (e.g. spatial patterns in household wealth and agricultural activity can make the influence of protected areas on nearby land prices harder to detect). Spatial relations are also important elements of comparative analyses across different case studies (Cumming 2011).

In SES research, spatial analysis and mapping have been widely used for the following: measuring changes in land use and land cover across landscapes; understanding abiotic, biotic and anthropogenic influences on ecosystems and exploring potential consequences of management actions; doing research on conservation planning and making decisions; and doing research that connects ecosystem services, land tenure and access to ecosystems and markets. Some published examples of social-ecological questions addressed by spatial approaches include:

- How does the success of a health-care intervention relate to household income in a fishing community? (Short et al. 2018)
- How should policy and management include animal movements? (Hays et al. 2019)
- How do governance institutions influence land-use change? (Holzhauer, Brown, and Rounsevell 2019)
- Does connectivity in marine reserve networks facilitate biodiversity conservation? (Magris et al. 2018)
- How does landscape structure affect ecosystem service delivery? (Ridding et al. 2018)

Brief description of key methods

The three dimensions of geographic space (longitude, latitude, elevation), plus the fourth dimension of time, provide a matrix in which we exist. Within this matrix, maps describe the coordinates of features or events in both space and time, providing a simplified record of reality. People create maps in a variety of ways and at scales ranging from molecular (e.g. 3D protein structures) to interstellar (e.g. star charts). Spatial analysis of any of these maps nonetheless has many shared ingredients, e.g. the use of coordinates, the calculation of basic spatial properties such as proximity or connectivity, and visualisation.

Autocorrelation is a particular concern in spatial analysis. It refers to the increased likelihood that values of a given variable at two points near each other in space will be similar to one another (Ord 2010). Most spatial datasets have autocorrelation. In some cases, autocorrelation is itself the variable of interest; in other cases, it is a nuisance variable that must be removed or factored out from the analysis, either statistically or through careful sampling. In SES research, spatial mapping and analysis typically use a GIS that provides an operational environment for spatial analyses, such as overlaying or intersecting polygons, extracting data from different data layers into a grid to create a comparable dataset, and smoothing or cleaning mapped data. In practice, GIS manipulation in a program like ArcGIS or Imagine is often a precursor to more intensive statistical analyses in other software packages (e.g. R, Matlab). Table 24.1 provides a summary of key applications of spatial mapping and analysis.

Limitations

Limits on spatial mapping and analysis are imposed by practical trade-offs between scale, processing time and information storage. The minimum mapping unit (MMU) is the size of the smallest feature that is reliably mapped by a given mapping approach. To distinguish between features requires a grain size (resolution) smaller than the MMU. Turner, Gardner and O'Neill (2001), for example, suggest that land-cover maps should have a grain at least three to five times smaller than the smallest patch and an extent three to five times larger than the

Table 24.1 Summary of key applications of spatial mapping and analysis

Main applications	*Description*	*References*
Remote sensing	Remotely sensed data typically take the form of pictures of the earth's surface. These data can be collected across a wide variety of different wavelengths of radiation (e.g. visible spectrum, ultraviolet, infra-red) and either actively (using radar or LIDAR) or passively (using sunlight, e.g. aerial photography or satellite platforms such as Landsat). After some processing to correct for errors, noise and distortions, the typical result is spatially explicit data at a consistent extent and resolution, e.g. a single Landsat Thematic Mapper (TM) image covers about 250 km² at a resolution of about 30 × 30 m (see also Chapter 25: Historical assessment and Chapter 31: Ecosystem service modelling).	*Key introductory text* Campbell and Wynne 2011 *Applications to SES* Jenerette et al. 2007; Brody et al. 2008; Lauer and Aswani 2008; Kennedy et al. 2009; Newton et al. 2009; Ament and Cumming 2016; Fernández-Giménez et al. 2018
Land-use and land-cover change analysis	Land cover describes the nature of different constituents of the earth's surface, such as forest, grassland, water or built environment. More specifically, land use refers to how people use a parcel of land, e.g. a forest (land cover) may be a conservation area or a managed forest used for timber (land use). Analysis of spatial patterns and changes over time in land-use and land-cover change (LULCC) are often used either as explanations or as response variables in SES analyses (see also Chapter 25: Historical assessment and Chapter 31: Ecosystem service modelling).	*Key introductory text* Lambin and Geist 2008 *Applications to SES* Veldkamp and Lambin 2001; Agarwal et al. 2002; Liu et al. 2007; Meyfroidt et al. 2018; Holzhauer, Brown, and Rounsevell 2019
Geostatistics	Geostatistics is a branch of statistics that developed in geology. It focuses on measuring the strength of spatial relationships. Geostatistics is used in SES analysis for understanding spatial pattern and autocorrelation, particularly the spatial scale of autocorrelation in point or continuous data, via semi-variograms, correlograms and measures of autocorrelation and dispersion such as Moran's I or Ripley's K (see also Chapter 18: Statistical analysis).	*Key introductory text* Isaaks and Srivastava 1989 *Applications to SES* Overmars, De Koning, and Veldkamp 2003; Mets, Armenteras, and Dávalos 2017; Fletcher and Fortin 2018

Main applications	*Description*	*References*
Species distribution models	A species distribution model uses observations of species of interest to estimate the probability of species occurrence in geographic space. It is also used as a tool for understanding influences on species occurrences and for linking drivers of species distributions (e.g. physiological tolerance) to distributions. It uses a wide range of statistical tools, e.g. multiple regression, MaxEnt and discriminant analysis. A species distribution model is useful for SES analyses in which the spatial mapping of animal or plant habitat is relevant, e.g. estimating ecosystem service provision by hunted or harvested wild species, or looking at the potential sustainability of harvesting. These models also provide useful null hypotheses for understanding the impacts of human use, e.g. if suitable habitat is unoccupied as a consequence of over-harvesting or pollution (see also Chapter 18: Statistical analysis).	*Key introductory text* Franklin 2010 *Applications to SES* Cumming and Van Vuuren 2006; Sherrouse, Semmens, and Clement 2014; Uden et al. 2015; Bonebrake et al. 2018
Telemetry	Telemetry uses trackable transmitters or global positioning system (GPS) units attached to animals or people to determine where they go. It originally used radio-frequency transmitting devices that were tracked using a handheld antenna. The most sophisticated approaches now use a lightweight solar-powered satellite platform terminal transmitter to transmit GPS data via satellite (see also Chapter 6: Ecological field data collection).	*Key introductory text* Hooten et al. 2017 *Applications to SES* Krause et al. 2013; Lin et al. 2018; Oppel et al. 2018; Hays et al. 2019
Home range, resource selection and utilisation density analysis	Home range, resource selection and utilisation density analysis uses telemetry data to map out which parts of a landscape are visited most frequently by a tracked organism and where key foraging areas and other vital elements occur. It is useful for establishing how, when and where animals and people use landscapes, and is relevant in studies of human–wildlife conflict (e.g. crop raiding in relation to human diurnal rhythms, interactions of people and carnivores, understanding locations of fishing effort) (see also Chapter 18: Statistical analysis).	*Key introductory text* Moorcroft and Lewis 2013 *Applications to SES* Bodin et al. 2006; Hebblewhite and Haydon 2010; Zetterberg, Mörtberg, and Balfors 2010; Iwamura et al. 2014

(Continued)

Table 24.1 (Continued)

Main applications	*Description*	*References*
Spatial conservation planning	Spatial conservation planning uses spatially explicit data layers of features of social, ecological or economic interest to select priority areas for conservation action. Software such as MARXAN can take a large number of individually mapped costs (e.g. property prices) and benefits (e.g. recorded occurrence of desirable habitat or species) and find an optimal spatial solution given a set of predefined constraints. Conservation planning tools have broader, but largely unexploited, applications in spatial optimisation in SES. MARXAN, for example, could easily be used to determine areas of greatest value (given mapped costs and benefits) for ecosystem service production (see also Chapter 29: Decision analysis based on optimisation).	*Key introductory text* Margules and Pressey 2000 *Applications to SES* Poiani et al. 2000; Possingham, Ball, and Andelman 2000; Pressey et al. 2007; Ban et al. 2013; Magris et al. 2017

largest patch. Maps with smaller grains usually have smaller extents. Obtaining a picture at finer grain size requires a closer or 'zoomed-in' lens; images that seek very high resolution at broad extents often run into a depth-of-field problem, which leads to blurring or very high levels of distortion at the edges. Also, information storage and processing demands increase rapidly with decreases in grain or increases in extent. Although Landsat Thematic Mapper images offer a good and widely used compromise (30×30 m grain and c. 250 km^2 extent) for mapping vegetation and land-cover types, they cannot generally be used to map individual tree canopies, roofs or vehicles.

Satellite data are also limited, at present, to system elements that can be readily observed from the sky. Remote-sensing approaches for features that are under water (e.g. coral reefs, seagrass beds) are challenging and require special technologies. Approaches that ignore three-dimensional structure can result in fundamental errors, such as mapping urban tree canopies as forests or shade-grown coffee as pristine rainforest. Spatial analysis may be limited by a lack of data for features – like people and small animals (e.g. human attitudes, household incomes or distributions of rodents) – that cannot be sampled via satellite. Similarly, land use cannot always be reliably deduced from land cover. Most countries have a national data provider, such as the British Ordnance Survey, which assembles critical spatial data (e.g. economic zones, census data, farming types, cadastral data), but the information demands of SES research can be high and it is not uncommon for SES researchers to embark on intensive spatial data-collection programmes of their own.

Resource implications

Spatial mapping and analysis cover a broad spectrum of methodological complexity. Data visualisation and simple measures of distance or proximity can be undertaken over the

Internet using freely available maps and software platforms provided by companies such as ESRI (esri.com/en-us/arcgis/products/index) and Google Maps (cloud.google.com/maps-platform). By contrast, Google's Earth Engine (earthengine.google.com) provides a sophisticated online platform that effectively gives the user free access to a mainframe computer and a range of datasets through a low-speed desktop interface (developers.google.com/earth-engine/datasets). This can reduce the time demands of tasks that would currently take two to three months on a contemporary desktop machine to two to three days. Large quantities of high-quality data are also freely available from many national-level governmental programmes, such as the United States Geological Survey (USGS) and NASA.

There is an increasing trend towards open-source, shareware packages (e.g. QGIS) and the addition of GIS capabilities to statistical platforms such as R and Matlab. At the time of writing, we would still advocate undertaking standard GIS operations (e.g. visualisation, cleaning datasets, merging and joining, extracting information from raster layers into vector layers) in a GIS environment and then exporting cleaned datasets, in the desired format, for advanced statistical analysis into R. For field data collection, a wide variety of GPS-enabled hardware and software is now available including, for example, mobile phone technologies, GPS wristwatches and custom handheld GPS devices. These generally offer good accuracy to resolutions down to 5–10 m, with some variation by location. For higher accuracy (e.g. submetre), a backpack GPS is still recommended.

Geographic information system analysis can be undertaken quite quickly and easily with relatively little training, and free online starter courses are available (e.g. via the ESRI website). A key point for SES research, and a particular trap for beginners, is that datasets should be critically examined and groundtruthed by using independent data from ground observations or field studies to verify datasets. 'Garbage in, garbage out' is a standard principle in GIS; using low-quality GIS data is easy and can rapidly produce advanced-looking maps. But if the data-collection protocol is weak, maps are poorly aligned and proper unbiased sampling approaches have not been observed (among other things), then the conclusions of the analysis will be unreliable.

New directions

Spatial mapping and analysis is a fast-moving research area that has progressed rapidly since the advent of modern computers and the Internet. New and exciting maps are increasingly being made available through new satellite platforms, new sensors and new platforms for carrying sensors (e.g. drones, solar-powered fliers, remotely operated vehicles (ROVs) and autonomous underwater vehicles (AUVs)). At the same time, data on human preferences and spatial movement patterns are increasingly becoming available via the widespread adoption and use of mobile phones, fitness devices and the Internet. Combining different data sources and data streams has huge and largely untapped potential for connecting human movement patterns and ecosystem access with human preferences (e.g. linking accommodation bookings, shopping data, entry to national parks, ecological interests and social network analysis via data from Internet search engines, mobile phone GPS and locations of connections made by mobile phone). Trasarti et al. (2015), for example, used mobile phone data to map movement patterns of people between Paris and surrounding rural areas over time, showing the role of key transport nodes. Technological advances are opening up new fields of analysis in SES research, such as participatory mapping to

Case study 24.1: Mapping a complex SES pattern for Germany

The objective of this research was to understand how socio-environmental conditions influence the pattern and distribution of multiple ecosystem services. It thus exemplifies a spatially explicit SES and shows how ecosystem services bundles and associated socio-environmental gradients interrelate across Germany.

Eleven spatially explicit ecosystem service indicators of provisioning ecosystem services (crop, livestock, wood production, clean water), regulating ecosystem services (nitrogen and flood retention, erosion control, pollination potential) and cultural ecosystem services (water and landscape recreation) in Germany were synthesised using self-organising maps (Agarwal and Skupin 2008; Mouchet et al. 2014). Eight types of ecosystem services bundles (SEBs) were characterised to varying degrees for provisioning, cultural and regulating/maintenance services, and summarised in a spatially explicit map. To relate this to socio-economic drivers, 18 covariates were used (e.g. price of drinking water, employees, ratio female/male, population density, etc. per district) to delineate socio-environmental clusters (Figure 24.2).

In Figure 24.1 the bar plots (so-called 'codebook vectors') show normalised values of the ecosystem services characterising each ecosystem services bundle, with zero representing the national average. The relative contribution of provisioning ecosystem services per ecosystem services bundle is indicated by the percentages next to the bar plots.

Figure 24.2 shows a mapped socio-environmental cluster (SEC). The bar plots show normalised values of covariates characterising each socio-environmental cluster, with zero representing the national average. Socio-environmental clusters are dominated either by socio-economic or environmental covariates. The relative contribution of these groups per socio-environmental cluster is indicated by the percentage next to the bar plots.

Overlaying these two maps provides information about the relationship between the spatial configuration of ecosystem services bundles and co-occurring socio-environmental clusters (Figure 24.3). The clustering method of self-organising maps incorporates properties of each grid cell as well as its location and thus accounts for spatial auto-correlations.

Whereas ecosystem services bundles that were dominated by provisioning ecosystem services were linked to regions with distinct environmental characteristics, cultural ecosystem services bundles were associated with areas where environmental and socio-economic gradients had equal importance. This regional concentration of specific provisioning services probably reflects the ongoing specialisation in land use and especially in agricultural production, which accelerated around 1950. A hotspot for tourism was found next to the shoreline in northern Germany, where high values for recreation relate to the scenic beauty of the sea. Large areas along the coastline in Germany have been designated as national parks and other protected areas, providing infrastructure for nature appreciation and protection of resting places

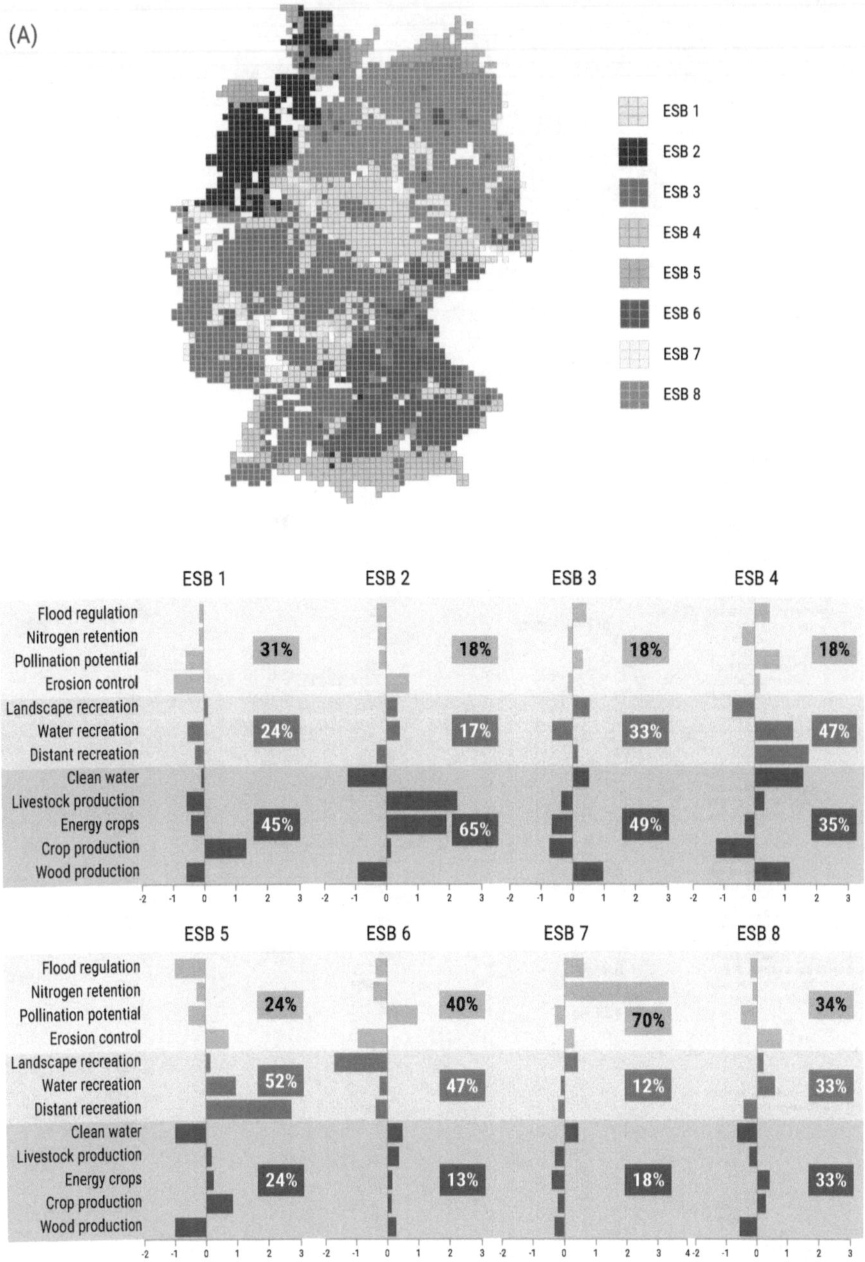

Figure 24.1 Ecosystem services bundles (ESBs) mapped in Germany (Dittrich et al. 2017; arcg.is/1C81jD) (© Ralf Seppelt)

(Continued)

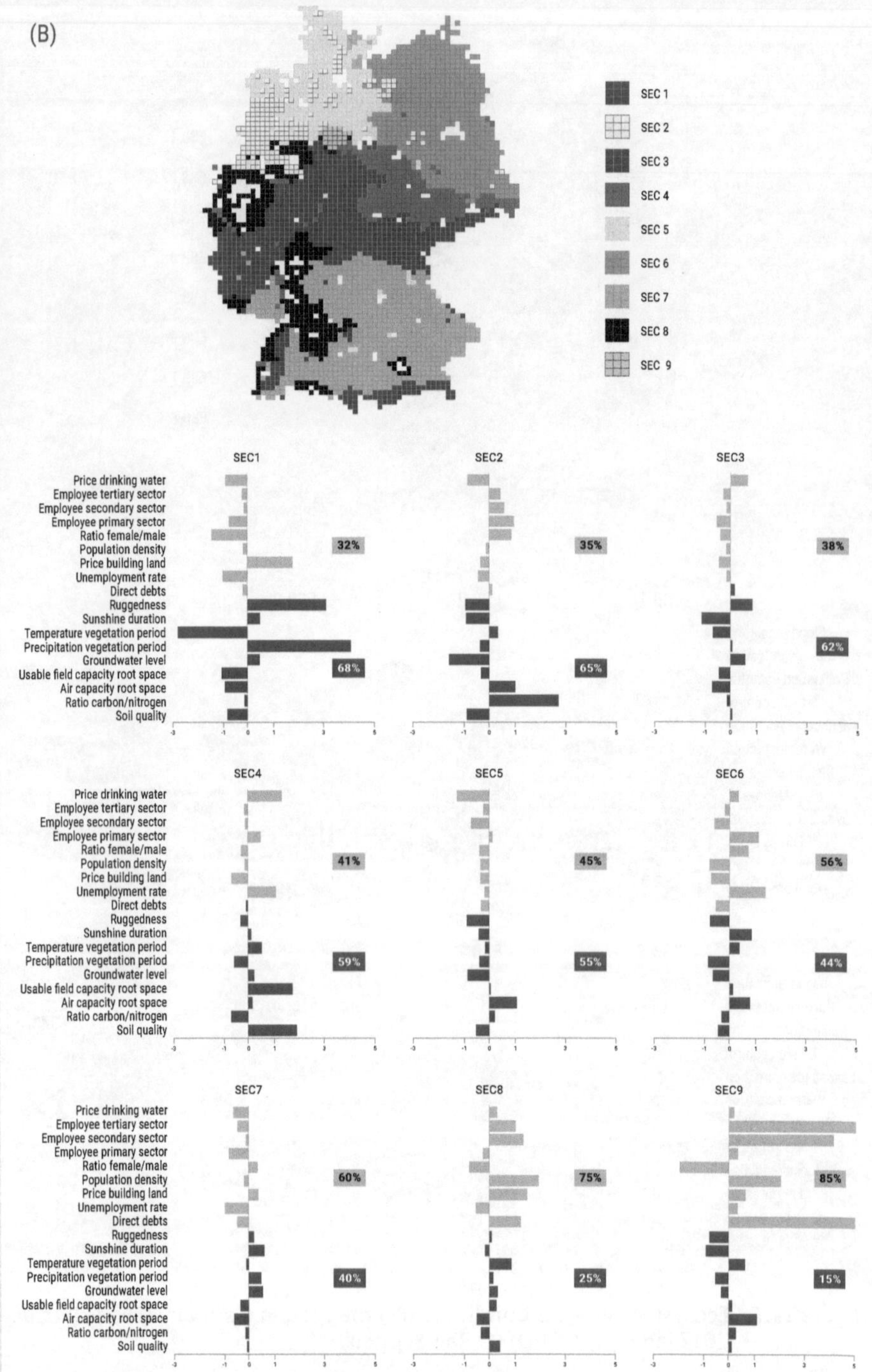

Figure 24.2 Mapped socio-environmental clusters (Dittrich et al. 2017; arcg.is/1Caaf50) (© Ralf Seppelt)

for migratory birds. It is also noteworthy that the Alpine Mountains are attractive for recreation.

Ecosystem services bundles dominated by cultural ecosystem services overlapped to a wide extent not only with socio-environmental clusters determined by environmental conditions but also with socio-economic conditions. For instance, ESB 8 in Figure 24.3, indicating multi-functional landscapes, overlapped mainly with the intermediate SEC 6, which is characterised by relatively equal levels of both socio-economic and environmental variables.

The rows in Figure 24.3 sum up to 100% and the circle sizes illustrate the extent of co-occurrence of the respective socio-economic drivers and ecosystem services bundles. The circles represent the dominance of environmental or socio-economic variables in characterising socio-environmental clusters, whereby colour intensity reflects the degree of dominance (dark = strong; light = weak).

In this case, the absence of pronounced environmental gradients may have hindered a specialisation in certain provisioning services and in turn also prevented known trade-offs with regulating/maintenance services. A spatial stratification of ecosystem services bundles indicated hotspots in which more detailed analysis is needed within national assessments.

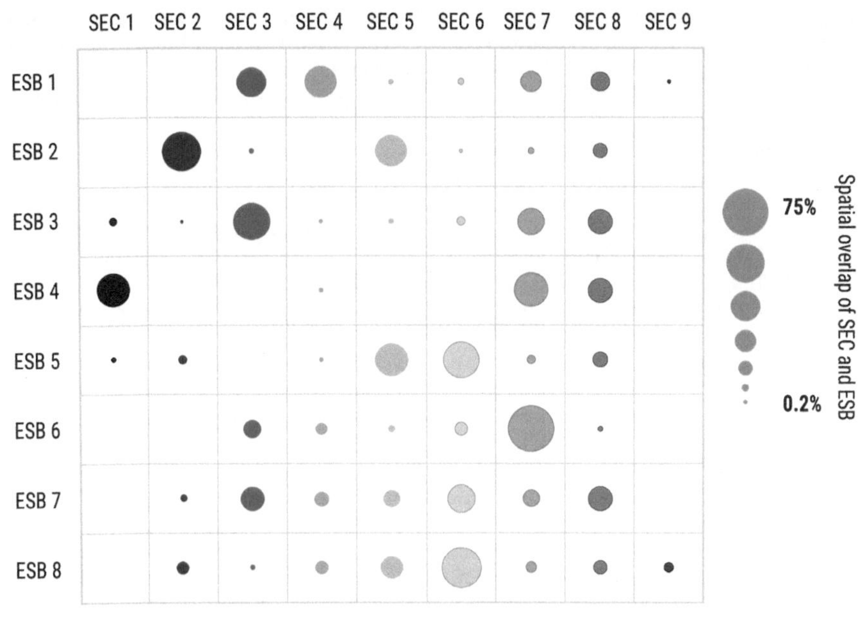

Figure 24.3 Results of spatial overlap analysis of each socio-environmental cluster (SEC) per ecosystem services bundle in percentage of area (Dittrich et al. 2017)

understand human–nature interactions or 'citizen science' as a tool for intensive data collection and monitoring ecosystem change (Hochachka et al. 2012).

We envisage that spatial mapping and analysis will continue as a rapid growth area in SES research for many years to come, as researchers explore and exploit the many insights it can offer. The field is ripe for the development of new and imaginative approaches to SES analysis that take advantage of new technologies to further develop and test theory.

Key readings

Cumming, G.S. 2011. *Spatial Resilience in Social-Ecological Systems.* Dordrecht: Springer.
Ellis, E.C. 2011. 'Anthropogenic Transformation of the Terrestrial Biosphere.' *Philosophical Transactions of the Royal Society A: Mathematical, Physical and Engineering Sciences* 369: 1010–1035.
Lambin, E.F., H.J. Geist, and E. Lepers. 2003. 'Dynamics of Land-Use and Land-Cover Change in Tropical Regions.' *Annual Review of Environment and Resources* 28: 205–241.
Levin, S.A. 1992. 'The Problem of Pattern and Scale in Ecology.' *Ecology* 73: 1943–1967.
Turner, M.G., R.H. Gardner, and R.V. O'Neill. 2015. *Landscape Ecology in Theory and Practice: Pattern and Process.* New York: Springer.

Acknowledgements

We thank Andreas Böhme, UFZ, Leipzig, Germany, for the implementation of online visualisation of results from Dittrich et al. (2017).

References

Agarwal, C., G.M. Green, J. Grove, T.P. Evans, and C.M. Schweik. 2002. 'A Review and Assessment of Land-use Change Models: Dynamics of Space, Time, and Human Choice.' General Technical Report NE-297. Forest Service, US Department of Agriculture.
Agarwal, P., and A. Skupin. 2008. *Self-organising Maps: Applications in Geographic Information Science.* Hoboken: John Wiley & Sons.
Ament, J.M., and G.S. Cumming. 2016. 'Scale Dependency in Effectiveness, Isolation, and Social-Ecological Spillover of Protected Areas.' *Conservation Biology* 30: 846–855.
Ban, N.C., M. Mills, J. Tam, C.C. Hicks, S. Klain, N. Stoeckl, M.C. Bottrill et al. 2013. 'A Social-Ecological Approach to Conservation Planning: Embedding Social Considerations.' *Frontiers in Ecology and the Environment* 11(4): 194–202.
Bodin, Ö., M. Tengö, A. Norman, J. Lundberg, and T. Elmqvist. 2006. 'The Value of Small Size: Loss of Forest Patches and Ecological Thresholds in Southern Madagascar.' *Ecological Applications* 16: 440–451.
Bonebrake, T.C., C.J. Brown, J.D. Bell, J.L. Blanchard, A. Chauvenet, C. Champion, I.C. Chen, T.D. Clark, R.K. Colwell, and F. Danielsen. 2018. 'Managing Consequences of Climate-driven Species Redistribution Requires Integration of Ecology, Conservation and Social Science.' *Biological Reviews* 93: 284–305.
Brody, S.D., S.E. Davis, W.E. Highfield, and S.P. Bernhardt. 2008. 'A Spatial-temporal Analysis of Section 404 Wetland Permitting in Texas and Florida: Thirteen Years of Impact along the Coast.' *Wetlands* 28: 107–116.
Campbell, J.B., and R.H. Wynne. 2011. *Introduction to Remote Sensing.* New York: Guilford Press.
Cord, A.F., K.A. Brauman, R. Chaplin-Kramer, A. Huth, G. Ziv, and R. Seppelt. 2017. 'Priorities to Advance Monitoring of Ecosystem Services using Earth Observation.' *Trends in Ecology and Evolution* 32: 416–428.
Cumming, G.S. 2011. *Spatial Resilience in Social-Ecological Systems.* New York: Springer.
Cumming, G.S., and D.P. van Vuuren. 2006. 'Will Climate Change Affect Ectoparasite Species Ranges?' *Global Ecology and Biogeography* 15: 486–497.
Dittrich, A., R. Seppelt, T. Václavík, and A.F. Cord. 2017. 'Integrating Ecosystem Service Bundles and Socio-environmental Conditions – A National Scale Analysis from Germany.' *Ecosystem Services* 28: 273–282.

Fernández-Giménez, M.E., G.R. Allington, J. Angerer, R.S. Reid, C. Jamsranjav, T. Ulambayar, K. Hondula, B. Baival, B. Batjav, and T. Altanzul. 2018. 'Using an Integrated Social-Ecological Analysis to Detect Effects of Household Herding Practices on Indicators of Rangeland Resilience in Mongolia.' *Environmental Research Letters* 13: 075010.

Fletcher, R., and M-J. Fortin. 2018. 'Spatial Dependence and Autocorrelation.' In *Spatial Ecology and Conservation Modeling: Applications with R*, 133–168. New York: Springer.

Franklin, J. 2010. *Mapping Species Distributions: Spatial Inference and Prediction*. Cambridge: Cambridge University Press.

Gardner, R.H., and D.L. Urban. 2007. 'Neutral Models for Testing Landscape Hypotheses.' *Landscape Ecology* 22: 15–29.

Geldmann, J., M. Barnes, L. Coad, I.D. Craigie, M. Hockings, and N.D. Burgess. 2013. 'Effectiveness of Terrestrial Protected Areas in Reducing Habitat Loss and Population Declines.' *Biological Conservation* 161: 230–238.

Hays, G.C., H. Bailey, S.J. Bograd, W.D. Bowen, C. Campagna, R.H. Carmichael, P. Casale, A. Chiaradia, D.P. Costa, and E. Cuevas. 2019. 'Translating Marine Animal Tracking Data into Conservation Policy and Management.' *Trends in Ecology and Evolution* 34(5): 459–473.

Hebblewhite, M., and D.T. Haydon. 2010. 'Distinguishing Technology from Biology: A Critical Review of the Use of GPS Telemetry Data in Ecology.' *Philosophical Transactions of the Royal Society B: Biological Sciences* 365: 2303–2312.

Hochachka, W.M., D. Fink, R.A. Hutchinson, D. Sheldon, W.K. Wong, and S. Kelling. 2012. 'Data-intensive Science Applied to Broad-scale Citizen Science.' *Trends in Ecology and Evolution* 27: 130–137.

Holzhauer, S., C. Brown, and M. Rounsevell. 2019. 'Modelling Dynamic Effects of Multi-scale Institutions on Land Use Change.' *Regional Environmental Change* 19: 733–746.

Hooten, M.B., D.S. Johnson, B.T. McClintock, and J.M. Morales. 2017. *Animal Movement: Statistical Models for Telemetry Data*. Boca Raton: CRC press.

Isaaks, E.H., and R.M. Srivastava. 1989. *An Introduction to Applied Geostatistics*. Oxford: Oxford University Press.

Iwamura, T., E.F. Lambin, K.M. Silvius, J.B. Luzar, and J.M. Fragoso. 2014. 'Agent-based Modeling of Hunting and Subsistence Agriculture on Indigenous Lands: Understanding Interactions between Social and Ecological Systems.' *Environmental Modelling and Software* 58: 109–127.

Jenerette, G.D., S.L. Harlan, A. Brazel, N. Jones, L. Larsen, and W.L. Stefanov. 2007. 'Regional Relationships between Surface Temperature, Vegetation, and Human Settlement in a Rapidly Urbanizing Ecosystem.' *Landscape Ecology* 22: 353–365.

Kennedy, R.E., P.A. Townsend, J.E. Gross, W.B. Cohen, P. Bolstad, Y.Q. Wang, and P. Adams. 2009. 'Remote Sensing Change Detection Tools for Natural Resource Managers: Understanding Concepts and Tradeoffs in the Design of Landscape Monitoring Projects.' *Remote Sensing of Environment* 113: 1382–1396.

Krause, J., S. Krause, R. Arlinghaus, I. Psorakis, S. Roberts, and C. Rutz. 2013. 'Reality Mining of Animal Social Systems.' *Trends in Ecology and Evolution* 28: 541–551.

Lambin, E.F., and H.J. Geist. 2008. *Land-use and Land-cover Change: Local Processes and Global Impacts*. New York: Springer.

Lauer, M., and S. Aswani. 2008. 'Integrating Indigenous Ecological Knowledge and Multi-spectral Image Classification for Marine Habitat Mapping in Oceania.' *Ocean and Coastal Management* 51: 495–504.

Lin, H.Y., C.J. Brown, R.G. Dwyer, D.J. Harding, D.T. Roberts, R.A. Fuller, S. Linke, and H.P. Possingham. 2018. 'Impacts of Fishing, River Flow and Connectivity Loss on the Conservation of a Migratory Fish Population.' *Aquatic Conservation: Marine and Freshwater Ecosystems* 28: 45–54.

Liu, J., T. Dietz, S.R. Carpenter, M. Alberti, C. Folke, E. Moran, A.N. Pell, P. Deadman, T. Kratz, and J. Lubchenco. 2007. 'Complexity of Coupled Human and Natural Systems.' *Science* 317: 1513–1516.

Magris, R.A., M. Andrello, R.L. Pressey, D. Mouillot, A. Dalongeville, M.N. Jacobi, and S. Manel. 2018. 'Biologically Representative and Well-connected Marine Reserves Enhance Biodiversity Persistence in Conservation Planning.' *Conservation Letters* 11(4): e12439.

Magris, R.A., R.L. Pressey, M. Mills, D.A. Vila-Nova, and S. Floeter. 2017. 'Integrated Conservation Planning for Coral Reefs: Designing Conservation Zones for Multiple Conservation Objectives in Spatial Prioritisation.' *Global Ecology and Conservation* 11: 53–68.

Margules, C.R., and R.L. Pressey. 2000. 'Systematic Conservation Planning. *Nature* 405: 243.

Mets, K.D., D. Armenteras, and L.M. Dávalos. 2017. 'Spatial Autocorrelation Reduces Model Precision and Predictive Power in Deforestation Analyses.' *Ecosphere* 8: e01824.

Meyfroidt, P., R.R. Chowdhury, A. de Bremond, E.C. Ellis, K-H. Erb, T. Filatova, R.D. Garrett et al. 2018. 'Middle-range Theories of Land System Change.' *Global Environmental Change* 53(August): 52–67. doi:10.1016/j.gloenvcha.2018.08.006.

Moorcroft, P.R., and M.A. Lewis 2013. *Mechanistic Home Range Analysis* (MPB-43). Princeton: Princeton University Press.

Mouchet, M.A., P. Lamarque, B. Martín-López, E. Crouzat, P. Gos, C. Byczek, and S. Lavorel. 2014. 'An Interdisciplinary Methodological Guide for Quantifying Associations between Ecosystem Services.' *Global Environmental Change* 28: 298–308.

Newton, A.C., R.A. Hill, C. Echeverria, D. Golicher, J.M.R. Benayas, L. Cayuela, and S.A. Hinsley. 2009. 'Remote Sensing and the Future of Landscape Ecology.' *Progress in Physical Geography* 33: 528–546.

Oppel, S., M. Bolton, A.P. Carneiro, M.P. Dias, J.A. Green, J.F. Masello, R.A. Phillips. E.Owen, P. Quillfeldt, and A. Beard. 2018. 'Spatial Scales of Marine Conservation Management for Breeding Seabirds.' *Marine Policy* 98: 37–46.

Ord, J.K. 2010. 'Spatial Autocorrelation: A Statistician's Reflections.' In *Perspectives on Spatial Data Analysis*, edited by L. Anselin and S.J. Rey, 165–180. New York: Springer.

Overmars, K.D., G. de Koning, and A. Veldkamp. 2003. 'Spatial Autocorrelation in Multi-scale Land Use Models.' *Ecological Modelling* 164: 257–270.

Poiani, K.A., B.D. Richter, M.G. Anderson, and H.E. Richter. 2000. 'Biodiversity Conservation at Multiple Scales: Functional Sites, Landscapes, and Networks.' *BioScience* 50: 133–146.

Possingham, H., I. Ball, and S. Andelman. 2000. 'Mathematical Methods for Identifying Representative Reserve Networks.' In *Quantitative Methods for Conservation Biology*, edited by S. Ferson and M. Burgman, 291–305. New York: Springer.

Pressey, R.L., M. Cabeza, M.E. Watts, R.M. Cowling, and K.A. Wilson. 2007. 'Conservation Planning in a Changing World.' *Trends in Ecology and Evolution* 22: 583–592.

Ridding, L.E., J.W. Redhead, T.H. Oliver, R. Schmucki, J. McGinlay, A.R. Graves, J. Morris, R.B. Bradbury, H. King, and J.M. Bullock. 2018. 'The Importance of Landscape Characteristics for the Delivery of Cultural Ecosystem Services.' *Journal of Environmental Management* 206: 1145–1154.

Sherrouse, B.C., D.J. Semmens, and J.M. Clement. 2014. 'An Application of Social Values for Ecosystem Services (SolVES) to Three National Forests in Colorado and Wyoming.' *Ecological Indicators* 36: 68–79.

Short, R., R. Gurung, M. Rowcliffe, N. Hill, and E.J. Milner-Gulland. 2018. 'The Use of Mosquito Nets in Fisheries: A Global Perspective.' *PLoS ONE* 13: e0191519.

Skellam, J.G. 1951. 'Random Dispersal in Theoretical Populations.' *Biometrika* 38: 196–218.

Trasarti, R., A-M. Olteanu-Raimond, M. Nanni, T. Couronné, B. Furletti, F. Giannotti, Z. Smoreda, and C. Ziemlicki. 2015. 'Discovering Urban and Country Dynamics from Mobile Phone Data with Spatial Correlation Patterns.' *Telecommunications Policy* 39: 347–362.

Turner, M.G., R.H. Gardner, and R.V. O'Neill. 2001. *Landscape Ecology in Theory and Practice: Pattern and Process.* New York: Springer.

Uden, D.R., C.R. Allen, D.G. Angeler, L. Corral, and K.A. Fricke. 2015. 'Adaptive Invasive Species Distribution Models: A Framework for Modeling Incipient Invasions.' *Biological Invasions* 17: 2831–2850.

Veldkamp, A., and E.F. Lambin. 2001. *Predicting Land-use Change.* Amsterdam: Elsevier.

Von Humboldt, A., and A. Bonpland. 1807 (reprint 2010*). Essay on the Geography of Plants*, edited by S.T. Jackson, 296. Chicago: University of Chicago Press.

Zetterberg, A., U.M. Mörtberg, and B. Balfors. 2010. 'Making Graph Theory Operational for Landscape Ecological Assessments, Planning, and Design.' *Landscape and Urban Planning* 95: 181–191.

Methods for analysing systems – system dynamics

25

Historical assessment

Sarah E. Gergel[1] *and Ruth H. Thurstan*[2]

[1]DEPARTMENT OF FOREST AND CONSERVATION SCIENCES, UNIVERSITY OF BRITISH COLUMBIA, VANCOUVER, CANADA
[2]CENTRE FOR ECOLOGY AND CONSERVATION, COLLEGE OF LIFE AND ENVIRONMENTAL SCIENCES, UNIVERSITY OF EXETER, PENRYN, UK

Key methods discussed in this chapter

Methods related to data obtained from sediment cores, archaeological/zooarchaeological materials, dendrochronology/sclerochronology, land surveys, historical aerial photography, satellite remote sensing, documentary sources, governmental data, interviews and oral histories

Connections to other chapters

Historical assessment methods may include participatory techniques (i.e. interviews and participatory mapping, see Chapters 7 and 8) or use biophysical information obtained from historical maps, satellite imagery (Chapter 24), sediment cores, tree rings and ancient artefacts. Historical analysis may involve discerning the content and key themes found in old documents and news sources, spatial mapping and analysis (Chapter 24), as well as quantitative statistical analysis of a government census or survey (Chapters 18 and 19).

Introduction

Historical assessment is the task of reconstructing the long-term dynamics of a social-ecological systems (SES) over time and may span years, centuries or millennia (Tomscha et al. 2016). Approaches can be qualitative or quantitative. While historical reconstruction may incorporate very direct measurements of phenomena, often more indirect indicators and proxies of a system must be used instead. The general goal of historical assessment is to determine if and how an SES has changed over time. Historical assessment can help us to understand baselines (e.g. initial starting conditions) (Morgan, Gergel, and Coops 2010) and provide context for recent changes (e.g. are recent changes large or small compared to historical changes?) (Rosenberg et al. 2005).

Historical assessment enjoys the benefits – as well as the challenges – of contributions from a variety of disciplines, a plurality of data sources and a diverse array of methodological approaches. Although historical assessment draws on many disciplines, historical ecology

 DOI: 10.4324/9781003021339-30

SUMMARY TABLE: HISTORICAL ASSESSMENT

DISCIPLINARY BACKGROUND	KNOWLEDGE TYPE
The methods in this chapter are derived from or have most commonly been used in: Historical Ecology, Environmental History, Palaeoecology, Climatology, Archaeology, Ethnography, Landscape Ecology, Anthropology, Environmental Social Sciences, Human Geography, Maritime History, Social History	*The methods in this chapter are primarily used to generate the following types of knowledge:* • Descriptive • Exploratory • Explanatory
RESEARCH APPROACH	**PURPOSE OF METHOD**
The methods in this chapter originate from or most commonly adopt the following research approaches: • Analytical/objective • Interpretive/subjective	*The most common purposes of using the methods in this chapter are:* • Data collection/generation • System understanding
TEMPORAL DIMENSION	**SYSTEMIC FEATURES AND PROCESSES**
The methods in this chapter are most commonly applied to the following temporal dimensions: • Recent past (post-1700s) • Pre-industrial revolution (pre-1700s)	*While most methods can do many things, the methods in this chapter are particularly good (i.e. go-to methods) for addressing the following:* • Social-ecological interactions over time • Adaptation and self-organisation • Regime shifts
SPATIAL DIMENSION	
The methods in this chapter are primarily either or both: • Non-spatial • Explicitly spatial *The methods in this chapter are most commonly applied at the following spatial scales:* • Local • Regional (provincial/state to continental) • Global • Multiple places/sites around the world	

Table 25.1 Summary of key data sources or types used in historical assessment

Data source or type	*Description*	*References*
Sediment cores	Sediment cores often contain floral and faunal remains, which, alongside dating and isotopic techniques, enable inferences about the ecological communities that existed in the past, changes that occurred and the potential drivers of change. Depending on the environment being sampled, the length and depth of the sediment core and the rate of sedimentation/erosion, sediment cores may be used to infer changes spanning decades to millions of years into the past. Methods include stable isotope analysis and core sediment composition or heavy-metal analysis to demonstrate the scale, direction and drivers of environmental change (see also Chapter 6: Ecological field data collection).	*Key introductory text* Last and Smol 2001 *Applications to SES* Jeffers, Nogué, and Willis 2015
Archaeological materials	Archaeological materials consist of artefacts, architectural remains and cultural landscapes, among many others. Archaeologists use both quantitative and semi-quantitative measures to compare changes in remains left behind. When combined with dating techniques and other sources, these approaches can be used to infer the timing and drivers of change for hundreds to tens of thousands of years in the past. Field methods include surveys (remote or in-person) and the excavation of sites. Post-excavation analysis typically includes dating, classifying or compositional analysis of artefacts and archaeological features.	*Key introductory text* Gebhard 2003 *Applications to SES* Crumley 2017
Zooarchaeological materials	Animal remains excavated from archaeological sites can be used to track changes in the presence, abundance and/or size of the species. These data can be used to infer social and ecological changes through time as far back as hundreds to tens of thousands of years ago. Analytical methods are diverse and include determining the age at death and biometry of animal remains, e.g. to identify the timing of commencement of animal domestication.	*Key introductory text* Gifford-Gonzalez 2018 *Applications to SES* Harland et al. 2016; Wallman 2018
Dendrochronology and sclerochronology	Both dendrochronology and sclerochronology reconstruct past environmental changes in climate and ecosystem productivity, which may reflect natural and anthropogenic causes.	*Key introductory texts* Butler and Schöne 2017; UWICER 2017

Data source or type	*Description*	*References*
	Dendrochronology examines tree rings over time spans of decades to centuries. These growth patterns reflect particular increments of time (i.e. seasonal, annual growth). Tree rings and scars can thus provide information on past climate and drought, cultural uses of trees (such as bark stripping), and traditional burning practices and fire suppression. Sclerochronology studies the bands of new growth laid down in invertebrate shell and coral remains. Similar to tree rings, the study and dating of these growth patterns can be used to reconstruct information on environmental and climatic changes over periods of decades to centuries. Methods include dating of samples to determine their age or date of death, followed by growth trend estimation and statistical reconstruction of spatial and temporal variations in, for example, climate anomalies, using rate of growth as a proxy.	*Applications to SES* Smith, Mackie, and Sumpter 2005; Stahle and Dean 2010
Land surveys	Historical maps, originally for surveying purposes, can be adapted for SES purposes ranging from mapping early transportation networks and vegetation change to carbon storage. In North America, General Land Office Notes and Land Surveys exist for the USA and Canada. In European countries these resources are even older, more extensive and phenomenally detailed. In some cases created by royal decree, these maps can be a source of information regarding cultural and economic priorities. Cartography is the study and practice of making maps. Georeferencing of chart features to geographic coordinates enables the comparison of charts and their features through time and space (the same principles apply for historical aerial photography).	*Key introductory text* Fuchs et al. 2015 *Applications to SES* Rhemtulla, Mladenoff, and Clayton 2009
Historical aerial photography	Historical photography can be used to map long-term patterns of landscape change from anthropogenic and/or biophysical causes. Historical photos can divulge 'baseline' or reference conditions preceding industrial expansion in some regions. Oblique imagery (captured perpendicular to the earth's surface) is routinely available for many parts of the world, beginning in the 1950s and occasionally in the 1930s. New sources of declassified 'cold war era' imagery are also emerging. For some nation states, the archival resources of a former colonial power or occupying country must be consulted in order to find historical imagery. Turn-of-the-century ground-based images are available in rare cases, which can represent historical vistas and viewsheds.	*Key introductory text* Morgan, Gergel, and Coops 2010 *Applications to SES* Coomes, Takasaki, and Rhemtulla 2011

(Continued)

Table 25.1 (Continued)

Data source or type	*Description*	*References*
Satellite remote sensing	Satellite imagery can be used to track long-term dynamics of agriculture, forests, rivers, glacial retreat, sea-level rise and urbanisation, to name a few. The primary workhorse is the Landsat series of satellites that have been providing repeat, continuous, freely available imagery at 30 m spatial resolution since the 1980s for many locations globally. Another primary source of information comes from a satellite sensor called Moderate Resolution Imaging Spectroradiometer (MODIS), which images the earth every one to two days. Although at coarser spatial resolution, the temporal frequency of MODIS makes it useful for climate change and phenological applications, such as tracking the timing of seasonal 'green-up' of vegetation and degradation of vegetation (see also Chapter 24: Spatial mapping and analysis).	*Key introductory text* Cohen and Goward 2004 *Applications to SES* Barbosa, Atkinson, and Dearing 2015; Eddy et al. 2017
Documentary sources	Documentary sources include government and non-government publications such as Commissions of Evidence, newspaper articles, popular media, and art and travel publications, among many other sources. Both quantitative and qualitative data may be of interest. An analysis may combine data from multiple sources, or multiple years from the same source. Particular care must be taken in interpretation to ensure that the context in which the data were originally created or presented is well understood and accounted for. Typical methods include source criticism, which evaluates the reliability, context and integrity of a source, and triangulation, which cross-references different sources or approaches to validate a finding.	*Key introductory texts* Hsieh and Shannon 2005; Braun and Clarke 2006 *Applications to SES* Thurstan, Buckley, and Pandolfi 2018; Thurstan et al. 2018
Governmental/ population census statistics/ health data	Governmental census and survey data can encompass a wide range of topics and may be collected for distinct foci such as population, health and sector-specific statistics. This information may be collected quarterly, annually or every 10 years, and at local, regional or national levels. Statistics may include social and ecological factors, e.g. landings of fish or numbers of fishers working from a specific port. By design, these datasets primarily emphasise quantitative information. These data are often accompanied by important textual explanations which not only provide important context but can even be further analysed directly. The methods of analysis can vary widely and will depend on the question being asked, the granularity of the data and the number of repeated data points.	*Key introductory text* Newsom, Jones, and Hofer 2012 *Applications to SES* Renard, Rhemtulla, and Bennett 2015

Data source or type	*Description*	*References*
Structured and unstructured interviews and oral histories	Retrospective analyses can be incorporated into interview methods by asking participants about their past observations, experiences and behaviours. These methods can also be combined with participatory mapping approaches to collect long-term spatial information. However, retrospective work should be undertaken with an appreciation of issues associated with recall bias and the shifting baselines syndrome, along with other potential issues associated with perception and memory recall. Methods such as timeline mapping and triangulation using other data sources can help to acknowledge and minimise biases of this nature (see Chapter 7 for a more detailed discussion of different interview methodologies and data types related to interviews).	*Key introductory text* Zusman 2010 *Applications to SES* Buckley et al. 2017; Selgrath, Gergel, and Vincent 2017

and environmental history play particularly important roles (McClenachan et al. 2015; see Table 25.1). As a result, the conceptual background, assumptions and level of quantitative and qualitative approaches behind any historical assessment can vary widely depending upon the researchers' disciplinary background, the sources available to them and the specific research question being examined. Interview techniques, for example, can provide rich detail on how people have used and influenced land- and seascapes (Selgrath, Gergel, and Vincent 2017). Archaeological and palaeo-ecological evidence can provide information on historical activities of humans along with the constraints under which they lived. New historical sources of geospatial information are becoming available as formerly classified mapping information becomes unclassified. Given their strengths, weakness and differences in spatio-temporal resolution, historical assessments are strongest when several methods are used in combination to build a more complete picture of long-term SES dynamics. To assess a recreational fishery, for example, Thurstan et al. (2018) used sources ranging from popular media articles, government statistical reports and early research surveys from the 19th century along with semi-structured interviews with current and retired fishers (in Thurstan et al. 2018, Table 2).

SES problems and questions

Historical assessment methods are routinely used to understand interactions between human and ecological communities and the outcomes of these interactions (Kittinger et al. 2015). Historical ecology is often used to study ecosystem change and to understand human use of resources through time. Historical ecology may also track the subsequent adaptation of human communities to environmental changes. Tackling these types of issues often includes the reconstruction of baseline ecosystem conditions and the characterisation of variation in periods with and without significant human impacts.

Some current topical challenges in historical social-ecological fields include the evaluation of shifting baselines (inter-generational differences in the perceptions of ecosystem baseline conditions), trade-offs among ecosystem services (the benefits that humans gain from nature and how these interact; Bennett, Peterson, and Gordon 2009) and shifts in the cultural importance of ecosystem services over time (Daniel et al. 2012). Trade-offs among ecosystem services have increasingly been evaluated over longer time frames (Renard, Rhemtulla, and Bennett 2015; Tomscha et al. 2016). Trade-offs resulting from historical agricultural expansion for food production, for example, may result in long-term problems for downstream water quality (Bennett, Peterson, and Gordon 2009).

Historical assessment methods must routinely assess similarities, differences and synergies between what is often termed 'academic' or 'scientific' knowledge and other forms of knowledge. These other forms of knowledge may include local ecological knowledge, traditional ecological knowledge and indigenous ecological knowledge. Overall, many academic scholars struggle to examine these issues in a way that fully captures the nuances, depth, significance and fundamental nature of indigenous knowledge and cultural ecosystem services, which includes how such perspectives are captured in archival sources (Todd 2016).

Because debates over what constitutes a reasonable baseline are notably fierce, any assumptions in this regard should be examined with deep care, consideration and humility by researchers. The conventional use of 'pre-European' contact as a baseline, for example, makes implicit assumptions about the limited role of indigenous peoples in shaping landscapes. However, recent academic research is attempting to do a better job of appreciating the much longer-term use and management of landscapes by indigenous communities. This understanding now includes active landscape management over thousands of years, indicative of complex social-ecological interactions (e.g. creation and management of clam gardens in North America of at least 2 000 years old (Jackley et al. 2016); and contemporary plant diversity of Amazonian forests driven by pre-Columbian agriculture (Levis et al. 2017)).

Brief description of key methods

For the sake of simplicity, we group historical methods by the general source of the information, loosely organised from oldest (most historical) to sources useful for more recent information. On a continuum from oldest to most recent, these categories span sediment cores, archaeological/zooarchaeological materials, dendrochronology/sclerochronology, land surveys,

Case study 25.1: Path dependencies can create poverty traps: long-term analysis of land use

Historical approaches can improve our understanding of path dependencies in SES. In a study examining poverty traps in Amazonia, Coomes, Takasaki and Rhemtulla (2011) examined whether a household's initial land holdings influenced subsequent land-use decisions, and whether these land-use trajectories resulted in greater poverty. To examine these questions, a combination of aerial photography and satellite imagery from the 1960s, the 1990s and 2007 was used to map land holdings and land cover. Special emphasis was placed on mapping forest age in order to capture differences between primary (intact) forest and agriculture land uses such as crops, orchards and fallow land. Household surveys of income and assets conducted in prior decades were also repeated to assess the contemporary status of households. Together, these datasets created a 30-year record of social-ecological change.

historical aerial photography, satellite remote sensing, documentary sources, governmental data, interviews and oral histories. Each of these is defined and explained in Table 25.1, which provides a general overview of the types of data and approaches useful in examining long-term SES. Examples from terrestrial, aquatic and marine SES are included (see Tomscha et al. 2016 for more detailed explanations of historical data, and Gergel and Turner 2017 and Chapter 24 of this book for user-friendly advice on using geospatial tools).

Limitations

Overall, historical assessment can suffer due to the degradation of information sources as one moves back in time. The loss of traditional cultures and languages worldwide has eroded deep traditions of oral histories in many regions, for example. Where subsequent generations recall different memories of SES components, shifting baselines may result, e.g. perceptions of what constitutes a 'large fish catch' may change across generations. In addition, individuals may forget events with the passing of time, making for less accurate recall of events in the deeper past. Poor image quality can reduce the utility of historical maps and photographs, especially when improperly archived (Morgan, Gergel, and Coops 2010). A comparison of the strengths and limitations of local ecological knowledge and remote sensing for long-term assessment is provided by Eddy et al. (2017).

Historical assessment routinely relies on post hoc (after the fact) analyses to infer patterns. Where a statistical approach is needed, it may be constrained by sample-size limitations or a lack of suitable controls (e.g. finding locations that have not been directly affected by human activities). Where sample sizes are limited, trends can be evaluated using a paired *t*-test (at two time periods), repeated measures ANOVA (e.g. three to five measurements over time) or Mann-Kendall tests for roughly 10 or more observations. While causation can potentially be inferred, it can rarely be definitively shown. In many cases mere correlations and associations may result. Finally, understanding the context of historical data (why it was written, who wrote it and who published the documents) is important when it comes to interpreting the

When tracked over three decades, two distinct land-use trajectories were linked to poverty traps. These paths to poverty originated from limitations in farmers' initial land holdings. The first poverty trap was associated with 'land-poor' farmers who focused on subsistence crops. Farmers who were initially 'land rich' were able to devote some of their land to orchard production. Orchards were beneficial in that they produce a higher-value product, but with a delay before planted trees bear fruit. In contrast, land-poor farmers were unable to devote much land to orchards that could only produce benefits in the future. Instead, land-poor farmers focused on immediate but lower-value subsistence crops, ultimately remaining trapped in poverty.

A second poverty trap arose with 'short fallows'. Fallowing is critical to replenish soil fertility prior to another round of cultivation. Farmers with limited land holdings were unable to keep their fallows out of production for very long. As such, their repeated cycles of cultivation with shorter fallows ultimately reduced soil fertility and overall crop production. These path dependencies would not have been evident without a longer-term historical perspective.

findings. The interpretation of historical sources and data may be heavily influenced by the disciplinary lens and personal experiences or values of the researcher.

Resource implications

The resource implications vary widely by method and whether primary or secondary data are used. Software is generally required for geospatial analysis of historical imagery. In some cases, data sources such as aerial photography must be purchased (Morgan, Gergel, and Coops 2010). Increasingly, extensive open-data repositories for satellite imagery from sources such as Landsat and MODIS satellites, covering much of the late 20th century, are revolutionising geospatial analysis capabilities. Storage space for data can be a problem, as can confidentiality and anonymity concerns requiring use of encrypted laptops. For participatory mapping, the translation of terms, and the use of terms such as 'degradation' which may not be used in the local language of an area can present real problems for survey implementation and interpretation. Ethics training for any interviews involving human research subjects, especially so for marginalised communities, should be undertaken. Finally, a variety of historical data sources can be labour and time intensive to convert into a usable form and may be impossible to access when historical reports are archived improperly or simply disposed of.

New directions

Historical assessments may comprise data and methods from multiple fields of research, but research is often still conducted through the lens of a single discipline. Interdisciplinary research between the humanities and social/natural sciences increasingly occurs and fostering this integration will make it possible to answer different, bolder and more innovative research questions. Looking ahead, long-term perspectives must be better incorporated into management and policy to guide efforts to reduce or halt ecosystem degradation or other negative long-term trends. This integration will be aided by technological improvements such as improved optical character recognition and increased access to archival sources via online platforms. Increasing access may enhance connections between individuals and groups of people to share their histories (e.g. transfer of knowledge from elders to the wider society). It may also increase knowledge-sharing platforms for marginalised members of society and so create opportunities to highlight and correct past inequities (e.g. support environmental justice movements).

Enabling 'big data' analyses of historical documents will also create the potential to move from singular place-based assessments to broader-scale comparative analyses. However, increased access to these sources and online platforms also comes with the potential for wider misuse or abuse of historical data and archived cultural memory, through the (intended or unintended) misinterpretation and communication of historical data and/or events.

Key readings

Crumley, C.L. 1994. *Historical Ecology: Cultural Knowledge and Changing Landscapes*. Seattle: University of Washington Press.

Kittinger J.N., L. McClenachan, K. Gedan, and L. Blight, eds. 2015. *Marine Historical Ecology in Conservation. Applying the Past to Manage for the Future*. San Francisco: University of California Press.

Máñez, K.S., and B. Poulsen. 'Of Seascapes and People: Multiple Perspectives on Oceans Past.' *Perspectives on Oceans Past* 1–10. doi:10.1007/978-94-017-7496-3_1.

Morgan, J.L., S.E. Gergel, and N.C. Coops. 2010. 'Aerial Photography: A Rapidly Evolving Tool for Ecological Management.' *BioScience* 60(1): 47–59. doi:10.1525/bio.2010.60.1.9.

Tomscha, S.A., I.J. Sutherland, D. Renard, S.E. Gergel, J.M. Rhemtulla, E.M. Bennett, L.D. Daniels, I.M.S. Eddy, and E.E. Clark. 2016. 'A Guide to Historical Data Sets for Reconstructing Ecosystem Service Change over Time.' *BioScience* 66(9): 747–762. doi:10.1093/biosci/biw086.

Acknowledgements

Sarah Gergel was supported by a Discovery Grant from NSERC, the National Science and Engineering Research Council of Canada. Ruth Thurstan was supported by the European Union's Horizon 2020 research and innovation programme under the Marie Skłodowska-Curie grant agreement (MarHIST No. 787671).

References

Barbosa, C.C. de A., P.M. Atkinson, and J.A. Dearing. 2015. 'Remote Sensing of Ecosystem Services: A Systematic Review.' *Ecological Indicators* 52: 430–443. doi:10.1016/j.ecolind.2015.01.007.

Bennett, E.M., G.D. Peterson, and L.J. Gordon. 2009. 'Understanding Relationships among Multiple Ecosystem Services.' *Ecology Letters* 12(12): 1394–1404. doi:10.1111/j.1461-0248.2009.01387.x.

Braun, V., and V. Clarke. 2006. 'Using Thematic Analysis in Psychology.' *Qualitative Research in Psychology* 3(2): 77–101. doi:10.1191/1478088706qp063oa.

Buckley, S.M., R.H. Thurstan, A. Tobin, and J.M. Pandolfi. 2017. 'Historical Spatial Reconstruction of a Spawning-Aggregation Fishery.' *Conservation Biology* 31(6): 1322–1332. doi:10.1111/cobi.12940.

Butler, P.G., and B.R. Schöne. 2017. 'New Research in the Methods and Applications of Sclerochronology.' *Palaeogeography, Palaeoclimatology, Palaeoecology* 465: 295–299. doi:10.1016/j.palaeo.2016.11.013.

Cohen, W.B., and S.N. Goward. 2004. 'Landsat's Role in Ecological Applications of Remote Sensing.' *BioScience* 54(6): 535. doi:10.1641/0006-3568(2004)054[0535:LRIEAO]2.0.co;2.

Coomes, O.T., Y. Takasaki, and J.M. Rhemtulla. 2011. 'Land-use Poverty Traps Identified in Shifting Cultivation Systems Shape Long-term Tropical Forest Cover.' *Proceedings of the National Academy of Sciences* 108(34): 13925–13930. doi:10.1073/pnas.1012973108.

Crumley, C.L. 2017. 'Historical Ecology and the Study of Landscape.' *Landscape Research* 42(Supplement 1). doi:10.1080/01426397.2017.1399994.

Daniel, T.C., A. Muhar, A. Arnberger, O. Aznar, J.W. Boyd, K.M.A. Chan, R. Costanza et al. 2012. 'Contributions of Cultural Services to the Ecosystem Services Agenda.' *Proceedings of the National Academy of Sciences of the United States of America* 109(23): 8812–8819. www.internationaleonline.org/research/decolonising_practices/54_decolonial_sensibilities_indigenous_research_and_engaging_with_archives_in_contemporary_colonial_canada.

Eddy, I.M.S., S.E. Gergel, N.C. Coops, G.M. Henebry, J. Levine, H. Zerriffi, and E. Shibkov. 2017. 'Integrating Remote Sensing and Local Ecological Knowledge to Monitor Rangeland Dynamics.' *Ecological Indicators* 82: 106–116. doi:10.1016/j.ecolind.2017.06.033.

Fuchs, R., P.H. Verburg, J.G.P.W. Clevers, and M. Herold. 2015. 'The Potential of Old Maps and Encyclopaedias for Reconstructing Historic European Land Cover/Use Change.' *Applied Geography* 59: 43–55. doi:10.1016/j.apgeog.2015.02.013.

Gebhard, R. 2003. 'Material Analysis in Archaeology.' *Hyperfine Interactions* 150(1–4): 1–5. doi:10.1023/b:hype.0000007175.85659.15.

Gergel, S.E., and M.G. Turner, eds. 2017. *Learning Landscape Ecology: A Practical Guide to Concepts and Techniques* (2nd ed). New York: Springer. doi:10.1007/978-1-4939-6374-4.

Gifford-Gonzalez, D. 2018. *An Introduction to Zooarchaeology*. New York: Springer. doi:10.1007/978-3-319-65682-3.

Harland, J., A.K.G. Jones, D.C. Orton, and J.H. Barrett. 2016. 'Fishing and Fish Trade in Medieval York: The Zooarchaeological Evidence.' In *Cod and Herring: The Archaeology and History of Medieval Sea Fishing*, edited by J.H. Barrett and D.C. Orton. Oxford: Oxbow Books.

Hsieh, H-F., and S.E. Shannon. 2005. 'Three Approaches to Qualitative Content Analysis.' *Qualitative Health Research* 15(9): 1277–1288. doi:10.1177/1049732305276687.

Jackley, J., L. Gardner, A.F. Djunaedi, and A.K. Salomon. 2016. 'Ancient Clam Gardens, Traditional Management Portfolios, and the Resilience of Coupled Human-Ocean Systems.' *Ecology and Society* 21(4): 20. doi:10.5751/ES-08747–210420.

Jeffers, E.S., S. Nogué, and K.J. Willis. 2015. 'The Role of Palaeoecological Records in Assessing Ecosystem Services.' *Quaternary Science Reviews* 112: 17–32. doi:10.1016/j.quascirev.2014.12.018.

Kittinger, J.N., L. McClenachan, K. Gedan, and L. Blight, eds. 2015. *Marine Historical Ecology in Conservation. Applying the Past to Manage for the Future.* San Francisco: University of California Press.

Last, W.M., and J.P. Smol. 2001. 'An Introduction to Basin Analysis, Coring, and Chronological Techniques Used in Paleolimnology.' In *Tracking Environmental Change Using Lake Sediments*, Developments in Paleoenvironmental Research, 1–5. doi:10.1007/0-306-47669-x_1.

Levis, C., F.R.C. Costa, F. Bongers, M. Peña-Claros, C.R. Clement, A.B. Junqueira, E.G. Neves et al. 2017. 'Persistent Effects of Pre-Columbian Plant Domestication on Amazonian Forest Composition.' *Science* 355(6328): 925–931.

McClenachan, L., A.B. Cooper, M.G. Mckenzie, and J.A. Drew. 2015. 'The Importance of Surprising Results and Best Practices in Historical Ecology.' *BioScience* 65: 932–939. doi:10.1093/biosci/biv100.

Morgan, J.L., S.E. Gergel, and N.C. Coops. 2010. 'Aerial Photography: A Rapidly Evolving Tool for Ecological Management.' *Bioscience* 60(1): 47–59. doi:10.1525/bio.2010.60.1.9.

Newsom, J.T., R.N. Jones, and S.M. Hofer, eds. 2012. *Multivariate Application Series, Volume 18. Longitudinal Data Analysis: A Practical Guide for Researchers in Aging, Health, and Social Sciences.* Abingdon: Routledge.

Renard, D., J.M. Rhemtulla, and E.M. Bennett. 2015. 'Historical Dynamics in Ecosystem Service Bundles.' *Proceedings of the National Academy of Sciences of the United States of America* 112(43): 13411–13416.

Rhemtulla, J.M., D.J. Mladenoff, and M.K. Clayton. 2009. 'Historical Forest Baselines Reveal Potential for Continued Carbon Sequestration.' *Proceedings of the National Academy of Sciences* 106(15): 6082–6087. doi:10.1073/pnas.0810076106.

Rosenberg, A.A., W.J. Bolster, K.E. Alexander, W.B. Leavenworth, A.B. Cooper, and M.G. McKenzie. 2005. 'The History of Ocean Resources: Modeling Cod Biomass Using Historical Records.' *Frontiers in Ecology and the Environment* 3(2): 84–90.

Selgrath, J.C, S.E. Gergel, and A.C.J. Vincent. 2017. 'Incorporating Spatial Dynamics Greatly Increases Estimates of Long-term Fishing Effort: A Participatory Mapping Approach.' *ICES Journal of Marine Science* 75(1): 210–220. doi:10.1093/icesjms/fsx108.

Smith, D.J., A.P. Mackie, and I.D. Sumpter. 2005. 'Building Quaksweaqwul: Dendroarchaeological Investigations at Kiix?in National Historic Site, Vancouver Island, Canada.' *Dendrochronologia* 22(3): 195–201. doi:10.1016/j.dendro.2005.04.004.

Stahle, D.W., and J.S. Dean. 2010. 'North American Tree Rings, Climatic Extremes, and Social Disasters.' *Dendroclimatology Developments in Paleoenvironmental Research* 11: 297–327. doi:10.1007/978-1-4020-5725-0_10.

Thurstan, R.H., Z. Brittain, D.S. Jones, E. Cameron, J. Dearnaley, and A. Bellgrove. 2018. 'Aboriginal Uses of Seaweeds in Temperate Australia: An Archival Assessment.' *Journal of Applied Phycology* 30: 1821–1832. doi:10.1007/s10811-017-1384-z.

Thurstan, R.H., S.M. Buckley, and J.M. Pandolfi. 2018. 'Trends and Transitions Observed in an Iconic Recreational Fishery Across 140 Years.' *Global Environmental Change* 52: 22–36. doi:10.1016/j.gloenvcha.2018.06.002.

Todd, Z. 2016. 'An Indigenous Feminist's Take on the Ontological Turn: "Ontology" is just another Word for Colonialism.' *Journal of Historical Sociology* 29(1): 4–22. doi:10.1111/johs.12124.

Tomscha, S.A., and S.E. Gergel. 2016. 'Ecosystem Service Trade-offs and Synergies Misunderstood without Landscape History.' *Ecology and Society* 21(1): 43. doi:10.5751/es-08345-210143.

Tomscha, S.A., I.J. Sutherland, D. Renard, S.E. Gergel, J.M. Rhemtulla, E.M. Bennett, L.D. Daniels, I.M.S. Eddy, and E.E. Clark. 2016. 'A Guide to Historical Data Sets for Reconstructing Ecosystem Service Change over Time.' *BioScience* 66(9): 747–762. doi:10.1093/biosci/biw086.

UWICER. 2017. *Dendrochronology Manual.* Ugyen Wangchuck Institute for Conservation and Environmental Research, Department of Forests and Park Services. Bogor: UWICER Press. researchgate.net/publication/325114690_Dendrochronological_Manual.

Wallman, D. 2018. 'Histories and Trajectories of Socio-ecological Landscapes in the Lesser Antilles: Implications of Colonial Period Zooarchaeological Research.' *Environmental Archaeology* 23(1): 13–22. doi:10.1080/14614103.2017.1345086.

Zusman, A. 2010. *Story Bridges: A Guide for Conducting Intergenerational Oral History Projects.* Walnut Creek: Left Coast Press. doi:10.4324/9781315419572.

26

Dynamical systems modelling

Steven J. Lade,[1,2] *John M. Anderies,*[3]
Paul Currie[4,5] *and Juan C. Rocha*[1,6]

[1]STOCKHOLM RESILIENCE CENTRE, STOCKHOLM UNIVERSITY, STOCKHOLM, SWEDEN
[2]FENNER SCHOOL OF ENVIRONMENT AND SOCIETY, THE AUSTRALIAN NATIONAL UNIVERSITY, CANBERRA, AUSTRALIA
[3]SCHOOL OF SUSTAINABILITY AND SCHOOL OF HUMAN EVOLUTION AND SOCIAL CHANGE, ARIZONA STATE UNIVERSITY, TEMPE, ARIZONA, USA
[4]ICLEI AFRICA, CAPE TOWN, SOUTH AFRICA
[5]CENTRE FOR SUSTAINABILITY TRANSITIONS, STELLENBOSCH UNIVERSITY, STELLENBOSCH, SOUTH AFRICA
[6]BEIJER INSTITUTE OF ECOLOGICAL ECONOMICS, THE ROYAL SWEDISH ACADEMY OF SCIENCES, STOCKHOLM, SWEDEN

Key methods discussed in this chapter

Causal loop diagrams, loop analysis, qualitative analysis of differential equations (including bifurcation analysis and stability analysis), numerical simulation of dynamical systems

Connections to other chapters

Systems scoping (Chapter 5) and participatory modelling and planning (Chapter 13) methods can help to construct the dynamical systems model. Statistical methods (Chapter 18) can be used to parameterise the model and/or test its outputs. The results of dynamical systems models are often used for futures analysis (Chapter 10) or scenario development (Chapter 11). Agent-based modelling (Chapter 28) is a closely related dynamical modelling method.

Introduction

Dynamical systems modelling provides a rigorous approach for studying how causal interactions within a social-ecological systems (SES) lead to dynamics at the system level. In studies of SES, dynamical systems models are generally used at an aggregated level, e.g. modelling total fish stocks or average harvester effort, rather than modelling the life cycles of individual fish or the effort level dynamics of individual harvesters, as might be done in an agent-based model (Chapter 28). Dynamical systems modelling provides conceptual, mathematical and computational tools to deal with key SES concepts such as feedbacks, non-linearity and regime shifts.

The modern notion of dynamical systems began with the work of Poincaré (1890; see Holmes 2007), who studied the qualitative dynamics generated by systems of non-linear differential equations. The advent of modern digital computing permitted advances in the numerical study of dynamical systems, most famously when Edward Lorenz, a meteorologist, discovered sensitivity to initial conditions, later named chaos, when performing computer simulations of air movement in the atmosphere (Lorenz 1963). Concepts such as chaos and

DOI: 10.4324/9781003021339-31

SUMMARY TABLE: DYNAMICAL SYSTEMS MODELLING	
DISCIPLINARY BACKGROUND	**KNOWLEDGE TYPE**
The methods in this chapter are derived from or have most commonly been used in: Mathematics, Physics, Ecology, Business	*The methods in this chapter are primarily used to generate the following types of knowledge:* • Exploratory • Explanatory
RESEARCH APPROACH	**PURPOSE OF METHOD**
The methods in this chapter originate from or most commonly adopt the following research approaches: • Analytical/objective • Collaborative/process	*The most common purposes of using the methods in this chapter are:* • System understanding • Stakeholder engagement and co-production • Policy/decision support
TEMPORAL DIMENSION	**SYSTEMIC FEATURES AND PROCESSES**
The methods in this chapter are most commonly applied to the following temporal dimensions: • Present (typically within the last 5–10 years) • Recent past (post-1700s) • Future **SPATIAL DIMENSION** *The methods in this chapter are primarily either or both:* • Non-spatial *The methods in this chapter are most commonly applied at the following spatial scales:* • Local • Regional (provincial/state to continental) • Global	*While most methods can do many things, the methods in this chapter are particularly good (i.e. go-to methods) for addressing the following:* • SES components and linkages • Social-ecological dependence and impact • Social-ecological interactions over time • Path dependency • Regime shifts • Evaluating policy options

bifurcations were mathematically formalised in the 1980s through work by researchers such as Guckenheimer and Holmes (1983). These and other dynamical systems concepts have influenced the development of many classical concepts used first in theoretical ecology and later in research on the resilience of SES, such as feedbacks, attractors, regime shifts, slow and fast variables, and definitions of resilience itself.

Computer-based tools for numerically solving dynamical systems were taken up by other academic fields for scenario development and planning (see Chapters 10 and 11). These modelling traditions were developed first in the military and later in industrial development, policy design and management sciences. In these fields, dynamical systems modelling is more commonly known as 'system dynamics'. A famous example of system dynamics is the World3 model developed by MIT professor Jay Forrester, which inspired the seminal book *Limits to Growth* (Meadows et al. 1972). The system dynamics tradition generally emphasises graphical methods for model construction and analysis, which are well suited to participatory settings (see Chapter 13). Causal loop diagrams, for example, provide a frequently used graphical representation of feedbacks.

Today, dynamical systems modelling is a highly interdisciplinary field and a core method for the study of social-ecological and other complex systems. Elements from both the system dynamics and mathematical dynamical systems traditions are widely used in SES research. Dynamical systems models of SES range from empirically parameterised models of a specific case (Elsawah et al. 2017) through to abstract models used to develop theory about SES dynamics (Lade et al. 2013). At the broadest level, dynamical systems modelling promotes a 'systems view' that sees elements of an SES as causally interconnected and interdependent.

SES problems and questions

Dynamical systems methods are used by individual researchers or collectively among groups of scientists and stakeholders for a wide range of purposes, including:

- Mapping the structure of causal relationships within a system (e.g. what are the social, ecological and social-ecological interactions within Lake Victoria fisheries? (Downing et al. 2014))
- Understanding how system-level dynamics result from causal relationships (e.g. what social and ecological mechanisms led to the collapse of the Baltic cod fishery? (Lade et al. 2015)). What poverty traps are produced by different poverty-environment relationships? (Lade et al. 2017). What mechanisms and drivers determine the state of a managed fire-driven rangeland system? (Anderies, Janssen, and Walker 2002))
- Predicting future dynamics of an SES in response to policy decisions and other drivers (e.g. how will natural perturbations and management decisions affect a coastal fishery? (Martone, Bodini, and Micheli 2017)). How will water supply and demand in the Australian Capital Territory respond to climate changes and management decisions? (Elsawah, McLucas, and Mazanov 2015))

A common use of dynamical systems models is to perform simulations that produce the model's behaviour over time. These simulations can be used to validate a model based on past dynamics and to predict not only a system's future dynamics but also the effects of different interventions or policies on future dynamics. However, there are many other concepts and methods in dynamical systems modelling that are well suited to studying research questions involving key SES concepts, e.g. resilience, feedbacks, attractors and regime shifts. These other methods and concepts, elaborated in Section 'Brief description of key methods', may even have triggered the original development of these SES concepts.

Dynamical systems approaches are also well suited to facilitating participatory processes (see Chapter 13). First, a complete model or model-under-development is a useful boundary object to show how each stakeholder or action contributes to system behaviour. Bringing stakeholders together to discuss a systemic problem challenges their preconceptions and tests their assumptions against larger system behaviour as elucidated in a model, often leading to changed actor and system behaviours. Second, testing the assumptions of the model builders and stakeholders as the model is built results in more robust causal links and therefore a better system model. Finally, once stakeholders are satisfied that the model displays accurate baseline behaviours, policy and intervention scenario options can be incorporated into the model and their results tested against these behaviours. Policy options could include environmental water provision or pumping restrictions in groundwater social-ecological models; subsidies, water pricing and regulations on the salt content of irrigation water; land management choices in irrigation agriculture models; and water supply and demand management options in public water-distribution models (Elsawah et al. 2017).

Brief description of key methods

In a dynamical systems model, dynamics result from interactions between different variables within the system. These variables could be properties of actors or groups (e.g. wealth or opinions), properties of species (e.g. population) or other biophysical quantities (e.g. temperature). Causal relationships between these variables could represent anything from biophysical laws to a stakeholder's belief about how the relationship operates. Methods for representing and analysing these relationships range from graphical representations that can be constructed by hand to computational simulations and analyses using formal mathematical methods (Table 26.1).

The first step in a dynamical systems analysis is often the construction of a causal loop diagram, which maps out causal relationships within the SES. The process of constructing the causal loop diagram can help to develop a shared understanding of the SES among multiple stakeholders (see Chapter 13). Especially in participatory settings, a causal loop diagram and its loop analysis may be the endpoint of the dynamical systems methodology.

Causal loop diagrams are well suited to identifying feedback loops in which a chain of causal mechanisms forms a closed loop of cause and effect. Feedback loops are generally classed as either 'reinforcing' or 'balancing'. In a reinforcing feedback loop, the initial changes to a variable are amplified by the feedback loop, generally leading to the system accelerating change. In a balancing feedback loop, the initial changes to a variable are counterbalanced by the feedback loop, generally leading to the system resisting change. 'Reinforcing' and 'balancing' feedback loops are also referred to as 'positive' and 'negative', respectively, where these terms are understood in a mathematical rather than normative sense as amplifying or dampening initial changes. In loop analyses, these feedback loops are identified to help explain system dynamics.

The foundation of any computational or mathematical study of a dynamical systems model is to formulate a set of differential equations (where time is continuous) or difference equations (where time is discrete, i.e. increases in steps). These equations specify how variables in the system change given information about their current states. They can be represented graphically using a stock and flow diagram, or mathematically in equation form.

Stock and flow diagrams represent causal relationships, like in causal loop diagrams, but also distinguish between 'stocks', which are variables that accumulate over time, and 'flows', which increase or decrease stocks over time. The level of water in a bathtub, for example, could be a stock and the flows of water in through the tap and out through the

Table 26.1 Summary of key methods used in dynamical systems modelling

Method	*Description*	*References*
Causal loop diagram	A causal loop diagram is a technique to map out the feedback structure of a system, identifying reinforcing or balancing feedback behaviours.	*Key introductory texts* Sterman 2000; Maani and Cavana 2007 *Applications to SES* Fazey et al. 2011; Hanspach et al. 2014; Pollard, Biggs, and Du Toit 2014; Kim et al. 2017
Loop analysis	Loop analysis infers possible stability properties of a system based on feedback loops in the causal loop diagram. It may involve comparison against reference archetypes or modes, or assessment of loop strength.	*Key introductory texts* Puccia and Levins 1985; Justus 2005 *Applications to SES* Downing et al. 2014; Martone, Bodini, and Micheli 2017; Abram and Dyke 2018
Qualitative analysis of differential equations	Bifurcation and stability analysis are powerful tools to determine the existence of and characterise dynamical patterns such as attractors and transitions between them (such as regime shifts).	*Key introductory texts* Strogatz 1994; Kuznetsov 2013 *Applications to SES* Anderies, Janssen, and Walker 2002; Anderies 2006; Lade et al. 2015; Lade et al. 2017; Tekwa et al. 2019
Numerical simulation of dynamical systems	Numerical simulation of dynamical systems models produces behaviour trends over time and allows the testing of scenarios. These models are implemented using differential or difference equations, but can also be represented graphically using stock and flow diagrams. Algorithms such as the Euler or Runge–Kutta methods numerically solve differential equations to produce the dynamics of variables. Researchers often use software with these algorithms built in.	*Key introductory texts* Sterman 2000; Butcher 2016 *Applications to SES* Cifdaloz et al. 2010; Elsawah et al. 2017 (see listed cases); Pizzitutti et al. 2017

plughole are 'flows'. Other variables, called 'dynamic variables' or 'intermediate variables', mediate causal relationships between different stocks or flows. The water temperature perceived by somebody in the bath, for example, could be an intermediate variable that is causally affected by the actual water temperature and causes a change, depending on the person's behaviour, in the rate of inflow of hot and cold water into the bath.

Underlying a stock and flow diagram is a set of differential or difference equations that some modelling approaches present directly.

- A paradigmatic ecological differential equation is the logistic equation (Verhulst 1845, 1847) for the growth of a population P,

$$\frac{dP}{dt} = rP\left(1 - \frac{P}{K}\right).$$

  Here dP/dt denotes the rate of change of the population P. The logistic equation assumes a rate of population growth proportional to the population size for a low population (with proportionality r), which unbounded would produce exponential growth. At higher population sizes, however, the growth of the population saturates at carrying capacity K.
- The replicator equation from evolutionary game theory is an example of a commonly used differential equation describing human behaviour (Cressman and Tao 2014). It states that the fraction of individuals f_i that follow strategy i changes over time at a rate

$$\frac{df_i}{dt} = (u_i - \bar{u}) f_i,$$

  where u_i is the payoff or utility associated with strategy i, and $\bar{u} = \sum_i u_i f_i$ is the population average payoff. The replicator equation assumes that individuals randomly encounter other individuals in a well-mixed population and switch strategies at a rate proportional to the difference between their payoffs (Cressman and Tao 2014).

While a theoretical pedigree such as that of the logistic equation or the replicator equation can help to build confidence in the choice of an equation, dynamical systems models are often also built from assumed or elicited knowledge about the causal relationships in an SES, e.g. from the causal loop diagrams described above (see Chapter 13). Statistical methods (see Chapter 18) can compare the fit of different forms of model equations to data, or produce non-parametric fits using interpolation.

After specifying the forms of relationships between variables through a stock and flow diagram or differential or difference equations, the final steps before a numerical simulation can be performed are to assign values to the model's parameters and initial values (also called 'initial conditions') to all stocks. Parameters, such as r and K in the logistic equation above, are quantities that are prescribed externally to the model and usually remain constant for the duration of the simulation. Parameter values can be chosen from previous knowledge about the SES, fitted to historical trends using statistical methods (see Chapter 18), or alternatively set to a range of different values to explore sensitivity of the model's dynamics to parameter values.

Once the model is fully specified, the difference or differential equations can be solved to produce time series of the variables in the systems. Numerical solution methods are readily available in many software products or can be programmed directly by the researcher. Symbolic mathematical solutions can be achieved only in rare cases. These time series can be used to project future states of the system, test the fit of the model to historical data or explore how the model's dynamics depend on changes to parameters within the model.

Beyond numerical simulation, a range of mathematical methods study the qualitative dynamics of dynamical systems. Here 'qualitative' refers to the differences between the various dynamical patterns that dynamical systems can display.

- Stability analysis can identify attractors, which are states that an SES approaches in its long-term dynamics. Attractors can be points (equilibrium points), stable oscillations (limit cycles) or more complicated geometric objects (strange attractors).

- State space diagrams and stability landscapes can graphically represent attractors and basins of attraction. A basin of attraction is the set of initial conditions whose dynamics approach the attractor. The resilience of SES has often been conceptualised using various aspects of the basin of attraction (Walker et al. 2004; Meyer 2016; Donges and Barfuss 2017). Stability landscapes, which are often sketched qualitatively, have a precise definition within dynamical systems theory (Strogatz 1994).
- Bifurcations are sudden, qualitative changes in system dynamics (where 'sudden' is with respect to changes in some external parameter). Regime shifts in ecological systems (Scheffer et al. 2001) and SES (Lade et al. 2013) are commonly associated with a type of bifurcation called the fold bifurcation in which an attractor disappears, forcing a transition to another attractor. Mathematical tools (such as normal forms) and computational tools (such as bifurcation continuation) can help to characterise bifurcations. There exists a large family of bifurcations (Kuznetsov 2013), although few of them have been studied in SES.
- Bifurcation diagrams can be used to graphically represent regime shifts and accompanying phenomena such as hysteresis (Scheffer et al. 2001). These diagrams plot how the stable and unstable states of a system depend on a chosen parameter. Hysteresis is a form of path dependence where increasing a parameter and then decreasing the parameter to its original value does not return the system to its original state. It commonly appears as an S-curve in bifurcation diagrams where the corners of the S are a pair of fold bifurcations (Scheffer et al. 2001).
- The importance of considering fast and slow variables is a common theme in resilience research (Biggs, Schlüter, and Schoon 2015). Whereas fast and slow variables are often identified heuristically, singular perturbation theory (Kuehn 2015) can formally decompose a model into fast and slow components. Crépin (2007), for example, decomposed a coral reef ecosystem model into fast variables (algae and herbivore biomass) and a slow variable (coral biomass).

Limitations

Dynamical systems models can be initially developed without quantitative data, relying on the mental models of model builders or stakeholders in participatory settings (see Chapter 13), or using previous literature and theories for theoretical models. This is useful for modelling in data-scarce environments, as system behaviour can be explored at a high level without data, and model accuracy can be improved as data become available. However, mental models require wide substantiation or validation by experts and their limitations must be acknowledged.

Large stock and flow models or simulation models can make detailed statements about the dynamics of the SES being studied. However, constructing these models is both data and time intensive, requiring validation of each stage as the modelling process continues in order to retain confidence in the system behaviour it is representing. Even with a rigorous validation process, it can be difficult to fully understand and analyse the operation of such a large model. Large and complicated system diagrams can also hinder communication and the development of shared system understanding.

At the other extreme of complexity, simple models can deliver transparent and generalisable insights but can be more challenging to develop than complex models. Simple models require difficult decisions about what simplifying assumptions one can make to maximise clarity, tractability, reproducibility and generalisability without giving up too much representativeness (the extent to which the model matches empirical reality, e.g. its capacity to predict outcomes). Simple models can make general statements about SES but are of limited use when analysing outcomes and presenting options for intervention in specific systems.

Dynamical systems methods focus on system-level phenomena, such as feedbacks, with less attention paid to the roles, agency and decision-making processes of individual actors in that system. Actor heterogeneity can be included, e.g. through variables representing different groups, or a variable that characterises the variability of a property across a population. Decision-making processes can be represented through causal relationships between the decision context and the consequences of a choice. However, these representations are generally less intuitive than in actor-focused methods such as agent-based modelling (see Chapter 28).

Resource implications

Causal loop diagrams can be sketched with pen and paper or created with graphical computer software such as Stella (iseesystems.com), Vensim (vensim.com), NetLogo (ccl.northwestern.edu/netlogo), Mental Modeler (mentalmodeler.org) or Insight Maker (insightmaker.com). In participatory modelling (see Chapter 13), an expert facilitator can equip stakeholders with the knowledge to understand and draw their own causal loop diagrams, giving them ownership over the process. This can result in novel outputs.

Most of the software listed above can also be used to construct a computational stock and flow model of dynamical systems. This process would usually require an expert system

Case study 26.1: Using generalised modelling to study the 1980s collapse of the Baltic cod fishery

Dynamical systems modelling can be used to investigate regime shifts in SES, such as the collapse of the Baltic cod fishery. In the mid- to late 1980s, the Baltic cod fishery suddenly changed from historically high cod biomass and catches to a sprat-dominant ecosystem with low cod abundance (Möllmann et al. 2009). Although the ecological causes and dynamics of this collapse have been well studied, the contribution of social processes to the collapse has received less attention. A dynamical systems modelling study (Lade et al. 2015) filled this research gap by investigating the cod collapse as a social-ecological phenomenon.

A group of ecological and social scientists with expertise on the Baltic cod fishery constructed a conceptual representation of the important variables and processes influencing the collapse of the cod-fishery SES (Figure 26.1). Quantitative and qualitative data on these variables and processes were assembled. Using a dynamical systems method called 'generalised modelling' (Lade and Niiranen 2017), these data were used to estimate the stability of the SES and its subsystems, the directions in which the system variables changed during the collapse, the dominant feedback loops and the possible effects of various policies. Among other results, the model showed that adaptive fisher behaviour temporarily stabilised an otherwise unstable ecological system and allowed the cod boom to persist longer than if the fishers had not adapted. Physical and ecological changes in the Baltic Sea, and increasing pressure from Swedish west coast fishers, eventually outweighed this adaptive capacity and drove the cod fishery into collapse.

dynamics modeller. The graphical interfaces in these software packages are useful for sharing model design and results with co-authors or stakeholders, but the capacity for sophisticated mathematical analyses by software packages that rely on graphical interfaces can be limited.

To deal directly with the representation and analysis of the dynamical systems as differential or difference equations, a basic level of algebra and calculus is required to understand and manipulate these equations. Some insight can be gained by symbolic manipulation, especially for systems with few variables, which can be performed by hand without computer assistance. However, even small models frequently have behaviour that is too complex for symbolic mathematical solutions. Any general purpose software such as C, Python, MATLAB or Julia can be used to numerically solve differential equations; this software often includes inbuilt equation solvers. The inbuilt package AUTO in the specialist free software XPP (Ermentrout 2007) and plugins for MATLAB such as MATCONT (Govaerts, Kuznetsov, and Sautois 2006) and GRIND (SparcS 2018) perform numerical bifurcation continuation to produce bifurcation diagrams.

New directions

One of the main challenges in constructing quantitative empirical models of SES is the level of knowledge and data about causal relationships that is required to fully specify and

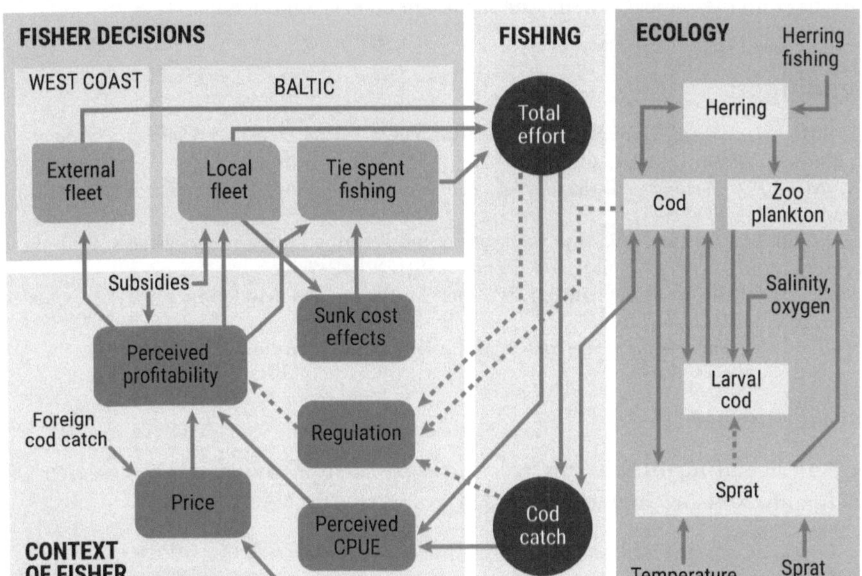

Figure 26.1 Overview of the components and causal relationships included in a social-ecological model of the 1980s Baltic cod fishery (Lade et al. 2015)

parameterise the model. 'Generalised modelling' is an emerging dynamical systems method for modelling SES that could deal with this challenge (Lade and Niiranen 2017). This modelling approach is useful in situations where the stability of the SES is of primary interest. Estimating stability requires less data than specifying a full simulation model. The generalised modelling procedure specifies exactly which data are required in order to estimate stability and can often incorporate qualitative data. Case study 26.1 summarises how generalised modelling was used to study the 1980s collapse of the Baltic cod fishery.

New developments in computational multi-scale modelling seek to embed outputs of one type of model as inputs to another. This allows system models, which typically operate at one level over one time scale and time step, to interact at multiple levels and times. A regime shift, for example, can be a long process over a wide territory that can be measured and modelled over decades. If the modeller wishes to understand how rapid urban expansion of a number of towns or seasonal resource-demand fluctuations, measured in years or months respectively, contributed to regime shifts, these models could be embedded for greater accuracy and decision support.

Finally, a broad range of dynamical systems theory that is currently unused by dynamical systems modelling in SES research could be used to characterise and explain SES phenomena. There are many more types of bifurcation than the fold bifurcation (Kuznetsov 2013) that is the basis for the regime shift concept. The concept of chaos, in which systems are highly sensitive to initial conditions, is rarely used. There are many dynamical patterns beyond equilibria, such as mixed-mode oscillations (Kuehn 2015), that display fast-slow dynamics reminiscent of the adaptive cycle (Holling and Gunderson 2002). These concepts could aid the analysis of transient dynamics in SES. Numerical methods (Colon, Claessen, and Ghil 2015; Van Strien et al. 2019) or formal statistical mechanics methods could also be used to apply concepts such as bifurcations from dynamical systems theory to agent-based models.

Key readings

Ford, A. 2010. *Modeling the Environment: An Introduction to System Dynamics Models of Environmental Systems* (2nd ed). Washington: Island Press.

Roussel, M. 2019. *Nonlinear Dynamics: A Hands-on Introductory Survey.* San Rafael: Morgan & Claypool.

Scheffer, M., and Carpenter, S.R. 2003. 'Catastrophic Regime Shifts in Ecosystems: Linking Theory to Observation.' *Trends in Ecology and Evolution* 18(12): 648–656.

Sterman, J. 2000. 'Business Dynamics: Systems Thinking and Modeling for a Complex World.' Boston: Irwin/McGraw-Hill.

Strogatz, S. 1994. *Nonlinear Dynamics and chaos.* Philadelphia: Westview.

Acknowledgements

Steven J. Lade and Juan C. Rocha received support from the Swedish Research Council Formas (grants 2014-589 and 942-2015-731, respectively).

References

Abram, J.J., and J.G. Dyke 2018. 'Structural Loop Analysis of Complex Ecological Systems.' *Ecological Economics* 154: 333–342.

Anderies, J.M. 2006. 'Robustness, Institutions, and Large-scale Change in Social-Ecological Systems: The Hohokam of the Phoenix Basin.' *Journal of Institutional Economics* 2(2): 133–155.

Anderies, J.M., M.A. Janssen, and B.H. Walker. 2002. 'Grazing Management, Resilience, and the Dynamics of a Fire-driven Rangeland System.' *Ecosystems* 5(1): 23–44.

Biggs, R., M. Schlüter, and M.L. Schoon, eds. 2015. *Principles for Building Resilience: Sustaining Ecosystem Services in Social-Ecological Systems*. Cambridge: Cambridge University Press.

Butcher, J.C. 2016. *Numerical Methods for Ordinary Differential Equations* (3rd ed). Chichester: John Wiley and Sons.

Cifdaloz, O., A. Regmi, J.M. Anderies, and A.A. Rodriguez. 2010. 'Robustness, Vulnerability, and Adaptive Capacity in Small-scale Social-Ecological Systems: The Pumpa Irrigation System in Nepal.' *Ecology and Society* 15(3): 39.

Colon, C., D. Claessen, and M. Ghil. 2015. 'Bifurcation Analysis of an Agent-based Model for Predator-Prey Interactions.' *Ecological Modelling* 317: 93–106.

Crépin, A-S. 2007. 'Using Fast and Slow Processes to Manage Resources with Thresholds.' *Environmental and Resource Economics* 36(2): 191–213.

Cressman, R., and Y. Tao. 2014. 'The Replicator Equation and Other Game Dynamics.' *Proceedings of the National Academy of Sciences* 111(Supplement 3): 10810–10817.

Donges, J.F., and W. Barfus. 2017. 'From Math to Metaphors and Back Again: Social-Ecological Resilience from a Multi-Agent-Environment Perspective.' *GAIA – Ecological Perspectives for Science and Society* 26(Supplement 1): 182–190.

Downing, A.S., E. van Nes, J. Balirwa, J. Beuving, P. Bwathondi, L.J. Chapman, I.J.M. Cornelissen et al. 2014. 'Coupled Human and Natural System Dynamics as Key to the Sustainability of Lake Victoria's Ecosystem Services.' *Ecology and Society* 19(4): 31.

Elsawah, S., A. McLucas, and J. Mazanov. 2015. 'Communicating about Water Issues in Australia: A Simulation/Gaming Approach.' *Simulation and Gaming* 46(6): 713–741.

Elsawah, S., S.A. Pierce, S.H. Hamilton, H. van Delden, D. Haase, A. Elmahdi, and A.J. Jakeman. 2017. 'An Overview of the System Dynamics Process for Integrated Modelling of Socio-Ecological Systems: Lessons on Good Modelling Practice from Five Case Studies.' *Environmental Modelling and Software* 93: 127–145.

Ermentrout, B. 2007. XPPAUT. *Scholarpedia* 2(1): 1399.

Fazey, I., N. Pettorelli, J. Kenter, D. Wagatora, and D. Schuett. 2011. 'Maladaptive Trajectories of Change in Makira, Solomon Islands.' *Global Environmental Change* 21(4): 1275–1289.

Govaerts, W., Y.A. Kuznetsov, and B. Sautois. 2006. 'MATCONT.' *Scholarpedia* 1(9): 1375.

Guckenheimer, J., and P. Holmes. 1983. *Nonlinear Oscillations, Dynamical Systems and Bifurcations of Vector Fields*. New York: Springer.

Hanspach, J., T. Hartel, A.I. Milcu, F. Mikulcak, I. Dorresteijn, J. Loos, H. von Wehrden et al. 2014. 'A Holistic Approach to Studying Social-Ecological Systems and its Application to Southern Transylvania.' *Ecology and Society* 19(4): 32.

Holling, C.S., and L.H. Gunderson. 2002. 'Resilience and Adaptive Cycles.' In *Panarchy: Understanding Transformations in Systems of Humans and Nature*, edited by L.H. Gunderson and C.S. Holling, 25–62. Washington: Island Press.

Holmes, P.H. 2007. 'History of Dynamical Systems.' *Scholarpedia* 2(5): 1843.

Justus, J. 2005. 'Qualitative Scientific Modeling and Loop Analysis.' *Philosophy of Science* 72: 1272–1286.

Kim, M., S. You, J. Chon, and J. Lee. 2017. 'Sustainable Land-use Planning to Improve the Coastal Resilience of the Social-Ecological Landscape.' *Sustainability* 9: 1086.

Kuehn, C. 2015. *Multiple Time Scale Dynamics*. Cham: Springer.

Kuznetsov, Y.A. 2013. *Elements of Applied Bifurcation Theory*. New York: Springer.

Lade, S.J., L.J. Haider, G. Engström, and M. Schlüter. 2017. 'Resilience Offers Escape from Trapped Thinking on Poverty Alleviation.' *Science Advances* 3(5): e1603043.

Lade, S.J., and S. Niiranen. 2017. 'Generalized Modeling of Empirical Social-Ecological Systems.' *Natural Resource Modeling* 30(3): e12129.

Lade, S.J., S. Niiranen, J. Hentati-Sundberg, T. Blenckner, W. Boonstra, K. Orach, M. Quaas, H. Österblom, and M. Schlüter. 2015. 'An Empirical Model of the Baltic Sea Reveals the Importance of Social Dynamics for Ecological Regime Shifts.' *Proceedings of the National Academy of Sciences* 112(35): 11120–11125.

Lade, S.J., A. Tavoni, S.A. Levin, and M. Schlüter. 2013. 'Regime Shifts in a Social-Ecological System.' *Journal of Theoretical Ecology* 6: 359–372.

Lorenz, E. 1963. 'Deterministic Nonperiodic Flow.' *Journal of the Atmospheric Sciences* 20(2): 130–141.

Maani, K.E., and R.Y. Cavana. 2007. *Systems Thinking, System Dynamics*. Auckland: Pearson.

Martone, R.G., A. Bodini, and F. Micheli. 2017. 'Identifying Potential Consequences of Natural Perturbations and Management Decisions on a Coastal Fishery Social-Ecological System using Qualitative Loop Analysis.' *Ecology and Society* 22(1): 34.

Meadows, D.H., D.L. Meadows, J. Randers, and W.W. Behrens III. 1972. *The Limits to Growth.* Washington: Potomac Associates.

Meyer, K. 2016. 'A Mathematical Review of Resilience in Ecology.' *Natural Resource Modeling* 29(3): 339–352.

Möllmann, C., R. Diekmann, B. Müller-Karulis, G. Kornilovs, M. Plikshs, and P. Axe. 2009. 'Reorganization of a Large Marine Ecosystem due to Atmospheric and Anthropogenic Pressure: A Discontinuous Regime Shift in the Central Baltic Sea.' *Global Change Biology* 15(6): 1377–1393.

Pizzitutti, F., S.J. Walsh, R.R. Rindfuss, R. Gunter, D. Quiroga, R. Tippett, and C.F. Mena. 2017. 'Scenario Planning for Tourism Management: A Participatory and System Dynamics Model Applied to the Galapagos Islands of Ecuador.' *Journal of Sustainable Tourism* 25(8): 1117–1137.

Poincaré, H.J. 1890. 'Sur le Problème des Trois Corps et les Équations de la Dynamique.' *Acta Mathematica* 13: 1–270.

Pollard, S., H. Biggs, and D.R. du Toit. 2014. 'A Systemic Framework for Context-based Decision Making in Natural Resource Management: Reflections on an Integrative Assessment of Water and Livelihood Security Outcomes following Policy Reform in South Africa.' *Ecology and Society* 19(2): 63.

Puccia, C.J., and R. Levins. 1985. *Qualitative Modeling of Complex Systems: An Introduction to Loop Analysis and Time Averaging.* Cambridge: Harvard University Press.

Scheffer, M., S. Carpenter, J.A. Foley, C. Folke, and B. Walker 2001. 'Catastrophic Shifts in Ecosystems.' *Nature* 413: 591–596.

SparcS (Synergy program for analyzing resilience and critical transitionS). 2018. *GRIND for MATLAB.* www.sparcs-center.org/grind.

Sterman, J. 2000. *Business Dynamics: Systems Thinking and Modeling for a Complex World.* Boston: Irwin/McGraw-Hill.

Strogatz, S. 1994. *Nonlinear Dynamics and Chaos.* Philadelphia: Westview.

Tekwa, E.W., E.P. Fenichel, S.A. Levin, and M.L. Pinsky. 2019. 'Path-dependent Institutions Drive Alternative Stable States in Conservation.' *Proceedings of the National Academy of Sciences* 116(2): 689–694.

Van Strien, M.J., S.H. Huber, J.M. Anderies, and A. Grêt-Regamey. 2019. 'Resilience in Social-Ecological Systems: Identifying Stable and Unstable Equilibria with Agent-based Models.' *Ecology and Society* 24(2): 8.

Verhulst, P-F. 1845. 'Recherches Mathématiques sur la Loi D'Accroissement de la Population.' *Nouv. mém. de l'Academie Royale des Sci. et Belles-Lettres de Bruxelles* 18: 1–41.

Verhulst, P-F. 1847. 'Deuxième Mémoire sur la Loi D'Accroissement de la Population.' *Mém. de l'Academie Royale des Sci., des Lettres et des Beaux-Arts de Belgique* 20: 1–32.

Walker, B., C.S. Holling, S.R. Carpenter, and A. Kinzig. 2004. 'Resilience, Adaptability and Transformability in Social-Ecological Systems.' *Ecology and Society* 9(2): 5.

27

State-and-transition modelling

Brandon Bestelmeyer,[1] *Maria Fernández-Giménez,*[2]
Bulgamaa Densambuu[3] *and Retta Bruegger*[4]

[1]USDA-ARS JORNADA EXPERIMENTAL RANGE, NEW MEXICO STATE UNIVERSITY, LAS CRUCES, NEW MEXICO, USA
[2]DEPARTMENT OF FOREST AND RANGELAND STEWARDSHIP, COLORADO STATE UNIVERSITY, FORT COLLINS, COLORADO, USA
[3]GREEN GOLD AND ANIMAL HEALTH PROJECT, SWISS AGENCY FOR DEVELOPMENT AND COOPERATION, ULAANBAATAR, MONGOLIA
[4]COLORADO STATE UNIVERSITY EXTENSION, GRAND JUNCTION, COLORADO, USA

Key methods discussed in this chapter

State-and-transition modelling

Connection to other chapters

State-and-transition model development relies on a suite of related methods. Interviews and participant observation (Chapter 7) and participatory data collection (Chapter 8) are used to gather local knowledge to describe states and identify the causes of transitions among states. Participatory modelling and planning (Chapter 13) can involve local users in the construction of models and the linkage of models to adaptive management practices. Historical assessment (Chapter 25) identifies the ecological states and processes that are considered as references for ecological assessment while ecological field data collection (Chapter 6) provides data on extant vegetation, soils and animal communities comprising states. Statistical analysis (Chapter 18), including multivariate and machine-learning techniques, is used to develop and provide empirical validation of concepts for ecological states and to quantify transitions. Spatial mapping and analysis (Chapter 24) is used to map ecological states for use in management. Ecosystem service modelling (Chapter 31) can be used to identify the suite of ecosystem services provided by alternative states and collections of states in a landscape. Finally, expert modelling (Chapter 16) methods can be used to predict the probability of state transitions.

Introduction

State-and-transition models are tools to explain the causes and consequences of ecosystem change (Bestelmeyer et al. 2017). These models are used in several ways. Most commonly, state-and-transition models are heuristic tools to explain the processes involved in ecosystem change. They can link to specific management strategies that cause or prevent specific ecosystem changes. These models are also used in scenario development through simulations of the effects of external drivers (such as climate) and management actions (such as prescribed burning) on vegetation. All of these uses are aimed at guiding the management of ecosystems and natural resources.

DOI: 10.4324/9781003021339-32

SUMMARY TABLE: STATE-AND-TRANSITION MODELLING	
DISCIPLINARY BACKGROUND	**KNOWLEDGE TYPE**
The methods in this chapter are derived from or have most commonly been used in: Vegetation Ecology, Rangeland Science, Social Science	*The methods in this chapter are primarily used to generate the following types of knowledge:* • Descriptive • Exploratory • Explanatory
RESEARCH APPROACH	**PURPOSE OF METHOD**
The methods in this chapter originate from or most commonly adopt the following research approaches: • Analytical/objective • Collaborative/process	*The most common purposes of using the methods in this chapter are:* • System understanding • Stakeholder engagement and co-production • Policy/decision support
TEMPORAL DIMENSION	**SYSTEMIC FEATURES AND PROCESSES**
The methods in this chapter are most commonly applied to the following temporal dimensions: • Present (typically within the last 5–10 years) • Recent past (post-1700s) • Future	*While most methods can do many things, the methods in this chapter are particularly good (i.e. go-to methods) for addressing the following:* • Regime shifts • Social learning • Evaluating policy options
SPATIAL DIMENSION	
The methods in this chapter are primarily either or both: • Non-spatial • Explicitly spatial *The methods in this chapter are most commonly applied at the following spatial scales:* • Local	

The state–transition concept was developed initially for arid rangelands as a flexible way of organising information about vegetation dynamics that draws on a wide range of concepts about ecosystem change (Westoby, Walker, and Noy-Meir 1989). State-and-transition models are generally consistent with concepts of resilience and regime shifts, acknowledging the potential for abrupt and persistent change in ecosystems (Briske et al. 2008; López et al. 2011). In state-and-transition models, ecosystems potentially exhibit multiple states, usually defined by changes in ecosystem structures and processes of interest. Distinctions among ecological states reflect differences in the ecosystem services provided, as well as the risks and opportunities for change in the provision of ecosystem services. Although state-and-transition models are sometimes linked to alternative stable-state theory (Petraitis 2013), in practice concepts for states are variable and based largely on management utility. States can be highly stable and resilient or be transient and change relatively easily and often. In most state-and-transition models, 'state' circumscribes both regimes and states within regimes, following Biggs et al. (2012). While state-and-transition models were initially conceived to link rangeland management to the emerging concepts of ecosystem non-equilibrium and catastrophic transitions (Walker and Westoby 2011), they have become widely used in many types of ecosystems (Hobbs and Suding 2009).

State-and-transition models have been developed and applied following four general approaches (see also Table 27.1).

1. **Conceptual state-and-transition models:** Most commonly, conceptual state-and-transition models (involving diagrams and text) are used by scientists to communicate the roles of drivers and feedbacks involved in state change (McGlathery et al. 2013). These models have been used in resilience assessments as part of the resilience planning process (Huber-Sannwald et al. 2012; Walker and Salt 2012). Conceptual state-and-transition models can include quantitative values and linkages of states and transitions to specific management recommendations (USDA 2019). Conceptual state-and-transition models with quantitative state criteria are being produced as part of government land management programmes in the USA, Mongolia and Argentina (Bestelmeyer et al. 2017).
2. **State-and-transition simulation models:** Conceptual models can be extended to simulation models. Simulation state-and-transition models use multi-temporal data (or best guesses) to estimate transition probabilities for broad land-cover states or plant communities and develop scenarios of ecosystem change under different management regimes (Zweig and Kitchens 2009; Bino et al. 2015; Perry et al. 2015; Daniel et al. 2016). Simulation state-and-transition models may be non-spatial or spatially explicit.
3. **'Process-based' state-and-transition models:** 'Process-based' state-and-transition models (closely related to simulation state-and-transition models) seek to quantify in greater detail how interacting factors influence transition probabilities (Bashari, Smith, and Bosch 2009).
4. **Ecosystem service-based state-and-transition models:** Conceptual state-and-transition models have been expanded to include information about the ecosystem services and economic values provided by states (Ritten et al. 2018).

In addition to these well-developed roles for state-and-transition models, there have been efforts towards social-ecological state-and-transition models (or regime shift models), in which feedbacks between social and ecological systems are used to define states or regimes (Easdale and López 2016; Wilcox et al. 2018).

State-and-transition models of all types are ideally co-developed by land managers and scientists for use in local decision-making (Kachergis et al. 2013). They are tools for enhancing the functioning of SES rather than for studying them (i.e. scientists as part of the SES rather than on the outside looking in). Collaboratively developed state-and-transition models have numerous benefits, including increased trust in and use of science in management, improved communication among participants from different organisations, and decreased conflict (Johanson and Fernández-Giménez 2015).

SES problems and questions

In the context of SES, state-and-transition models foster a community-level understanding of how ecosystems function and respond to management actions, particularly when there is a lack of understanding or disagreement about why ecosystems change. Community-level understanding of ecosystem function underpins collaborative efforts to promote the sustainable provision of desired ecosystem services tailored to specific parts of a land- or seascape. Several specific problems are addressed by state-and-transition models (Yates and Hobbs 1997; Bestelmeyer et al. 2010; Karl, Herrick, and Browning 2012; Kachergis et al. 2013). First, they are used to stratify the landscape according to variations in ecological potential (the plant communities which a site can possibly support) and to identify management and restoration targets (e.g. deciding what plant communities to try to restore at a particular site). Second, they are used to assess the risk of degradation and identify proactive measures to avoid it. These measures could include early-warning indicators used in grazing management or strategies for managing fire frequency in a landscape. Third, they specify constraints to, and opportunities for, desirable transitions based on a knowledge of ecological processes. The success of seeding to restore desired plants, for example, may depend on factors such as soil type, climate or rates of soil erosion. Fourth, state-and-transition models can link to specific intervention strategies that promote desirable transitions, such as a specific seed mixture that is suitable for a site. Finally, these models are used in the design and interpretation of monitoring programmes used to evaluate the success of management by specifying, for example, the plant community benchmarks against which monitoring data are evaluated.

Brief description of key methods

State-and-transition models have been co-developed by scientists, land managers and resource users in many settings (Chambers et al. 2014; Bruegger et al. 2016; Tarrasón et al. 2016). These models may or may not be used directly in resource management decisions, but reflect and influence the mental models underlying those decisions.

Best practices in developing all state-and-transition models include three steps:

1. **Define the spatial extent:** State-and-transition models should be grounded to specific land areas, which helps to avoid confusing inherent differences in ecological potential with state transitions by focusing users' attention on usefully comparable land areas. Soil, landform, and ecoregional land classifications and maps can be used to organise multiple state-and-transition models pertaining to distinct land areas. Recent advances in machine-learning-assisted digital mapping can utilise widely available global spatial datasets to create these maps. Land classifications should reflect how local people and resource users classify and distinguish land types (Bestelmeyer et al. 2009; Duniway, Bestelmeyer, and Tugel 2010; Spiegal, Bartolome, and White 2016; Maynard et al. 2019).
2. **Conceptual state-and-transition model development:** Literature review, historical records and concepts developed from semi-structured interviews of key informants are used to develop initial models for spatial areas. Ideally, workshops with collaborators are

used to introduce state-and-transition modelling concepts and refine state-and-transition models based on local and expert knowledge, using a combination of break-out and large-group discussions. Group field visits to discuss states and transitions are invaluable. Key uncertainties for testing are identified at the end of the initial workshop(s) (Knapp, Fernández-Giménez, and Kachergis 2010; Knapp et al. 2011; Kachergis et al. 2013).

3. **Model testing and refinement:** Draft state-and-transition models are used as a basis for hypothesis testing and evaluation by users. Vegetation and soil (inventory) data collected by scientists can be used to quantify state characteristics and test for differences

Table 27.1 Summary of key applications of state-and-transition modelling

Main applications	*Description*	*References*
Conceptual state-and-transition models	Conceptual or descriptive state-and-transition models are produced in a variety of ways, but involve narratives and graphical descriptions of states and transitions between states. These transitions typically involve natural drivers, management actions and feedback processes. Descriptions of states and transitions can also include quantitative values.	*Key introductory texts* Bestelmeyer et al. 2010; Bruegger et al. 2016 *Applications to SES* Barrio et al. 2018; Peinetti et al. 2019
State-and-transition simulation models – non-spatial	Conceptual models are extended to quantitative models by (a) defining discrete states (e.g. by using multivariate analysis of community data), (b) defining transitions (e.g. by using multi-temporal data, stratified sampling in space with space-for-time substitution assumptions, expert estimates), and (c) simulating scenarios of change (e.g. using transition matrix models and Monte Carlo methods featuring probabilistic or deterministic transitions between states).	*Key introductory text* Daniel et al. 2016 *Applications to SES* Zweig and Kitchens 2009; Bino et al. 2015
State-and-transition simulation models – spatial	These models are similar to non-spatial state-and-transition simulation models, but run for multiple spatial cells that can incorporate spatial variation in conditional transition probabilities and spatial processes such as dispersal.	*Key introductory text* Daniel et al. 2016 *Applications to SES* Perry et al. 2015; Miller et al. 2017
Process-based state-and-transition models	Processes involved in transitions are quantified to produce a probabilistic model of cause and effect that can be updated over time with new knowledge (a Bayesian network; see also Chapter 16: Expert modelling). Probabilistic transition estimates include uncertainty about transitions.	*Key introductory text* Bashari, Smith, and Bosch 2009 *Applications to SES* Rumpff et al. 2011
Ecosystem service-based state-and-transition models	Once conceptual or simulation-based state-and-transition models are sufficiently developed, additional information about states and transitions can be presented. This includes the multiple ecosystem services and economic values provided by states and model-based 'value-added' information on processes of interest in states (e.g. wind erosion). This is an active area of research.	*Key introductory text* Brown and MacLeod 2011 *Applications to SES* Webb, Herrick, and Duniway 2014; Eastburn et al. 2017; Ritten et al. 2018

between alternative states. Field data are combined with local knowledge or spatial data on past management treatments to test ideas about deterministic transitions. Experimental monitoring can also provide tests of transitions over longer time frames. Workshops are used to discuss evidence and revise models iteratively. This step is, ideally, never fully completed, as collaborative groups constantly refine state-and-transition models based on new knowledge (Young et al. 2014; Bruegger et al. 2016; Porensky et al. 2016; Arterburn et al. 2018; Jamiyansharav et al. 2018; Tipton et al. 2018).

Conceptual state-and-transition model development and dissemination can be supported by systematic formats that enable state-and-transition models to be included in a database and linked to other computational tools. The ecosystem dynamics interpretive tool (EDIT) (edit.jornada.nmsu.edu) is a database for housing state-and-transition models linked to land classifications and spatial data and making these models available via the web and mobile devices. Application programming interfaces (APIs) allow state-and-transition model data to be linked to a variety of web and mobile applications. A globally accessible version of EDIT (editglobal.org) is in development.

Table 27.1 categorises and summarises the general types of state-and-transition models that have been produced and gives introductory references on state-and-transition model development methods and example applications.

Case study 27.1: State-and-transition models for management of Mongolian rangelands

The country of Mongolia has developed a system of conceptual rangeland state-and-transition models with quantification of key state characteristics (Figure 27.1). These state-and-transition models are coupled with monitoring and community-based rangeland management across the country. Communal, rangeland-based livestock production is a dominant land use, an important source of livelihood across Mongolia, and an equally important element of national identity. Rangeland management, however, changed dramatically with the transition from socialism to a free-market economy in the early 1990s. Privatisation of livestock coupled with a collapse of government support and an influx of new herders has led to ever-increasing livestock numbers and weakly coordinated management (Ulambayar and Fernández-Giménez 2019). There have been widespread reports of rangeland degradation (Addison et al. 2012; Eckert et al. 2015). Nonetheless, tools for assessing the true nature of rangeland change and responding to it did not exist.

State-and-transition model development combined with monitoring and support for community-based rangeland management began in 2008 via cooperation of an international donor organisation (Swiss Agency for Development and Cooperation's 'Green Gold' programme), the Mongolian government and US scientists. Mongolia is fortunate to have government-supported technical staff associated with local governments. Government staff are able to carry out monitoring, use state-and-transition models and participate in community-based rangeland management. Starting in 2009, the Mongolian Green Gold programme scientists were trained in monitoring, model

Limitations

The development of state-and-transition models that are useful for the management of SES is limited primarily by information available on states and transitions and the resources to support collaborative development efforts. Conceptual state-and-transition models developed at broad scales by scientists, usually in one-time efforts, can be useful for education, but they lack participatory and feedback elements that increase the quality of and trust in the models. In addition, people must be willing to use the state-and-transition models. A sense that state-and-transition models could be a waste of time or only add burdensome regulations may thwart development efforts even when resources are available. A lack of data and information on ecosystem responses to management is often a critical limitation. An absence of data or accurate local knowledge will yield state-and-transition models of limited complexity and predictive value, to such an extent that these models are ignored.

Resource implications

Successful conceptual model development requires human and financial resources to support interviews and workshops, literature review and legacy data compilation, new data collection and analysis, and the production of documents. Leaders should foremost have good facilitation

development methods and database management by US Department of Agriculture's Jornada Experimental Range scientists. In 2011, rangeland assessment and monitoring procedures, based on techniques used by US government agencies, were established and Green Gold scientists trained over 400 technicians.

Green Gold scientists and technicians conducted an inventory of vegetation and soils at over 600 sites across Mongolia. These measurements provided an empirical basis for developing state-and-transition models. In addition, workshops were conducted to elicit local knowledge about reference conditions and the presumed causes of vegetation change, and to identify informative sites for additional inventory.

A national core group was established to oversee state-and-transition model development. The core group comprised experienced plant community ecologists representing different ecoregions across Mongolia alongside representatives of science and land management agencies. The core group designated 22 distinct land classes by grouping together finer-grained soil variations (called 'ecological site groups'), for each of which a state-and-transition model was developed. Other tasks included: (a) reviewing published materials to establish reference conditions and causes of state change, (b) working in close collaboration with Green Gold scientists to develop and revise state-and-transition models, and (c) performing outreach activities to encourage the adoption of models by local government and herder cooperatives.

A primary objective of the model development effort was to specify rangeland management strategies to maintain or recover perennial grasses. In contrast to the pre-existing narrative, monitoring data interpreted via state-and-transition

(Continued)

models indicated that many sites were not significantly altered from reference conditions and that the majority of them could be restored with changes to grazing management. The core group expanded the state-and-transition models to contain detailed information about recommended stocking rates and grazing deferment periods, tailored to the objectives of either maintaining a current state or recovering a former state. Recommendations and expectations were linked to specific vegetation-cover indicators that could be monitored as part of community-based rangeland management.

In Figure 27.1, plant composition and key plant species are described for each state. At the bottom the estimated above-ground plant biomass yield and resilient carrying capacity (RCC) are displayed, indicating the number of Mongolian sheep units that can be grazed to maintain or improve plant community composition and productivity. Tables (not shown) contain more information about each state and the various transitions.

Today, agencies are using state-and-transition models as part of community-based rangeland management across Mongolia to plan grazing and resting periods. Positive changes in vegetation are occurring where coordinated grazing management has been implemented. In addition, responsible agencies are using state-and-transition models to interpret national-scale monitoring data (at 1 516 monitoring sites) for periodic reporting of rangeland trends to the public via national news organisations. A recently established non-governmental organisation, the National Federation of Pasture User Groups, serves as a coordinating body that promotes the use of state-and-transition models across agencies and herder organisations.

skills. Leaders or key participants should have a background in natural resource ecology (including the precise measurement of natural resources of interest (e.g. Herrick et al. 2017 for rangelands)), geographic information systems (GIS) and statistical analysis, and participatory science approaches. No specific software or techniques are required, although multivariate analysis options in the R programming language are often used (e.g. vegan, labdsv packages).

New directions

The linkage of state-and-transition models to multiple ecosystem services, economic modelling and structured decision-making frameworks are promising recent directions (Fraser et al. 2017; Ritten et al. 2018). In addition, there have been great strides in using machine-learning algorithms on spatial data to generate maps of ecological conditions and state changes (Jones et al. 2018). Maps of states can greatly enhance the use of state-and-transition models for landscape management. State-and-transition models have also been incorporated into interactive computer-based games to help landowners and students learn about concepts of states and thresholds, and to explore the consequences of hypothetical management decisions in a risk-free environment (Ritten et al. 2011). However, all these advances are predicated on high-quality conceptual state-and-transition models.

Figure 27.1 An example state-and-transition model for sandy loam alluvial fan soils in the dry steppe of central-eastern Mongolia (Densambuu et al. 2018)

Key readings

Bestelmeyer, B.T., A. Ash, J.R. Brown, B. Densambuu, M. Fernández-Giménez, J. Johanson, M. Levi et al. 2017. 'State and Transition Models: Theory, Applications, and Challenges.' In *Rangeland Systems: Processes, Management and Challenges*, edited by D.D. Briske, 303–345. Cham: Springer.

Briske, D.D., B.T. Bestelmeyer, T.K. Stringham, and P.L. Shaver. 2008. 'Recommendations for Development of Resilience-based State-and-transition Models.' *Rangeland Ecology & Management* 61(4): 359–367.

Knapp, C.N., M. Fernández-Giménez, and E. Kachergis. 2010. 'The Role of Local Knowledge in State-and-Transition Model Development.' *Rangelands* 32(6): 31–36.

References

Addison, J., M. Friedel, C. Brown, J. Davies, and S. Waldron. 2012. 'A Critical Review of Degradation Assumptions Applied to Mongolia's Gobi Desert.' *Rangeland Journal* 34(2): 125–137.

Arterburn, J.R., D. Twidwell, W.H. Schacht, C.L. Wonkka, and D.A. Wedin. 2018. 'Resilience of Sandhills Grassland to Wildfire During Drought.' *Rangeland Ecology & Management* 71(1): 53–57.

Barrio, I.C., D.S. Hik, J. Thórsson, K. Svavarsdóttir, B. Marteinsdóttir, and I.S. Jónsdóttir. 2018. 'The Sheep in Wolf's Clothing? Recognizing Threats for Land Degradation in Iceland Using State-and-transition Models.' *Land Degradation & Development* 29(6): 1714–1725.

Bashari, H., C. Smith, and O.J.H. Bosch. 2009. 'Developing Decision Support Tools for Rangeland Management by Combining State and Transition Models and Bayesian Belief Networks.' *Agricultural Systems* 99: 23–34.

Bestelmeyer, B.T., A. Ash, J.R. Brown, B. Densambuu, M. Fernández-Giménez, J. Johanson, M. Levi et al. 2017. 'State and Transition Models: Theory, Applications, and Challenges.' In *Rangeland Systems: Processes, Management and Challenges*, edited by D.D. Briske, 303–345. Cham: Springer.

Bestelmeyer, B.T., K. Moseley, P.L. Shaver, H. Sanchez, D.D. Briske, and M.E. Fernández-Giménez. 2010. 'Practical Guidance for Developing State-and-transition Models.' *Rangelands* 32(6): 23–30.

Bestelmeyer, B.T., A.J. Tugel, G.L. Peacock, D.G. Robinett, P.L. Shaver, J.R. Brown, J.E. Herrick, H. Sanchez, and K.M. Havstad. 2009. 'State-and-transition Models for Heterogeneous Landscapes: A Strategy for Development and Application.' *Rangeland Ecology & Management* 62(1): 1–15.

Biggs, R., T. Blenckner, C. Folke, L. Gordon, A. Norström, M. Nyström, and G. Peterson. 2012. 'Regime Shifts.' In *Encyclopedia of Theoretical Ecology*, edited by A. Hastings and L. Gross, 609–617. Berkeley: University of California Press.

Bino, G., S.A. Sisson, R.T. Kingsford, R.F. Thomas, and S. Bowen. 2015. 'Developing State and Transition Models of Floodplain Vegetation Dynamics as a Tool for Conservation Decision-Making: A Case Study of the Macquarie Marshes Ramsar Wetland.' *Journal of Applied Ecology* 52(3): 654–664.

Briske, D.D., B.T. Bestelmeyer, T.K. Stringham, and P.L. Shaver. 2008. 'Recommendations for Development of Resilience-based State-and-transition Models.' *Rangeland Ecology & Management* 61(4): 359–367.

Brown, J., and N. MacLeod. 2011. 'A Site-based Approach to Delivering Rangeland Ecosystem Services.' *Rangeland Journal* 33(2): 99–108.

Bruegger, R.A., M.E. Fernández-Giménez, C.Y. Tipton, J.M. Timmer, and C.L. Aldridge. 2016. 'Multistakeholder Development of State-and-transition Models: A Case Study from Northwestern Colorado.' *Rangelands* 38(6): 336–341.

Chambers, J.C., R.F. Miller, D.I. Board, D.A. Pyke, B.A. Roundy, J.B. Grace, E.W. Schupp, and R.J. Tausch. 2014. 'Resilience and Resistance of Sagebrush Ecosystems: Implications for State and Transition Models and Management Treatments.' *Rangeland Ecology & Management* 67(5): 440–454.

Daniel, C.J., L. Frid, B.M. Sleeter, and M-J. Fortin. 2016. 'State-and-transition Simulation Models: A Framework for Forecasting Landscape Change.' *Methods in Ecology and Evolution* 7(11): 1413–1423.

Densambuu, B., T. Indree, A. Battur, and S. Sainnemekh. 2018. *State and Transition Models of Mongolian Rangelands*. Ulaanbaatar: Agency for Land Management, Geodesy, and Cartography.

Duniway, M.C., B.T. Bestelmeyer, and A. Tugel. 2010. 'Soil Processes and Properties that Distinguish Ecological Sites and States.' *Rangelands* 32(6): 9–15.

Easdale, M.H., and D.R. López. 2016. 'Sustainable Livelihoods Approach through the Lens of the State-and-transition Model in Semi-Arid Pastoral Systems.' *The Rangeland Journal* 38(6): 541–551.

Eastburn, D.J., A.T. O'Geen, K.W. Tate, and L.M. Roche. 2017. 'Multiple Ecosystem Services in a Working Landscape.' *PLoS ONE* 12(3): e0166595.

Eckert, S., F. Hüsler, H. Liniger, and E. Hodel. 2015. 'Trend Analysis of MODIS NDVI Time Series for Detecting Land Degradation and Regeneration in Mongolia.' *Journal of Arid Environments* 113: 16–28.

Fraser, H., L. Rumpff, J.D.L. Yen, D. Robinson, and B.A. Wintle. 2017. 'Integrated Models to Support Multiobjective Ecological Restoration Decisions.' *Conservation Biology* 31(6): 1418–1427.

Herrick, J.E., J.W. van Zee, S.E. McCord, E.M. Courtright, J.W. Karl, and L.M. Burkett. 2017. *Monitoring Manual for Grassland, Shrubland and Savanna Ecosystems Volume I: Core Methods* (2nd ed). Las Cruces: USDA-ARS Jornada Experimental Range.

Hobbs, R.J., and K.N. Suding. 2009. *New Models for Ecosystem Dynamics and Restoration*. Washington: Island Press.

Huber-Sannwald, E., M.R. Palacios, J.T.A. Moreno, M. Braasch, R.M.M. Peña, J.G. de A. Verduzco, and K.M. Santos. 2012. 'Navigating Challenges and Opportunities of Land Degradation and Sustainable Livelihood Development in Dryland Social-Ecological Systems: A Case Study from Mexico.' *Philosophical Transactions of the Royal Society B: Biological Sciences* 367(1606): 3158–3177.

Jamiyansharav, K., M.E. Fernández-Giménez, J.P. Angerer, B. Yadamsuren, and Z. Dash. 2018. 'Plant Community Change in Three Mongolian Steppe Ecosystems 1994–2013: Applications to State-and-transition Models.' *Ecosphere* 9(3): e02145.

Johanson, J., and M. Fernández-Giménez. 2015. 'Developers of Ecological Site Descriptions Find Benefits in Diverse Collaborations.' *Rangelands* 37(1): 14–19.

Jones, M.O., B.W. Allred, D.E. Naugle, J.D. Maestas, P. Donnelly, L.J. Metz, J. Karl et al. 2018. 'Innovation in Rangeland Monitoring: Annual, 30 m, Plant Functional Type Percent Cover Maps for U.S. Rangelands, 1984–2017.' *Ecosphere* 9(9): e02430.

Kachergis, E.J., C.N. Knapp, M.E. Fernández-Giménez, J.P. Ritten, J.G. Pritchett, J. Parsons, W. Hibbs, and R. Roath. 2013. 'Tools for Resilience Management: Multidisciplinary Development of State-and-Transition Models for Northwest Colorado.' *Ecology and Society* 18(4): 39.

Karl, J.W., J.E. Herrick, and D.M. Browning. 2012. 'A Strategy for Rangeland Management Based on Best Available Knowledge and Information.' *Rangeland Ecology & Management* 65(6): 638–646.

Knapp, C.N., M.E. Fernández-Giménez, and E. Kachergis. 2010. 'The Role of Local Knowledge in State-and-transition Model Development.' *Rangelands* 32(6): 31–36.

Knapp, C.N., M. Fernández-Giménez, E. Kachergis, and A. Rudeen. 2011. 'Using Participatory Workshops to Integrate State-and-Transition Models Created with Local Knowledge and Ecological Data.' *Rangeland Ecology & Management* 64(2): 158–170.

López, D.R., L. Cavallero, M.A. Brizuela, and M.R. Aguiar. 2011. 'Ecosystemic Structural-functional Approach of the State and Transition Model.' *Applied Vegetation Science* 14(1): 6–16.

Maynard, J., T.W. Nauman, S.W. Salley, M. Duniway, B. Bestelmeyer, C. Talbot, and J. Brown. 2019. 'Digital Mapping of Ecological Land Units Using a Nationally Scalable Modeling Framework.' *Soil Science Society of America Journal* 83: 666–686.

McGlathery, K.J., M.A. Reidenbach, P. D'Odorico, S. Fagherazzi, M.L. Pace, and J.H. Porter. 2013. 'Nonlinear Dynamics and Alternative Stable States in Shallow Coastal Systems.' *Oceanography* 26(3): 220–231.

Miller, B.W., A.J. Symstad, L. Frid, N.A. Fisichelli, and G.W. Schuurman. 2017. 'Co-Producing Simulation Models to Inform Resource Management: A Case Study from Southwest South Dakota.' *Ecosphere* 8(12): e02020.

Peinetti, H.R., B.T. Bestelmeyer, C.C. Chirino, A.G. Kin, and M.E.F. Buss. 2019. 'Generalized and Specific State-and-transition Models to Guide Management and Restoration of Caldenal Forests.' *Rangeland Ecology & Management*, 72(2): 230–236.

Perry, G.L.W., J.M. Wilmshurst, J. Ogden, and N.J. Enright. 2015. 'Exotic Mammals and Invasive Plants Alter Fire-Related Thresholds in Southern Temperate Forested Landscapes.' *Ecosystems* 18(7): 1290–12305.

Petraitis, P. 2013. *Multiple Stable States in Natural Ecosystems*. Oxford: Oxford University Press.

Porensky, L.M., K.E. Mueller, D.J. Augustine, and J.D. Derner. 2016. 'Thresholds and Gradients in a Semi-arid Grassland: Long-term Grazing Treatments Induce Slow, Continuous and Reversible Vegetation Change.' *Journal of Applied Ecology* 53(4): 1013–1022.

Ritten, J., M. Fernández-Giménez, E. Kachergis, W. Hibbs, and J. Pritchett. 2011. 'Do Livestock and Ecosystem Services Compete? A State-and-transition Approach.' In *Range Beef Cow Symposium*, 298. http://digitalcommons.unl.edu/rangebeefcowsymp/298.

Ritten, J., M.E. Fernández-Giménez, J. Pritchett, E. Kachergis, and W. Bish. 2018. 'Using State and Transition Models to Determine the Opportunity Cost of Providing Ecosystem Services.' *Rangeland Ecology & Management* 71(6): 737–752.

Rumpff, L., D.H. Duncan, P.A. Vesk, D.A. Keith, and B.A. Wintle. 2011. 'State-and-transition Modelling for Adaptive Management of Native Woodlands.' *Biological Conservation* 144(4): 1224–1236.

Spiegal, S., J.W. Bartolome, and M.D. White. 2016. 'Applying Ecological Site Concepts to Adaptive Conservation Management on an Iconic Californian Landscape.' *Rangelands* 38(6): 365–370.

Tarrasón, D., F. Ravera, M.S. Reed, A.J. Dougill, and L. Gonzalez. 2016. 'Land Degradation Assessment through an Ecosystem Services Lens: Integrating Knowledge and Methods in Pastoral Semi-Arid Systems.' *Journal of Arid Environments* 124: 205–213.

Tipton, C.Y., T.W. Ocheltree, K.E. Mueller, P. Turk, and M.E. Fernández-Giménez. 2018. 'Revision of a State-and-transition Model to Include Descriptions of State Functional Attributes.' *Ecosphere* 9(5): e02201.

Ulambayar, T., and M.E. Fernández-Giménez. 2019. 'How Community-based Rangeland Management Achieves Positive Social Outcomes in Mongolia: A Moderated Mediation Analysis.' *Land Use Policy* 82: 93–104.

USDA. 2019. 'Ecological site R042XB012NM: Sandy.' *EDIT – Ecosystem Dynamics Interpretive Tool*. Washington, DC: USDA. https://edit.jornada.nmsu.edu/catalogs/esd/042X/R042XB012NM.

Walker, B., and D. Salt. 2012. *Resilience Practice: Building Capacity to Absorb Disturbance and Maintain Function*. Washington: Island Press.

Walker, B., and M. Westoby. 2011. 'States and Transitions: The Trajectory of an Idea, 1970–2010.' *Israel Journal of Ecology & Evolution* 57(1–2): 17–22.

Webb, N.P., J.E. Herrick, and M.C. Duniway. 2014. 'Ecological Site-based Assessments of Wind and Water Erosion: Informing Accelerated Soil Erosion Management in Rangelands.' *Ecological Applications* 24(6): 1405–1420.

Westoby, M., B. Walker, and I. Noy-Meir. 1989. 'Opportunistic Management for Rangelands Not at Equilibrium.' *Journal of Range Management* 42(4): 266–274.

Wilcox, B.P., A. Birt, S.R. Archer, S.D. Fuhlendorf, U.P. Kreuter, M.G. Sorice, W.J.D. van Leeuwen, and C.B. Zou. 2018. 'Viewing Woody-plant Encroachment through a Social-Ecological Lens.' *Bioscience* 68(9): 691–705.

Yates, C.J., and R.J. Hobbs. 1997. 'Woodland Restoration in the Western Australian Wheatbelt: A Conceptual Framework Using a State and Transition Model.' *Restoration Ecology* 5(1): 28–35.

Young, D., H.L. Perotto-Baldivieso, T. Brewer, R. Homer, and S.A. Santos. 2014. 'Monitoring British Upland Ecosystems with the Use of Landscape Structure as an Indicator for State-and-transition Models.' *Rangeland Ecology & Management* 67(4): 380–388.

Zweig, C.L., and W.M. Kitchens. 2009. 'Multi-state Succession in Wetlands: A Novel Use of State and Transition Models.' *Ecology* 90(7): 1900–1909.

28

Agent-based modelling

Maja Schlüter,[1] *Emilie Lindkvist,*[1]
Nanda Wijermans[1] *and Gary Polhill*[2]

[1]STOCKHOLM RESILIENCE CENTRE, STOCKHOLM UNIVERSITY, STOCKHOLM, SWEDEN
[2]THE JAMES HUTTON INSTITUTE, ABERDEEN, UK

Key methods discussed in this chapter

Agent-based modelling

Connections to other chapters

This method connects to participatory modelling (Chapter 8), as agent-based models are often developed and/or used in participatory processes. Agent-based models can also be used to support scenario development (Chapter 11) and futures analysis (Chapter 10), or to develop serious games (Chapter 12). The design and parameterisation of an agent-based model can be informed by knowledge and data collected through systems scoping (Chapter 5), interviews and surveys (Chapter 7), or ecological field data collection (Chapter 6). Agent-based models are generally analysed by running many simulations which generate synthetic data that can then be processed with statistical methods (Chapter 18). Agent-based models can also be combined with dynamical systems modelling (Chapter 26) in hybrid models.

Introduction

Agent-based modelling is a computational method that emerged in the early 1970s simultaneously in several fields, particularly in complexity, economic, sociological and computer sciences (distributed artificial intelligence) (Hare and Deadman 2004). An agent-based model (often referred to as ABM) is a computer program composed of autonomous agents, i.e. agents whose behaviour is not centrally controlled, who are diverse and interact with one another and their environment. The program is simulated over time: at each time step or event, agents take decisions and act based on their internal state and/or in response or anticipation of other agents' behaviours or changes in the environment. The microlevel actions and interactions of many agents give rise to macrolevel patterns and dynamics which are typically the focus of analysis. The micro- and macrolevel outcomes of individual actions can modify the internal state of the agents and prompt them to change their behaviour in the subsequent time steps. The agents thus adapt to the contexts they jointly create, which is a key

DOI: 10.4324/9781003021339-33

SUMMARY TABLE: AGENT-BASED MODELLING

DISCIPLINARY BACKGROUND

The methods in this chapter are derived from or have most commonly been used in:

Complexity Science, Computer Science/ Distributed Artificial Intelligence

KNOWLEDGE TYPE

The methods in this chapter are primarily used to generate the following types of knowledge:

- Descriptive
- Exploratory
- Explanatory

RESEARCH APPROACH

The methods in this chapter originate from or most commonly adopt the following research approaches:

- Analytical/objective
- Collaborative/process

PURPOSE OF METHOD

The most common purposes of using the methods in this chapter are:

- System understanding
- Stakeholder engagement and co-production
- Policy/decision support

TEMPORAL DIMENSION

The methods in this chapter are most commonly applied to the following temporal dimensions:

- Present (typically within the last 5–10 years)
- Recent past (post-1700s)
- Pre-industrial revolution (pre-1700s)
- Future

SPATIAL DIMENSION

The methods in this chapter are primarily either or both:

- Non-spatial
- Explicitly spatial

The methods in this chapter are most commonly applied at the following spatial scales:

- Local
- Regional

SYSTEMIC FEATURES AND PROCESSES

While most methods can do many things, the methods in this chapter are particularly good (i.e. go-to methods) for addressing the following:

- SES components and linkages
- Diversity
- Multiple scales and levels or cross-level interactions
- Social-ecological interactions over time
- Path dependency
- Adaptation and self-organisation
- Evaluating policy options
- Exploring uncertainty

characteristic of complex adaptive systems. The evolution of the system over time can result in system-level outcomes that stabilise in some form of dynamic equilibrium or continuously change indefinitely in a chaotic or a regularly fluctuating manner.

The conceptual foundation of agent-based modelling is the theory of complex adaptive systems (Arthur, Durlauf, and Lane 1997). Agent-based modelling is one of the key methods to study emergent phenomena, i.e. system-level patterns or behaviours that cannot be explained by the individual components alone, but arise from interactions among individual agents who adapt and learn about their environment. In that sense, agent-based modelling supports the development of generative explanations: explanations that specify the microlevel interactions of heterogeneous agents that bring about a macrolevel phenomenon of interest (Epstein 2006). While agent-based models have been primarily used to study the emergence of system-level outcomes from microlevel interactions of agents, ultimately both system-level and microlevel behaviour matter as they influence each other.

Agent-based models have been used to represent and study many complex adaptive systems, from immune systems, social systems and ecosystems to social-ecological systems (SES). In models of SES, the agents often represent individual or collective actors (e.g. fishers, households, organisations) or biological organisms (e.g. fish, fish populations or livestock). The social environment comprises social structures (e.g. social networks or neighbourhoods). The biophysical environment represents natural resources or ecosystems that are used or affected by the behaviour of the agents (e.g. patches of land, fishing areas, forest patches or landscapes). The agents are characterised by their properties and behaviours, which may vary among agent types (e.g. fishers who use different fishing styles) or within one type (e.g. a farmer with more or less wealth). The agents are often also characterised by their location in space or in a social or ecological network.

One of the first and most well-known agent-based models of social phenomena was Sakoda–Schelling's model of segregation (Sakoda 1971; Schelling 1971). One of the earliest applications of agent-based modelling to SES is Lansing and Kremer's model of the emergence of patterns of rice-paddy irrigation in Bali (Lansing and Kremer 1993) and the model of a fishery developed by Bousquet et al. (1993). The use of agent-based modelling in support of participatory processes has a long tradition in natural resource management (Bousquet et al. 1999; Bousquet and Le Page 2004). As agent-based modelling has gained momentum in the last decades, it has been applied in more diverse fields, including sociology, psychology, political science, economics, philosophy, ecology, natural resource use, land-system science and SES research (Schulze et al. 2017; Gotts et al. 2018). Agent-based models can be developed for many different purposes, ranging from exploring, understanding, explaining, predicting, communicating or illustrating to an analogy or a mediator for social interaction between diverse stakeholders or disciplinary researchers (Edmonds et al. 2019).

SES problems and questions

Several characteristics of agent-based modelling make it an interesting and important method for SES research: (a) its focus on the change of an SES over time from mutual adaptations of agents and their environments, (b) its ability to generate emergent system-level outcomes from microlevel interactions and macrolevel feedbacks, thus enabling the study of an SES as a complex adaptive system, (c) its ability to represent the diversity and heterogeneity of human and non-human actors as well as the spatial characteristics of an SES, and (d) its capacity to serve as a virtual laboratory in which researchers and stakeholders can experiment with an SES to explore possible consequences of interventions or identify and test causal relationships that underlie an emergent phenomenon.

Agent-based modelling is applied to both theoretical and empirical questions. Theoretical models can be conceptual or stylised with the aim to explore or understand key mechanisms that determine the behaviour of an SES. Empirical models can be built for a particular system (e.g. a fishery or a landscape) using data and knowledge collected in the field through empirical methods (see Chapters 5 and 7) and/or participatory processes (see Chapter 13) (Janssen and Ostrom 2006; Smajgl and Barreteau 2014). The degree of realism of an agent-based model is tied to the purpose of the model. Agent-based models of SES are often developed for one of the following purposes: (a) to explore or explain the emergence of social-ecological outcomes and understand the evolution of the SES over time, (b) to assess the impact of a new policy or disturbance on an SES that is understood as a complex adaptive system, including potential unintended consequences, and (c) to support a participatory process that aims to enhance problem understanding and co-develop problem solutions.

Typical questions related to enhancing the understanding of emergent outcomes and system change over time include:

- What land-use patterns emerge from land-use choices of farmers, as influenced by formal or informal institutions, social networks or the behaviour of other farmers? (Parker and Meretsky 2004; Manson et al. 2016)
- How robust is cooperation for shared natural resource use to environmental variability? (Schlüter, Tavoni, and Levin 2016)
- What are the implications of pastoralist behavioural diversity for the sustainability of a rangeland? (Dressler et al. 2018)
- How important is actors' diversity for the resilience of SES to global change? (Grêt-Regamey, Huber, and Huber 2019)

Typical questions related to the use of agent-based models for policy assessment include:

- How will a fuel policy affect poverty patterns and future land use and land cover in a region in Indonesia? (Smajgl and Bohensky 2011)
- What are the benefits of spatial cooperative harvesting schemes in a marine fishery? (Gutierrez et al. 2017)
- How can we design policies that incentivise farmers to conserve biodiversity? (Gimona and Polhill 2011; Polhill, Gimona, and Gotts 2013)

The aim of the use of agent-based modelling in participatory processes is to disclose and discuss different interpretations of the system and stakeholder objectives, develop a shared system understanding, support joint deliberation and social learning, and develop management solutions (see also Chapter 13). Joint model development requires stakeholders to make their assumptions explicit and stimulates discussions about what is important. The process of specifying a model, particularly in an inter- or transdisciplinary team or in participatory processes with stakeholders, is often as insightful as the model outcomes themselves and can lead to new questions for further research. Applications of agent-based modelling in participatory processes include:

- The use of an agent-based model in support of a process to resolve a conflict over sharing water resources between two communities along a river (Gurung, Bousquet, and Trébuil 2006)
- The use of an agent-based model in a participatory process for analysing a fishery and developing a management plan (Worrapimphong et al. 2010)

Recent work has highlighted the potential of agent-based modelling for studying systemic change such as regime shifts (Filatova, Polhill, and Van Ewijk 2016; Polhill et al. 2016). Land-system science has made extensive use of the potential of agent-based modelling to represent interactions among actors in space, e.g. a farmer making a decision based on her neighbour's actions (Happe, Kellermann, and Balmann 2006; Bert et al. 2011; Villamor et al. 2014). Land-use agent-based models typically represent the landscape as a grid of cells with different land-use types or a spatially explicit representation from a geographic information system (Crooks, Heppenstall, and Malleson 2018). Many agent-based models in land-system science focus on land-use and land-cover change or policy scenario analysis, with a focus on explicit representation of spatial dynamics, psychologically plausible decision-making and biophysical and social interactions. Many agent-based models of natural resource management focus on the sustainable exploitation of resources such as fish (Burgess et al. 2020; Lindkvist et al. 2020), water or forests, given the diversity of resource users, social dilemmas, social interactions and the dynamics of the resource. Other areas of application of agent-based modelling are human–wildlife interactions or tourism in protected areas. Agent-based modelling has also been extensively used to study the collapse of past societies such as the Anasazi or the Maya, particularly how it may have been caused by environmental or social change (Janssen 2009; Heckbert 2013).

Brief description of key methods

The process of developing and using an agent-based model includes several steps: model design, model building, model testing, simulation experiments or scenario analysis, and communication of results. In general, the development of an agent-based model starts with identifying the purpose and research question the model should address. This guides the development of the model, particularly the selection of the agents, their behaviours and the social and ecological environments in which they (inter-)act. The assumptions underlying these choices can be: (a) based on qualitative and quantitative empirical studies of a particular place, such as a farming landscape, (b) co-developed with researchers and/or stakeholders in a participatory process, (c) based on stylised facts derived from expert knowledge and the literature, or (d) based on theory. Often a mix of sources is used because no single source provides all the understanding and evidence needed.

Once the model has been designed and implemented in computer code, the virtual SES can be simulated by running the software. After being thoroughly tested and validated, it can then be deployed to simulate different scenarios or run experiments, for instance to assess the impact of policies, changes in selected parameter values, or the presence or absence of certain processes. The simulations produce large amounts of simulated data about the model system, which are then analysed to study outcomes of interest using statistical methods or machine-learning algorithms.

Various methods can be applied to each of the steps outlined above. These include: (a) methods to collect and analyse qualitative or quantitative empirical data, elicit expert or stakeholder knowledge (Bharwani 2006), or identify theories to inform the design of the structure of the model (what to include and what not) and behavioural rules governing the agents' behaviour (Smajgl et al. 2011; Smajgl and Barreteau 2014; Schlüter et al. 2017), (b) methods to determine parameter values and assess the sensitivity of the model to parameter choices (Thiele, Kurth, and Grimm 2014), (c) methods of experimental design to develop and analyse scenarios (Lorscheid, Heine, and Meyer 2012), and (d) statistical, pattern-recognition and mathematical methods to analyse, validate and represent simulation data and outcomes

(Lee et al. 2015; Martin and Thomas 2016; Thomas, Lloyd, and Skeldon 2016). The applicability of the different methods and the type of analysis and validation needed depend on the purpose of the model and the context of model development.

Model development is often an iterative process, looping between the design of the model and a case study or theoretical foundations. Given the interdisciplinary and complex nature of SES problems, the agent-based modelling activity is ideally embedded in a larger interdisciplinary research process where multiple methods are being combined (Schlüter, Müller, and Frank 2019; Schlüter et al. 2019). A model and its results should always be interpreted in light of the underlying assumptions, the model purpose and the intended use of the model (Schlüter, Müller, and Frank 2019). A careful process of choosing the assumptions and transparency in communicating model assumptions is thus essential (Schlüter et al. 2014; Gotts et al. 2018). The agent-based modelling community has developed several protocols to facilitate good and standardised model communication (Grimm et al. 2006; Schmolke et al. 2010; Müller et al. 2013).

Table 28.1 summarises a few of the most common uses of agent-based modelling in SES research.

Table 28.1 Summary of key applications of agent-based modelling

Main applications	*Description*	*References*
Exploring emergent SES outcomes and dynamics	A model is developed and used to explore system-level outcomes and system trajectories that emerge from interactions between actors, elements of ecosystems and their social and biophysical environments.	*Applications to SES* Carpenter and Brock 2004; Wilson, Yan, and Wilson 2007; Evans and Kelley 2008; Heckbert 2013; Castilla-Rho et al. 2017
Policy assessment	A model is used to assess possible outcomes of different policy scenarios, e.g. the effect of introducing an individual transferable quota on the sustainability of fisheries, or of a fuel subsidy on poverty.	*Applications to SES* Smajgl and Bohensky 2011; Sun and Müller 2013
Assessment of response of SES to environmental or social change	A model is used to explore how an SES may respond to environmental or social change, such as climate change impacts or changes in market conditions.	*Applications to SES* Melbourne-Thomas et al. 2011; Klein, Barbier, and Watson 2017
Explaining emergent SES phenomena	A model is developed and used to identify social-ecological mechanisms that produced an observed outcome of interest.	*Applications to SES* Schill et al. 2016; Lindkvist, Basurto, and Schlüter 2017; Plank et al. 2017
As boundary object in participatory processes	A model is co-developed and/or used as boundary object to facilitate a process of social learning and co-developing solutions to problems in a particular case or for integration of different (disciplinary) understandings.	*Applications to SES* Castella, Trung, and Boissau 2005; Gurung, Bousquet, and Trébuil 2006; Worrapimphong et al. 2010; Forrester et al. 2014

Limitations

Agent-based models are very flexible and can incorporate many details of the real world in a disaggregate manner. This is a great advantage because important aspects of an SES, such as diversity, spatial arrangements and heterogeneity, can be represented. These aspects are usually difficult to model with aggregate approaches such as dynamical systems modelling. The realism of agent-based models also makes it easier for the modeller to communicate them to non-modellers. This flexibility, however, also poses a challenge as choices need to be made about what to include or exclude in the model. There are no general guidelines because these choices are context-specific and model-purpose-specific and influenced by the backgrounds, knowledge and experience of those involved in building the model. Including too many details or too much heterogeneity leads to very complex models that are difficult to analyse and that pose challenges for transparency and reproducibility (Kremmydas, Athanasiadis, and Rozakis 2018). Complex agent-based models of a particular SES that are developed to explore responses of the SES to interventions or other external changes, however, can still work well if they are validated with empirical data. Complex models are more problematic when the aim of the model is to explain how a specific outcome came about.

Agent-based models need to be thoroughly analysed to build confidence in the model, i.e. to ascertain that a result is not an artefact of the model structure or parameter values, to test their validity, and to assess the robustness of model results to uncertainties about the structure of the model or parameter values. A common method to deal with uncertainty of modelling choices and parameter values is to conduct sensitivity and uncertainty analyses. Uncertainty that arises from the context-dependence of human decision-making and behaviour is often modelled using stochastic variables, e.g. a behavioural choice is modelled with a certain probability. Because of the complex, multi-level and evolutionary nature of agent-based models, standard mathematical tools, such as identification of equilibrium points and stability analysis, are not applicable. Instead, model behaviour needs to be systematically explored.

Recent developments in statistical methods as well as advances in mathematical analysis or approximation of agent-based models (Martin and Thomas 2016; Thomas, Lloyd, and Skeldon 2016) can significantly enhance model analysis. At the same time, agent-based models provide the opportunity to analyse the evolution of the SES over time when it is out of equilibrium, examine the implications of random events, and study feedbacks and path dependencies. All this is difficult to do with other modelling approaches. Analysis of the transient behaviour of SES using agent-based models is a current research frontier (see Section 'New directions').

Resource implications

The most commonly used platform for developing agent-based models in ecology and SES research is NetLogo (ccl.northwestern.edu/netlogo). NetLogo is easy to learn and has a simple user interface that has made it popular for both teaching and scientific applications of agent-based modelling (see textbooks by Railsback and Grimm 2012 and Janssen 2020). There are alternative platforms (reviewed by Kravari and Bassiliades 2015), as well as the option of writing all the software to implement the simulation yourself. Most agent-based models can be run on personal computers; however, large-scale model output analyses and more complex or empirical agent-based models typically use larger machines or clusters. There is a growing repository of agent-based models of SES at comses.net.

The development of an SES agent-based model requires programming skills and skills in identifying and synthesising knowledge and data to develop a representation of an SES that can answer a research question. The modeller will need data analysis skills to analyse large amounts of simulation data using software, such as R, and skills to visualise the data. There are an increasing number of courses on agent-based modelling in SES or related fields that teach many of these skills. If the combination of skills seems daunting, there is the option of working in small collaborative teams to design, implement and analyse the results of agent-based models. Agent-based modelling is in any case a cross-disciplinary activity (Squazzoni 2010).

Whereas building a first prototype agent-based model can be done quickly, constructing a full model and testing and analysing it thoroughly is often time consuming. The different phases of building and using an agent-based model require different amounts of time depending on

Case study 28.1: The establishment and persistence of cooperative forms of self-governance in small-scale fisheries in north-west Mexico

This example is taken from the work of Lindkvist, Basurto and Schlüter (2017), who developed an agent-based model to study the self-governance of small-scale fisheries in north-west Mexico. The aim of this agent-based model is to explore and explain why non-cooperative forms of self-governance, such as fisher–trader relationships, dominate many small-scale fishing communities in north-west Mexico. In particular, it was developed to investigate how the heterogeneity of fishers' reliability, organisational characteristics of different self-governance forms (e.g. loyalty and trust) and the fish population, through their dynamic interactions, may explain observed differences in self-governance arrangements. The researchers developed an agent-based model of an archetypical

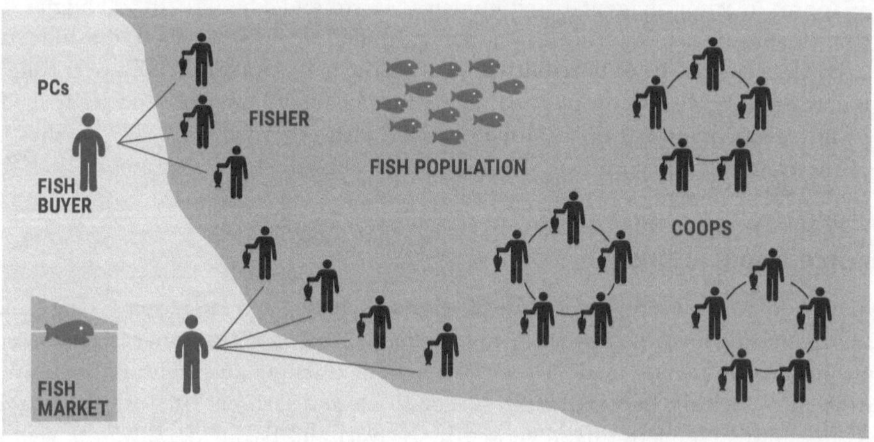

Figure 28.1 Key elements of the stylised agent-based model of a small-scale fishery (Lindkvist, Basurto, and Schlüter 2017)

the skills of the researcher(s) and the problem at hand. The development of a research question that is suitable to be addressed with an agent-based model and the collection and synthesis of the data that will inform the model structure are often as time consuming as the programming of the model itself.

New directions

New interesting fields of application of agent-based modelling include understanding social-ecological regime shifts (Filatova, Polhill, and Van Ewijk 2016; Martin, Schlüter, and Blenckner 2020) or societal transformations (Holtz et al. 2015; Köhler et al. 2018) from a complex adaptive systems perspective. Recent agent-based modelling of regime shifts has focused on

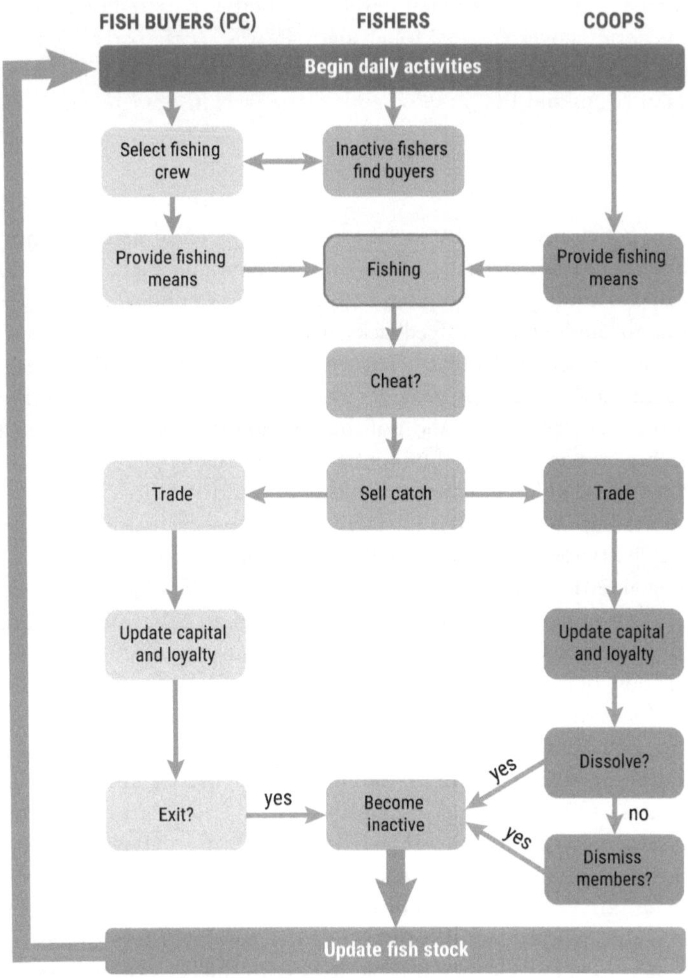

Figure 28.2 Flowchart of actions and interactions of the three key types of agents in the model (Lindkvist, Basurto, and Schlüter 2017)

(Continued)

small-scale fishery with cooperative and non-cooperative self-governance forms (Figures 28.1 and 28.2). The aim of the model is to answer the research question, 'Under which conditions can cooperative forms of self-governance establish and survive?' It captures key insights and hypotheses from in-depth fieldwork in north-west Mexico about fishers' day-to-day fishing and trading, the activities of fish buyers and cooperatives, as well as factors that influence their performance (Basurto et al. 2020). It represents two ubiquitous forms of self-governance: hierarchical non-cooperative arrangements between fishers and fish buyers, such as patron–client relationships (PCs), versus non-hierarchical, cooperative arrangements among fishers, such as fishing cooperatives (coops).

The model reveals that initial levels of trust and the diversity among fishers matter for different self-governance arrangements to be established and persist, and should therefore be taken into account when developing better, targeted policies for improved small-scale fisheries governance. Model analysis was particularly helpful to unravel an intriguing interplay between the macro- (the fish stock, the population of fishers, the numbers of coops and PCs), meso- and microlevels (fishers' reliability and loyalty),

investigating how policy responses by social actors affect the transient dynamics of an ecosystem while it moves towards an attractor (Martin, Schlüter, and Blenckner 2020). Agent-based modelling provides as yet largely unexplored opportunities to understand how an SES pathway unfolds over time through emergent feedbacks and resulting path dependencies or shifts in the direction or strength of feedbacks that determine the future development path of the system. Furthermore, agent-based modelling provides exciting opportunities for identifying social-ecological mechanisms (i.e. the human and non-human actors, interactions and causal processes) that generate emergent phenomena, thus contributing to the development of explanations and middle-range theories of SES phenomena (Schlüter et al. 2019).

Another important new direction is the inclusion of more realistic representations of human decision-making, particularly its embeddedness in changing social and biophysical environments (Schlüter et al. 2017; Huber et al. 2018; Schill et al. 2019, Wijermans et al. 2020). Agent-based models can also provide a method for including human behaviour in models of the Earth system, which is an important research frontier in Earth-system science (Müller-Hansen et al. 2017). This will allow better accounting for the diversity of human motivations and behaviours in SES beyond *Homo oeconomicus* and the implications of these motivations and behaviours for sustainability, e.g. by leading to unexpected outcomes of policies or management.

New methodological directions in support of the use of agent-based modelling to study the dynamics of SES as complex adaptive systems include the combination of social or social-ecological networks with agent-based modelling (Manson et al. 2016; Dobson et al. 2019), the development of multi-level models that incorporate interactions across scales (Lippe et al. 2019) as well as hybrid models that combine different modelling approaches, such as a dynamical systems representation of an ecosystem with an agent-based representation of human and societal interactions with the ecosystem (Martin and Schlüter 2015). New methodological developments regarding human behaviour include the combination of agent-based models with behavioural experiments (Janssen and Baggio 2016; Schill et al. 2016) and novel formalisations of social science theories, and empirical evidence of the diversity of human behaviour in agent-based models (Wijermans et al. 2020).

which would have been impossible to study empirically. The community-level patterns of fishers' reliability and loyalty influence the cheating behaviour of individual fishers and ultimately the resulting organisational membership at the mesolevel. The survival of an organisation at the mesolevel, however, was also influenced by the state of the fish population, which resulted from the harvesting activities of all existing PCs and coops in the community at the macrolevel. Finally, the organisational membership also influenced the macrolevel patterns that determine the predominance of either PCs or coops in a community. It is the interplay among these three levels that explains the dominance of PCs or coops at the macrolevel.

In Figure 28.2, each box represents an activity that is a stylised representation of observations from fisheries in north-west Mexico, where the different types of self-governance forms have slightly different ways of operating. Cheating is a key activity, which involves fishers selling their catch to a different organisation than the one they belong to. Fish buyers can select their crew and also exit the fishery if their capital is too low, whereas coops will dismiss members and dissolve if overall loyalty or capital is too low.

Key readings

Crooks, A., A. Heppenstall, and N. Malleson. 2018. 'Agent-based Modeling.' In *Comprehensive Geographic Information Systems*, edited by B. Huang, 218–243. Amsterdam: Elsevier.

Gilbert, G.N. 2008. *Agent-based Models.* Thousand Oaks: Sage.

Heckbert, S., T. Baynes, and A. Reeson. 2010. 'Agent-based Modeling in Ecological Economics.' *Annals of the New York Academy of Sciences* 1185(1): 39–53. doi:10.1111/j.1749-6632.2009.05286.x.

Janssen, M. 2020. *Introduction to Agent-based Modeling.* https://intro2abm.com.

Railsback, S.F., and V. Grimm. 2012. *Agent-based and Individual-based Modeling – A Practical Introduction.* Princeton: Princeton University Press. www.railsback-grimm-abm-book.com.

Acknowledgements

Maja Schlüter received support from the European Research Council (ERC) under the European Union's Horizon 2020 research and innovation programme (grant agreement no. 682472 – MUSES). Emilie Lindkvist was funded by the Swedish Research Council (Dnr 2018-05862) and the US NSF Coupled Natural and Human Systems Program (DEB-1632648). Gary Polhill's work is funded by the Scottish Government, Horizon 2020 and the Research Council of Norway.

References

Arthur, W.B., S. Durlauf, and D.A. Lane. 1997. 'Process and Emergence in the Economy.' In *The Economy as an Evolving Complex System II*, edited by W.B. Arthur, S. Durlauf, and D.A. Lane, 14. Reading: Addison-Wesley. http://tuvalu.santafe.edu/~wbarthur/Papers/ADL_Intro.pdf.

Basurto, X., A. Bennett, E. Lindkvist, and M. Schlüter. 2020. 'Governing the Commons beyond Harvesting: An Empirical Illustration from Fishing.' *PLoS ONE* 15(4): e0231575. doi:10.1371/journal.pone.0231575.

Bert, F.E., G.P. Podestá, S.L. Rovere, Á.N. Menéndez, M. North, E. Tatara, C.E. Laciana, E. Weber, and F.R. Toranzo. 2011. 'An Agent Based Model to Simulate Structural and Land Use Changes in Agricultural Systems of the Argentine Pampas.' *Ecological Modelling* 222(19): 3486–3499. doi:10.1016/j.ecolmodel.2011.08.007.

Bharwani, S. 2006. 'Understanding Complex Behavior and Decision Making Using Ethnographic Knowledge Elicitation Tools (KnETs).' *Social Science Computer Review* 24(1): 78–105. doi:10.1177/0894439305282346.

Bousquet, F., O. Barreteau, C. Le Page, C. Mullon, and J. Weber. 1999. 'An Environmental Modelling Approach. The Use of Multiagents Simulations.' In *Advances in Environmental and Ecological Modelling*, edited by F. Blasco and A. Weill, 113–122. Amsterdam: Elsevier.

Bousquet, F., C. Cambier, P. Morand, J. Quensiere, and C.A.P. Mullon. 1993. 'Simulating the Interaction between a Society and a Renewable Resource.' *Journal of Biological Systems* 1: 199–214.

Bousquet, F., and C. Le Page. 2004. 'Multi-agent Simulations and Ecosystem Management: A Review.' *Ecological Modelling* 176(3–4): 313–332. doi:10.1016/j.ecolmodel.2004.01.011.

Burgess, M.G., E. Carrella, M. Drexler, R.L. Axtell, R.M. Bailey, J.R. Watson, R.B. Cabral et al. 2020. 'Opportunities for Agent-based Modelling in Human Dimensions of Fisheries.' *Fish and Fisheries*. doi:10.1111/faf.12447.

Carpenter, S.R., and W.A. Brock. 2004. 'Spatial Complexity, Resilience, and Policy Diversity: Fishing on Lake-rich Landscapes.' *Ecology and Society* 9(1): 8.

Castella, J.C., T.N. Trung, and S. Boissau. 2005. 'Participatory Simulation of Land-use Changes in the Northern Mountains of Vietnam: The Combined Use of an Agent-based Model, a Role-playing Game, and a Geographic Information System.' *Ecology and Society* 10(1): 27. doi:10.5751/ES-01328-100127.

Castilla-Rho, J.C., R. Rojas, M.S. Andersen, C. Holley, and G. Mariethoz. 2017. 'Social Tipping Points in Global Groundwater Management.' *Nature Human Behaviour* 1(9): 640–649. doi:10.1038/s41562-017-0181-7.

Crooks, A., A. Heppenstall, and N. Malleson. 2018. 'Agent-based Modeling.' In *Comprehensive Geographic Information Systems*, edited by B. Huang, 218–243. Amsterdam: Elsevier.

Dobson, A.D.M., E. de Lange, A. Keane, H. Ibbett, and E.J. Milner-Gulland. 2019. 'Integrating Models of Human Behaviour between the Individual and Population Levels to Inform Conservation Interventions.' *Philosophical Transactions of the Royal Society B: Biological Sciences* 374(1781). doi:10.1098/rstb.2018.0053.

Dressler, G., J. Groeneveld, C.M. Buchmann, C. Guo, N. Hase, J. Thober, K. Frank, and B. Müller. 2018. 'Implications of Behavioral Change for the Resilience of Pastoral Systems – Lessons from an Agent-based Model.' *Ecological Complexity*. doi:10.1016/j.ecocom.2018.06.002.

Edmonds, B., V. Grimm, R. Meyer, C. Montañola, P. Ormerod, H. Root, and F. Squazzoni. 2019. 'Different Modelling Purposes.' *Journal of Artificial Societies and Social Simulation* 22(3): 30.

Epstein, J.M. 2006. *Growing Artificial Societies: Social Science from the Bottom Up*. Princeton: Princeton University Press.

Evans, T.P., and H. Kelley. 2008. 'Assessing the Transition from Deforestation to Forest Regrowth with an Agent-based Model of Land Cover Change for South-central Indiana (USA).' *Geoforum, Conversations Across the Divide* 39(2): 819–832. doi:10.1016/j.geoforum.2007.03.010.

Filatova, T., J.G. Polhill, and S. van Ewijk. 2016. 'Regime Shifts in Coupled Socio-environmental Systems: Review of Modelling Challenges and Approaches.' *Environmental Modelling & Software* 75: 333–347. doi:10.1016/j.envsoft.2015.04.003.

Forrester, J., R. Greaves, H. Noble, and R. Taylor. 2014. 'Modeling Social-Ecological Problems in Coastal Ecosystems: A Case Study.' *Complexity* 19(6): 73–82. doi:10.1002/cplx.21524.

Gimona, A., and J.G. Polhill. 2011. 'Exploring Robustness of Biodiversity Policy with a Coupled Metacommunity and Agent-based Model.' *Journal of Land Use Science* 6(2/3): 175–193.

Gotts, N.M., G.A.K. van Voorn, J.G. Polhill, E. de Jong, B. Edmonds, G.J. Hofstede, and R. Meyer. 2018. 'Agent-based Modelling of Socio-Ecological Systems: Models, Projects and Ontologies.' *Ecological Complexity*. doi:10.1016/j.ecocom.2018.07.007.

Grêt-Regamey, A., S.H. Huber, and R. Huber. 2019. 'Actors' Diversity and the Resilience of Social-Ecological Systems to Global Change.' *Nature Sustainability* 2(4): 290–297. doi:10.1038/s41893-019-0236-z.

Grimm, V., U. Berger, F. Bastiansen, S. Eliassen, V. Ginot, J. Giske, J. Goss-Custard et al. 2006. 'A Standard Protocol for Describing Individual-based and Agent-based Models.' *Ecological Modelling* 198(1–2): 115–126.

Gurung, T.R., F. Bousquet, and G. Trébuil. 2006. 'Companion Modeling, Conflict Resolution, and Institution Building: Sharing Irrigation Water in the Lingmuteychu Watershed, Bhutan.' *Ecology and Society* 11(2): 36.

Gutierrez, N.L., P. Halmay, R. Hilborn, A.E. Punt, and S. Schroeter. 2017. 'Exploring Benefits of Spatial Cooperative Harvesting in a Sea Urchin Fishery: An Agent-based Approach.' *Ecosphere* 8(7): e01829. doi:10.1002/ecs2.1829.

Happe, K., K. Kellermann, and A. Balmann. 2006. 'Agent-based Analysis of Agricultural Policies: An Illustration of the Agricultural Policy Simulator AgriPoliS, its Adaptation and Behavior.' *Ecology and Society* 11(1): 49. doi:10.5751/ES-01741-110149.

Hare, M., and P. Deadman. 2004. 'Further towards a Taxonomy of Agent-based Simulation Models in Environmental Management.' *Mathematics and Computers in Simulation* 64: 25–40.

Heckbert, S. 2013. 'MayaSim: An Agent-based Model of the Ancient Maya Social-Ecological System.' *Journal of Artificial Societies and Social Simulation* 16(4). doi:10.18564/jasss.2305.

Holtz, G., F. Alkemade, F. de Haan, J. Köhler, E. Trutnevyte, T. Luthe, J. Halbe et al. 2015. 'Prospects of Modelling Societal Transitions: Position Paper of an Emerging Community.' *Environmental Innovation and Societal Transitions* 17: 41–58. doi:10.1016/j.eist.2015.05.006.

Huber, R., M. Bakker, A. Balmann, T. Berger, M. Bithell, C. Brown, A. Grêt-Regamey et al. 2018. 'Representation of Decision-making in European Agricultural Agent-based Models.' *Agricultural Systems* 167: 143–160. doi:10.1016/j.agsy.2018.09.007.

Janssen, M.A. 2009. 'Understanding Artificial Anasazi.' http://jasss.soc.surrey.ac.uk/12/4/13.html.

Janssen, M. 2020. *Introduction to Agent-based Modeling*. https://intro2abm.com.

Janssen, M.A., and J.A. Baggio. 2016. 'Using Agent-based Models to Compare Behavioral Theories on Experimental Data: Application for Irrigation Games.' *Journal of Environmental Psychology* 46: 106–115. doi:10.1016/j.jenvp.2016.04.003.

Janssen, M.A., and E. Ostrom. 2006. 'Empirically Based, Agent-based Models.' *Ecology and Society* 11(2): 37.

Klein, E.S., M.R. Barbier, and J.R. Watson. 2017. 'The Dual Impact of Ecology and Management on Social Incentives in Marine Common-pool Resource Systems.' *Royal Society Open Science* 4(8): 170740. doi:10.1098/rsos.170740.

Köhler, J., F. de Haan, G. Holtz, K. Kubeczko, E. Moallemi, G. Papachristos, and E. Chappin. 2018. 'Modelling Sustainability Transitions: An Assessment of Approaches and Challenges.' *Journal of Artificial Societies and Social Simulation* 21(1): 8.

Kravari, K., and N. Bassiliades. 2015. 'A Survey of Agent Platforms.' *Journal of Artificial Societies and Social Simulation* 18(1): 11.

Kremmydas, D., I.N. Athanasiadis, and S. Rozakis. 2018. 'A Review of Agent Based Modeling for Agricultural Policy Evaluation.' *Agricultural Systems* 164: 95–106. doi:10.1016/j.agsy.2018.03.010.

Lansing, J.S., and J.N. Kremer. 1993. 'Emergent Properties of Balinese Water Temple Networks: Coadaptation on a Rugged Fitness Landscape.' *American Anthropologist* 95(1): 97–114. doi:10.1525/aa.1993.95.1.02a00050.

Lee, J-S., T. Filatova, A. Ligmann-Zielinska, B. Hassani-Mahmooei, F. Stonedahl, I. Lorscheid, A. Voinov, Gary Polhill, Z. Sun, and D.C. Parker. 2015. 'The Complexities of Agent-based Modeling Output Analysis.' *Journal of Artificial Societies and Social Simulation* 18(4). doi:10.18564/jasss.2897.

Lindkvist, E., X. Basurto, and M. Schlüter. 2017. 'Micro-level Explanations for Emergent Patterns of Self-governance Arrangements in Small-scale Fisheries – A Modeling Approach.' *PloS ONE* 12(4): e0175532.

Lindkvist, E., N. Wijermans, T.M. Daw, B. Gonzalez-Mon, A. Giron-Nava, A.F. Johnson, I. van Putten, Xavier Basurto, and Maja Schlüter. 2020. 'Navigating Complexities: Agent-based Modeling to Support Research, Governance, and Management in Small-scale Fisheries.' *Frontiers in Marine Science* 6. doi:10.3389/fmars.2019.00733.

Lippe, M., M. Bithell, N. Gotts, D. Natalini, P. Barbrook-Johnson, C. Giupponi, M. Hallier et al. 2019. 'Using Agent-based Modelling to Simulate Social-Ecological Systems across Scales.' *GeoInformatica*. doi:10.1007/s10707-018-00337-8.

Lorscheid, I., B-O. Heine, and M. Meyer. 2012. 'Opening the "Black Box" of Simulations: Increased Transparency and Effective Communication through the Systematic Design of Experiments.' *Computational and Mathematical Organization Theory* 18(1): 22–62. doi:10.1007/s10588-011-9097-3.

Manson, S.M., N.R. Jordan, K.C. Nelson, and R.F. Brummel. 2016. 'Modeling the Effect of Social Networks on Adoption of Multifunctional Agriculture.' *Environmental Modelling & Software* 75: 388–401. doi:10.1016/j.envsoft.2014.09.015.

Martin, R., and M. Schlüter. 2015. 'Combining System Dynamics and Agent-based Modeling to Analyze Social-Ecological Interactions – An Example from Modeling Restoration of a Shallow Lake.' *Frontiers in Environmental Science* 3. doi:10.3389/fenvs.2015.00066.

Martin, R., M. Schlüter, and T. Blenckner. 2020. 'The Importance of Transient Social Dynamics for Restoring Ecosystems beyond Ecological Tipping Points.' *Proceedings of the National Academy of Sciences* 117(5): 2717–2722. doi:10.1073/pnas.1817154117.

Martin, R., and S.A. Thomas. 2016. 'Analyzing Regime Shifts in Agent-based Models with Equation-free Analysis.' http://scholarsarchive.byu.edu/iemssconference/2016/Stream-B/54.

Melbourne-Thomas, J., C.R. Johnson, P. Perez, J. Eustache, E.A. Fulton, and D. Cleland. 2011. 'Coupling Biophysical and Socioeconomic Models for Coral Reef Systems in Quintana Roo, Mexican Caribbean.' *Ecology and Society* 16(3): 23. doi:10.5751/ES-04208-160323.

Müller, B., F. Bohn, G. Dreßler, J. Groeneveld, C. Klassert, R. Martin, M. Schlüter, J. Schulze, H. Weise, and N. Schwarz. 2013. 'Describing Human Decisions in Agent-based Models – ODD + D, an Extension of the ODD Protocol.' *Environmental Modelling & Software* 48: 37–48. doi:10.1016/j.envsoft.2013.06.003.

Müller-Hansen, F., M. Schlüter, M. Mäs, R. Hegselmann, J.F. Donges, J.J. Kolb, K. Thonicke, and J. Heitzig. 2017. 'How to Represent Human Behavior and Decision Making in Earth System Models? A Guide to Techniques and Approaches.' *Earth Systems Dynamics Discussions*, 1–53.

Parker, D.C., and V. Meretsky. 2004. 'Measuring Pattern Outcomes in an Agent-based Model of Edge-effect Externalities Using Spatial Metrics.' *Agriculture, Ecosystems & Environment* 101(2–3): 233–250. doi:10.1016/j.agee.2003.09.007.

Plank, M.J., J. Kolding, R. Law, H.D. Gerritsen, and D. Reid. 2017. 'Balanced Harvesting Can Emerge from Fishing Decisions by Individual Fishers in a Small-scale Fishery.' *Fish and Fisheries* 18(2): 212–225. doi:10.1111/faf.12172.

Polhill, J.G., T. Filatova, M. Schlüter, and A. Voinov. 2016. 'Modelling Systemic Change in Coupled Socio-environmental Systems.' *Environmental Modelling & Software* 75: 318–332. doi:10.1016/j.envsoft.2015.10.017.

Polhill, J.G., A. Gimona, and N.M. Gotts. 2013. 'Nonlinearities in Biodiversity Incentive Schemes: A Study Using an Integrated Agent-based and Metacommunity Model.' *Environmental Modelling & Software*, Thematic Issue 45: 74–91. doi:10.1016/j.envsoft.2012.11.011.

Railsback, S.F., and V. Grimm. 2012. *Agent-based and Individual-based Modeling – A Practical Introduction*. Princeton: Princeton University Press. www.railsback-grimm-abm-book.com.

Sakoda, J.M. 1971. 'The Checkerboard Model of Social Interaction.' *The Journal of Mathematical Sociology* 1(1): 119–132. doi:10.1080/0022250X.1971.9989791.

Schelling, T.C. 1971. 'Dynamic Models of Segregation.' *The Journal of Mathematical Sociology* 1(2): 143–186. doi:10.1080/0022250X.1971.9989794.

Schill, C., J.M. Anderies, T. Lindahl, C. Folke, S. Polasky, J.C. Cárdenas, A-S. Crépin, M.A. Janssen, J. Norberg, and M. Schlüter. 2019. 'A More Dynamic Understanding of Human Behaviour for the Anthropocene.' *Nature Sustainability* 2(12): 1075–1082. doi:10.1038/s41893-019-0419-7.

Schill, C., N. Wijermans, M. Schlüter, and T. Lindahl. 2016. 'Cooperation Is Not Enough – Exploring Social-Ecological Micro-foundations for Sustainable Common-pool Resource Use.' *PLoS ONE* 11(8): e0157796. doi:10.1371/journal.pone.0157796.

Schlüter, M., A. Baeza, G. Dressler, K. Frank, J. Groeneveld, W. Jager, M.A. Janssen et al. 2017. 'A Framework for Mapping and Comparing Behavioural Theories in Models of Social-Ecological Systems.' *Ecological Economics* 131: 21–35. doi:10.1016/j.ecolecon.2016.08.008.

Schlüter, M., J. Hinkel, P.W.G. Bots, and R. Arlinghaus. 2014. 'Application of the SES Framework for Model-based Analysis of the Dynamics of Social-Ecological Systems.' *Ecology and Society* 19(1): 36. doi:10.5751/ES-05782-190136.

Schlüter, M., B. Müller, and K. Frank. 2019. 'The Potential of Models and Modeling for Social-Ecological Systems Research: The Reference Frame ModSES.' *Ecology and Society* 24(1): 31. doi:10.5751/ES-10716-240131.

Schlüter, M., K. Orach, E. Lindkvist, R. Martin, N. Wijermans, Ö. Bodin, and W.J. Boonstra. 2019. 'Toward a Methodology for Explaining and Theorizing about Social-Ecological Phenomena.' *Current Opinion in Environmental Sustainability* 39: 44–53. doi:10.1016/j.cosust.2019.06.011.

Schlüter, M., A. Tavoni, and S. Levin. 2016. 'Robustness of Norm-driven Cooperation in the Commons.' *Proceedings of the Royal Society B: Biological Sciences* 283(1822): 20152431. doi:10.1098/rspb.2015.2431.

Schmolke, A., P. Thorbek, D.L. DeAngelis, and V. Grimm. 2010. 'Ecological Models Supporting Environmental Decision Making: A Strategy for the Future.' *Trends in Ecology & Evolution* 25(8): 479–486. doi:10.1016/j.tree.2010.05.001.

Schulze, J., B. Müller, J. Groeneveld, and V. Grimm. 2017. 'Agent-based Modelling of Social-Ecological Systems: Achievements, Challenges, and a Way Forward.' *Journal of Artificial Societies and Social Simulation* 20(2). doi:10.18564/jasss.3423.

Smajgl, A., and O. Barreteau. 2014. *Empirical Agent-based Modelling – Challenges and Solutions: Volume 1, The Characterisation and Parameterisation of Empirical Agent-based Models.* New York: Springer. doi:10.1007/978-1-4614-6134-0_1.

Smajgl, A., and E. Bohensky. 2011. 'Behaviour and Space in Agent-based Modelling: Poverty Patterns in East Kalimantan, Indonesia.' *Environmental Modelling & Software.* doi:10.1016/j.envsoft.2011.10.014.

Smajgl, A., D.G. Brown, D. Valbuena, and M.G.A. Huigen. 2011. 'Empirical Characterisation of Agent Behaviours in Socio-Ecological Systems.' *Environmental Modelling & Software* 26(7): 837–844. doi:10.1016/j.envsoft.2011.02.011.

Squazzoni, F. 2010. 'The Impact of Agent-based Models in the Social Sciences after 15 Years of Incursions.' *History of Economic Ideas* 18: 197–233.

Sun, Z., and D. Müller. 2013. 'A Framework for Modeling Payments for Ecosystem Services with Agent-based Models, Bayesian Belief Networks and Opinion Dynamics Models.' *Environmental Modelling & Software* 45: 15–28. doi:10.1016/j.envsoft.2012.06.007.

Thiele, J.C., W. Kurth, and V. Grimm. 2014. 'Facilitating Parameter Estimation and Sensitivity Analysis of Agent-based Models: A Cookbook Using NetLogo and R.' *Journal of Artificial Societies and Social Simulation* 17(3): 11.

Thomas, S.A., D.J.B. Lloyd, and A.C. Skeldon. 2016. 'Equation-free Analysis of Agent-based Models and Systematic Parameter Determination.' *Physica A: Statistical Mechanics and Its Applications* 464: 27–53. doi:10.1016/j.physa.2016.07.043.

Villamor, G.B., Q.B. Le, U. Djanibekov, M. van Noordwijk, and P.L.G. Vlek. 2014. 'Biodiversity in Rubber Agroforests, Carbon Emissions, and Rural Livelihoods: An Agent-based Model of Land-use Dynamics in Lowland Sumatra.' *Environmental Modelling & Software* 61: 151–165. doi:10.1016/j.envsoft.2014.07.013.

Wijermans, N., W.J. Boonstra, K. Orach, J. Hentati-Sundberg, and M. Schlüter. 2020. 'Behavioural Diversity in Fishing – Towards a Next Generation of Fishery Models.' *Fish and Fisheries* 21(5): 872–890. doi:10.1111/faf.12466.

Wilson, J., L. Yan, and C. Wilson. 2007. 'The Precursors of Governance in the Maine Lobster Fishery.' *Proceedings of the National Academy of Sciences* 104(39): 15212–15217.

Worrapimphong, K., N. Gajaseni, C. Lee Page, and F. Bousquet. 2010. 'A Companion Modeling Approach Applied to Fishery Management.' *Environmental Modelling & Software* 25(11): 1334–1344. doi:10.1016/j.envsoft.2010.03.012.

Methods for analysing systems – directly informing decision-making

29

Decision analysis based on optimisation

Anne-Sophie Crépin[1] *and Stephen Polasky*[2]

[1]BEIJER INSTITUTE OF ECOLOGICAL ECONOMICS, THE ROYAL SWEDISH ACADEMY OF SCIENCES, STOCKHOLM, SWEDEN
[2]UNIVERSITY OF MINNESOTA, MINNEAPOLIS, MINNESOTA, USA

Key methods discussed in this chapter

Mathematical programming, optimal control theory, game theory, decision theory, cost-benefit analysis, multi-criteria decision analysis

Connections to other chapters

Methods for data generation and systems scoping (Chapters 5–8), futures analysis (Chapter 10), scenario development (Chapter 11), dynamical systems modelling (Chapter 26), state-and-transition modelling (Chapter 27), agent-based modelling (Chapter 28) and other methods for analysing systems (Chapters 30–32) can be used to inform decision processes. Controlled behavioural experiments (Chapter 21) can help to evaluate potential impacts of decisions. The methods in this chapter can help to model people's behaviour in futures analysis, scenario development, dynamical systems modelling and agent-based modelling.

Introduction

Decision analysis is a systematic approach to evaluating information about alternative choices, when multiple options are possible, with many possible outcomes and different trade-offs. In social-ecological systems (SES), multiple types of decisions (policy, management, private, other) – all with different objectives – influence the social, economic and ecological dimensions, making it hard to compare across alternatives. Decision analysis can analyse these situations and their impacts on individual actors, society and the rest of the system.

The objective of a decision can, for example, be related to maximising measures of human well-being ('utility') or reaching a particular target, such as remaining below a maximum level of pollution, reducing inequality or conserving biodiversity. These decisions also involve multiple constraints, e.g. the desire to remain within a given budget, or the physical restrictions posed by particular ecosystem dynamics. Selecting from several possible decisions requires specific criteria for assessment. Here we focus on optimisation as a criterion.

The problem of optimisation has been of interest for centuries in mathematics. Optimisation methods are used in a wide array of disciplines including decision science, economics, engineering,

DOI: 10.4324/9781003021339-35

SUMMARY TABLE: DECISION ANALYSIS BASED ON OPTIMISATION

DISCIPLINARY BACKGROUND	KNOWLEDGE TYPE
The methods in this chapter are derived from or have most commonly been used in: Mathematics, Engineering, Economics, Physics, Political Sciences, Sociology, Biology	*The methods in this chapter are primarily used to generate the following types of knowledge:* • Descriptive • Exploratory • Explanatory • Prescriptive
RESEARCH APPROACH	**PURPOSE OF METHOD**
The methods in this chapter originate from or most commonly adopt the following research approaches: • Analytical/objective	*The most common purposes of using the methods in this chapter are:* • System understanding • Policy/decision support
TEMPORAL DIMENSION	**SYSTEMIC FEATURES AND PROCESSES**
The methods in this chapter are most commonly applied to the following temporal dimensions: • Present (typically within the last 5–10 years) • Recent past (post-1700s) • Pre-industrial revolution (pre-1700s) • Future	*While most methods can do many things, the methods in this chapter are particularly good (i.e. go-to methods) for addressing the following:* • SES components and linkages • Social-ecological dependence and impact • Evaluating policy options
SPATIAL DIMENSION	
The methods in this chapter are primarily either or both: • Non-spatial • Explicitly spatial *The methods in this chapter are most commonly applied at the following spatial scales:* • Local • Regional (provincial/state to continental) • Global • Multiple places/sites around the world	

mathematics, political science, psychology and sociology. These methods build on early attempts (sometimes dating back to ancient Greece) to solve problems related to risk management, resource allocation and strategic interactions, which occur in all SES. Strategic interactions are likely to influence people's own objectives and how they expect others to react. Methods to analyse strategic interactions assess people's expected actions and how their interactions are likely to influence the SES outcome. Here we focus on game theory, which studies the interactions of multiple optimising agents. This method was already documented in the early 18th century.

Decision analysis methods in SES typically use models to predict how various choices made by actors affect the evolution of the SES and how the system in turn affects the individual or collective objective. Among those methods, optimisation methods then search through possible choices to find the one that generates the highest score for the objective. Game theory searches instead for the equilibrium likely to occur from people's interactions. Decision analysis in the context of SES must incorporate social and ecological components with feedbacks among components that accurately capture the system's dynamic behaviour (Polasky et al. 2011). At least four aspects of these dynamics substantially influence the choice of the method that would be appropriate for decision analysis:

1. **Who is making the decision?** A centralised decision process implies that someone has exclusive power to make a decision or that all agents agree to abide by a group decision. The decision-maker could, for example, be a national government, an individual allocating their own resources or a community managing a common resource. In contrast, a decentralised decision process involves several people, all with different objectives and criteria for assessing their objective. While optimisation methods are most useful for studying centralised decision processes, game theory analyses the different actors' decision options and the outcomes of decentralised processes.
2. **How much information is available?** The amount of information available is important, in particular whether there is full certainty (complete knowledge), risk (the outcome is unknown, but all possible outcomes and their probabilities are known) or uncertainty (some outcomes and/or their probabilities are unknown). Risk situations require stochastic optimisation methods, which are particular methods that maximise some expected value of the objective (e.g. expected utility). In contrast, uncertainty requires the use of methods such as futures analysis (Chapter 10), scenario development (Chapter 11) or resilience assessments (Chapter 14). Information may not be evenly distributed among agents. Economic theories of decisions with imperfect information for problems of moral hazard (e.g. an insured person taking more risk than an uninsured one), principal agents (e.g. a person taking action on behalf of another) or asymmetric information (e.g. a seller of a product knowing more about its quality than a buyer) can use game theory to inform policy design and studies of power relationships and decisions in the social part of SES.
3. **Is there a temporal dimension?** Some decision analyses focus on one-shot decisions and use so-called static optimisation methods, whereas others incorporate processes that evolve with time and use dynamic optimisation methods. Mathematical programming is a static optimisation method, while optimal control methods are dynamic. Time can be modelled as a continuous or discrete process, with each process requiring different optimisation technologies. Modelling strategic interactions also entails using different tools depending on the timing of the different individuals' decisions. People can act simultaneously (static games), after one another (sequential games), once (one-shot games), repeatedly (repeated games), or take into account dynamic processes (dynamic games) and evolution (evolutionary games).

4. **Are there any spatial patterns?** Most decision analysis methods were developed without explicit spatial considerations. However, optimisation and game theory can be adapted to account for spatial heterogeneity and spatial interactions when decisions made in one place affect the net benefits of making decisions somewhere else (Costello and Polasky 2008; Polasky et al. 2008, 2014; Smith, Sanchirico, and Wilen 2009; Brock and Xepapadeas 2010).

Optimisation and game theory rely on a number of important assumptions. Varying some of these assumptions contributes to theory development in these fields. The study of SES, in particular slow and delayed time processes, for example, has led to borrowing methods from mathematics, such as perturbation theory (Crépin 2007), or new ways of modelling risk with delays (Crépin and Nævdal 2020). The field of behavioural economics has grown in response to the observation that, in many situations, people do not behave in a way consistent with rational choice, i.e. choices that are in their perceived best interest. Researchers at the boundary between psychology and economics have introduced new explanations such as bounded rationality and prospect theory to explain these deviations (Shogren and Taylor 2008).

Another important assumption in optimisation methods is that the systems being studied are well behaved, which leads to a single optimum equilibrium. Research on economies of scales identified the existence of two possible optimal equilibria and an indifference point where either equilibrium could be an optimal target for management (Skiba 1978) depending on initial conditions. Complexity economics (Arthur 1999) emerged partly in response to observations that many of the systems studied were complex adaptive systems with possibly multiple and complex attractors, rather than a unique optimal equilibrium.

SES problems and questions

Decision analysis is essential to modelling social dynamics in SES models, exploring and testing SES behaviour, and comparing the outcomes of different policies under various conditions. Optimisation methods can help to answer questions such as the following: 'where should various interventions occur?' (Polasky et al. 2008); 'what would be the optimal release of nutrients in a lake given that users value both water quality and agricultural production on the shore?' (Mäler, Xepapadeas, and De Zeeuw 2003); 'how much fish should we harvest in a coral reef?' (Crépin 2007). Game theory can address questions such as the following: 'what would be the outcome in the lake if the users did not cooperate?' (Mäler, Xepapadeas and De Zeeuw 2003); 'if farmers did not cooperate, how many animals would each farmer allow to graze on a common grassland?' (Crépin and Lindahl 2009). Cost-benefit analysis and multi-criteria decision analysis can help to answer questions such as 'what investment should we make?' and 'which alternative should we choose among those possible?'.

Case study 29.1 illustrates these different options to inform decisions in a community managing a lake. Optimisation and game theory can also enrich other methods that are often used in the study of SES, in particular bio-economic modelling and agent-based models (Chapter 28), where they contribute to testing the impacts of a wide array of different behaviours and assumptions on the outcome of the model.

Brief description of key methods

Decision analysis includes many methods that can be used for various purposes. Those selected in Table 29.1 represent the variety of methods using optimisation, which take into

Table 29.1 Summary of key methods used in decision analysis based on optimisation

Method	*Description*	*References*
Mathematical programming	Mathematical methods are used to solve optimisation problems when decision is centralised, full information is available and time does not matter. Some of these methods (e.g. Lagrange and Kuhn–Tucker methods) address situations when constraints (e.g. a budget that must be fully spent or a pollution level that cannot be transgressed) limit the range of possible decisions.	*Key introductory text* Intriligator 2002 *Applications to SES* Robinson, Williams, and Albers 2002; Watson et al. 2011
Game theory	Game theory involves mathematical methods to study the outcomes of strategic interactions, usually between rational decision-makers. A fully defined game specifies the players of the game (e.g. all users of a lake), the information and actions available to each player at each decision point (e.g. to fish or not), the order in which actions can be taken (who gets to fish when), and the pay-offs of each outcome. In the simplest solution concept (called the Nash equilibrium), each actor maximises their expected utility given the equilibrium behaviour of other actors. More complex games incorporate aspects like time (repeated, differential and evolutionary games) and asymmetric information between players.	*Key introductory text* Myerson 2013 *Applications to SES* Mäler, Xepapadeas, and De Zeeuw 2003; Crépin and Lindahl 2009; Diekert 2012; Tavoni, Schlüter, and Levin 2012
Optimal control theory	Optimal control theory is a set of mathematical methods to solve optimisation problems when decision is centralised, full information is available and some variables change over time. The methods aim to identify the values of a variable (control) that can be manipulated (e.g. the size of a harvest) to optimise a particular objective of a system (e.g. the sum of profits from a fishery over time). The dynamics of some system variable (e.g. how the fish stock changes over time) limit the range of possible decisions along with other constraints (e.g. how much fish should be left at the end of the period). These methods include calculus of variation, dynamic programming (also useful for computer programming) and Pontryagin's maximum principle.	*Key introductory text* Kamien and Schwartz 2012 *Applications to SES* Crépin 2007; Diekert et al. 2010; Quaas et al. 2013; Ashander et al. 2019

Method	*Description*	*References*
Decision theory	Decision theory is a collection of decision methods that incorporate risk and uncertainty. With risk, the objective can be expressed as an expected value, and stochastic optimisation methods can be applied. With uncertainty, different decision rules are available depending on whether the decision-maker puts more weight on precaution, confidence in information about the relative likelihood of outcomes, robustness of choice, or flexibility in the timing of the decision. Decision rules emerging from this process could, for example, aim to minimise regrets associated with a decision, minimise potential losses or maximise best possible outcomes.	*Key introductory text* Morgan and Henrion 1990 *Applications to SES* Polasky, De Zeeuw, and Wagener 2011; Cox 2012
Cost-benefit analysis	Cost-benefit analysis (CBA) is a systematic approach to determine whether an investment is sound by comparing the net present value of different investments. CBA identifies the flow of all the costs and benefits over time associated with a particular project. A CBA should contain a comparison between one and more particular development projects and the status quo.	*Key introductory text* Boadway 2016 *Applications to SES* Bateman et al. 2003; Pearce, Atkinson, and Mourato 2006; Wegner and Pascual 2011
Multi-criteria decision analysis	Multi-criteria decision analysis (MCDA) uses formal approaches that take explicit account of multiple criteria in exploring decisions. MCDA applies in particular when the criteria to take into account are sufficiently important for the outcome of the decision and some of them may conflict with one another. There is typically not a unique optimal solution to this problem; instead, the goal may be to find the preferred alternative among several or to find all solutions that would not sacrifice at least one dimension.	*Key introductory text* Mendoza and Martins 2006 *Applications to SES* Karjalainen et al. 2013

account three of the four important aspects mentioned in the introduction (i.e. who is making the decision, the amount of information available, the temporal dimension). Although these methods were not specifically developed with spatial variations, it is possible to incorporate such heterogeneity – but then often at the expense of more complex solutions.

Mathematical programming solves the simplest optimisation problems if decision-making is centralised, there is full certainty and time does not influence the outcome. Many daily decisions in SES emerge from the strategic interactions of multiple stakeholders who do not necessarily cooperate, such as a small number of firms competing with one another or users of a common-pool fishery. Game theory aims to study these types of strategic interactions when the decision is decentralised. Optimal control theory covers optimisation methods over time, with centralised decisions and full certainty, whereas decision theory focuses on optimisation methods when some information is missing.

In contrast to these theoretical methods, cost-benefit analysis and multi-criteria decision analysis are of a more practical nature. They apply to centralised decisions and are useful for choosing among multiple project alternatives when more detailed information is available (Polasky and Binder 2012). Cost-benefit analysis measures all have a monetary impact and compare alternatives in terms of net benefits (benefits minus costs) to find optimal policy or management choices. Multi-criteria decision analysis methods were developed to structure, analyse and solve decision problems involving multiple criteria measured in different metrics.

Limitations

Despite their powerful and broad applicability, decision analysis methods are based on simplified models that limit the ability to fully represent complex SES dynamics. Optimisation and game theory methods both focus on equilibria, but system dynamics outside of any equilibrium are also important in the study of SES. Decision analysis methods that explicitly incorporate time variations can be used to analyse dynamics outside of equilibrium, but this usually requires computer simulations even for relatively simple problems. There is currently no example of an optimisation or game theory model that simultaneously incorporates all typical characteristics of SES as a complex adaptive system (Levin et al. 2013). Such models would be extremely hard if not impossible to fully analyse. The information obtained may not be meaningful because the results could be very sensitive to initial conditions and model assumptions.

Different assumptions about preferences, how they trade off against one another and in time, and different decision criteria can result in widely different recommendations about possible actions. Individual decision-making involves subjective tastes and preferences. Decision processes involving several people require weighting the preferences of different individuals against one another, and trading off values between today and the future. Uncertainty and the choice of appropriate decision criteria generate yet another layer of subjectivity. Any outcome from a decision analysis therefore needs to be carefully considered and understood in terms of the assumptions it relies on. Sensitivity analyses are an important component of any analysis to explore how the optimal outcome changes given different assumptions.

Cost-benefit analysis and multi-criteria decision analysis tend to neglect aspects of decisions that are difficult to quantify and value, compared to easily quantifiable variables. Data of sufficient resolution for both natural and social variables are often not available and information about causal relationships related to behavioural, social, economic, ecological and technical aspects is generally lacking. The difference in timing when the costs and benefits are estimated can generate decision biases because the cost occurring today is easier to estimate than the uncertain future costs and benefits.

Game theory is particularly challenging to apply when studying decisions in larger groups. With more than two players, it becomes difficult to represent all possibilities. To address this issue, one could study the behaviour of an average player or simulate many potential outcomes to identify the distribution of these outcomes given a set of potential responses and initial conditions. Agent-based models (Chapter 28) are probably better suited to representing the outcomes of strategic interactions between many heterogeneous agents.

Resource implications

Using decision analysis methods based on optimisation requires mathematical training. Basic mathematics is enough to solve the simplest mathematical programs, cost-benefit analyses and games. However, many of the methods studied hinge on advanced mathematics to model and analyse multiple kinds of dynamic, stochastic and spatial processes. Cost-benefit analysis and multi-criteria decision analysis also require using methods (such as valuation methods) to quantify expected changes and find ways to compare these changes to one another.

Problems with SES dynamics are so complex that they typically do not have any analytical solutions. Instead, numerical analysis can identify solution properties through computer simulations using, for example, Monte Carlo methods, which replicate the studied system using a large number of different parameter values. Model investigations over the whole range of possible parameter values require extensive computing power to perform numerous iterations. Careful model calibration to real data, with good quality data, can limit the necessary number of iterations.

While it is straightforward to program discrete time models, continuous time models are more problematic because computers can only handle discrete time steps. The Runge–Kutta methods were developed for the purpose of approximating the solutions to ordinary differential equations. These methods are usually available as a standard package of the most commonly used software for mathematical programming and numerical analysis.

New directions

The methods presented in this chapter contribute to SES analysis but were not designed for this purpose. Further progress should broaden these methods to account for the complex adaptive systems properties of SES (Levin et al. 2013). More interaction with the field of complexity economics (Arthur 1999, 2006) could be fruitful, including the study of dynamics outside of equilibrium and more heterogeneous agents.

Substantial effort has been made to incorporate elements of complexity in standard optimisation and game theory tools. With the appropriate assumptions, these methods can help to analyse regime shifts (Grass et al. 2008), spatial heterogeneity (Brock and Xepapadeas 2010), diversity (Crépin, Norberg, and Mäler 2011), and fast and slow variables (Crépin 2007). Resource economics has investigated management with multiple attractors for a variety of ecosystems (e.g. forests: Crépin 2003; lakes: Mäler, Xepapadeas, and De Zeeuw 2003; coral reefs and fisheries: Crépin 2007). Recent developments have also focused on the role of spatial heterogeneity and spatial interactions (Polasky et al. 2008; Smith, Sanchirico, and Wilen 2009; Brock and Xepapadeas 2010, 2019; Epanchin-Niell and Wilen 2012), diversity (Van der Heide, Van den Bergh, and Van Ierland 2005), learning (Peterson, Carpenter, and Brock 2003), uncertainty about critical thresholds (Polasky, De Zeeuw, and Wagener 2011) and other aspects relevant to SES.

Agent-based models, resilience assessments and scenario planning often only include poor representation of the economic incentives that influence the behaviour of different agents or groups of agents (Crépin 2019). These approaches should start using the methods presented in this chapter in a more systematic way. Some progress in this direction includes a framework for integrated ecosystem-based management that aims to merge ecosystem-based management, economic decision approaches and an SES perspective, accounting for aspects that influence system resilience (Crépin et al. 2017).

The impacts of uneven distribution of information among people have been widely studied using the methods presented in this chapter. However, these studies usually neglect further complexity. It could be fruitful to revisit key studies of complex strategic human

Case study 29.1: Decision analysis and the management of North Temperate Lakes in Wisconsin, USA

The North Temperate Lakes (NTL) Long-Term Ecological Research (LTER) site studies the ecology of seven lakes in Wisconsin, USA. It focuses on how biophysical setting, climate and changing land use and land cover interact to shape lake characteristics and dynamics over time. Figure 29.1, for example, shows Lake Mendota and Lake Monona. Both lakes have regular algal blooms, due primarily to nutrient run-off from dairy manure and eroding agricultural fields. The aim of the project, which started in 1981, was to study long-term change and ecosystem regime shifts and assess their potential causes.

Using long-term monitoring and large-scale experimentation, researchers could identify multiple drivers of ecosystem change such as extreme events and land-use

Figure 29.1 The isthmus of Madison, Wisconsin, with Lake Mendota in the foreground and Lake Monona in the background (© Eric Booth 2014)

behaviour by placing them in an SES context or to introduce agents with various degrees of information and complex strategic behaviour into agent-based models.

Another blind spot concerns strategic interactions when regime shifts occur in a spatial context. Earlier results investigating strategic considerations and behavioural responses with regime shifts show a very rich set of strategic responses compared to when there are no regime shifts (Schill, Lindahl, and Crépin 2015). The same holds true for investigations relating to the optimisation of resource use with regime shifts and spatial heterogeneity (Brock and Xepapadeas 2019). Integrating all these dimensions is likely to provide novel insights.

changes. These could lead to altered nutrient inflows, which could, beyond a certain threshold, trigger a regime shift. This knowledge led to simplified mathematical models of lakes. Other scientists (e.g. Mäler, Xepapadeas, and De Zeeuw 2003) could then use these simplified models as constraints in models of decision analysis. Optimal control models could answer the question 'What is the optimal amount of nutrient release from agriculture to a lake?' and differential game models would answer the questions 'What will be the release of nutrients to the lake if land users do not cooperate?' and 'Can we design a tax on nutrients that will achieve the optimal amount of nutrient release even if land users do not cooperate?' (Mäler, Xepapadeas, and De Zeeuw 2003; Kossioris et al. 2011).

These studies showed, for example, that for some initial conditions it might be optimal to release so many nutrients into the lake that its water would become turbid, while for other initial conditions the lake water should remain clear of nutrients. Indeed, the outcome would depend on how the value of having a clear lake is traded off against the value of being able to release nutrients (e.g. to grow crops on the shore). If the community did not manage to cooperate, they would reach a suboptimal outcome, which could trigger a regime shift to a turbid lake although it would have been optimal to keep the lake clear. The studies also showed that a tax scheme designed to reach an optimal nutrient release when the community does not cooperate would be extremely complicated and probably difficult to implement.

An advantage of using decision analysis based on optimisation in a dynamic context is that one can characterise not only the long-term equilibrium outcomes but also the complete path leading to these outcomes. Doing so revealed that while the long-term optimal equilibrium could in theory be achieved using a constant tax, the path leading to this outcome was suboptimal and contained discontinuities. This triggered discussion about whether such a tax would be successful at all.

While decision theory methods applied to simpler decision problems are able to reveal general rules and patterns, it is much more difficult to identify these patterns when decision analysis is applied to a more complex problem like the management of lakes that can exhibit regime shifts. The models used in this context were tractable only because they were substantially simplified (one differential equation expressing the nutrient dynamics). If the system is more complex, the solution to these types of problems must rely on advanced computational techniques. However, for lake management, the general results seem to be robust in more complex models that also incorporate a slow mud dynamics equation in addition to the nutrient dynamics (Grass, Xepapadeas, and De Zeeuw 2017).

Key readings

Grass, D., J.P. Caulkins, G. Feichtinger, G. Tragler, and D.A. Behrens. 2008. *Optimal Control of Nonlinear Processes*. Berlin: Springer.

Schmedders, K., and K.L. Judd. 2014. *Handbook of Computational Economics*, Volume 3. Amsterdam: Elsevier.

Seierstad, A., and K. Sydsæter. 1987. *Optimal Control Theory with Economic Applications*. Amsterdam: North Holland.

Acknowledgements

We thank the Kjell and Märta Beijer Foundation for financial support.

References

Arthur, W.B. 1999. 'Complexity and the Economy.' *Science* 284(5411): 107–109.

Arthur, W.B. 2006. 'Out-of-equilibrium Economics and Agent-based Modeling.' *Handbook of Computational Economics* 2: 1551–1564.

Ashander, J., L.C. Thompson, J.N. Sanchirico, and M.L. Baskett. 2019. 'Optimal Investment to Enable Evolutionary Rescue.' *Theoretical Ecology* 12: 165–177.

Bateman, I.J., A.A. Lovett, J.S. Brainard, and D.W. Pearce. 2003. *Applied Environmental Economics: A GIS Approach to Cost-benefit Analysis.* Cambridge: Cambridge University Press.

Boadway, R. 2016. 'Cost-benefit Analysis.' In *The Oxford Handbook of Well-being and Public Policy*, edited by M.D. Adler and M. Fleurbaey, 47–81. New York: Oxford University Press.

Booth, E. 2014. *Water Sustainability and Climate Project.* https://wsc.limnology.wisc.edu.

Brock, W., and A. Xepapadeas. 2010. 'Pattern Formation, Spatial Externalities and Regulation in Coupled Economic–Ecological Systems.' *Journal of Environmental Economics and Management* 59(2): 149–164.

Brock, W., and A. Xepapadeas. 2019. 'Regional Climate Change Policy under Positive Feedbacks and Strategic Interactions.' *Environmental and Resource Economics* 72(1): 51–75.

Costello, C., and S. Polasky. 2008. 'Optimal Harvesting of Stochastic Spatial Resources.' *Journal of Environmental Economics and Management* 56(1): 1–18.

Cox, L.A. Jr. 2012. 'Community Resilience and Decision Theory Challenges for Catastrophic Events.' *Risk Analysis: An International Journal* 32(11): 1919–1934.

Crépin, A-S. 2003. 'Multiple Species Boreal Forests – What Faustmann Missed.' *Environmental and Resource Economics* 26(4): 625–646.

Crépin, A-S. 2007. 'Using Fast and Slow Processes to Manage Resources with Thresholds.' *Environmental and Resource Economics* 36(2): 191–213.

Crépin, A-S. 2019. 'Complexity, Resilience and Economics.' In *Global Challenges, Governance, and Complexity: Applications and Frontiers*, edited by V. Galaz, 166–187. Cheltenham: Edward Elgar.

Crépin, A-S., Å. Gren, G. Engström, and D. Ospina. 2017. 'Operationalising a Social-Ecological System Perspective on the Arctic Ocean.' *Ambio* 46(Supplement 3): S475–S485.

Crépin, A-S., and T. Lindahl. 2009. 'Grazing Games: Sharing Common Property Resources with Complex Dynamics.' *Environmental and Resource Economics* 44: 29–46.

Crépin, A-S., and E. Nævdal. 2020. 'Inertia in Risk: Improving Economic Models of Catastrophes.' *Scandinavian Journal of Economics* 122(4): 1259–1285. doi:10.1111/sjoe.12381.

Crépin, A-S., J. Norberg, and K.G. Mäler. 2011. 'Coupled Economic-ecological Systems with Slow and Fast Dynamics – Modelling and Analysis Method.' *Ecological Economics* 70(8): 1448–1458.

Diekert, F.K. 2012. 'The Tragedy of the Commons from a Game-theoretic Perspective.' *Sustainability* 4(8): 1776–1786.

Diekert, F.K., D.Ø. Hjermann, E. Nævdal, and N.C. Stenseth. 2010. 'Spare the Young Fish: Optimal Harvesting Policies for North-East Arctic Cod.' *Environmental and Resource Economics* 47(4): 455–475.

Epanchin-Niell, R.S., and J.E. Wilen. 2012. 'Optimal Spatial Control of Biological Invasions.' *Journal of Environmental Economics and Management* 63(2): 260–270.

Grass, D., J.P. Caulkins, G. Feichtinger, G. Tragler, and D.A. Behrens. 2008. *Optimal Control of Nonlinear Processes.* Berlin: Springer.

Grass, D., A. Xepapadeas, and A. de Zeeuw. 2017. 'Optimal Management of Ecosystem Services with Pollution Traps: The Lake Model Revisited.' *Journal of the Association of Environmental and Resource Economists* 4(4): 1121–1154.

Intriligator, M.D. 2002. *Mathematical Optimization and Economic Theory.* Philadelphia: Society for Industrial and Applied Mathematics.

Kamien, M.I., and N.L. Schwartz. 2012. *Dynamic Optimization: The Calculus of Variations and Optimal Control in Economics and Management.* North Chelmsford: Courier Corporation.

Karjalainen, T.P., P.M. Rossi, P. Ala-Aho, R. Eskelinen, K. Reinikainen, B. Kløve, M. Pulido-Velasquez, and H. Yang. 2013. 'A Decision Analysis Framework for Stakeholder Involvement and Learning in Groundwater Management.' *Hydrology and Earth System Sciences* 17: 1–13.

Kossioris, G., M. Plexousakis, A. Xepapadeas, and A. de Zeeuw. 2011. 'On the Optimal Taxation of Common-pool Resources.' *Journal of Economic Dynamics and Control* 35(11): 1868–1879.

Levin, S., T. Xepapadeas, A-S. Crépin, J. Norberg, A. de Zeeuw, C. Folke, T. Hughes et al. 2013. 'Social-Ecological Systems as Complex Adaptive Systems: Modeling and Policy Implications.' *Environment and Development Economics* 18(2): 111–132.

Mäler, K.G., A. Xepapadeas, and A. de Zeeuw. 2003. 'The Economics of Shallow Lakes.' *Environmental and Resource Economics* 26(4): 603–624.

Mendoza, G.A., and H. Martins. 2006. 'Multi-criteria Decision Analysis in Natural Resource Management: A Critical Review of Methods and New Modelling Paradigms.' *Forest Ecology and Management* 230(1): 1–22.

Morgan, M.G., and M. Henrion. 1990. *Uncertainty: A Guide to Dealing with Uncertainty in Quantitative Risk and Policy Analysis*. Cambridge: Cambridge University Press.

Myerson, R.B. 2013. *Game Theory*. Cambridge: Harvard University Press.

Pearce, D., G. Atkinson, and S. Mourato. 2006. *Cost-benefit Analysis and the Environment: Recent Developments*. Paris: OECD.

Peterson, G.D., S.R. Carpenter, and W.A. Brock. 2003. 'Uncertainty and the Management of Multistate Ecosystems: An Apparently Rational Route to Collapse.' *Ecology* 84(6): 1403–1411.

Polasky, S., and S. Binder. 2012. 'Valuing the Environment for Decision-making.' *Issues in Science and Technology* 28(4): 53–62.

Polasky, S., S.R. Carpenter, C. Folke, and B. Keeler. 2011. 'Decision-making under Great Uncertainty: Environmental Management in an Era of Global Change.' *Trends in Ecology and Evolution* 26(8): 398–404.

Polasky, S., A. de Zeeuw, and F. Wagener. 2011. 'Optimal Management with Potential Regime Shifts.' *Journal of Environmental Economics and Management* 62(2): 229–240.

Polasky, S., D. Lewis, A. Plantinga, and E. Nelson. 2014. 'Implementing the Optimal Provision of Ecosystem Services.' *Proceedings of the National Academy of Sciences* 111(17): 6248–6253.

Polasky, S., E. Nelson, J. Camm, B. Csuti, P. Fackler, E. Lonsdorf, C. Montgomery et al. 2008. 'Where to Put Things? Spatial Land Management to Sustain Biodiversity and Economic Returns.' *Biological Conservation* 141(6): 1505–1524.

Quaas, M.F., T. Requate, K. Ruckes, A. Skonhoft, N. Vestergaard, and R. Voss. 2013. 'Incentives for Optimal Management of Age-structured Fish Populations.' *Resource and Energy Economics* 35(2): 113–134.

Robinson, E.J., J.C. Williams, and H.J. Albers. 2002. 'The Influence of Markets and Policy on Spatial Patterns of Non-Timber Forest Product Extraction.' *Land Economics* 78(2): 260–271.

Schill, C., T. Lindahl, and A-S. Crépin. 2015. 'Collective Action and the Risk of Ecosystem Regime Shifts: Insights from a Laboratory Experiment.' *Ecology and Society* 20(1): 48.

Shogren, J.F., and L.O. Taylor. 2008. 'On Behavioral-environmental Economics.' *Review of Environmental Economics and Policy* 2(1): 26–44.

Skiba, A.K. 1978. 'Optimal Growth with a Convex-concave Production Function.' *Econometrica* 46(3): 527–539.

Smith, M.D., J.N. Sanchirico, and J.E. Wilen. 2009. 'Economics of Spatial-dynamic Processes: Applications to Renewable Resource Use.' *Journal of Environmental Economics and Management* 57(1): 104–121.

Tavoni, A., M. Schlüter, and S. Levin. 2012. 'The Survival of the Conformist: Social Pressure and Renewable Resource Management.' *Journal of Theoretical Biology* 299: 152–161.

Van der Heide, C.M., J.C. van den Bergh, and E.C. van Ierland. 2005. 'Extending Weitzman's Economic Ranking of Biodiversity Protection: Combining Ecological and Genetic Considerations.' *Ecological Economics* 55(2): 218–223.

Watson, J.E., H.S. Grantham, K.A. Wilson, and H.P. Possingham. 2011. 'Systematic Conservation Planning: Past, Present and Future.' In *Conservation Biogeography*, edited by R.J. Ladle and R.J. Whittaker, 136–160. Hoboken: John Wiley & Sons.

Wegner, G., and U. Pascual. 2011. 'Cost-benefit Analysis in the Context of Ecosystem Services for Human Well-being: A Multidisciplinary Critique.' *Global Environmental Change* 21(2): 492–504.

30

Flow and impact analysis

Lisa Deutsch[1,2] *and Max Troell*[1,3]

[1]STOCKHOLM RESILIENCE CENTRE, STOCKHOLM UNIVERSITY, STOCKHOLM, SWEDEN
[2]NORDIC INSTITUTE FOR LATIN AMERICAN STUDIES, STOCKHOLM UNIVERSITY, STOCKHOLM, SWEDEN
[3]BEIJER INSTITUTE OF ECOLOGICAL ECONOMICS, THE ROYAL SWEDISH ACADEMY OF SCIENCES, STOCKHOLM, SWEDEN

Key methods discussed in this chapter

Physical trade flows, multi-regional input-output analysis, environmentally extended multi-regional input-output analysis, environmental footprints, Life Cycle Assessment, energy return on investment, multi-scale integrated analysis of societal and ecosystem metabolism, global commodity chain analysis

Connections to other chapters

Methods for flow and impact analysis can require input from, be used as input into, be combined with and/or contain similar data as several other methods in this book. Environmental footprints and Life Cycle Assessment (LCA) studies, for example, need the ecological data generated from methods in ecological field data collection (Chapter 6) to convert material flows into environmental footprints and establish environmental relevance and impact for LCAs. Physical trade flows can be used as input into comparative case study analysis (Chapter 20), statistical analysis (Chapter 18), dynamical systems modelling (Chapter 26) and livelihood analysis (Chapter 32). Moreover, network analysis (Chapter 23) can be combined with physical trade flows to discern social-ecological linkages across scales. Also, related or similar information can be found in historical assessments (Chapter 25), while environmental footprints and some LCA impact categories are related to spatial mapping and analysis (Chapter 24).

Introduction

Flow and impact analysis is mainly used to measure and monitor how ecosystems are linked to and support human well-being and are affected by human–nature interactions. Flows are composed of inputs (e.g. energy and material resources), outputs (products and services produced) and wastes (e.g. emissions associated with production). In the conceptual causal chain from natural resource flows to impacts, extraction, production and emissions flows affect various biophysical structures or processes. This in turn has effects

DOI: 10.4324/9781003021339-36

SUMMARY TABLE: FLOW AND IMPACT ANALYSIS

DISCIPLINARY BACKGROUND	KNOWLEDGE TYPE
The methods in this chapter are derived from or have most commonly been used in: Human, Political and Systems Ecology, Cultural and Physical Geography, Physics, Geology, Chemistry, Hydrology, Ecological Economics, Political Economy, Sociology	*The methods in this chapter are primarily used to generate the following types of knowledge:* • Descriptive • Explanatory • Prescriptive
RESEARCH APPROACH	**PURPOSE OF METHOD**
The methods in this chapter originate from or most commonly adopt the following research approaches: • Analytical/objective • Interpretive/subjective	*The most common purposes of using the methods in this chapter are:* • Data collection/generation • System understanding • Policy/decision support
TEMPORAL DIMENSION	**SYSTEMIC FEATURES AND PROCESSES**
The methods in this chapter are most commonly applied to the following temporal dimensions: • Present (typically within the last 5–10 years) • Recent past (post-1700s) **SPATIAL DIMENSION** *The methods in this chapter are primarily either or both:* • Non-spatial • Explicitly spatial *The methods in this chapter are most commonly applied at the following spatial scales:* • Local • Regional (provincial/state to continental) • Global • Multiple places/sites around the world	*While most methods can do many things, the methods in this chapter are particularly good (i.e. go-to methods) for addressing the following:* • Social-ecological dependence and impact • Multiple scales and levels or cross-level interactions • Social-ecological interactions over time • Evaluating policy options

(e.g. reduction of ecosystem function and provision of ecosystem services) that result in various impacts on both ecological and social systems (e.g. human well-being). Studies can be performed over an entire product life cycle, from production and processing to consumption and waste. However, not all studies encompass the entire cycle, including impact analysis.

This large and diverse group of methods shares a common family tree with roots in both social and natural science (Haberl et al. 2019). In the natural sciences, methods arose from a variety of disciplinary fields, e.g. biology, systems ecology and hydrology (Falkenmark 2003). The aim was to collect empirical data on fundamental biophysical processes and stocks (e.g. energy: Odum 1971), nutrient cycles (nitrogen: Galloway et al. 2014) and other 'embodied' natural resources or affected systems tied to the impacts of human activities on the biosphere. These studies did not focus on natural cycles, but on resource flows both directly and indirectly related to human activities. Deutsch et al. (2010), for example, quantified global hydrological flows to give context to the share of embodied water in livestock production for human consumption.

Studies based in social science originally focused on the role of resource use for societal development, e.g. industrial metabolism (Ayres 1994). Methods also arose in different disciplines as researchers analysed, for example, unequal ecological exchange (Martinez-Alier 2002) and global commodity chains (Gereffi 2018) based in world systems theory (Hopkins and Wallerstein 1977). Today, flow and impact studies are common in the interdisciplinary fields of ecological economics (Jansson 1991), human ecology (Rees 1992), political-economic sociology (Hirschman 1980) and political ecology (Peterson 2000).

Flow and impact studies include a large variety and number of different methods that can be deployed at multiple spatial scales to track the accelerating use of natural and social resources. There are methods focused at farm level (e.g. water use for animal feed (Ran et al. 2013)), at the national scale (e.g. the ecological footprint of Swedish land use for food consumption (Deutsch and Folke 2005)), at regional level (e.g. the environmental footprint of EU consumption of biomass for non-food purposes (Bruckner et al. 2019)) and at the global scale (e.g. 24 indicators depicting the Great Acceleration (Steffen et al. 2015a); nine proposed planetary boundaries (Steffen et al. 2015b)).

Studies can use single methods or metrics, such as household incomes, corporate market values, GDP, human appropriation of land, net primary productivity, fisheries catch or energy use, to name just a few. Studies also use methods that combine several metrics to generate multi-dimensional analyses of multiple resource flows into and out of socio-economic systems and emissions (e.g. carbon footprints). Methods can show where resources are used and/or emitted (e.g. multi-regional input-output models) and what the impacts are (e.g. impact categories in LCAs). Moreover, there are methods that are now capable of accounting for upstream flows. This enables the analysis of not only the entire supply chain and product life cycle but also the unintended or indirect spill-over effects into other systems than those included as direct producers, processors and consumers in the studied social-ecological systems (SES) (Godar et al. 2015).

This chapter describes some of the main research approaches and gives examples of methods used for flow and impact analysis. The majority of methods generate descriptive accounting studies for decision support. Flow and impact methods can track natural resource and human capital flows both directly and indirectly related to human activities, and associated direct and indirect impacts. Studies can track changes over time and space and simultaneously at different scales. Some studies are change-oriented and explicitly intend to communicate information (e.g.

environmental footprint and LCA-based eco-labelling/certification) and even prescribe preferred options (environmental footprint and LCA). Whereas the methods are mostly analytical, there is certainly some subjective interpretation (e.g. LCA has an 'interpretation' phase where information from the life cycle inventory and/or the life cycle impact assessment is evaluated).

SES problems and questions

Flow and impact analysis methods can have widely differing goals and uses, from global-level scanning (planetary boundaries), to equitable sharing of resources between nations (environmental footprints), to product-level management tools (LCAs). They are mainly used to support decision- and policymaking and monitoring efforts. They can also be used in different phases of problem solving: (a) problem identification and agenda setting, by influencing the worldviews of consumers and policymakers (e.g. environmental footprints), (b) design of policy tools or policies (e.g. an LCA of public-school food procurement), and (c) monitoring, for feedbacks and learning on the impacts of different production methods (e.g. an LCA of salmon farming).

The following are examples of problems and questions typically addressed by some of the main methods discussed in this chapter:

- Are producers and consumers able to escape local biophysical ecosystem limits using trade? (e.g. can cities maintain food security with food trade? (physical trade flows: Porter et al. 2014))
- What is the level of diversity of animal feed supply sources for Thai jumbo shrimp aquaculture production? (physical trade flows: Deutsch et al. 2007; also see Case study 30.1)
- Are consumers able to dislocate the environmental impacts of their consumption to production sites abroad? And who is affected globally? (physical trade flows: Godar et al. 2015)
- Can water-scarce countries 'save' water at global scale by importing products from countries where water is abundant? (environmental footprints: Hoekstra and Mekkonen 2012)
- How do global value chains affect a country's own environment? (environmental footprints: Deutsch and Folke 2005)
- What are the greenhouse gas emissions per kilogram of fish produced in an intensive salmon aquaculture production system? (LCA: Newton and Little 2018)
- Which part of the product life cycle of salmon aquaculture production requires the most energy use: manufacture of feed, production site or processing? (LCA: Pelletier et al. 2009)
- Which product is most sustainable based on predefined sustainability criteria? (LCA)

Brief description of key methods

The largest group of methods used for flow and impact analyses falls into the category of material- and energy-flow accounting (MEFA) methods. This group includes material-flow accounting approaches that do not incorporate energy, and energy-flow accounting methods that focus specifically on energy flows. Material- and energy-flow accounting methods can assess the correlation between physical and monetary growth, e.g. if and how water use is related to GDP. Material-flow accounting approaches track biophysical material flows both directly and indirectly related to human activities. Energy-flow accounting illuminates the essential role that energy plays in our economy.

There are two broad ways to track material and energy flows. The 'bottom-up' approach uses physical flows, whereas the 'top-down' approach traces resource flows through economic sectors with monetary input-output tables. Six common MEFA methods are described in this chapter:

1. Physical trade flows (PTF) typically map and link resource flows in material and spatial terms and can help to discern levels of dependence on different ecosystems for production and consumption.
2. Large international input-output (IO) databases enable multi-regional input-output (MRIO) analyses between nations and can account for bilateral trade relationships between several countries.
3. Environmentally extended multi-regional input-output (EE-MRIO) models complement monetary sectoral data with environmental and social indicators.
4. Environmental footprints (EF) can quantify levels of dependence on different types of ecosystem support (e.g. freshwater), illuminate support that is located in different places and linked through trade, and balance local, national and global resource budgets.
5. Life Cycle Assessment (LCA) quantifies environmental impacts associated with products, processes or activities; the main environmental impacts investigated in LCA are greenhouse global warming potential, water use, energy consumption, acidification and abiotic depletion.
6. Energy return on investment (EROI) is the ratio of the amount of usable energy obtained to the amount of energy required to obtain that energy, where the focus is not on the total amount of primary energy used, but on the energy gained by society.

Another type of flow and impact method is multi-scale integrated analysis of societal and ecosystem metabolism (MuSIASEM). This integrated accounting method simultaneously represents the metabolic pattern of social, economic, socio-metabolic and biophysical flows (e.g. labour, value added, energy use) and their interrelations with the complex system (society) interacting with its environment, e.g. the water-energy-food nexus.

Lastly, there are several 'global chain methods' that can analyse international trade and production networks of a specific commodity and map how people, places and processes are interlinked in the global economy. Global commodity chain (GCC) analysis maps what resources are produced and consumed, who governs, wields power and accumulates capital within the entire commodity chain from production to consumption, and where these processes occur. Table 30.1 provides a summary of key methods used in flow and impact analysis.

Whereas most material-flow and energy-flow accounting studies cover very recent years, with some exceptions from the 1900s (Krausmann et al. 2009) and the 1960s, many global chain studies are historical (Marichal, Topik, and Frank 2006; Topik 2008). Furthermore, some studies have used a single method but many studies combine multiple methods and indicators in an effort to yield more comprehensive inventories and impact assessments. Physical trade flows, for example, can be used to quantify the material flows needed for input into environmental footprint calculations of land areas, which could then be used as an input into LCA inventories.

Table 30.1 Summary of key methods used in flow and impact analysis

Methods	*Description*	*References*
Physical trade flows	Physical trade flow (PTF) methods can trace the size and composition of traded goods, i.e. imports and exports. Methods exclude domestic production and total consumption or use. PTF focuses only on goods moved in/outside relevant borders (e.g. nation, city). PTF can use: • National and international physical trade flow data • Adjusted PTF data (e.g. with bilateral trade flow matrices) • Input-output tables (alone or as input into multi-regional input-output models) • A hybrid of sources, especially if one source is incomplete	*Key introductory text* Kastner, Kastner, and Nonhebel 2011 *Applications to SES* Deutsch et al. 2007; Porter et al. 2014; Godar et al. 2015; Gephart et al. 2017; trase.earth
Multi-regional input-output anlysis Environmentally extended multi-regional input-output analysis	Today, large international input-output (IO) databases enable multi-regional input-output (MRIO) analyses between nations and can account for bilateral trade relationships between several countries. MRIO can use: • Physical input-output models (e.g. based on FAOSTAT) • Monetary input-output models (e.g. EXIOBASE) • A hybrid of sources, especially if one source is incomplete (e.g. FABIO) Environmentally extended multi-regional input-output (EE-MRIO) models can complement monetary sectoral data with environmental and social indicators.	*Key introductory text* Haberl et al. 2019 *Applications to SES* Kastner et al. 2014; Stadler et al. 2018 (EXIOBASE 3); Bruckner et al. 2019 (FABIO); Kummu et al. 2020
Environmental footprints	Environmental footprints (EF) calculate the biological capacity needed to produce materials and/or assimilate waste from a given population's consumption of products (i.e. domestic production + imports – (re-)exports). Environmental footprints are calculated in two main ways: • Bottom-up based on physical flows • Top-down using (MR)IO Material flows are then combined with other methods, e.g. hydrological data on crop-water use to quantify the freshwater needed to generate a product or service. There are several footprints (e.g. ecological footprint of land areas, water footprint/virtual water, nitrogen and carbon footprints).	*Key introductory text* Wiedmann and Lenzen 2018 *Applications to SES* Deutsch and Folke 2005 (land and sea areas); Hoekstra and Mekkonen 2012 (water); Galloway et al. 2014 (nitrogen)

(*Continued*)

Table 25.1 (Continued)

Methods	*Description*	*References*
Life Cycle Assessment	Life Cycle Assessment (LCA) aims to be a comprehensive impact assessment tool for researchers and policymakers. LCA begins at the 'cradle' (point of primary resource extraction) of a product or service and extends along the supply chain to encompass all life-cycle stages. LCA can identify and assess the environmental impacts associated with a product, process or activity by quantifying raw materials, energy and waste released into the air, water and soil. Chosen impact categories vary widely, but commonly include resource depletion and emissions-related environmental concerns and toxicological potentials (e.g. greenhouse global warming potential, water footprint, energy consumption, acidification, abiotic depletion). Unlike many other biophysical accounting tools, LCA is internationally standardised (ISO 14040-14044).	*Key introductory text* Curran 2012 *Applications to SES* Pelletier et al. 2009; Henriksson et al. 2018; Newton and Little 2018
Energy return on investment	Energy return on investment (EROI) is the ratio of the amount of usable energy (i.e. the ratio of energy obtained to the amount of energy required to obtain that energy). The main focus is not the total amount of primary energy used, but the energy gained by society.	*Key introductory text* Odum 1971 *Applications to SES* Hall 2011; Pelletier et al. 2011
Multi-scale integrated analysis of societal and ecosystem metabolism	Multi-scale integrated analysis of societal and ecosystem metabolism (MuSIASEM) integrates social, economic and socio-metabolic and biophysical flows (e.g. labour, value added, energy use). Although it can use some of the MEFA methods, studies are typically context specific.	*Key introductory text* Gerber and Scheidel 2018 *Applications to SES* Giampietro and Bukkens 2014
Global commodity chains	Global commodity chain (GCC) methods can map not only what resources are produced and consumed and where, but also who governs, wields power and accumulates wealth or capital within the entire commodity chain, from production to commercialisation and consumption. GCC also identifies the external institutional context in which the chain operates.	*Key introductory text* Gereffi 2018 *Applications to SES* Topik 2008; Gereffi 2018; World Bank 2019

Limitations

Flow and impact analysis methods alone do not fully integrate human and ecological dimensions and do not account for many interdependencies between social-ecological processes. The most obvious shortcoming is that social-ecological researchers performing flow and impact studies must often use methods that were not fundamentally designed to analyse SES. Instead, they often use and adapt methods with a singular approach, e.g. an approach designed for a capitalist market system that solely tracks indicators of economic value (GDP, market value, commodity prices).

Moreover, data that are most readily available for use in flow and impact studies may not be available at the appropriate social and ecological scales. The specific location of where resources originate matters for ecosystems and societies, particularly in terms of understanding who benefits from resource extraction, who governs resource use and what ecosystem effects are experienced. To be able to discern the social-ecological effects of soybean production in the Brazilian *Cerrado* biome, for example, exports, ownership, land-cover changes, tax policies and company profits must all be traceable to enable the linking of specific products to specific agro-ecosystems and particular actors such as land owners, land managers, service contractors, processors or multi-national companies supplying seeds. Most nations do not track flows and impacts at subnational level and states allow corporations to hinder traceability of products to landscapes, producers, inputs and profits as proprietary knowledge. Thus, if data are not available or accessible, studies must upscale (or downscale) data to another level with unknown relevance, e.g. most studies of food consumption at the city level are based on adjusted national-level consumption statistics (Porter et al. 2014).

Social-ecological systems researchers often combine several methods and data sources to adequately reflect the current complexity of flows and the multi-dimensional nature of effects (e.g. how to combine toxicity studies related to chemical use and pollution at the individual insect species level with national-level data on volumes of pesticide use). To modify data for relevancy, researchers also use additional methods (e.g. expert interviews) to gather additional data (e.g. particular species of fish) or conversion factors (e.g. city-level consumption). This is time and resource intensive (e.g. combining LCA studies (ecological effects of different production methods) with MRIO is very data intensive) and requires expertise in several methods.

If methods or data cannot be meaningfully modified, it places significant limitations on the conclusions that can be drawn. For instance, although environmental footprints have successfully illuminated the scale of the social-ecological impacts of the human enterprise for the general public and decision-makers, it is doubtful whether certain flows and impacts (e.g. the use of chemicals and biodiversity loss) can actually be meaningfully converted into environmental footprints (e.g. land or sea areas). As for LCA, while modern software with built-in inventory databases and impact assessment methods has simplified the LCA process, the rigour of these models is highly dependent on data quality. The use of the generic data available in many public and commercial life-cycle inventory databases may therefore provide a starting point for scoping analyses, but more context-specific data are required for robust modelling of specific production systems and technologies. Moreover, there is a need for standardisation of methodologies with respect to key model assumptions, scope and allocation methods.

Data availability is a major challenge for many flow and impact studies. In addition to the need for social-ecological relevance and the unavailability of data at relevant scales discussed above, there are general issues of: (a) data quality, (b) access fees for quality data, (c) restricted access, and (d) a paucity of historical data. Data on waste and emissions are also highly fragmentary, thus limiting studies that wish to include these or to get a full mass balance accounting (Krausmann et al. 2017).

There are many national statistical bureaus and quite a few international sources for data. The United Nations (UN) Food and Agriculture Organization has several free online physical trade flow databases based on self-reported national data, e.g. FAOSTAT and FISHBASE. The UN Environment Programme (UNEP) has a material flow accounting database. The OECD has an input-output database, and the World Bank, EUROSTAT and COMTRADE also have databases. Although these sources are sometimes inaccurate, they are

Case study 30.1: The dependence of intensive shrimp and salmon aquaculture production in Thailand and Norway on access to fishmeal imports from South American marine ecosystems

Seafood is the most popular food commodity traded in the world, with aquaculture (or farmed seafood) being the fastest growing food sector globally (Troell et al. 2014). In fact, every other bite of fish we take today is from aquaculture. Jumbo shrimp and salmon are two of the most economically valuable aquaculture products (by weight) and Thailand and Norway are dominant producer countries of shrimp and salmon, respectively. Although non-fish substitutes are emerging, both of these products still depend on fishmeal as a key feed ingredient.

In a study by Deutsch et al. (2007), global production, trade and consumption of fishmeal for shrimp and salmon aquaculture production were traced from 1980 to 2000 as the aquaculture industries developed in Thailand and Norway. Despite the two nations having very different socio-economic and cultural backgrounds, social-ecological resource bases and geographic locations, and producing two entirely different products, there were some significant similarities:

- Shrimp and salmon aquaculture use the same fishmeal in feed pellets. Moreover, these are similar to chicken and pig feeds, so aquaculture can now be likened to 'aquatic livestock'.
- Thailand and Norway were both able to increase production (i.e. escape the limitations of own local fishing waters) and expand production and export through access to global markets and trade.
- There was no 'north-south' difference in supply sources of fishmeal. Shrimp farms in Thailand imported fishmeal from the same marine ecosystem as Norwegian salmon farms.

The study revealed that key products and producers are highly dependent on the same marine ecosystem, namely the south-east Pacific Ocean, for their key input, fishmeal. Using FISHBASE, fishmeal use and physical trade flows were tied to actual species of wild fish and linked to the locations of marine ecosystems needed to produce them, and tied to agrofood production systems (aquaculture). Thus, the physical trade flows study illuminated invisible marine subsidies in feed production and patterns of ecosystem support and flows.

The fact that the south-east Pacific Ocean (via Peru and Chile) supplies much of the world with fishmeal is an economic and ecological vulnerability, with both a reduced diversity of suppliers and increased pressure on a single marine area. Most people do not realise the dependence of aquaculture on fishmeal derived from distant marine ecosystems. Furthermore, although volumes of fish imports were traced, the study could not separate fishmeal use in aquaculture from other animal production systems

often the only available or free data sources. More accurate, high-resolution data are gathered by private consultants and trade associations, but have (sometimes exorbitant) access fees. Thus, for some studies, there may be no or limited access to data on major actors with the most power in the global chain. Furthermore, since free-trade zones exist, some trade data completely lose

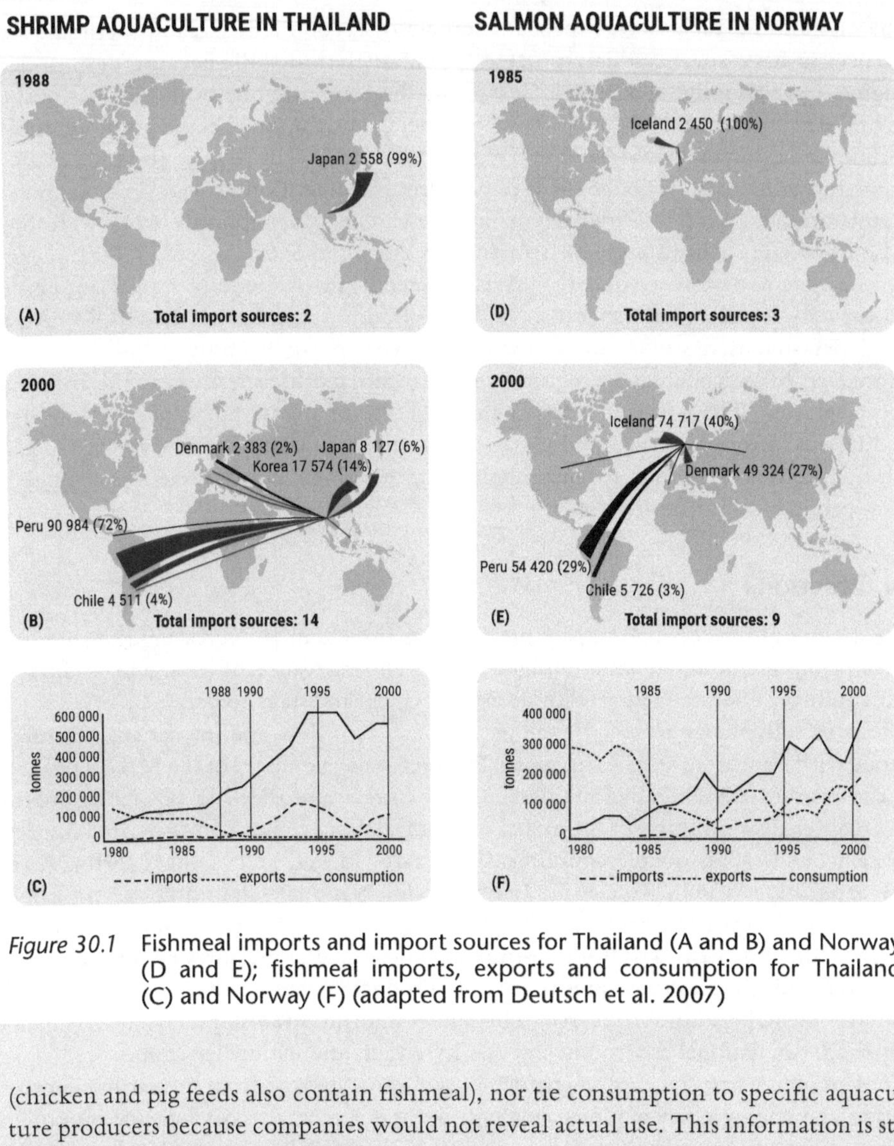

Figure 30.1 Fishmeal imports and import sources for Thailand (A and B) and Norway (D and E); fishmeal imports, exports and consumption for Thailand (C) and Norway (F) (adapted from Deutsch et al. 2007)

(chicken and pig feeds also contain fishmeal), nor tie consumption to specific aquaculture producers because companies would not reveal actual use. This information is still not available today (Fry et al. 2016).

Figure 30.1 shows the imported fishmeal and import sources for Thailand in 1988 (A) and 2000 (B) and for Norway in 1985 (D) and 2000 (E); and fishmeal imports, exports and consumption from 1980 to 2000 for Thailand (C) and Norway (F). Fishmeal amounts are in metric tonnes and numbers in parentheses denote percentages of total imports.

their origin, e.g. links to production systems. The current institutional framework for trade, i.e. the World Trade Organization, does not allow trade restrictions on the basis of production methods and data have therefore not been traced in this way. Another widespread problem is a lack of historical data prior to the 1960s, especially at larger scales, in digital format or in several languages, e.g. major databases like FAOSTAT start in 1961.

Resource implications

All the methods in Table 30.1, except for some global commodity chain methods, require researchers to have advanced data-processing and analytical skills because desired datasets are digital, large and highly detailed. Certain methods require one or more areas of specific expertise (e.g. an ecological economist who understands the links between biophysical and economic flows, such as MuSIASEM). To gather data may thus require multiple skills from the same researcher or a skilled team of researchers and practitioners. It takes time, resources and trust to build an interdisciplinary team of researchers that not only work well together but also transcend their disciplines, truly inform one another and reach a joint understanding of the system. The sheer quantity of data needed can also require significant computer server capacity for data processing (e.g. adjusted physical trade flows, EE-MRIO). All these detailed, data-intensive studies are resource intensive and can be costly.

Moreover, historical analyses require rigorous source criticism on how the information has been selected. These analyses may require the use of a wide range of literature dealing with different historical epochs and places, and representing divergent perspectives and research traditions (e.g. agronomy, anthropology, archaeology, climate science, ecology, economics, geography and/or history).

New directions

Future directions for flow and impact methods are threefold: (a) increasing the comprehensiveness of the methods, (b) improving access to and relevance of data, and (c) increasing transdisciplinary efforts together with decision-makers and stakeholders.

Efforts to fully integrate social-ecological aspects into flow and impact analysis methods continue with some noteworthy successes. The recent development of the MRIO framework with environmentally extended multi-regional input-output (EE-MRIO) tables provides a more comprehensive linking of the global economy, labour inputs and associated impacts on ecosystems, and avoids double counting (Cabernard, Pfister, and Hellweg 2019). A recent policy application is Policy-Relevant Indicators for National Consumption and Environment (PRINCE) (prince-project.se). PRINCE links Swedish national input-output tables with EXIOBASE (an EE-MRIO) to estimate environmental pressures of consumption and production, and allocates those pressures to 60 'product groups' to show where in the world the environmental pressures occurred. The model informs Swedish policy development and monitoring, e.g. national environmental quality targets and national accounts.

Similarly, LCA has, since its emergence in the 1970s, moved from primarily being a tool for waste and energy-efficiency management to a more general eco-efficiency measurement decision-support framework. New developments in the LCA ISO standard include additional environmental as well as economic and social aspects (e.g. human health), and deepening its analytic capacity (e.g. considering behavioural aspects in the inventory modelling). However, it is debatable whether there is a limit to how comprehensive some methods or approaches can be.

Improvements in access to and the relevance of data are increasing the capability to trace flows and impacts at disaggregated levels spatially, temporally and physically. A notable application based on cooperation between researchers and corporate stakeholders is Spatially Explicit Information on Production to Consumption Systems (SEI-PCS) (Godar et al. 2015). This application traces material-flow analysis via the individual companies that export and import specific commodities. SEI-PCS data have been used by the Trase Initiative (Trase 2018) to map supply chains from the local subnational production regions through trading companies all the way to import nations (e.g. Brazilian soybeans, Indonesian palm oil).

To date, flow and impact methods have not been widely developed in collaboration with decision-makers or societal stakeholders. However, inherently subjective aspects have recently been explicitly acknowledged (Pelletier, Bamber, and Brandão 2019) and even encouraged (Einarsson and Cederberg 2019) (e.g. the choice of weighting factors in LCA) in recognition of the need for guidance and transparency with regard to prioritisations that take place in decision-making. There is growing interest in exploring the benefits of a transdisciplinary approach that could address the need for combining methods and engaging skilled teams of researchers and practitioners in flow and impact studies.

Key readings

Goedkoop, M., R. Heijungs, M.A.J. Huijbregts, A. de Schryver, J. Struijs, and R. van Zelm. 2009. 'ReCiPe 2008 – A Life Cycle Impact Assessment Method which Comprises Harmonised Category Indicators at the Midpoint and the Endpoint Level.' Report I: Characterisation. file:///C:/Users/User/Downloads/ReCiPe_main_report_MAY_20131.pdf.

Kastner, T., M. Kastner, and S. Nonhebel. 2011. 'Tracing Distant Environmental Impacts of Agricultural Products from a Consumer Perspective.' *Ecological Economics* 70: 1032–1040.

Krausmann, F., S. Gingrich, N. Eisenmenger, K-H. Erb, H. Haberl, and M. Fischer-Kowalski. 2009. 'Growth in Global Materials Use, GDP and Population During the 20th Century.' *Ecological Economics* 68(10): 2696–2705.

Marichal, C., S. Topik., and Z. Frank. 2006. 'Commodity Chains and Globalization in Historical Perspective.' In *From Silver to Cocaine: Latin American Commodity Chains and the Building of The World Economy, 1500–2000*, edited by S. Topik, Z.L. Frank, and C. Marichal, 1–24. Durham: Duke University Press.

Monfreda, C., M. Wackernagel, and D. Deumling. 2004. 'Establishing National Natural Capital Accounts Based on Detailed Ecological Footprint and Biological Capacity Assessments.' *Land Use Policy* 21(3): 231–246. www.sciencedirect.com/science/article/B6VB0-4BNVWBB-1/2/e5a72c38dc783550ac0b0a6fd625ad4e.

Acknowledgements

The authors would like to thank Michelle Audouin for getting this chapter started and the critical enthusiasm of our discussions. Also, thanks are due to Tiina Häyhä for her advice on disentangling the multitude of methods. Lisa Deutsch's work was partially financed by the Swedish Research Council grant 'Agrofood Globalization through the Lens of the Soybean'.

References

Ayres, R.U. 1994. *Industrial Metabolism: Theory and Policy, the Greening of Industrial Ecosystems*. Washington: National Academy Press.

Bruckner, M., T. Häyhä, S. Giljum, V. Maus, G. Fischer, S. Tramberend, and J. Börner. 2019. 'Quantifying the Global Cropland Footprint of the European Union's Non-Food Bioeconomy.' *Environmental Research Letters* 14(4): 45011. doi:10.1088/1748-9326/ab07f5.

Cabernard, L., S. Pfister, and S. Hellweg. 2019. 'A New Method for Analyzing Sustainability Performance of Global Supply Chains and Its Application to Material Resources.' *Science of The Total Environment* 684: 164–177. doi:10.1016/j.scitotenv.2019.04.434.

Curran, M.A., ed. 2012. *Life Cycle Assessment Handbook: A Guide for Environmentally Sustainable Products*. Hoboken: John Wiley & Sons. doi:10.1002/9781118528372.

Deutsch, L., M. Falkenmark, L.J. Gordon, J. Rockström, and C. Folke. 2010. 'Water-mediated Ecological Consequences of Intensification and Expansion of Livestock Production.' In *Livestock in a Changing Landscape*, Volume 1, edited by H. Steinfeld, H. Mooney, F. Schneider, and L. Neville, 97–110. Washington, DC: Island Press. www.fao.org/3/a-am074e.pdf.

Deutsch, L., and C. Folke. 2005. 'Ecosystem Subsidies to Swedish Food Consumption from 1962 to 1994.' *Ecosystems* 8(5): 512–528. doi:10.1007/s10021-005-0035-4.

Deutsch, L., S. Gräslund, C. Folke, M. Huitric, N. Kautsky, M. Troell, and L. Lebel. 2007. 'Feeding Aquaculture Growth through Globalization: Exploitation of Marine Ecosystems for Fishmeal.' *Global Environmental Change* 17(2): 238–249. doi:10.1016/j.gloenvcha.2006.08.004.

Einarsson, R., and C. Cederberg. 2019. 'Is the Nitrogen Footprint Fit for Purpose? An Assessment of Models and Proposed Uses.' *Journal of Environmental Management* 240: 198–208. doi:10.1016/j.jenvman.2019.03.083.

Falkenmark, M. 2003. 'Freshwater as Shared between Society and Ecosystems: From Divided Approaches to Integrated Challenges.' *Philosophical Transactions of the Royal Society B: Biological Sciences* 358(1440): 2037–2049. doi:10.1098/rstb.2003.1413.

Fry, J.P., D.C. Love, G.K. MacDonald, P.C. West, P.M. Engstrom, K.E. Nachman, and R.S. Lawrence. 2016. 'Environmental Health Impacts of Feeding Crops to Farmed Fish.' *Environment International* 91: 201–214. doi:10.1016/j.envint.2016.02.022.

Galloway, J.N., W. Winiwarter, A. Leip, A.M. Leach, A. Bleeker, and J.W. Erisman. 2014. 'Nitrogen Footprints: Past, Present and Future.' *Environmental Research Letters* 9(11): 115003. doi:10.1088/1748-9326/9/11/115003.

Gephart, J.A., L. Deutsch, M.L. Pace, M. Troell, and D.A. Seekell. 2017. 'Shocks to Fish Production: Identification, Trends, and Consequences.' *Global Environmental Change* 42. doi:10.1016/j.gloenvcha.2016.11.003.

Gerber, J-F., and A. Scheidel. 2018. 'In Search of Substantive Economics: Comparing Today's Two Major Socio-metabolic Approaches to the Economy – MEFA and MuSIASEM.' *Ecological Economics* 144: 186–194. doi:10.1016/j.ecolecon.2017.08.012.

Gereffi, G. 2018. *Global Value Chains and Development – Redefining the Contours of 21st Century Capitalism*. New Delhi: Cambridge University Press.

Giampietro, M., and S.G. Bukkens. 2014. 'The Multi-scale Integrated Analysis of Societal and Ecosystem Metabolism.' In *Resource Accounting for Sustainability Assessment*, edited by M. Giampietro, R. Aspinall, J. Ramos-Martin, and S.G.F. Bukkens, 33–43. Abingdon: Routledge.

Godar, J.U., U.M. Persson, E.J. Tizado, and P. Meyfroidt. 2015. 'Towards More Accurate and Policy Relevant Footprint Analyses: Tracing Fine-scale Socio-environmental Impacts of Production to Consumption.' *Ecological Economics* 112: 25–35. doi:10.1016/j.ecolecon.2015.02.003.

Haberl, H., D. Wiedenhofer, S. Pauliuk, F. Krausmann, D.B. Müller, and M. Fischer-Kowalski. 2019. 'Contributions of Sociometabolic Research to Sustainability Science.' *Nature Sustainability* 2(3): 173–184. doi:10.1038/s41893-019-0225-2.

Hall, C.A.S. 2011. 'Introduction to Special Issue on New Studies in EROI (Energy Return on Investment).' *Sustainability* 3(10): 1773–1777. doi:10.3390/su3101773.

Henriksson, P.J.G., B. Belton, K. Murshed-E-Jahan, and A. Rico. 2018. 'Measuring the Potential for Sustainable Intensification of Aquaculture in Bangladesh Using Life Cycle Assessment.' *Proceedings of the National Academy of Sciences of the United States of America* 115(12): 2958–2963. doi:10.1073/pnas.1716530115.

Hirschman, A. 1980 (originally published 1945). *National Power and the Structure of Foreign Trade*. Berkeley: University of California Press.

Hoekstra, A.Y., and M.M. Mekonnen. 2012. 'The Water Footprint of Humanity.' *Proceedings of the National Academy of Sciences of the United States of America* 109(9): 3232–3237. doi:10.1073/pnas.1109936109.

Hopkins, T.K., and I. Wallerstein. 1977. 'Patterns of Development of the Modern World-System.' *Review (Fernand Braudel Center)* 1(2): 111–145.

Jansson, A.M. 1991. 'Ecological Consequences of Long-term Landscape Transformations in Relation to Energy Use and Economic Development.' In *Linking the Natural Environment and the Economy*, edited by C. Folke and T. Kåberger, 97–110. London: Kluwer Academic Publishers.

Kastner, T., M. Kastner, and S. Nonhebel. 2011. 'Tracing Distant Environmental Impacts of Agricultural Products from a Consumer Perspective.' *Ecological Economics* 70(6): 1032–1040.

Kastner, T., A. Schaffartzik, N. Eisenmenger, K-H. Erb, H. Haberl, and F. Krausmann. 2014. 'Cropland Area Embodied in International Trade: Contradictory Results from Different Approaches.' *Ecological Economics* 104: 140–144. doi:10.1016/j.ecolecon.2013.12.003.

Krausmann, F., S. Gingrich, N. Eisenmenger, K-H. Erb, H. Haberl, and M. Fischer-Kowalski. 2009. 'Growth in Global Materials Use, GDP and Population during the 20th Century.' *Ecological Economics* 68(10): 2696–2705.

Krausmann, F., H. Schandl, N. Eisenmenger, S. Giljum, and T. Jackson. 2017. 'Material Flow Accounting: Measuring Global Material Use for Sustainable Development.' *Annual Review of Environment and Resources* 42(1): 647–765. doi:10.1146/annurev-environ-102016-060726.

Kummu, M., P. Kinnunen, E. Lehikoinen, M. Porkka, C. Queiroz, E. Röös, M. Troell, and C. Weil. 2020. 'Interplay of Trade and Food System Resilience: Gains on Supply Diversity over Time at the Cost of Trade Independency.' *Global Food Security* 24: 100360. doi:10.1016/j.gfs.2020.100360.

Marichal, C., S. Topik., and Z. Frank. 2006. 'Commodity Chains and Globalization in Historical Perspective.' In *From Silver to Cocaine: Latin American Commodity Chains and the Building of The World Economy*, 1500–2000, edited by S. Topik, Z.L. Frank, and C. Marichal, 1–24. Durham: Duke University Press.

Martinez-Alier, J. 2002. *The Environmentalism of the Poor: A Study of Ecological Conflicts and Valuation*. Cheltenham: Edward Elgar.

Newton, R.W., and D.C. Little. 2018. 'Mapping the Impacts of Farmed Scottish Salmon from a Life Cycle Perspective.' *The International Journal of Life Cycle Assessment* 23(5): 1018–1029. doi:10.1007/s11367-017-1386-8.

Odum, H.T. 1971. Environment, Power, and Society. New York: Wiley-Interscience.

Pelletier, N., E. Audsley, S. Brodt, T. Garnett, P. Henriksson, A. Kendall, K.J. Kramer, D. Murphy, T. Nemecek, and M. Troell. 2011. 'Energy Intensity of Agriculture and Food Systems.' *Annual Review of Environment and Resources* 36(1): 223–246. doi:10.1146/annurev-environ-081710-161014.

Pelletier, N., N. Bamber, and M. Brandão. 2019. 'Interpreting Life Cycle Assessment Results for Integrated Sustainability Decision Support: Can an Ecological Economic Perspective Help Us to Connect the Dots?' *The International Journal of Life Cycle Assessment* 24(9): 1580–1586. doi:10.1007/s11367-019-01612-y.

Pelletier, N., P. Tyedmers, U. Sonesson, A. Scholz, F. Ziegler, A. Flysjo, S. Kruse, B. Cancino, and H. Silverman. 2009. 'Not All Salmon Are Created Equal: Life Cycle Assessment (LCA) of Global Salmon Farming Systems.' *Environmental Science & Technology* 43(23): 8730–8736. doi:10.1021/es9010114.

Peterson, G. 2000. 'Political Ecology and Ecological Resilience: An Integration of Human and Ecological Dynamics.' *Ecological Economics* 35(3): 323–336. doi:10.1016/S0921-8009(00)00217-2.

Porter, J.R., R. Dyball, D. Dumaresq, L. Deutsch, and H. Matsuda. 2014. 'Feeding Capitals: Urban Food Security and Self-provisioning in Canberra, Copenhagen and Tokyo.' *Global Food Security* 3(1). doi:10.1016/j.gfs.2013.09.001.

Ran, Y., L. Deutsch, M. Lannerstad, and J. Heinke. 2013. 'Rapidly Intensified Beef Production in Uruguay: Impacts on Water-Related Ecosystem Services.' *Aquatic Procedia* 1(0): 77–87. doi:10.1016/j.aqpro.2013.07.007.

Rees, W.E. 1992. 'Ecological Footprints & Appropriated Carrying Capacity.' *Environment & Urbanization* 4(2): 121–130.

Stadler, K., R. Wood, T. Bulavskaya, C-J. Södersten, M. Simas, S. Schmidt, A. Usubiaga et al. 2018. 'EXIOBASE 3: Developing a Time Series of Detailed Environmentally Extended Multi-regional Input-output Tables.' *Journal of Industrial Ecology* 22(3): 502–515. doi:10.1111/jiec.12715.

Steffen, W., W. Broadgate, L. Deutsch, O. Gaffney, and C. Ludwig. 2015a. 'The Trajectory of the Anthropocene: The Great Acceleration.' *Anthropocene Review* 2(1). doi:10.1177/2053019614564785.

Steffen, W., K. Richardson, J. Rockström, S.E. Cornell, I. Fetzer, E.M. Bennett, R. Biggs, S.R. Carpenter, W. de Vries, and C.A. de Wit. 2015b. 'Planetary Boundaries: Guiding Human Development on a Changing Planet.' *Science* 347(6223): 1259855.

Topik, S. 2008. 'Historicizing Commodity Chains: Five Hundred Years of the Global Coffee Commodity Chain.' In *Frontiers of Commodity Chain Research*, edited by J. Bair, 37–62. Stanford: Stanford University Press.

Trase. 2018. The Trase Yearbook 2018. Sustainability in Forest-Risk Supply Chains: Spotlight on Brazilian Soy. https://yearbook2018.trase.earth.

Troell, M., R.L. Naylor, M. Metian, M. Beveridge, P.H. Tyedmers, C. Folke, K.J. Arrow et al. 2014. 'Does Aquaculture Add Resilience to the Global Food System?' *Proceedings of the National Academy of Sciences* 111(37): 13257–13263.

Wiedmann, T., and M. Lenzen. 2018. 'Environmental and Social Footprints of International Trade.' *Nature Geoscience* 11(5): 314–321. doi:10.1038/s41561-018-0113-9.

World Bank. 2019. *World Development Report 2020: Trading for Development in the Age of Global Value Chains*. Washington: World Bank.

31

Ecosystem service modelling

Maike Hamann,[1,2,3] Justin A. Johnson,[3,4*] Tomas Chaigneau,[5] Rebecca Chaplin-Kramer,[3,6] Lisa Mandle[6] and Jesse T. Rieb[7]*

[1]CENTRE FOR SUSTAINABILITY TRANSITIONS, STELLENBOSCH UNIVERSITY, STELLENBOSCH, SOUTH AFRICA
[2]HUMPHREY SCHOOL OF PUBLIC AFFAIRS, UNIVERSITY OF MINNESOTA, MINNEAPOLIS, MINNESOTA, USA
[3]NATURAL CAPITAL PROJECT, INSTITUTE ON THE ENVIRONMENT, UNIVERSITY OF MINNESOTA, SAINT PAUL, MINNESOTA, USA
[4]DEPARTMENT OF APPLIED ECONOMICS, UNIVERSITY OF MINNESOTA, SAINT PAUL, MINNESOTA, USA
[5]ENVIRONMENT AND SUSTAINABILITY INSTITUTE, UNIVERSITY OF EXETER, PENRYN, UK
[6]NATURAL CAPITAL PROJECT, WOODS INSTITUTE FOR THE ENVIRONMENT, STANFORD UNIVERSITY, STANFORD, CALIFORNIA, USA
[7]DEPARTMENT OF NATURAL RESOURCE SCIENCES, MCGILL UNIVERSITY, SAINTE-ANNE-DE-BELLEVUE, CANADA
[*]THESE AUTHORS CO-LED THIS CHAPTER

Key methods discussed in this chapter

Decision-support modelling packages: Integrated valuation of ecosystem services and trade-offs (InVEST), artificial intelligence for ecosystem services (ARIES), Co$ting Nature/WaterWorld

Related technical models and frameworks: Integrated assessment models, general equilibrium models, Lund–Potsdam–Jena dynamic global vegetation model, Life Cycle Assessment models

Connections to other chapters

Ecosystem service modelling connects to a number of other methods, since ecosystem service assessments usually form part of a larger decision-support initiative. Systems scoping (Chapter 5) may lay the foundation for an ecosystem service assessment, while ecosystem service models may be developed through participatory modelling and planning (Chapter 13), and feed directly into scenario development (Chapter 11) or a livelihood and vulnerability analysis (Chapter 32). The results of ecosystem service modelling are usually mapped spatially (Chapter 24) and often support decision analysis based on optimisation (Chapter 29).

Introduction

The study of social-ecological systems (SES) is mainly concerned with understanding the interactions between people and nature. Ecosystem services represent an important subset of these interactions. They are the benefits from nature that support and fulfil human life

DOI: 10.4324/9781003021339-37

SUMMARY TABLE: ECOSYSTEM SERVICE MODELLING	
DISCIPLINARY BACKGROUND	**KNOWLEDGE TYPE**
The methods in this chapter are derived from or have most commonly been used in: Ecology, Resource Economics, Computational Geography, Systems Dynamics, Computer Science, Information Science	*The methods in this chapter are primarily used to generate the following types of knowledge:* • Explanatory • Prescriptive
RESEARCH APPROACH	**PURPOSE OF METHOD**
The methods in this chapter originate from or most commonly adopt the following research approaches: • Analytical/objective • Collaborative/process	*The most common purposes of using the methods in this chapter are:* • System understanding • Stakeholder engagement and co-production • Policy/decision support
TEMPORAL DIMENSION	**SYSTEMIC FEATURES AND PROCESSES**
The methods in this chapter are most commonly applied to the following temporal dimensions: • Present (typically within the last 5–10 years) • Recent past (post-1700s) • Future	*While most methods can do many things, the methods in this chapter are particularly good (i.e. go-to methods) for addressing the following:* • SES components and linkages • Social-ecological dependence and impact • Evaluating policy options
SPATIAL DIMENSION	
The methods in this chapter are primarily either or both: • Explicitly spatial *The methods in this chapter are most commonly applied at the following spatial scales:* • Local • Regional (provincial/state to continental) • Global • Multiple places/sites around the world	

(Millennium Ecosystem Assessment 2005; Díaz et al. 2015). Coastal habitats such as mangroves, for example, act as a buffer for people and infrastructure along coastlines against the impact of storms. Fields, forests and oceans provide food in the form of crops, game and fish. Natural landscapes around the world are important parts of people's cultural and spiritual identities. Functioning, healthy ecosystems are necessary for the production of ecosystem services, but often some sort of human input or action is also required to enhance the provision of these services and their contribution to well-being, such as forest management or cropland irrigation. Ecosystem services are therefore co-produced by people and nature (Reyers et al. 2013; Palomo et al. 2016).

The concept of ecosystem services has its roots in economics and ecology. One of the first studies to collate information on the economic value of a range of 'nature's services' was published in the late 1970s (Westman 1977), starting a trend of valuing services provided by ecosystems (i.e. natural capital) in a way that made them comparable to human-made goods and services (i.e. manufactured and financial capital). In ecology, the first mention of the term 'ecosystem services' occurred in a 1981 book on species extinctions and their consequences (Ehrlich and Ehrlich 1981). By the end of the 20th century, ecosystem services were formally defined as the 'conditions and processes through which natural ecosystems, and the species that make them up, sustain and fulfil human life' (Daily 1997). Since then, the concept of ecosystem services has moved beyond economics and ecology to become a widely used interdisciplinary approach within sustainability research (Abson et al. 2014).

A variety of methods can be used to quantify ecosystem services, depending on the type of ecosystem service, the spatial and temporal scale, and the scientific question or management decision being considered. Sometimes ecosystem service provision can be measured directly, either through field measurements (methods for generating data, Chapters 6–8) or through remote sensing of variables that are highly correlated with ecosystem service provision (Chapter 24). For instance, all the wheat produced on a field can be weighed, all the fish caught in a lake can be counted, and visits to parks can be tallied. However, direct measurement is usually only an option for a limited area over a limited time and tends to favour services that provide a physical product or have a market value. Often, ecosystem services have not been quantified for a location of interest, or they cannot be measured directly (such as the contribution of natural vegetation to maintaining water quality or regulating floods). Furthermore, decision-makers often need to understand changes in ecosystem service provision under different management scenarios or future conditions. In these cases, ecosystem services need to be modelled or estimated.

One option for modelling ecosystem services is the 'value transfer' or 'benefit transfer' approach, where the per-unit-area provision of ecosystem services from one area is transposed onto another area using a proxy variable such as land cover (see e.g. De Groot et al. 2012). Although this method is simple to use, it assumes similar social and ecological drivers of ecosystem service provision in both locations. Results from this type of approach should therefore be interpreted with caution (Eigenbrod et al. 2010).

Another approach is to construct process-based ecosystem service models, based on knowledge about ecosystem processes and functions that produce benefits for humans (so-called 'ecological production functions'). Process-based ecosystem service models typically translate geospatial inputs (such as land cover, vegetation and soil types, climate and topography) into estimates of an ecosystem service over space and/or time. This translation process occurs within small, standardised area units, such as pixels or parcels. Sometimes, the results for one unit influence the calculations for the next unit, creating a quantification of ecosystem service 'flow' between pixels or parcels. Vegetation that retains soil in one pixel will change the amount of sediment flowing into downhill pixels, for example.

While decision-making often involves spatial planning, non-spatial process-based simulation models can also be used to explore the implications of different management decisions for ecosystem service provision, or to explore future scenarios. The outcomes of these models are not necessarily quantitative predictions but rather a qualitative understanding of how different components of an SES interact and connect to one another to provide ecosystem services. Moreno et al. (2014), for example, used a participatory modelling approach to explore the factors affecting ecosystem service provision in protected areas, and found that it helped decision-makers learn about complex systems and identify opportunities for improving ecosystem service management.

SES problems and questions

Ecosystem service models help to reveal nature's benefits, especially when these benefits would otherwise be 'invisible' in decision-making (Daily 1997; Guerry et al. 2015). Usually, a variety of different ecosystem services are modelled to map and quantify the provision of multiple ecosystem services in a landscape. Ideally, this helps decision-makers to understand the impact of a decision (such as a development plan or land-management strategy) on multiple ecosystem services, and avoids unexpected outcomes where efforts to enhance the provision of one service inadvertently reduce the provision of another (Arkema et al. 2015). Decision-support packages such as InVEST, ARIES and Co$ting Nature have been designed to facilitate the modelling of multiple ecosystem services, and may assist in processes such as decision analysis based on optimisation (Chapter 29).

Another key application of ecosystem service models is the identification of important areas for conservation or restoration, based on an area's high level of ecosystem service provision (or potential provision) (Naidoo et al. 2008; Mandle et al. 2017; Nel et al. 2017). This approach can also identify which parts of a landscape supply ecosystem services that play a crucial role in supporting local livelihoods (Malmborg et al. 2018) and thus inform a livelihood and vulnerability analysis (Chapter 32). Ecosystem service modelling, and ecosystem service assessments more broadly, provide insights into the diverse ways that ecosystem services contribute to human well-being. These contributions to well-being may, in turn, influence environmental behaviour and the stewardship of natural resources, thereby affecting future ecosystem service provision in a complex set of social-ecological feedbacks (Masterson et al. 2019).

At a global scale, ecosystem service models are increasingly incorporated into other modelling frameworks, such as integrated assessment models and general equilibrium models, to investigate the impact of changes in land use, climate or commodity prices on ecosystems and the services they provide (e.g. Johnson et al. 2020).

Finally, ecosystem services are inherently linked to issues of equity and inclusivity. The provision of ecosystem service benefits depends not only on ecosystems and the functions they perform but also on their location relative to people who might benefit, and people's access to the services (Keeler et al. 2019a, b). Innovative approaches to ecosystem service assessments track how and where environmental changes affect specific beneficiaries, including impacts on the health or livelihoods of poor and marginalised communities (Arkema et al. 2013; Mandle et al. 2015; Chaigneau et al. 2018). These assessments can be used to compare the equity implications of different development options (Mandle et al. 2016).

Brief description of key methods

A diverse range of tools and models have been developed to assess ecosystem services (Bagstad et al. 2013; Neugarten et al. 2018). In Table 31.1, we briefly describe a selection of key decision-support modelling packages with a documented, software-based user interface that allows non-experts to run process-based, spatially explicit ecosystem service models. In addition, we describe several technical models and frameworks that are not strictly ecosystem service models, but are often used to generate ecosystem service-specific results in practice.

Table 31.1 Summary of key methods used in ecosystem service modelling

Method	*Description*	*References*
DECISION-SUPPORT MODELLING PACKAGES		
Integrated valuation of ecosystem services and trade-offs	Integrated valuation of ecosystem services and trade-offs (InVEST) is a suite of over 20 ecosystem service production function models. Typical inputs include land-use and land-cover (LULC) maps, climate data, topographic data and soil data. From these inputs, InVEST applies functions from peer-reviewed literature (e.g. the revised universal soil loss equation) to the input data and calculates ecosystem service provision. Each model is calculated separately, but most analyses then combine results for multiple models. InVEST is open source and developed in the programming language Python.	*Key introductory text* Kareiva et al. 2011 *Applications to SES* Mandle et al. 2017; Chaplin-Kramer et al. 2019
Artificial intelligence for ecosystem services	Artificial intelligence for ecosystem services (ARIES) is an artificially intelligent modelling platform that chooses which models to run in response to a user query, based on available spatial data and several decision rules. Methodologically, ARIES focuses on specifying how or where individuals benefit from the flows of ecosystem services from sources to sinks. ARIES is open source, but running the model relies on non-open-source tools (e.g. k.LAB).	*Key introductory text* Villa et al. 2014 *Applications to SES* Martínez-López et al. 2019
Co$ting Nature/ WaterWorld	Co$ting Nature and WaterWorld are web-based tools used to estimate terrestrial and freshwater ecosystem services. The models are built on default base data, such as soil type or precipitation, which allow ecosystem service calculations at detailed spatial scales of 1 km or 1 ha resolution. Co$ting Nature includes 13 services, such as hazard mitigation, nature-based tourism and timber supply. The code is not open source and requires payment for full functionality.	*Key introductory text* Mulligan 2012 *Applications to SES* Mulligan et al. 2013

Method	Description	References
TECHNICAL MODELS AND FRAMEWORKS		
Integrated assessment models	Integrated assessment models (IAMs) consider how changes in climate or the biosphere affect human activities, and vice versa. Key inputs include population growth, consumption patterns and climate change. IAMs are coarse in spatial resolution (grid cells of 30–110 km) but detailed in sectoral information (e.g. crop production figures) and have many explicit links between humans and the environment (e.g. through water scarcity or nutrient cycling).	*Key introductory text* Stanton, Ackerman, and Kartha 2009 *Applications to SES* Van Vuuren et al. 2015
Economic models that include ecosystem services	Several important economic models have been extended to report ecosystem service outcomes. These include general equilibrium models that allow a change in the system to affect other system components. Typically, these models track how economic changes affect land-use patterns. These in turn are translated into impacts on ecosystem services such as carbon storage and nutrient retention. In addition, a large body of literature has developed that models interactions between climate change and the economy, predicting greenhouse gas emissions as a function of economic activity while also tracking how the change in climate causes economic damage.	*Key introductory texts* Hertel 1997; Nordhaus 2017 *Applications to SES* Arndt et al. 2011; Meyfroidt et al. 2013
Lund–Potsdam–Jena dynamic global vegetation model	The Lund–Potsdam–Jena (LPJ) dynamic global vegetation model, along with other similar global vegetation models, provides detailed information on plant growth, mortality, soil interactions and other biophysical components of the model. Core parts of the LPJ, along with extensions of the model, report ecosystem service-specific results, such as soil carbon, water run-off or other factors. There is less interaction between people and the ecosystem in this type of model, with the exception of highly detailed agricultural models.	*Key introductory text* Sitch et al. 2003 *Applications to SES* Metzger et al. 2008
Life Cycle Assessment models	Life Cycle Assessment (LCA) models attempt to quantify the full environmental impact of a product through the many stages of its life, including the collection of input resources, assembly, usage and disposal. Many research institutes and private consultants have developed highly detailed databases of the impacts that different products have on the environment, although relatively few have focused on ecosystem service-specific impacts. One notable exception by Chaplin-Kramer et al. (2017) assesses how land-use and ecosystem service changes prompted by expanded production can also be included in life-cycle calculations (see also Chapter 30: Flow and impact analysis).	*Key introductory text* Curran 1996 *Applications to SES* Chaplin-Kramer et al. 2017

Limitations

Ecosystem service models are constrained by data quality. The general computing adage of 'garbage in, garbage out' also holds true for ecosystem service modelling. Beyond data quality, the quantity of data may be a limiting factor. The amount of data to be processed depends on the scale at which an ecosystem service model is applied: as extent increases or resolution becomes finer, computation time increases and it becomes more challenging to find suitable data (see also Chapter 24). In addition, many ecosystem service models rely on information or input assumptions that are not constant across a large extent, thus requiring different inputs for different subregions. Because these factors all increase modelling time and effort, it is important for researchers to identify and work at the optimal scale to answer their specific research questions.

Ecosystem service models can be used to express the value of nature in a diversity of ways, including mental health benefits (Bratman et al. 2015), self-reported importance (Martín-López et al. 2012) or the number of people affected by changes in ecosystem service provision (Olander et al. 2018; Keeler et al. 2019a). In addition, ecosystem service models are often used to quantify benefits in economic or monetary terms. It is important to note that economic value is only a small part of the value of nature and that ecosystem service models can provide a more complete and holistic set of values. However, even when ecosystem services are not given a monetary value, the ecosystem service concept still implies that the value of nature lies in its utility to people. Although the hope is that focusing on people's well-being will help promote nature's value in typical, economically driven decision-making, this 'commodification' of nature goes against the philosophical and spiritual values of many communities, cultures and worldviews (Gomez-Baggethun and Ruiz-Perez 2011; Díaz et al. 2018).

A related key limitation of current ecosystem service models is their relatively poor performance in capturing non-market and cultural values (Chan et al. 2016; Small, Munday, and Durance 2017), with the possible exception of nature-based recreation and tourism (e.g. Wood et al. 2013). Another challenge in capturing non-economic benefits is illustrated by the difficulties (most notably the lack of data availability) faced by modellers to 'disaggregate' the impact of ecosystem services for different groups of people, based on their access to and need for the service (Daw et al. 2011). For details on how many of these limitations are currently being addressed, refer to Section 'New directions' of this chapter.

Resource implications

Ecosystem service models can require substantial investments in learning to use the modelling software, and in pre-processing data into the correct formats. In addition, model outputs often need further processing and visualisation, which may require proficiency in geographic information system (GIS) software or geoprocessing code. Licences for widely used GIS programs can be very expensive, and high-resolution input data (e.g. satellite imagery) may be costly to acquire.

When involving stakeholders, ecosystem service modelling faces the same time and resource constraints, and must follow the same strict ethical research guidelines as any other participatory modelling process (Chapter 13). Since many ecosystem service models, including the ones outlined in this chapter, rely on Western scientific knowledge and technical expertise, they can appear obscure to stakeholders who rely on other kinds of knowledge systems. This lack of transparency may hinder community buy-in and reduce the effectiveness of ecosystem service models in assisting decision-making processes. This is why a thoughtful and respectful approach to using these models is needed, especially in participatory settings.

New directions

As the popularity of ecosystem services as a tool for addressing social-ecological challenges grows, ecosystem service models are continually changing and developing to better provide the types of information that practitioners require. For instance, improvements in remote-sensing technology are enabling the direct observation of more ecosystem services across wide areas and at fine spatial and temporal resolution (Ramirez-Reyes et al. 2019). Yet a number of research frontiers for advancing ecosystem service models remain (Rieb et al. 2017).

The first frontier addresses the shortcoming that current ecosystem service tools typically model multiple ecosystem services using separate models, as a 'snapshot' at a single point in time, and at the scale of a pixel or patch. Models that better represent interactions between multiple ecosystem services and their spatial and temporal dynamics would help decision-makers to predict the full implications of management actions for multiple ecosystem services across a landscape and into the future. Advances in remote-sensing technology and data availability have an important role to play in enhancing dynamic ecosystem service modelling, as well as moving beyond categorical land-cover inputs to more continuous and nuanced (i.e. realistic) input data.

The second frontier concerns the need for ecosystem service models to move beyond merely quantifying the biophysical supply of services to tracking changes in human well-being for different groups of beneficiaries. This includes adopting more meaningful valuation metrics that capture non-use and relational values of nature. An ecosystem service modelling approach that is explicit about beneficiaries, their needs and how changes in ecosystem services either enhance or diminish their well-being is crucial to understanding the role that ecosystem services can play in poverty alleviation and achieving the UN Sustainable Development Goals. Here, conventional models could be supplemented with more exploratory approaches, such as toy models and the use of scenario planning techniques (Chapter 11), to identify trade-offs between groups of people and how they benefit from ecosystem services, both now and in future scenarios (Daw et al. 2015).

The third frontier involves better understanding and modelling the implications of using different types of non-natural capital (e.g. technology, infrastructure and institutions) to produce or distribute ecosystem services. A study of shellfish fisheries in Spain and Portugal, for example, observed multiple ways of co-producing this ecosystem service using different amounts of equipment, active management and human labour, and found that the type of co-production affected the quantity of shellfish produced, as well as trade-offs with other ecosystem services (Outeiro et al. 2017). Key to this research area is understanding the difference that nature makes, relative to and accounting for the other kinds of capital. This frontier also involves understanding the burdens that are placed on ecosystems in one location by trade and decision-making in other, often far-away, parts of the world (Pascual et al. 2017; Schröter et al. 2018). It is especially important to consider global-scale flows of ecosystem services as countries work to develop ecosystem service accounting systems and metrics such as 'green GDP' (Li and Fang 2014).

Other frontiers for ecosystem service modelling lie at the intersection of different model types and the exploration of previously unexplored landscapes. Combining computable general equilibrium models like GTAP with ecosystem service models like InVEST, for example, will help to improve our understanding of cross-scale linkages between regional

Case study 31.1: Balancing conservation and agricultural expansion in the Volta watershed

Ecosystem service modelling can help decision-makers to understand potential trade-offs between development and conservation. Johnson et al. (2019) assessed how agricultural expansion may affect ecosystem service provision in the Volta watershed in Burkina Faso and Ghana. The watershed is critical to human livelihoods and well-being for a variety of reasons. Among other things, it supports the Akosombo Dam, which is a vital source of hydroelectric power. Other local livelihood strategies depend on the Volta River and surrounding ecosystems for brick-making and low-capital, low-intensity irrigated agriculture (Figure 31.1A and B). The basin also includes a particularly wide range of ecosystem types due to a strong precipitation gradient, from very dry, Sahara-like conditions in the north to extremely wet and lush ecosystems in the south.

A transnational agency, the Volta Basin Authority (VBA), is responsible for managing water withdrawals and other aspects of dam management, many of which have the potential to cause transboundary disputes. Recently, environmental degradation in the watershed has had detrimental impacts on ecosystem service provision. An increase in soil erosion, for example, has led to increasing sedimentation of reservoirs (in both the Akosombo Dam and smaller, run-of-river dams in the north). In response, the VBA committed to a strategic action programme aimed at preventing further environmental degradation and protecting ecosystem services (World Bank 2018). However, food security is extremely important to policymakers in this region, and thus any conservation plan needed to also consider impacts on food production.

Johnson et al. (2019) considered three different conservation strategies and assessed which lands should be protected in order to meet the VBA's dual goals of protecting ecosystem service provision and food security. InVEST models for sediment retention, phosphorus and nitrogen retention, water yield and carbon storage were run for the basin under the different strategies. The results showed that targeted conservation actions could attain much improved ecosystem service provision over the business-as-usual approach, while still meeting food-security goals through agricultural expansion. The study was limited by regional data availability and had to rely on global datasets for many components. It was also challenging to express the aggregate of multiple ecosystem services in a clear and appropriate way. In the end, an equally weighted overall ecosystem service score was created for the region based on the model results, and used to identify priority areas for conservation (Figure 31.1C).

This case study illustrates that ecosystem service modelling can highlight both conflicts and synergies between development and conservation goals in a spatially explicit manner, and provide decision-makers with a range of options to consider.

or global market or policy changes and local-level ecosystem service provision (Johnson et al. 2020). In addition, the increasing availability of fine-scale land-cover data down to the resolution of individual trees has opened up avenues to explore urban landscapes and the ecosystem service provision by 'green infrastructure' (such as street trees, parks,

Figure 31.1 Making bricks in the Volta watershed (A), irrigated agriculture drawing from the Volta (B), and combining scores for five ecosystem services into a single conservation value metric (C) (Johnson et al. 2019)

green roofs and community gardens). However, the dense assemblage of people of different socio-economic and cultural backgrounds in cities makes it particularly important to consider ecosystem service beneficiaries, as well as the equity and justice implications of changes in ecosystem service provision (Keeler et al. 2019b).

Key readings

Díaz S., U. Pascual, M. Stenseke, B. Martín-López, R.T. Watson, Z. Molnár, R. Hill et al. 2018. 'Assessing Nature's Contributions to People.' *Science* 359(6373): 270–272.

Millennium Ecosystem Assessment. 2005. *Ecosystems and Human Well-being: Synthesis*. Washington: Island Press.

Neugarten, R.A., P.F. Langhammer, E. Osipova, K.J. Bagstad, N. Bhagabati, S.H. Butchart, N. Dudley et al. 2018. *Tools for Measuring, Modelling, and Valuing Ecosystem Services: Guidance for Key Biodiversity Areas, Natural World Heritage Sites, and Protected Areas*. Gland: IUCN.

Reyers, B., R. Biggs, G.S. Cumming, T. Elmqvist, A.P. Hejnowicz, and S. Polasky. 2013. 'Getting the Measure of Ecosystem Services: A Social-Ecological Approach.' *Frontiers in Ecology and the Environment* 11(5): 268–273.

Rieb, J.T., R. Chaplin-Kramer, G.C. Daily, P.R. Armsworth, K. Böhning-Gaese, A. Bonn, G.S. Cumming et al. 2017. 'When, Where, and How Nature Matters for Ecosystem Services: Challenges for the Next Generation of Ecosystem Service Models.' *BioScience* 67: 820–833.

References

Abson, D.J., H. von Wehrden, S. Baumgärtner, J. Fischer, J. Hanspach, W. Härdtle, H. Heinrichs et al. 2014. 'Ecosystem Services as a Boundary Object for Sustainability.' *Ecological Economics* 103: 29–37.

Arkema, K., G. Guannel, G. Verutes, S.A. Wood, A. Guerry, M. Ruckelshaus, P. Kareiva et al. 2013. 'Coastal Habitats Shield People and Property from Sea-level Rise and Storms.' *Nature Climate Change* 3: 913.

Arkema, K., G. Verutes, S.A. Wood, C. Clarke-Samuels, S. Rosado, M. Canto, A. Rosenthal et al. 2015. 'Embedding Ecosystem Services in Coastal Planning Leads to Better Outcomes for People and Nature.' *Proceedings of the National Academy of Sciences of the United States of America* 112(24): 7390–7395.

Arndt, C., K. Strzepeck, F. Tarp, J. Thurlow, C. Fant IV, and L. Wright. 2011. 'Adapting to Climate Change: An Integrated Biophysical and Economic Assessment for Mozambique.' *Sustainability Science* 6(1): 7–20.

Bagstad, K., D.J. Semmens, S. Waage, and R. Winthrop. 2013. 'A Comparative Assessment of Decision-support Tools for Ecosystem Services Quantification and Valuation.' *Ecosystem Services* 5: 27–39.

Bratman, G., J.P. Hamilton, K.S. Hahn, G. Daily, and J.J. Gross. 2015. 'Nature Experience Reduces Rumination and Subgenual Prefrontal Cortex Activation.' *Proceedings of the National Academy of Sciences* 112(28): 8567–8572.

Chaigneau, T., S. Coulthard, K. Brown, T.M. Daw, and B. Schulte-Herbrüggen. 2018. 'Incorporating Basic Needs to Reconcile Poverty and Ecosystem Services.' *Conservation Biology* 33: 655–664.

Chan, K.M.A., P. Balvanera, K. Benessaiah, M. Chapman, S. Díaz, E. Gómez-Baggethun, R. Gould et al. 2016. 'Opinion: Why Protect Nature? Rethinking Values and the Environment.' *Proceedings of the National Academy of Sciences* 113(6): 1462–1465.

Chaplin-Kramer, R., R.P. Sharp, C. Weil, E.M. Bennett, U. Pascual, K. Arkema, K.A. Brauman et al. 2019. 'Global Modeling of Nature's Contributions to People.' *Science* 366(6462): 255–258.

Chaplin-Kramer, R., S. Sim, P. Hamel, B. Bryant, R. Noe, C. Mueller, G. Rigarlsford et al. 2017. 'Life Cycle Assessment Needs Predictive Spatial Modelling for Biodiversity and Ecosystem Services.' *Nature Communications* 8: 15065.

Curran, M.A. 1996. *Environmental Life-cycle Assessment*. New York: McGraw-Hill.

Daily, G. 1997. *Nature's Services*. Washington: Island Press.

Daw, T., K. Brown, S. Rosendo, and R. Pomeroy. 2011. 'Applying the Ecosystem Services Concept to Poverty Alleviation: The Need to Disaggregate Human Well-being.' *Environmental Conservation* 38(4): 370–379.

Daw, T.M., S. Coulthard, W.W.L. Cheung, K. Brown, C. Abunge, D. Galafassi, G.D. Peterson, T.R. McClanahan, J.O. Omukoto, and L. Munyi. 2015. 'Evaluating Taboo Trade-offs in Ecosystems Services and Human Well-being.' *Proceedings of the National Academy of Sciences* 112(22): 6949–6954.

De Groot, R., L. Brander, S. van der Ploeg, R. Costanza, F. Bernard, L. Braat, M. Christie et al. 2012. 'Global Estimates of the Value of Ecosystems and their Services in Monetary Units.' *Ecosystem Services* 1(1): 50–61.

Díaz, S., S. Demissew, J. Carabias, C. Joly, M. Lonsdale, N. Ash, A. Larigauderie et al. 2015. 'The IPBES Conceptual Framework–Connecting Nature and People.' *Current Opinion in Environmental Sustainability* 14: 1–16.

Díaz, S., U. Pascual, M. Stenseke, B. Martín-López, R.T. Watson, Z. Molnár, R. Hill et al. 2018. 'Assessing Nature's Contributions to People.' *Science* 359(6373): 270–272.

Ehrlich, P.R., and A.H. Ehrlich. 1981. *Extinction: The Causes and Consequences of the Disappearance of Species*. New York: Random House.

Eigenbrod, F., P.R. Armsworth, B.J. Anderson, A. Heinemeyer, S. Gillings, D.B. Roy, C.D. Thomas, and K.J. Gaston. 2010. 'The Impact of Proxy-based Methods on Mapping the Distribution of Ecosystem Services.' *Journal of Applied Ecology* 47(2): 377–385.

Gomez-Baggethun, E., and M. Ruiz-Perez. 2011. 'Economic Valuation and the Commo-dification of Ecosystem Services.' *Progress in Physical Geography* 35(5): 613–628.

Guerry, A.D., S. Polasky, J. Lubchenco, R. Chaplin-Kramer, G.C. Daily, R. Griffin, M. Ruckelshaus et al. 2015. 'Natural Capital and Ecosystem Services Informing Decisions: From Promise to Practice.' *Proceedings of the National Academy of Sciences* 112(24): 7348–7355.

Hertel, T.W. 1997. *Global Trade Analysis: Modeling and Applications*. Cambridge: Cambridge University Press.

Johnson, J.A., U. Baldos, T. Hertel, J. Liu, C. Nootenboom, S. Polasky, and T. Roxburgh. 2020. 'Global Futures: Modelling the Global Economic Impacts of Environmental Change to Support Policy-making.' Technical Report, January 2020. London: WWF UK. www.wwf.org.uk/globalfutures.

Johnson, J.A., S.K. Jones, S.L.R. Wood, R. Chaplin-Kramer, P.L. Hawthorne, M. Mulligan, D. Pennington, and F.A. DeClerck. 2019. 'Mapping Ecosystem Services to Human Well-being: A Toolkit to Support Integrated Landscape Management for the SDGs.' *Ecological Applications* 29(8): e01985. doi:10.1002/eap.1985.

Kareiva, P., H. Tallis, T.H. Ricketts, G.C. Daily, and S. Polasky. 2011. *Natural Capital: Theory and Practice of Mapping Ecosystem Services*. New York: Oxford University Press.

Keeler, B., B. Dalzell, J. Gourevitch, P. Hawthorne, K. Johnson, and R. Noe. 2019a. 'Putting People on the Map Improves the Prioritization of Ecosystem Services.' *Frontiers in Ecology and the Environment* 17(3): 151–156.

Keeler, B.L., P. Hamel, T. McPhearson, M. Hamann, M.L. Donahue, K.A.M. Prado, K.K. Arkema et al. 2019b. 'Social-Ecological and Technological Factors Moderate the Value of Urban Nature.' *Nature Sustainability* 2: 29–38.

Li, G., and C. Fang. 2014. 'Global Mapping and Estimation of Ecosystem Services Values and Gross Domestic Product: A Spatially Explicit Integration of National "Green GDP" Accounting.' *Ecological Indicators* 46: 293–314.

Malmborg, K., H. Sinare, E.E. Kautsky, I. Ouedraogo, and L.J. Gordon. 2018. 'Mapping Regional Livelihood Benefits from Local Ecosystem Services Assessments in Rural Sahel.' *PLoS ONE* 13(2): e0192019.

Mandle, L., J. Douglass, J.S. Lozano, R.P. Sharp, A.L. Vogl, D. Denu, T. Walschburger, and H. Tallis. 2016. 'OPAL: An Open-Source Software Tool for Integrating Biodiversity and Ecosystem Services into Impact Assessment and Mitigation Decisions.' *Environmental Modelling & Software* 84: 121–133.

Mandle, L., H. Tallis, L. Sotomayor, and A. Vogl. 2015. 'Who Loses? Tracking Ecosystem Service Redistribution from Road Development and Mitigation in the Peruvian Amazon.' *Frontiers in Ecology and the Environment* 13(6): 309–315.

Mandle, L., S. Wolny, N. Bhagabati, H. Helsingen, P. Hamel, R. Bartlett, A. Dixon et al. 2017. 'Assessing Ecosystem Service Provision under Climate Change to Support Conservation and Development Planning in Myanmar.' *PLoS ONE* 12(9): e0184951.

Martínez-López, J., K.J. Bagstad, S. Balbi, A. Magrach, B. Voigt, I. Athanasiadis, M. Pascual, S. Willcock, and F. Villa. 2019. 'Towards Globally Customizable Ecosystem Service Models.' *Science of the Total Environment* 650: 2325–2336.

Martín-López, B., I. Iniesta-Arandia, M. García-Llorente, I. Palomo, I. Casado-Arzuaga, D.G. del Amo, E. Gómez-Baggethun et al. 2012. 'Uncovering Ecosystem Service Bundles through Social Preferences.' *PLoS ONE* 7(6): e38970.

Masterson, V.A., S. Vetter, T. Chaigneau, T.M. Daw, O. Selomane, M. Hamann, G.Y. Wong, V. Mellegård, M. Cocks, and M. Tengö. 2019. 'Revisiting the Relationships between Human Well-being and Ecosystems in Dynamic Social-Ecological Systems: Implications for Stewardship and Development.' *Global Sustainability* 2: e8.

Metzger, M.J., D. Schröter, R. Leemans, and W. Cramer. 2008. 'A Spatially Explicit and Quantitative Vulnerability Assessment of Ecosystem Service Change in Europe.' *Regional Environmental Change* 8(3): 91–107.

Meyfroidt, P., E.F. Lambin, K. Erb, and T.W. Hertel. 2013. 'Globalization of Land Use: Distant Drivers of Land Change and Geographic Displacement of Land Use.' *Current Opinion in Environmental Sustainability* 5(5): 438–444.

Millennium Ecosystem Assessment. 2005. *Ecosystems and Human Well-being: Synthesis.* Washington: Island Press.

Moreno, J., I. Palomo, J. Escalera, B. Martín-López, and C. Montes. 2014. 'Incorporating Ecosystem Services into Ecosystem-based Management to Deal with Complexity: A Participative Mental Model Approach.' *Landscape Ecology* 29(8): 1407–1421.

Mulligan, M. 2012. 'WaterWorld: A Self-parameterising, Physically Based Model for Application in Data-poor but Problem-rich Environments Globally.' *Hydrology Research* 44(5): 748.

Mulligan, M., J.R. Rubiano, S. Burke, and A. van Soesbergen. 2013. 'Water Security in Amazonia.' Report for Global Canopy Programme and International Center for Tropical Agriculture as Part of the Amazonia Security Agenda Project.

Naidoo, R., A. Balmford, R. Costanza, B. Fisher, R.E. Green, B. Lehner, T.R. Malcolm, and T.H. Ricketts. 2008. 'Global Mapping of Ecosystem Services and Conservation Priorities.' *Proceedings of the National Academy of Sciences* 105(28): 9495–9500.

Nel, J.L., D.C. le Maitre, D.J. Roux, C. Colvin, J.S. Smith, L.B. Smith-Adao, A. Maherry, and N. Sitas. 2017. 'Strategic Water Source Areas for Urban Water Security: Making the Connection between Protecting Ecosystems and Benefiting from Their Services.' *Ecosystem Services* 28: 251–259.

Neugarten, R.A., P.F. Langhammer, E. Osipova, K.J. Bagstad, N. Bhagabati, S.H.M. Butchart, N. Dudley et al. 2018. *Tools for Measuring, Modelling, and Valuing Ecosystem Services: Guidance for Key Biodiversity Areas, Natural World Heritage Sites, and Protected Areas.* Gland: IUCN.

Nordhaus, W.D. 2017. 'Evolution of Assessments of the Economics of Global Warming: Changes in the DICE Model, 1992–2017.' NBER Working Paper No. 23319. Cambridge: National Bureau of Economic Research.

Olander, L.P., R.J. Johnston, H. Tallis, J. Kagan, L.A. Maguire, S. Polasky, D. Urban, J. Boyd, L. Wainger, and M. Palmer. 2018. 'Benefit Relevant Indicators: Ecosystem Services Measures that Link Ecological and Social Outcomes.' *Ecological Indicators* 85: 1262–1272.

Outeiro, L., E. Ojea, J.G. Rodrigues, A. Himes-Cornell, A. Belgrano, Y. Liu, E. Cabecinha et al. 2017. 'The Role of Non-natural Capital in the Co-production of Marine Ecosystem Services.' *International Journal of Biodiversity Science, Ecosystem Services & Management* 13(3): 35–50.

Palomo, I., M.R. Felipe-Lucia, E.M. Bennett, B. Martín-López, and U. Pascual. 2016. 'Disentangling the Pathways and Effects of Ecosystem Service Co-Production.' *Advances in Ecological Research* 54: 245–283.

Pascual, U., I. Palomo, W.A. Adams, K.M.A. Chan, T.M. Daw, E. Garmendia, E. Gómez-Baggethun et al. 2017. 'Off-stage Ecosystem Service Burdens: A Blind Spot for Global Sustainability.' *Environmental Research Letters* 12(7): 75001.

Ramirez-Reyes, C., K.A. Brauman, R. Chaplin-Kramer, G.L. Galford, S.B. Adamo, C.B. Anderson, C. Anderson et al. 2019. 'Reimagining the Potential of Earth Observations for Ecosystem Service Assessments.' *Science of the Total Environment* 665: 1053–1063.

Reyers, B., R. Biggs, G.S. Cumming, T. Elmqvist, A.P. Hejnowicz, and S. Polasky. 2013. 'Getting the Measure of Ecosystem Services: A Social-Ecological Approach.' *Frontiers in Ecology and the Environment* 11(5): 268–273.

Rieb, J.T., R. Chaplin-Kramer, G.C. Daily, P.R. Armsworth, K. Böhning-Gaese, A. Bonn, G.S. Cumming et al. 2017. 'When, Where, and How Nature Matters for Ecosystem Services: Challenges for the Next Generation of Ecosystem Service Models.' *BioScience* 67(9): 820–833.

Schröter, M., T. Koellner, R. Alkemade, S. Arnhold, K.J. Bagstad, K. Erb, K. Frank et al. 2018. 'Interregional Flows of Ecosystem Services: Concepts, Typology and Four Cases.' *Ecosystem Services* 31: 231–241.

Sitch, S., B. Smith, I.C. Prentice, A. Arneth, A. Bondeau, W. Cramer, J.O. Kaplan et al. 2003. 'Evaluation of Ecosystem Dynamics, Plant Geography and Terrestrial Carbon Cy-cling in the LPJ Dynamic Global Vegetation Model.' *Global Change Biology* 9(2): 161–185.

Small, N., M. Munday, and I. Durance. 2017. 'The Challenge of Valuing Ecosystem Services that Have No Material Benefits.' *Global Environmental Change* 44: 57–67.

Stanton, E.A., F. Ackerman, and S. Kartha. 2009. 'Inside the Integrated Assessment Models: Four Issues in Climate Economics.' *Climate and Development* 1(2): 166–184.

Van Vuuren, D.P., M. Kok, P.L. Lucas, A.G. Prins, R. Alkemade, M. van den Berg, L. Bouwman et al. 2015. 'Pathways to Achieve a Set of Ambitious Global Sustainability Objectives by 2050: Explorations Using the IMAGE Integrated Assessment Model.' *Technological Forecasting and Social Change* 98: 303–323.

Villa, F., K. Bagstad, B. Voigt, G. Johnson, R. Portela, M. Honzák, and D. Batker. 2014. 'A Methodology for Adaptable and Robust Ecosystem Services Assessment.' *PLoS ONE* 9(3): e91001.

Westman, W. 1977. 'How Much Are Nature's Services Worth?' *Science* 197(1969): 960–963.

Wood, S.A., A.D. Guerry, J.M. Silver, and M. Lacayo. 2013. 'Using Social Media to Quantify Nature-based Tourism and Recreation.' *Scientific Reports* 3: 1–7.

World Bank. 2018. 'Project Information Document (Appraisal Stage) – Volta River Basin Strategic Action Programme Implementation – P149969.' Washington: World Bank.

32

Livelihood and vulnerability analysis

Charlie Shackleton,[1] Kate Schreckenberg,[2] Sheona Shackleton[3] and Marty Luckert[4]

[1]DEPARTMENT OF ENVIRONMENTAL SCIENCE, RHODES UNIVERSITY, MAKHANDA, SOUTH AFRICA
[2]READER IN DEVELOPMENT GEOGRAPHY, KING'S COLLEGE, LONDON, UK
[3]AFRICAN CLIMATE AND DEVELOPMENT INITIATIVE, UNIVERSITY OF CAPE TOWN, CAPE TOWN, SOUTH AFRICA
[4]DEPARTMENT OF RESOURCE ECONOMICS & ENVIRONMENTAL SOCIOLOGY, UNIVERSITY OF ALBERTA, EDMONTON, CANADA

Key methods discussed in this chapter

Sustainable livelihood analysis, vulnerability analysis

Connections to other chapters

Livelihood analysis and vulnerability analysis are integrative approaches and consequently draw on a variety of methods to collect and analyse primary and secondary data covered in other chapters. Core ones include systems scoping (Chapter 5), ecological field data collection (Chapter 6), interviews and surveys (Chapter 7), participatory data collection (Chapter 8), action research (Chapter 15), statistical analysis (Chapter 18), qualitative content analysis (Chapter 19), comparative case study analysis (Chapter 20), institutional analysis (Chapter 22) and spatial mapping and analysis (Chapter 24).

Introduction

The origins of livelihood and vulnerability analyses can be traced back through a number of disciplines, each vested in subtly different ideologies and prescriptions about rural development and land use. Economists were interested in rural incomes and poverty outcomes through the use of land and the resources it produces; ecologists were concerned with the environmental sustainability of the same land and resources; sociologists sought to foster more equitable outcomes in terms of access to land and resources by different groups, genders and the power relations associated with these; and development planners considered strategies to simultaneously optimise land productivity, employment, markets and human development outcomes. Scoones (2009) outlines how the bridges between disciplines and ideologies evolved from the mid-1980s onwards towards more people-centred and 'holistic' policies and tools, laying the foundation for both livelihood analysis and vulnerability analysis.

Livelihood analysis concretised in the early 1990s with the appearance of the Chambers and Conway (1992) working paper. This paper provided the first definition of 'a livelihood' and when it is deemed sustainable, i.e. 'when it can cope with and recover from stresses and

 DOI: 10.4324/9781003021339-38

SUMMARY TABLE: LIVELIHOOD AND VULNERABILITY ANALYSIS

DISCIPLINARY BACKGROUND

The methods in this chapter are derived from or have most commonly been used in:

Interdisciplinary

KNOWLEDGE TYPE

The methods in this chapter are primarily used to generate the following types of knowledge:

- Exploratory
- Explanatory

RESEARCH APPROACH

The methods in this chapter originate from or most commonly adopt the following research approaches:

- Analytical/objective
- Interpretive/subjective
- Collaborative/process

PURPOSE OF METHOD

The most common purposes of using the methods in this chapter are:

- Data collection/generation
- Stakeholder engagement and co-production
- Policy/decision support

TEMPORAL DIMENSION

The methods in this chapter are most commonly applied to the following temporal dimensions:

- Present (typically within the last 5–10 years)
- Recent past (post-1700s)
- Future

SPATIAL DIMENSION

The methods in this chapter are primarily either or both:

- Explicitly spatial

The methods in this chapter are most commonly applied at the following spatial scales:

- Local
- Multiple places/sites around the world

SYSTEMIC FEATURES AND PROCESSES

While most methods can do many things, the methods in this chapter are particularly good (i.e. go-to methods) for addressing the following:

- Diversity
- Social-ecological dependence and impact
- Adaptation and self-organisation
- Evaluating policy options

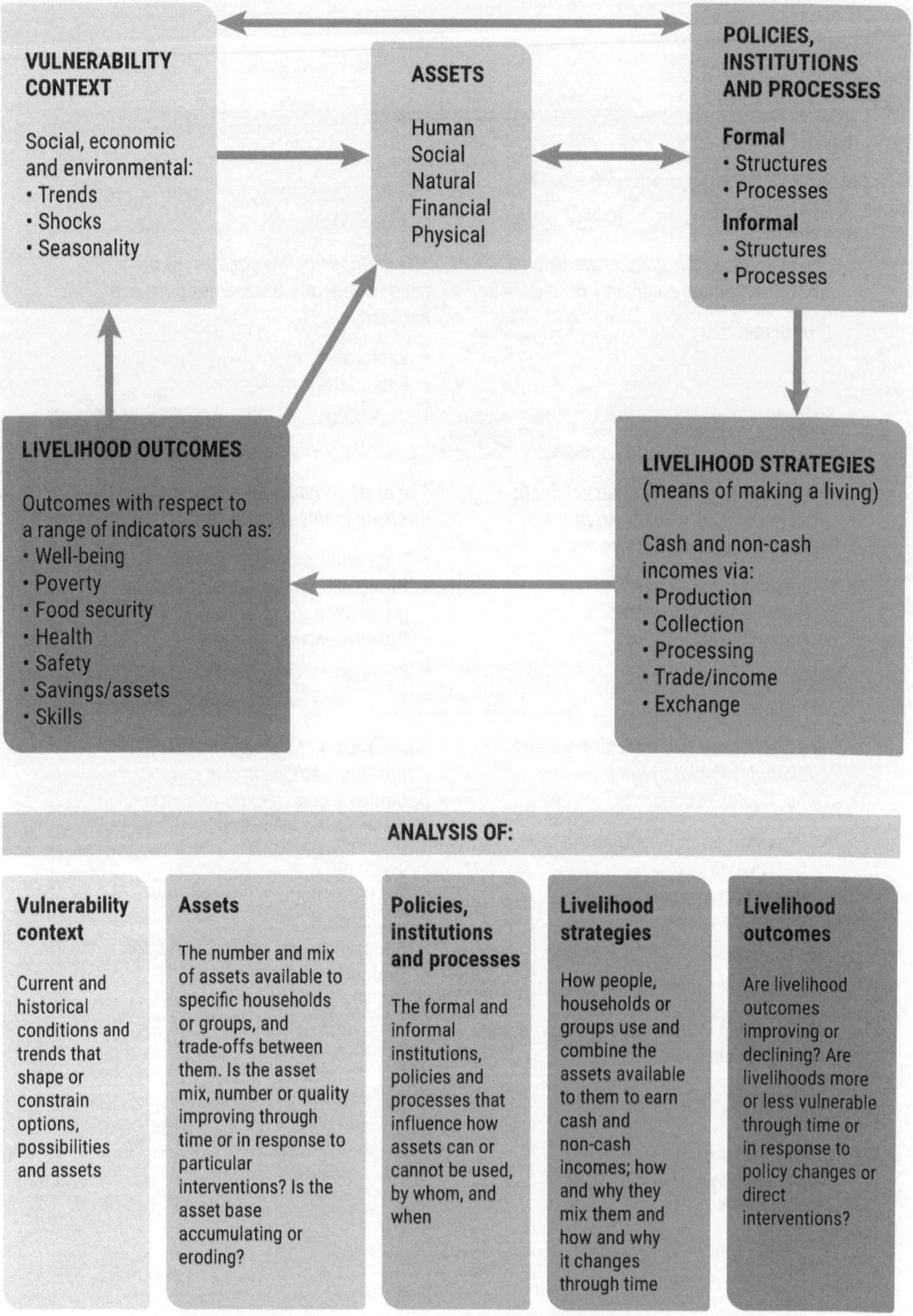

Figure 32.1 The sustainable livelihoods framework (adapted from the originals by Carney 1998 and Scoones 1998) (© Charlie Shackleton)

shocks and maintain or enhance its capabilities and assets both now and in the future, while not undermining the natural resource base' (Carney 1998). Sallu, Twyman and Stringer (2010) point out that this aligns strongly with the current concepts of vulnerability and resilience. These early beginnings laid the foundation for the sustainable livelihoods framework (SLF) (Figure 32.1) to emerge a few years later, commonly depicted as an input-output model or diagram. The inputs were five asset classes and the outputs a suite of livelihood strategies that resulted in given livelihood outcomes in particular contexts, mediated by a range of local and higher-level institutions. Although it originated in response to rural development and poverty issues, the framework has also been found useful in urban settings (e.g. Farrington, Ramasut, and Walker 2002; Rakodi 2002).

Vulnerability analysis followed a similar trajectory of a merging of ideas and philosophies from multiple disciplines over a very similar period (Fuchs 2009). Adger (2006) describes how social, geographical and ecological scientists formed different schools that each applied this kind of analysis in their own way. However, there was a gradual and steady convergence and today it is difficult to conceive that there had ever been a separation. Most vulnerability analyses now consider both exposure to hazards and the underlying, often structural and contextual, factors or causes that make some groups or people more vulnerable to these hazards than others. In contrast to livelihood analysis, the multi-disciplinary parentage of vulnerability analysis has resulted in numerous vulnerability assessment frameworks and tools; some, but not all, of which emphasise either social or ecological aspects rather than the two simultaneously. Many of the widely applied participatory tools for human vulnerability analysis have been developed by practitioners in international NGOs concerned with assisting communities to adapt to climate change and its impacts (Füssel and Klein 2006). Increasingly, these tools are much more responsive to local context than more deductive index-based approaches (Vogel et al. 2007). Vulnerability analysis is also increasingly linked to notions of resilience (Vogel et al. 2007; Nguyen et al. 2016) because increased resilience more often than not results in reduced vulnerability.

There are strong commonalities and a sharing of core elements between sustainable livelihood analysis and most vulnerability analysis approaches (e.g. Fraser et al. 2011), including understanding livelihood activities, assets and access to these, institutions and multiple shocks and stresses. The evolution of sustainable livelihood analysis and vulnerability analysis as a merging of multiple disciplinary approaches to consider the diversity and complexity of rural lives and poverty predates and resonates with the development of social-ecological thinking and methods. These analyses are therefore particularly useful tools in social-ecological systems (SES) research dealing with important development and sustainability questions. Their core strengths of being people centred, context specific, local level, inter- and transdisciplinary, unequivocally linking people and the natural environment and explicitly recognising diversity and multiple outcomes, are widely sought in SES research.

SES problems and questions

Livelihood analysis and vulnerability analysis draw upon a wide diversity of qualitative and quantitative data-collection methods which provide the information necessary to understand how individuals, households and communities: (a) make a living, (b) generate cash and non-cash incomes, (c) sustain themselves, their assets and networks within a given socio-economic and ecological context, (d) respond to short-term and long-term stressors and drivers, and (e) adapt to and cope with changing contexts and circumstances. These approaches seek to describe and analyse diversity among households and their members with respect to activities, assets, livelihood strategies and ultimately the livelihood outcomes, and pressures that shape vulnerability and sustainability.

The sustainable livelihoods approach and its associated sustainable livelihoods framework are widely used for organising data, information and insights obtained via multiple empirical methods to analyse the sustainability (or not) of current livelihoods. It differentiates livelihood strategies and livelihood outcomes. Livelihood strategies are the activities pursued to obtain cash and non-cash incomes and build the asset portfolio of the household or community, whereas livelihood outcomes are the ultimate benefits (or not) of engaging in a particular suite of strategies, such as increased well-being, reduced poverty or increased vulnerability. Understanding the vulnerability context of livelihoods is a core aspect of the sustainable livelihoods framework. However, there are also a host of more prescribed vulnerability assessment (vulnerability analysis) frameworks and tools (especially emerging from the climate change research field), which are applied independently of the sustainable livelihoods framework and that operate at varying scales, from global to local.

Sustainable livelihood analysis is useful in guiding a full and integrated analysis of the current situation regarding incomes and various 'capital' stocks that people can access and use. Common questions in sustainable livelihood analysis include:

- Who makes use of what natural resources in the local environment?
- Why?
- For what benefits?
- How is use or access controlled and by which institutions?

Questions about dynamics are also relevant, such as:

- How do households, particular groups or communities, cope or adapt in the face of temporary or longer-term changes in resource supply or access?
- How does livelihood diversification or trade-offs between strategies or between assets improve or undermine human well-being?
- How vulnerable are particular groups to specific stressors and how might that affect livelihood options and outcomes?

With respect to vulnerability analysis, questions of this nature can help to highlight the role of different natural resources in mediating some of the impacts of climate change on livelihoods. These questions can also reveal the potential for enhancing resilience through ecosystem-based adaptation.

Brief description of key methods

Both livelihood analysis and vulnerability analysis require holistic inter- and transdisciplinary perspectives and approaches. Methods and tools that provide data on the social, economic and ecological dimensions of livelihoods and vulnerability are required, e.g. household interviews, focus group discussions, oral narratives/histories (all in Chapter 7), participatory appraisal tools, e.g. seasonal calendars, timelines, resource and income ranking, participatory mapping, gender roles (Chapter 8), resource and ecological inventories (Chapter 6) and mapping (Chapter 24). Livelihood and vulnerability analysis may also include resource valuation, income determination, asset quantification, social network analysis, participatory modelling and planning, spatial analysis (for risk/

Table 32.1 Summary of key methods used in livelihood and vulnerability analysis

Method	*Description*	*References*
Sustainable livelihood analysis	Sustainable livelihood analysis is an integrative analysis of the vulnerability context, the institutions and the assets available to groups of interest and how these shape the livelihood strategies adopted, and ultimately the livelihood outcomes.	*Key introductory texts* Farrington et al. 1999 (ODI Natural Resource Perspectives 42); Serrat 2008 (Asian Development Bank); Valdés-Rodríguez and Pérez-Vázquez 2011 (Tropical and Subtropical Agroecosystems 14) *Applications to SES* Campbell et al. 2002; Sallu, Twyman, and Stringer 2010; Masunungure and Shackleton 2018; Östberg et al. 2018
Vulnerability analysis	Vulnerability analysis is primarily used to determine the expected impacts, risks and adaptive capacity of a region, sector, social group or person to the effects of climate change and other interacting stresses. It involves several methods, from indicator- or proxy-based methods to geographic information systems (GIS) and mapping, multiple-stressor-based methods, participatory approaches, narratives and stories, and household surveys. The purpose of the assessment and scale usually determines the type of method used. There is no single 'best' approach or method and all have their advantages and disadvantages. Participatory approaches have mainly been developed and applied by international NGOs such as CARE, the Red Cross, ActionAid and Oxfam. Quantitative approaches using indices are more in the realm of economists. All approaches should incorporate information on who/what is vulnerable to what (climate plus other stressors) and when; the causes of vulnerability across spatial and temporal scales (causal models, problem trees); factors influencing vulnerability outcomes for what or whom (barriers, assets, knowledge, policies and institutions – often provided through livelihood analysis); and actions or responses, including the potential for maladaptation ultimately increasing vulnerability.	*Key introductory texts* Frameworks Turner et al. 2003; Schröter, Polsky, and Patt 2005; Davis, Waagsaether, and Methner 2017 Tools and approaches (especially participatory tools) csir.co.za/documents/csir-global-change-ebook.pdf; letsrespondtoolkit.org/vulnerability-assessment; mediation-project.eu/platform/tbox/participatory_vulnerability_and_capacity_assessments.html; ifrc.org/Global/Publications/disasters/vca/vca-toolbox-en.pdf (International Red Cross - VAC – vulnerability and capacity assessment); youtube.com/watch?v=Fv5vE2vxYwY; actionaid.org.uk/sites/default/files/doc_lib/108_1_participatory_vulnerability_analysis_guide.pdf (ActionAid UK)

vulnerability assessments), and composite indices of vulnerability, risk and coping strategies (e.g. food security measures). Table 32.1 provides a summary of key methods used in livelihood and vulnerability analysis.

Limitations

Given that the sustainable livelihoods framework is used by disciplinary and interdisciplinary researchers, it will inevitably not be able to meet all needs at all times. However, the nature and extent of certain assumptions related to or shortcomings of the framework are contested. The merits of the debates (and whether we agree with them or not) will not be considered here. Instead, we aim to make readers aware of some of the debates about the robustness or weaknesses of the sustainable livelihoods framework (Hobley 2001; Adato and Meinzen-Dick 2002; Bryceson 2002; Toner 2003; Serrat 2008; Morse, McNamara, and Acholo 2009; Scoones 2009), which they can investigate further if required.

The references in Table 32.1 cover the commentary on the sustainable livelihoods framework in terms of:

- Insufficient acknowledgement of power relations in the framework, especially political power, to the extent that some argue that power relations should be included as an additional asset class in the asset pentagon

Case study 32.1: Livelihoods, change and vulnerability in rural Botswana

Appreciating the nature of local livelihoods and the available assets households can draw on to construct their livelihoods is crucial in understanding broader patterns of poverty, social justice and whether development intervention or policies are required. It also provides insights into the efficacy of local institutions and household strategies in responding to shocks and vulnerability. These were the core concerns explored by Sallu, Twyman and Stringer (2010) in two villages in arid Botswana, when they sought to identify the factors that in 'isolation and combination push livelihoods along particular "trajectories" towards vulnerability or resilience'. Trends and livelihood trajectories were investigated over a 30-year period.

The researchers adopted a rich mixed-methods approach to gather the necessary information and data, including household questionnaires about livelihood activities and resources used, participatory time lines, oral histories, focus group discussions, vegetation and wild animal surveys, and analysis of remote-sensing images. The different methods allowed a substantive process of triangulation across findings. Qualitative data analysis was undertaken via thematic analysis and iterative reflexivity allowing for inductive interpretation. The more quantitative data were analysed using standard statistics.

The low rainfall and generally dystrophic, albeit patchy, soils at the two sites limited the types and intensity of some livelihood options, such as cropping. Livelihoods at both sites were significantly contingent on local landscapes and biodiversity resources. Thus, livestock husbandry and collection of non-timber forest products (such as wild foods, medicines and construction materials) were considered key livelihood

- Its focus at the local level, which compromises its ability to be useful at higher spatial scales and in the face of global drivers of change
- Its strength in analysing the current, local-level situation, which results in a weakness for considering longer-term shifts in rural economies or institutions
- Its context-specific nature, which makes it hard to use in a comparative manner (and generalise across sites)
- The use of the words 'assets' and 'capitals', which embeds and overemphasises economic considerations
- The fact that markets are crucial to most rural livelihoods but are not explicit in the sustainable livelihoods framework
- The fact that the sustainable livelihoods framework is frequently erroneously reduced to just the asset pentagon, which makes it very difficult to operationalise as there is no single suite of accepted tools to do so

Some of the limitations of current vulnerability analysis approaches from an SES perspective relate to their (a) neglect of other livelihood stressors that may interact with climate hazards, (b) lack of attention to ecosystem health and ecosystem services delivery, (c) lack of clarity regarding which quantitative measures to select and apply from the large range used to date, and (d) neglect of scale issues and local context, and of social differentiation and intersectionality.

strategies for most households. However, households engaged in more than one livelihood strategy, such as small-scale cropping or vegetable gardening, hunting and small businesses around local needs, and a few had some form of salaried employment. Considering patterns across households, the researchers identified three broad clusters, which they labelled 'accumulators', 'diversifiers' and 'dependents'.

However, the broader environmental context had undergone marked changes over the previous 30 years, such that some livelihood strategies were no longer as successful as they used to be. Key contextual changes were: (a) an intense and prolonged drought in the mid-1980s, (b) the more frequent late onset of the rainy season, now approximately one month later than previously, (c) increased variability in the mean annual rainfall, (d) the loss of flood recession cultivation sites with the drying up of Lake Xau, and (e) land degradation. These translated into marked changes in livelihood strategies and outcomes for some households, and less so for others. Generally, an increase in household vulnerability was associated with a decline in or loss of one or more of the following: (a) access to local natural resources, (b) livestock, and (c) diversity of livelihood strategies. Households in the so-called 'dependent' cluster were the most vulnerable, followed by 'accumulators' if their primary livelihood strategy was placed at risk by changes in the local or broader context.

The study concluded that the findings had 'highlighted the importance of formal and informal institutions in building resilience and the need for increased effort to ensure [that] the most vulnerable households have access to a diversity of assets' (Sallu, Twyman, and Stringer 2010). Overall, the application of several quantitative and qualitative data-collection and analytical methods, framed within the current and historical context, allowed the researchers to develop deep insight into the livelihood trajectories and vulnerability of these village communities.

Resource implications

Seeking to understand local livelihoods and vulnerability is resource intensive in terms of time (for focus group discussions, participatory exercises, household surveys, key informant interviews, mapping and resource inventories). These activities require sufficient budget for extended periods of field research to cover transport, field accommodation and supplies, local guides or assistants, and perhaps translators and interpreters. However, there is generally no need for expensive field equipment or sophisticated software. Some participatory mapping tools may require GIS or hardcopy aerial photographs. Otherwise, most data can be captured on field sheets or electronic devices or, if focus groups are held, on flipcharts. Videos and voice recorders can be useful but are not mandatory. As per the norm, refreshments should be served at group meetings that last longer than two hours. It may also be necessary to pay for the use of a particular venue and provide transport to the venue. Careful consideration must be given to research ethics because some people or households may engage in what are deemed by some regulatory authorities as illegal or undesirable activities (such a poaching, selling of protected species or use of land to which they don't have formal, recognised access).

Data analysis can take various forms depending on the precise research questions asked or emphasised, as well as the methods used. Quantitative surveys, for example, require different resources and skills than participatory data. For quantitative surveys, data access to and experience in using spreadsheets are useful. Depending on the objectives of the study, this could extend into statistical analysis using any of a number of software packages. Qualitative content analysis can be achieved via several different means such as thematic or content analysis, grounded theory, oral histories, narrative analysis and the like, depending on the researcher's theoretical or philosophical position (see Chapter 19). There is increasing use of software to assist in the analysis of qualitative data such as NVivo, Atlas.ti and QDA Miner.

New directions

Although livelihood analysis and vulnerability analysis have been around for a long time, these approaches and tools are being used in cutting-edge SES work, which fosters new applications and refinements of the tools themselves. Although five asset classes are integral to the sustainable livelihoods framework, most studies do not quantify each class equally but tend to focus on one or two more than the others. Historically, natural and financial capital have received the most attention, but there are renewed efforts regarding the development and application of indices of social or human capital. This is fostering some innovative uses of social network analysis within livelihood and vulnerability framings. Providing more equal attention to all asset classes reinvigorates debates about their substitutability.

Another key area is the application of the tools in longitudinal studies to better understand how livelihoods and vulnerability are changing and for whom, and what is driving the changes. This inevitably leads towards the merging of livelihood and vulnerability approaches with scenarios (Chapter 11). Furthermore, there is an increasing realisation of the need to disaggregate vulnerability assessments and livelihood strategies and outcomes between specific groups, such as by gender, by size of land holding, by wealth classes, by proximity to certain resources and so on. Livelihood and vulnerability approaches are also finding greater application in urban contexts in terms of urban residents' reliance on urban green infrastructure for a variety of capitals, or to cope with or develop resilience in times of heightened vulnerability.

Key readings

Adger, W.N. 2006. 'Vulnerability.' *Global Environmental Change*, 16: 268–281.

Angelsen, A., H.O. Larsen, J.F. Lund, C. Smith-Hall, and S. Wunder. 2011. *Measuring Livelihoods and Environmental Dependence: Methods for Research and Fieldwork*. London: Earthscan.

De Lange, H.J., S. Sala, M. Vighi, and J.H. Faber. 2010. 'Ecological Vulnerability in Risk Assessment: A Review and Perspectives.' *Science of the Total Environment* 408: 3871–3879.

Scoones, I. 2009. 'Livelihoods Perspectives and Rural Development.' *The Journal of Peasant Studies* 36: 171–196.

Turner, B.L., R.E. Kasperson, P.A. Matsone, J. McCarthy, R.W. Corell, L. Christensen, N. Eckley et al. 2003. 'A Framework for Vulnerability Analysis in Sustainability Science.' *Proceedings of the National Academy of Science* 100: 8074–8079.

Acknowledgements

Charlie Shackleton's contribution was supported by the South African Research Chairs Initiative of the Department of Science and Innovation and the National Research Foundation of South Africa (grant 84379). Any opinion, finding, conclusion or recommendation expressed in this material is that of the authors and the NRF does not accept any liability in this regard.

References

Adato, M., and R. Meinzen-Dick. 2002. 'Assessing the Impact of Agricultural Research on Poverty Using the Sustainable Livelihoods Framework.' Environment and Production Technology Division Discussion Paper No. 89. Washington: International Food Policy Research Institute.

Adger, W.N. 2006. 'Vulnerability.' *Global Environmental Change* 16: 268–281.

Bryceson, D. 2002. 'Mutliplex Livelihoods in Rural Africa: Recasting the Terms and Conditions of Gainful Employment. *Journal of Modern African Studies* 40: 1–28.

Campbell, B.M., S. Jeffry, W. Kozanayi, M. Luckert, M. Mutamba, and C. Zindi. 2002. *Household Livelihoods in Semi-arid Regions: Options and Constraints*. Bogor: CIFOR.

Carney, D. 1998. 'Implementing the Sustainable Rural Livelihoods Approach.' In *Sustainable Rural Livelihoods: What Contribution Can We Make?* edited by D. Carney. London: DFID.

Chambers, R., and G. Conway. 1992. 'Sustainable Rural Livelihoods: Practical Concepts for the 21st Century.' IDS Discussion Paper, 296. Brighton: Institute of Development Studies.

Davis, C.L., K. Waagsaether, and N. Methner. 2017. 'Development of a Risk Assessment Methodology and Vulnerability Indices.' In *Climate Change over South Africa: From Trends and Projected Changes to Vulnerability Assessments and the Status Quo of National Adaptation Strategies*. South Africa's 3rd National Communication to UNFCCC. Pretoria: Department of Environmental Affairs. www.environment.gov.za/sites/default/files/reports/draftsouthafricas3rdnationalcommunication_unfccc2017.pdf.

Farrington, J., D. Carney, C. Ashley, and C. Turton. 1999. 'Sustainable Livelihoods in Practice: Early Applications of Concepts in Rural Areas.' *ODI Natural Resources Perspectives* 42.

Farrington, J., T. Ramasut, and J. Walker. 2002. 'Sustainable Livelihoods Approaches in Urban Areas: General Lessons, with Illustrations from Indian Examples.' ODI Working Paper.

Fraser, E.D., A. Dougill, K. Hubacek, C. Quinn, J. Sendzimir, and M. Termansen. 2011. 'Assessing Vulnerability to Climate Change in Dryland Livelihood Systems: Conceptual Challenges and Interdisciplinary Solutions.' *Ecology and Society* 16(3): 3. doi:10.5751/ES-03402-160303.

Fuchs, S. 2009. 'Susceptibility versus Resilience in Mountain Hazards in Austria: Paradigms of Vulnerability Revisited.' *Natural Hazards & Earth Systems Science* 9: 337–352.

Füssel, H-M., and R.J. Klein. 2006. 'Climate Change Vulnerability Assessments: An Evolution of Conceptual Thinking.' *Climate Change* 75: 301–329.

Hobley, M. 2001. *Unpacking the PIP Box*. Somerset: Hobley, Shields & Associates.

Masunungure, C., and S.E. Shackleton. 2018. 'Exploring Long-term Livelihood and Landscape Change in Two Semi-arid Sites in Southern Africa: Drivers and Consequences for Social-Ecological Vulnerability.' *Land* 7(2): 50. doi:10.3390/land7020050.

Morse, S., N. McNamara, and M. Acholo. 2009. 'Sustainable Livelihood Approach: A Critical Analysis of Theory and Practice.' Geographical Paper No. 189. Reading: University of Reading. www.reading.ac.uk/web/files/geographyandenvironmentalscience/GP189.pdf.

Nguyen, T.T., J. Bonetti, K. Rodgers, and C.D. Woodroffe. 2016. 'Indicator-based Assessment of Climate-change Impacts on Coasts: A Review of Concepts, Methodological Approaches and Vulnerability Indices.' *Ocean and Coastal Management* 123: 18–43.

Östberg, W., O. Howland, J. Mduma, and D. Brockington. 2018. 'Tracing Improving Livelihoods in Rural Africa Using Local Measures of Wealth: A Case Study from Central Tanzania, 1991–2016.' *Land* 7(2): 44. doi:10.3390/land7020044.

Rakodi, C. 2002. 'A Livelihoods Approach Conceptual Issues and Definitions.' In *Urban Livelihoods: A People-centred Approach to Reducing Poverty*, edited by C. Rakodi and T. Lloyd-Jones, 3–22. London: Earthscan.

Sallu, S.M., C. Twyman, and L.C. Stringer. 2010. 'Resilient or Vulnerable Livelihoods? Assessing Livelihood Dynamics and Trajectories in Rural Botswana.' *Ecology & Society* 15(4): 3. www.ecologyandsociety.org/vol15/iss4/art3.

Schröter, D., C. Polsky, and A.G. Patt. 2005. 'Assessing Vulnerabilities to the Effects of Global Change: An Eight Step Approach.' *Mitigation and Adaptation Strategies for Global Change* 10: 573–596.

Scoones, I. 1998. 'Sustainable Rural Livelihoods: A Framework for Analysis.' Working Paper 72. Brighton: Institute for Development Studies.

Scoones, I. 2009. 'Livelihoods Perspectives and Rural Development.' *The Journal of Peasant Studies* 36: 171–196.

Serrat, O. 2008. 'The Sustainable Livelihoods Approach.' *Knowledge Solutions*. Manila: Asian Development Bank.

Toner, A. 2003. 'Exploring Sustainable Livelihoods Approaches in Relation to Two Interventions in Tanzania.' *Journal of International Development* 15: 771–781.

Turner, B.L., R.E. Kasperson, P.A. Matsone, J. McCarthy, R.W. Corell, L. Christensen, N. Eckley et al. 2003. 'A Framework for Vulnerability Analysis in Sustainability Science.' *Proceedings of the National Academy of Science* 100: 8074–8079.

Valdés-Rodríguez, O.A., and A. Pérez-Vázquez. 2011. 'Sustainable Livelihoods: An Analysis of the Methodology.' *Tropical and Subtropical Agroecosystems* 14: 91–99.

Vogel, C., S.C. Moser, R.E. Kasperson, and G.D. Dabelko. 2007. 'Linking Vulnerability, Adaptation and Resilience Science to Practice: Pathways, Players and Partnerships.' *Global Environmental Change* 17: 349–364.

Part 3

33

Synthesis and emerging frontiers in social-ecological systems research methods

Maja Schlüter,[1] Reinette Biggs,[1,2] Hayley Clements,[2] Alta de Vos,[3] Kristine Maciejewski[2] and Rika Preiser[2]

[1]STOCKHOLM RESILIENCE CENTRE, STOCKHOLM UNIVERSITY, STOCKHOLM, SWEDEN
[2]CENTRE FOR SUSTAINABILITY TRANSITIONS, STELLENBOSCH UNIVERSITY, STELLENBOSCH, SOUTH AFRICA
[3]DEPARTMENT OF ENVIRONMENTAL SCIENCE, RHODES UNIVERSITY, MAKHANDA, SOUTH AFRICA

Introduction

Social-ecological systems (SES) research is a rapidly emerging new research domain within the broader emerging area of sustainability science. It is largely a problem-driven and action-oriented field, motivated by the immense sustainability and equity-related challenges facing society (see Chapter 1). Social-ecological systems research is based on an understanding that SES are complex adaptive systems (CAS), where social and ecological dynamics are deeply intertwined, and give rise to features and problems that cannot be understood or addressed by studying these dimensions in isolation (see Chapter 2). The field draws on and combines methods from both natural and social sciences, and combines quantitative and qualitative approaches. As such, SES research is characterised by epistemological and methodological pluralism, which is challenging for those entering the field and has complex implications for the research process, methods and ethical considerations to be taken into account in SES research (see Chapter 3).

This book aims to clarify and synthesise this plurality by providing an introduction to SES research (Part 1), and the diversity of methods currently used in the field (Part 2). The aim of this final chapter (Part 3) is to provide a synthesis of the current landscape of SES methods, critically reflect on the methods with respect to their ability to address systemic features of SES and discuss some of the most common methodological challenges associated with the complex adaptive and intertwined nature of SES. Based on this synthesis, we identify methodological gaps and discuss novel methods and method combinations that may help to address these gaps and move the field forward.

The current landscape of SES methods

Building on the summaries at the start of each chapter in Part 2, this section presents patterns of method characteristics and systemic features found across the set of methods presented in this book. Although we had multiple rounds of discussion with authors to identify the key characteristics of the methods and the systemic features they most commonly address, it is

DOI: 10.4324/9781003021339-40

important to note that these assessments remain somewhat subjective and could potentially be interpreted in other ways. The synthesis presented here is thus intended to identify broad trends, but not to draw specific conclusions with respect to individual methods. In addition, while we refer to 'methods' in the summary matrices (Tables 33.1 and 33.2) and in the text, most chapters (and thus rows in the matrices) include multiple methods, which may differ in their key characteristics and abilities. As such, a key characteristic may be reflective of a specific method in a chapter, and not of all methods in that chapter.

Synthesis of key characteristics of SES methods

Each method or method group was characterised at the beginning of a chapter according to key dimensions such as the type of knowledge it generates, its purpose, and the spatial and temporal scales commonly addressed. Looking across the methods discussed in this book, there are more groups of methods originating from and grounded in analytical or objective **research approaches** than in subjective or collaborative approaches (Table 33.1). There are, however, also many methods that are based on and allow for subjective and collaborative approaches. Some methods can be applied in different ways depending on the purpose and research question, which explains why they are suited to multiple approaches. The higher frequency of analytical approaches may be a result of early SES work coming more from the natural sciences, particularly ecology and geography, using established methods in those fields (Janssen et al. 2006). Participatory approaches and co-production of knowledge and action are more recent developments connected to the challenge of linking knowledge to action in contexts where decision stakes and uncertainty are high, and knowledge is diverse and contested (Wyborn et al. 2019; Caniglia et al. 2020; Norström et al. 2020). The abundance of co-production methods is in line with a growing movement away from expert-driven knowledge to community/stakeholder knowledge and the aim to ensure the societal relevance of research. In addition to developing their own novel methods, collaborative methods often build on methods that were originally developed for analytical purposes, such as scenario development or agent-based modelling.

Social-ecological systems research methods are most commonly used for exploratory research, with explanatory and descriptive knowledge also being common **knowledge types**. Many methods are used for more than one of these knowledge types. The abundance of methods for exploratory and descriptive research may be because SES research works with systems where the linkages between social and ecological systems and the resulting system behaviour are highly uncertain or unknown and, contrary to established disciplines, there is little existing knowledge or theory to draw on. Much SES research, particularly in the initial years, has been frontier work. Exploratory methods allow for making connections that have not been made before, and better accounting for the complexity of SES by being less restrictive in a priori defining what is in focus. Finally, exploratory methods are very suitable for informing or being applied in processes of knowledge co-production or effecting change because they allow for working with different understandings of a system. One can, for example, use different framings or assumptions to explore the possible effects of a management measure using dynamical systems modelling or a scenario analysis method.

Different methods may be based on different theories of change, i.e. their use is based on different views about how to effect change in SES. Whereas there are many methods that are used in SES research for the broader **purpose** of policy and decision support, only few of them generate prescriptive knowledge, such as optimal management strategies (e.g. spatial mapping and analysis, decision analysis based on optimisation). To some extent this may be due to the history of the field and the focus on exploration described above, but it may also be partly due to the inherent, irreducible uncertainty of SES (Polasky et al. 2011; Nuno, Bunnefeld, and Milner-Gulland 2014). Understanding SES as complex adaptive systems entails acknowledging that outcomes of interventions cannot be fully predicted and that uncertainty is an inherent aspect of SES. The reason for this is that SES interactions are always embedded in and shaped by temporal and spatial contexts. The best we can do is therefore to explore different possibilities for the evolution of SES under different conditions. Methods for knowledge co-production and methods such as modelling and scenario planning are particularly useful in this regard because they enable us to explore the range of possible outcomes and the uncertainty associated with them, and provide insight into processes that influence the emergence of different pathways. Based on analyses of this nature, we can make 'judicious suggestions' about potential actions and policies to influence SES outcomes, but cannot offer blueprint-type procedures for ensuring specific SES outcomes (Boulton, Allen, and Bowman 2015). The large number of methods that are used for the purpose of policy/decision support while not aiming at prescription indicates that methods are often used to enhance understanding of the features and processes of SES with the aim to provide valuable insights that can inform management and governance.

Some methods can be used for multiple purposes when applied in different ways. Scenario development in global assessments, for example, is used to integrate different models and data to develop policy recommendations (Kok et al. 2017; Rosa et al. 2017). Local-scale scenarios often involve participatory processes that are used to enhance stakeholder engagement and legitimacy in decision-making (Oteros-Rozas et al. 2015). Dynamical systems, agent-based modelling and state-and-transition modelling can be used for system understanding, stakeholder engagement or policy support. The purpose of applying these methods will, however, influence who is involved in model design, analysis and interpretation, and the processes used to specify the research question, determine the model structure, validate the model and interpret results.

Since SES researchers are often interested in understanding or effecting change over time, time is an important factor. Studying change over time and the temporal characteristics of SES is, however, challenging, as we can also see in the analysis of systemic features of SES (see the next section). In terms of **temporal dimensions**, most methods are primarily used to study the present or recent past, and only five methods are typically used to research the 'deep' (i.e. pre-industrial) past. Whereas this can to some extent be an artefact of how methods were grouped in our analysis, it possibly reflects that SES research focuses more on the present and the future than the past. This focus may to some extent be explained by the urgency of pressing sustainability problems and the rapid changes the world experiences that direct focus to the now and the future.

The majority of methods can be used in a spatially explicit way, although this is not always done. In terms of **spatial scales**, all methods are appropriate and typically used for local-scale research, with many also appropriate for research across multiple places. Fewer methods are appropriate for regional- and particularly global-scale research. This may be indicative of a focus of SES research on the local scale, with studies focusing on larger-scale SES processes and dynamics only increasing in recent years. To date, there are few methods for measuring and analysing how local-scale processes affect the global level and vice versa, i.e. for exploring cross-scale processes. This may be related to a lack of conceptual frameworks and theories to address cross-scale interactions, which is an area of active SES research (e.g. the telecoupling framework (Liu et al. 2018)). Table 33.1 provides a summary of the key characteristics of the methods covered in Part 2 of the handbook.

Table 33.1 Summary of the key characteristics of the methods covered in Part 2 of the handbook

	Research approach			Knowledge type				Purpose of method				Temporal dimension				Spatial dimension					
	Analytical/objective	Interpretive/subjective	Collaborative/process	Descriptive	Exploratory	Explanatory	Prescriptive	Data collection/generation	System understanding	Stakeholder engag. and coprod.	Policy/decision support	Present	Recent past	Pre-industrial revolution	Future	Explicitly spatial	Non-spatial	Local	Regional	Global	Multiple places/sites
METHODS FOR DATA GENERATION AND SYSTEMS SCOPING																					
5. Systems Scoping	■	■	■	■	■			■	■	■	■	■	■			■	■	■			
6. Ecological Field Data Collection	■			■	■			■				■	■			■		■	■		■
7. Interviews and Surveys	■	■	■	■	■	■		■	■		■	■	■		■		■	■			■
8. Participatory Data Collection		■	■	■	■			■		■		■	■		■	■	■	■			■
METHODS FOR KNOWLEDGE CO-PRODUCTION AND EFFECTING SYSTEM CHANGE																					
9. Facilitated Dialogues		■	■		■	■				■	■	■	■		■		■	■	■		■
10. Futures Analysis	■	■	■	■	■				■	■	■	■	■		■	■	■	■	■	■	■
11. Scenario Development	■	■	■		■				■	■	■	■	■		■	■	■	■	■	■	■
12. Serious Games		■	■		■	■			■	■	■	■	■		■		■	■			
13. Participatory Modelling	■		■	■	■	■			■	■	■	■	■		■	■	■	■			■
14. Resilience Assessment		■	■	■	■			■	■	■	■	■	■		■	■	■	■	■		■
15. Action Research		■	■		■	■				■		■					■	■	■		
METHODS FOR ANALYSING SYSTEMS – System components and linkages																					
16. Expert Modelling	■	■		■	■	■			■	■	■	■	■		■	■	■	■	■		■
17. Data Mining and Pattern Recognition	■			■	■			■	■		■	■	■			■	■	■	■	■	■
18. Statistical Analysis	■			■	■	■			■			■	■	■	■	■	■	■	■	■	■
19. Qualitative Content Analysis		■		■	■	■			■		■	■	■		■		■	■	■	■	■
20. Comparative Case Study Analysis	■	■		■	■	■			■			■	■			■		■			■
21. Controlled Behavioural Experiments	■					■		■	■			■					■	■			■
22. Institutional Analysis	■	■		■	■	■			■			■	■			■	■	■			■
23. Network Analysis	■			■	■	■			■			■	■				■	■	■		
24. Spatial Mapping and Analysis	■			■	■	■	■	■	■	■	■	■	■	■	■	■		■	■	■	■
METHODS FOR ANALYSING SYSTEMS – System dynamics																					
25. Historical Assessment	■	■		■	■	■		■	■				■	■		■	■	■	■	■	■
26. Dynamical Systems Modelling	■		■		■	■			■	■	■	■	■		■		■	■	■	■	
27. State-and-transition Modelling	■		■	■	■	■			■	■	■	■	■		■	■	■	■			
28. Agent-based Modelling	■		■	■	■	■			■	■	■	■	■	■	■	■	■	■	■		
METHODS FOR ANALYSING SYSTEMS – Directly informing decision-making																					
29. Decision Analysis based on Optimisation	■			■	■	■	■		■		■	■	■	■	■	■	■	■	■	■	■
30. Flow and Impact Analysis	■	■		■		■	■	■	■		■	■	■			■	■	■	■	■	■
31. Ecosystem Service Modelling	■		■			■	■		■	■	■	■	■		■	■		■	■	■	■
32. Livelihood and Vulnerability Analysis	■	■	■		■	■		■		■	■	■	■		■	■		■			■
	22	15	15	20	25	21	4	11	23	16	19	27	26	5	18	19	23	28	18	11	21

Ability of methods to address different systemic features of SES

The methods discussed in this book have different abilities when it comes to investigating systemic features of SES, but some features are more frequently addressed than others (Table 33.2). The fact that a method has not been selected for a particular feature does not, however, mean that it is not potentially suitable for addressing that feature. Expanding the scope of a method beyond its current use is in some cases an interesting method frontier (see Section 'Methodological gaps and frontiers').

There are many methods for capturing system components and one-way interactions (e.g. social-ecological components, diversity, social-ecological dependence), with far fewer methods capable of capturing processes and dynamics over time (e.g. path dependence, adaptation and self-organisation), or multiple scales and the interactions of these scales. A few methods address multiple SES features. These are often methods that have been developed more recently, specifically for studying or effecting change in complex adaptive systems (e.g. agent-based modelling, network analysis). By contrast, other methods are particularly good at addressing only a few selected features (e.g. institutional analysis and comparative case study analysis, both particularly suited to understanding and explaining collective action). Just because a method only captures a few systemic features does not mean it is a poor method for SES research. Conversely, a highly flexible method that can be used to study or support many different SES features or processes may not necessarily always be the best method for a specific feature or process. It is important to consider the limitations of methods, in terms of not only what they can capture on their own but also where several methods are needed in combination to capture different SES features.

Most methods that address system dynamics are either co-production methods (such as scenario and futures analysis) or modelling approaches, with the exception of historical profiling and methods that include time-series analysis such as spatial mapping and analysis. One of the reasons for the dominance of modelling and co-production methods in the study of SES dynamics is probably the difficulty of collecting data and analysing processes over time in empirical studies. Co-production methods are, not surprisingly, the go-to method for social learning and for facilitating processes of transformation. In contrast, collective action and collaborative governance are both addressed by co-production and by systems analysis methods, reflecting different aims: supporting processes of collective action and governance versus studying the factors and conditions that enable collective action.

Classical decision-support tools are most often used for addressing social-ecological dependence and informing policy options, but several features are consistently overlooked by this group of methods, including power relations, uncertainty, path dependence, social learning and collective action. Our analysis also suggests that methods that focus on analysing system components and linkages are more likely to address power relations, but are not suitable for understanding path dependency. Methods that analyse system dynamics, while able to address the role of power, rarely do so, but are among the few methods that address path dependency. Generally, methods used for analysing systems are not listed as methods of choice for exploring or supporting social learning or transformation, although some methods (e.g. modelling approaches) are often used in participatory processes with the aim to support social learning and transformation (see Chapter 13: Participatory modelling).

Table 33.2 Summary of the systemic features that methods covered in Part 2 of the handbook most commonly address

	SES components and linkages	Diversity	SE dependence and impact	Power relations	Multiple scales or cross-level interaction	SE interactions over time	Path dependency	Adaptation and self-organisation	Regime shifts	Transformation	Social learning	Collective action and collab. governance	Evaluating policy options	Exploring uncertainty	Total
METHODS FOR DATA GENERATION AND SYSTEMS SCOPING															
5. Systems Scoping	■	■		■	■						■	■			6
6. Ecological Field Data Collection	■	■	■												3
7. Interviews and Surveys	■	■	■	■											4
8. Participatory Data Collection	■		■	■							■				4
METHODS FOR KNOWLEDGE CO-PRODUCTION AND EFFECTING SYSTEM CHANGE															
9. Facilitated Dialogues	■	■		■	■	■				■	■	■		■	9
10. Futures Analysis			■	■		■	■			■	■			■	7
11. Scenario Development			■			■	■			■	■		■	■	7
12. Serious Games										■	■	■	■		4
13. Participatory Modelling		■	■								■	■	■		5
14. Resilience Assessment	■				■			■	■	■	■	■		■	8
15. Action Research		■		■						■	■	■	■	■	7
METHODS FOR ANALYSING SYSTEMS – System components and linkages															
16. Expert Modelling	■		■		■				■				■	■	6
17. Data Mining and Pattern Recognition	■	■							■						3
18. Statistical Analysis	■	■	■						■					■	5
19. Qualitative Content Analysis	■			■						■			■		4
20. Comparative Case Study Analysis										■		■			2
21. Controlled Behavioural Experiments						■		■				■			3
22. Institutional Analysis	■											■			2
23. Network Analysis	■			■								■			3
24. Spatial Mapping and Analysis	■	■				■			■						4
METHODS FOR ANALYSING SYSTEMS – System dynamics															
25. Historical Assessment						■		■	■						3
26. Dynamical Systems Modelling	■		■			■	■		■				■		6
27. State-and-transition Modelling									■		■		■		3
28. Agent-based Modelling	■	■			■	■	■	■					■	■	8
METHODS FOR ANALYSING SYSTEMS – Directly informing decision-making															
29. Decision Analysis based on Optimisation	■		■										■		3
30. Flow and Impact Analysis			■		■	■							■		4
31. Ecosystem Service Modelling	■		■										■		3
32. Livelihood and Vulnerability Analysis		■	■					■					■		4
Total	17	11	13	8	6	9	4	5	8	8	10	10	13	8	

Methodological challenges and practical limitations of current SES methods

Analysing or engaging with SES that are characterised by features of complexity and social-ecological intertwinedness poses considerable methodological challenges (Chapter 2). Each method or method group has different strengths and limitations which are discussed individually in the chapters of Part 2. Here we reflect on difficulties and challenges that cut across methods and method applications. We also summarise some of the most common practical difficulties that researchers should consider when planning to use a particular method.

Conceptual and methodological challenges of doing SES research

Many of the methods presented in this handbook have their roots in disciplines whose conceptual foundations may not align with those of SES research. Methods adopted from disciplines such as ecology, economics, anthropology and geography may have been developed for different questions and purposes than those common to SES research, and may be based on assumptions and worldviews that are not compatible with a view of SES as intertwined complex adaptive systems. Methods that are based on worldviews that assume linear causation, stability or independent, fully rational actors, for instance, may be problematic, particularly when their applicability and the validity of results depend on whether these underlying assumptions hold (Ferraro, Sanchirico, and Smith 2019). There can also be a mismatch between a chosen conceptual approach and the way methods are applied to operationalise it. An example is the concept of 'ecosystem services'. Although the concept itself is grounded in an intertwined view of people and nature (Fischer and Eastwood 2016), some ecosystem service models (Chapter 31) are strongly rooted in reductionist economic assumptions, which goes against the worldviews of many communities and cultures and that of SES as intertwined complex adaptive systems (Gómez-Baggethun and Ruiz-Pérez 2011; Lele et al. 2013; Díaz et al. 2015). Such methods are still useful for SES research, but their limitations and fit with a particular research question or transdisciplinary activity, and the consequences of their use, need to be carefully considered (Jahn, Bergmann, and Keil 2012; Popa, Guillermin, and Dedeurwaerdere 2015; Popa and Guillermin 2017). Below we discuss five common challenges of dealing with the complex adaptive and social-ecological intertwined nature of SES highlighted in the method chapters in Part 2.

1. **Defining system boundaries for an analysis or activity:** The radically open nature of SES means that there is no objectively real 'inside' and 'outside' the system (see Chapter 2). Nevertheless, when studying or effecting change in a complex system, one needs to make choices about what or whom to include and at what level or scale. Choices about where to draw the system 'boundary' are not always obvious (Preiser et al. 2018). Often these choices are determined by the worldviews, frameworks and theories that underlie a method or method application (e.g. institutional analysis (Chapter 22)) or methodological limitations (e.g. controlled behavioural experiments (Chapter 21)); sometimes they stem from the experience and intuition of those involved in applying the method (e.g. agent-based modelling (Chapter 28)), or are co-constructed by a group of researchers and/or stakeholders (e.g. participatory modelling (Chapter 13)). In this context, it becomes critical to be transparent about how these choices have been made and to reflect on their possible consequences. The importance of transparency about what

to include in defining system boundaries has been highlighted as a critical challenge in the chapters on participatory data collection (Chapter 8), ecological data collection (Chapter 6), interviews (Chapter 7), agent-based modelling (Chapter 28) and flow and impact analysis (Chapter 30).

2. **Dealing with disciplinary biases and accounting for diversity of views:** Many SES methods, particularly those for data generation and analysing systems, originate in a social or natural science discipline. When used within their respective discipline or field, key assumptions underlying a method are commonly known and accepted. Once a method is used outside its field of origin, however, this cannot be taken as given. In interdisciplinary contexts, transparency and reflexivity are thus critical for several reasons. First, choices made during method application can be heavily influenced by biases, histories and the contemporary context of researchers and participants (a challenge highlighted for instance in the chapters on interviews and surveys (Chapter 7) and historical assessment (Chapter 25)), as well as technical limitations of a method such as limitations in the number of variables that can be considered (see Chapter 21 on controlled behavioural experiments). Second, a researcher's framing and disciplinary lens impacts the interpretation of results (a challenge highlighted for instance by the chapters on historical assessment (Chapter 25) and scenarios (Chapter 11)). Third, results should always be evaluated in light of the assumptions underlying the analysis (highlighted in the chapter on agent-based modelling (Chapter 28)). In addition, most methods do not inherently require researchers to account for social differentiation (highlighted in the chapters on vulnerability analysis (Chapter 32) and interviews (Chapter 7)) and participatory research can be hard to conduct in a way that includes marginalised voices and non-scientific knowledge systems (highlighted in the chapters on systems scoping (Chapter 5) and participatory modelling (Chapter 13)). Drawing system boundaries and analysing results or engaging with a system is thus a profoundly ethical endeavour that requires transparency about underlying viewpoints, reflexivity, as well as careful consideration and discussion of ethical dilemmas that may arise during the research process.
3. **Dealing with context dependence:** Disciplines and their associated methods have different views about the relevance of contextual factors or processes for understanding, exploring or predicting SES outcomes. Social-ecological systems functions are contingent on context (Chapter 2) and these contexts are dynamic, i.e. they are shaped by and shape interactions in SES (Schill et al. 2019). Although many methods in SES research are used at the local scale and many researchers take pains to account for context, context-dependent methodological challenges are nevertheless pervasive. Moreover, given the cross-scale nature of SES, context extends beyond the local. Every process and action in an SES is embedded in and the result of a particular context, e.g. the behaviour of a resource user emerges from the individual's experiences, motivations, aims and her relations with her social-ecological environment (Kaaronen 2017; Raymond, Giusti, and Barthel 2018). The data collected in a research activity, the choices made when applying a method and the interpretations of results are similarly dependent on the context in which they were created. This challenge has been highlighted by many chapters dealing with methods such as interviews (Chapter 7), participatory data collection (Chapter 8), facilitated dialogues (Chapter 9), scenarios (Chapter 11), participatory modelling (Chapter 13), institutional analysis (Chapter 22), network analysis (Chapter 23), historical assessment

(Chapter 25), agent-based modelling (Chapter 28) and livelihood and vulnerability analysis (Chapter 32).

4. **Accounting for power relations:** Being able to critically engage with often subtle and hidden power relations and how they shape our understanding of phenomena is very important in order to understand how certain groups of people or organisations ascribe value to, for example, some natural resources or certain practices, and how these meanings shape SES interactions and stewardship practices. Understanding power relations is not easy. They are often hidden because they are ingrained in the identity of a group or individual, and find expression in language forms, how we dress, what we value as important and how we make judgements about certain actions and attitudes (Foucault 1982; Bourdieu 1991). Several method chapters highlighted the challenge and lack of accounting for power relationships. Not acknowledging the role of power and politics or how power influences decisions in the research process can limit the diversity of knowledge or actors that are taken into account when conceptualising a system (highlighted by Chapter 5 on systems scoping). It may also influence the legitimacy of research. Actors who hold social, political or economic power may not always accept discussing or playing together on a level playing field (highlighted by Chapter 12 on serious games). Some methods have been criticised for not acknowledging or focusing enough on power relations in their frameworks (see Chapter 32 on livelihood and vulnerability analysis) or analysis (see Chapter 22 on institutional analysis).

   In general, methods that investigate and analyse multiple genres, intertextual relationships and the tension between how structure and agency are co-constituted (Giddens 1984) are well placed to reflect on power relations. These include some of the chapters that have highlighted the lack of accounting for power relations as a key challenge (e.g. Chapter 7 on interviews and surveys; Chapter 19 on qualitative content analysis, and methods that enable the co-production of knowledge and systemic change such as Chapter 9 on facilitated dialogues, Chapter 10 on futures analysis, Chapter 13 on participatory modelling and Chapter 15 on action research). These methods, when used appropriately, provide the possibility of engaging with the stories, narratives, discourses, visions and myths that construct the ways in which people make sense and ascribe meaning to their place in this world. There is much potential, however, to expand the use of these methods for addressing power relations in SES research.

5. **Dealing with complex causation:** Social-ecological systems are characterised by complex causation and continuous change, which poses immense challenges for analysis and action (see Chapter 2). Few methods are able to deal with the complex interactions across spatial and temporal scales that give rise to the feedbacks, path dependencies and time lags that shape the emergent pathways and outcomes of SES. In addition, context sensitivity of social-ecological processes and the fact that agent- and system-level processes affect one another in various ways make identifying or untangling causal relationships difficult, if not impossible. Many quantitative methods for causal inference cannot deal with social-ecological feedbacks, non-linearities, emergence or multiple interacting causes, which limits their ability to address complex causation (Levin et al. 2012; Meyfroidt 2016; Preiser et al. 2018; De Vos, Biggs, and Preiser 2019). Some methods, such as narrative analysis and qualitative content analysis (Chapter 19) or facilitated dialogues (Chapter 9), are good at highlighting non-linear and cross-scale relationships.

However, understanding how these relationships drive cause and effect or bring about emergent patterns of behaviour is a far more complex task (Levin et al. 2012; Chapter 27 on state-and-transition modelling).

Challenges of dealing with complex causation have been mentioned across many chapters (e.g. Chapter 7 on interviews and surveys, Chapter 22 on institutional analysis, Chapter 29 on decision analysis based on optimisation, Chapter 18 on statistical analysis and Chapter 26 on dynamical systems modelling). Many methods are limited by a lack of knowledge about possible causal processes and conditions that may have brought about a particular phenomenon of interest or underlie a particular problem. Some methods cannot address causality, or need to be combined with other methods to be able to do so (e.g. meta-analysis with an in-depth case study). Others, such as serious games (Chapter 12), behavioural experiments (Chapter 21), simple dynamical systems models (Chapter 26) or decision analysis (Chapter 29), can only incorporate a small set of variables and processes, which limits their ability to address complex causation. Others again can include many variables and connections (e.g. statistical methods (Chapter 18) or agent-based modelling (Chapter 28)) but run the risk of becoming a 'black box' that is difficult to analyse, validate and communicate. Similar to other choices during a research process, the way researchers study causation in SES is influenced by the purpose of a study (e.g. to understand, explain or predict), researchers' interests and backgrounds, and also practical considerations. A researcher who aims to provide policy support may, for instance, focus on those causes that can be manipulated and try to assess their effect on outcomes. The challenge is then to understand how these causes play out within the broader network of SES relations.

Practical limitations and resource challenges

In addition to significant conceptual and methodological challenges, most SES research also faces significant practical and resource limitations. Social-ecological systems research can be more time intensive than disciplinary research: it takes time to develop a complex adaptive systems mindset within a research team or a group of stakeholders, and many of the methods used in knowledge co-production processes require time-consuming trust-building and iterative engagement processes (Lang et al. 2012; Angelstam et al. 2013; Norström et al. 2020). Much SES research is carried out in collaborative endeavours, often across disciplines and knowledge systems, which requires openness, epistemological agility (Haider et al. 2018), communication and facilitation skills. Unfortunately, existing funding and institutional contexts are often not conducive to the collaborative research and action processes needed in SES research. It can be difficult to find funding for the longer project durations needed in inter- or transdisciplinary projects; proposals are too often still categorised and judged within disciplinary silos, and it may be difficult to recruit and train students and early career researchers because of (disciplinary) constraints within the educational system.

Furthermore, many methods require advanced technical skills, such as statistical, programming and modelling skills or facilitation experience. This is particularly challenging when methods are combined and a researcher or team needs to acquire multiple skills that might be quite diverse. When combining ethnographic research with agent-based modelling, for instance, the researcher will need to be able to engage with the rich details of a particular context while at the same time abstracting this rich knowledge into a model. Another critical issue is data availability and quality. Obtaining data on social and ecological aspects at comparable and relevant spatial and temporal scales can be challenging (see Chapter 30 on flow and impact analysis), which limits the integration of different datasets (e.g. scenarios). Poor quality and availability of data may also limit the extent to which certain modelling approaches can be used and accepted by non-modellers (see Chapter 27 on state-and-transition modelling). Existing datasets or mined data may also not be representative of the population of interest (see Chapter 17 on data mining and pattern recognition; Chapter 18 on statistical analysis; Chapter 24 on spatial mapping and analysis). Data availability is particularly problematic for studies of change over time, such as dynamic models and historical analyses, and studies that need spatially explicit data (see Chapter 24 on spatial mapping and analysis).

Methodological gaps and frontiers

The above synthesis highlights the diversity of methods used in SES research today. In general, methods that aim to support policy or action are well represented. There are many methods that are well suited to exploration of the present and future of SES at local scales and that can support stakeholder engagement and knowledge co-production. Whereas the majority of methods are used to study interactions within one scale, there are fewer methods for studying dynamics and cross-scale interactions. Our synthesis also highlights methodological challenges rooted in the complex, adaptive and social-ecologically intertwined nature of SES, such as accounting for social-ecological feedbacks, emergence and complex dynamics that push many traditional methods to their limits.

In this section, we highlight methodological gaps related to addressing key features of SES and to supporting the SES research approach, co-production processes and knowledge synthesis. We present ways to address these gaps and point towards some emerging methods and methodological frontiers in the field. Social-ecological systems research is a rather young, interdisciplinary field. The development of novel methods, the use of existing methods in novel ways, the introduction of methods from other disciplines and the development of new combinations of methods all present exciting ongoing research frontiers.

Methods that account for emergence, cross-scale interactions and social-ecological intertwinedness

Methods for understanding and navigating emergence

Social-ecological systems research has relatively few methods to study the complex and intertwined social-ecological processes that give rise to emergent novel properties, phenomena or behaviours of SES or to foster the capacity to navigate them. In particular, there are few methods to identify and study how the system unfolds over time and how its pathway is shaped by local adaptations, non-linear feedbacks, path dependencies and chance. The capacity to appreciate and navigate emergence is essential for transformation because it allows identifying different kinds of opportunities based on an improved understanding of complex dynamics and ways to deal with unpredictability, uncontrollability and contestation (Moore et al. 2018).

A number of methods go some way towards addressing complex dynamics and unfolding processes, but there is much potential for further development. Case study research has been instrumental in developing narratives of how transformations succeed through the cross-scale interactions of actors, networks and structural features of an SES (e.g. Gelcich et al. 2010;

Moore et al. 2014; Herrfahrdt-Pähle et al. 2020). Network analysis, a method most commonly associated with providing a snapshot in time, can also be used to capture dynamism through time (Ryan and D'Angelo 2018), or to look at multiple time periods (Yletyinen et al. 2018; Zhao et al. 2018). Combinations of network approaches with agent-based modelling are a promising method frontier for understanding how the interplay of structure and agency influences system-level outcomes, such as the effects of a conservation intervention (Dobson et al. 2019). Dynamic modelling approaches are in general well suited to studying change of SES over time, but their potential for studying transient dynamics, coevolution and processes of emergence has not yet been fully realised. On the empirical side, process tracing (Beach and Pedersen 2013) is gaining attention as a method to study the historical processes that may have produced an outcome of interest, such as a trap (Boonstra and De Boer 2014) or an environmental policy (Orach, Schlüter, and Österblom 2017). New advances have also been proposed to better understand the coevolution of institutions and SES, such as the combined IAD-SES framework, the institutional grammar tool and the power of polycentric governance approach (Epstein et al. 2020). Finally, long-term social-ecological research sites may provide time series of social-ecological processes that will help understand patterns of dynamic interactions and their effects on the SES (Bretagnolle et al. 2019).

The conceptual and methodological challenges of emergence and complex causality have recently received attention in various subfields of sustainability science, such as land system science, ecological economics and Earth system science (Meyfroidt 2016; Carlson et al. 2018; Ferraro, Sanchirico, and Smith 2019; Runge et al. 2019). Novel methods such as advances in time-series analysis (convergent cross-mapping (CCM), Sugihara et al. 2012) or Bayesian score-based approaches (Chickering 2002) have been proposed for data-rich contexts. At the same time, authors highlight the need for multi-method approaches and triangulation because individual methods all have their limitations and biases that need to be carefully navigated (see Section 'Advances in multi- or mixed-methods approaches'). Statistical methods, for instance, are based on the assumption of absence of interference (i.e. the effect of manipulating one part of the system does not depend on changes in other parts of the system), which is highly unlikely in SES given social-ecological feedbacks (Ferraro, Sanchirico, and Smith 2019). Most importantly, there is always a need for expert knowledge about the system to guide interpretation of the results, and for recognising the assumptions and limitations of the method used. Beyond quantitative methods, qualitative methods can shed light on complex causal processes in individual cases. Biesbroek, Dupuis and Wellstead (2017), for instance, argue for mechanism-based approaches and the use of process tracing to unravel the complex causal mechanisms underlying adaptive governance (see an example in Sieber, Biesbroek, and De Block 2018).

Methods accounting for multiple scales/levels or cross-level interactions

Most disciplines and research fields focus on a selected level or scale, such as the individual, community or societal levels; or local, regional or global scales. The associated methods are often particularly suitable for that level or scale, and may be incompatible with others. Moreover, technical limitations may constrain the level of complexity and hence the number of levels or scales a method can address. Together, these characteristics limit the ability of methods to address multi-scale, multi-level and cross-level dynamics. It may be difficult, for example, for methods suited to studying systems at the local scale (e.g. institutional analysis (Chapter 22), livelihood and vulnerability analysis (Chapter 32)) to include cross-scale drivers, and most methods have difficulty accounting for dynamic interactions across scales (De Vos, Biggs, and Preiser 2019). Agent-based modelling, network analysis and GIS mapping are examples of methods that have the potential to address cross-scale interactions and are already doing so (e.g. Guerrero, Mcallister, and Wilson 2015; Maciejewski and Cumming 2016; Miyasaka et al. 2017; Lippe et al. 2019; Cumming and Dobbs 2020). In this context, multi-method approaches also become very important (see Section 'Advances in multi- or mixed-methods approaches'). The challenge of cross-scale and cross-level interactions calls for research in inter- and transdisciplinary teams that use multi-scalar entry points when collecting information while at the same time engaging in partnerships to account for emergent properties, feedbacks and non-linearities at and across scales so that various system facets can be connected to one another (Pricope et al. 2020).

Methods to overcome dichotomies and account for social-ecological intertwinedness

Whereas the need to better integrate the social and the ecological in SES research is increasingly recognised (see e.g. Fischer et al. 2015; Guerrero et al. 2018), doing so poses particularly difficult methodological challenges. As discussed in Chapter 2, conceptualising SES as co-constituted by social-ecological relations requires an ontology that does not separate social and ecological, culture and nature, subject and object (Hertz, Mancilla García, and Schlüter 2020). Methods rooted in either the social or the natural sciences are, however, often based on such dichotomies, which limit their ability to address social-ecological intertwinedness. A method that requires working with distinct social and ecological entities that exist independently from one another cannot account for the creation of novel SES elements through continuously interacting social and ecological processes. The same applies to measuring SES. Indicators for biodiversity conservation and human well-being, for instance, are largely developed separately and often viewed in opposition to one another, which makes it impossible to conceive of human and ecological well-being as an interrelated system (Caillon et al. 2017). In addition, most methods are prone to focusing more on particular elements, actors or processes of SES from either the social or the ecological realm. An example is flow and impact analysis (Chapter 30). Despite having developed from both the social and the natural sciences, SES researchers performing flow and impact analysis often use methods adapted from particular disciplines (e.g. economics), which limits their ability to integrate human and ecological dimensions, or account for interdependencies in SES processes.

Attempts to overcome dichotomies range from developing a framework that puts interactions between human and non-human actors at the centre of analysis (Schlüter et al. 2019a), to the use of relational approaches (West et al. 2020), to methods such as radical empiricism that are based on process-relational ontologies (Mancilla García, Hertz, and Schlüter 2020). Process-relational approaches encourage careful questioning and rebuilding of the concepts used to study or engage with SES, thus making it possible to overcome dichotomies (Mancilla García, Hertz, and Schlüter 2020; West et al. 2020). Other fields can also be sources for ideas and concepts that help overcome dichotomies. Ecofeminism, for instance, draws on how women understand their connection with nature and uses this embodied knowledge as a motivation and justification for introducing new notions of intertwinedness, partnership, agency, care and stewardship (Merchant 2018). Contemporary scholars draw on ideas of post-humanism and new materialism in which the role and agency of humans is not elevated above the agency of non-living and transient beings. Novel notions such as 'sympoeisis' suggest that there is a radical kind of entangled reciprocity between all living and non-living beings and that the world as such comes about because of a kind of 'making with'. As described by the leading 'multi-species' feminist, Donna Haraway (Haraway 2018), sympoeisis ascribes a kind of intertwinedness where all are ultimately connected to one another in ways that the specificity and proximity of connections matter. And it is the nature of the relations that emerge from the interactions that brings about structures and 'ways of being and becoming' in this world. Experiential methods, ritual practices, facilitating knowledge co-production and immersive practices can allow researchers and stakeholders to experience these modes of being co-constituted in a relational way, and can bring about a deeper awareness of the intertwined nature of SES as complex adaptive systems.

Methods to support knowledge co-production and reflexive research processes

Arts-based and other creative approaches to support knowledge co-production

Methods addressing the complexity of SES interactions and challenges are generally not good at offering clear-cut or rationally deduced directives about the best actions or interventions to effect desired SES changes. Sense-making processes are often the most appropriate way of initiating action in SES, and methods that allow multiple perspectives and voices to be included tend to be most effective. Methods that support knowledge co-production processes are good at facilitating these processes of joint sense-making. Many of the knowledge co-production methods discussed in this book (Chapters 9–15) are at the forefront of methodological development in this respect. These methods foster broader engagement with understanding diverse values and ethical imperatives regarding what is considered as desirable and just change.

One aspect that has recently gained increased attention is the importance of drawing on creativity as a resource for facilitating knowledge co-production and engagement processes. Both art and science provide avenues for inquiry and communication, impacting different audiences through the generation of a multiplicity of diverse narratives and modes of representation. Art has the ability to convey the complexity of SES intertwinedness in experimental and experiential mediums and platforms and can generate shifts in social perceptions and behaviours that can provide complementary pathways for SES knowledge co-production and engagement (Born and Barry 2010). Art–science collaborations provide a means for artists, scientists and societal stakeholders to discover new ways to convey their understanding of SES interactions to others, and provide an open platform to juxtapose potentially conflicting and contradictory perspectives (Galafassi et al. 2018; Paterson et al. 2020). Art–science approaches are increasingly used to inform scientific and public literacy and engagement concerning sustainability challenges (Eldred 2016; Angeler, Alvarez-Cobelas, and Sánchez-Carrillo 2018) and foster more embodied and experiential participation in social-ecological research projects (see Chapter 8 on participatory data collection; Chapter 15 on action research). Novel examples include data sonification, an approach that allows converting scientific data into music (Angeler, Alvarez-Cobelas, and Sánchez-Carrillo 2018) and poetic inquiry, an approach that can encourage researcher reflexivity, disrupt hierarchies and humanise research by centring on participants' lived experience (Fernández-Giménez, Jennings, and Wilmer 2019). Another example is the use of performance, e.g. through forum theatre or role-playing games, to generate empathy, engage with a range of emotions and explore how participants can collectively find solutions to a shared problem (Brown, Seo, and Rounsevell 2019).

Methods to monitor the impact of research and action

There are relatively few methods to assess the impacts of research and action in SES. Standard 'key performance index' appraisals do poorly when it comes to evaluating whether social learning or reflective and iterative processes of collaborative knowledge co-creation or change-making have taken place or not in an SES. Reflecting on lessons learnt and building this into futures planning is an essential part of action-oriented SES research, and monitoring and evaluation provide important data and experiences that contribute to this learning (Morris and Lawrence 2010). Monitoring, evaluation and learning (MEL) processes stimulate and capture shared learning and are particularly relevant to SES research and programmes designed with a systems orientation. RESILIM-O, a USAID-funded programme focused on building resilience in the Olifants River Basin in South Africa (AWARD 2017), for example, was based on the premise that the environmental and social challenges are complex and interdependent. The programme therefore used an evaluation approach that relied on 'systemic social learning' and 'learning together what is not yet known', through interactive, participatory and open-ended methods that included institutions at multiple levels. This meant that involvement of multiple activities and role-plays was seen as critical in assessing project outcomes. The purpose of the MEL approach in this programme included accountability, not only for the funders but for all stakeholders involved in the process of building resilience in the region – to communicate success stories and areas that need attention; to guide strategic planning on which projects should continue, change or stop; and to provide internal learning among the project partners, and external learning among development partners and external stakeholders (AWARD 2017).

Methods to support reflexivity

The importance of more reflexive modes of engaging with SES, in both research processes and knowledge co-production, has been a central theme throughout this book (see Chapters 1–3). Reflexivity is particularly important in view of multiple understandings of SES: to position one's research; to ensure consistency within one's research approach; to enable collaboration within diverse groups of scientists, practitioners or stakeholders; to make explicit and deal with biases, including understanding and communicating how one's personal biases may affect results and their interpretation; and to ensure that ethical aspects are taken into account, such as how inclusive the research is (or not) and whether diverse viewpoints have been considered. Despite the growing awareness of the need for reflexivity, there is still a lack of tools to support processes of reflexive engagement with SES. A few recent developments include a toolbox for philosophical dialogue, which is a set of questions to help identify and address philosophical disparities and commonalities across a group of researchers (Eigenbrode et al. 2007), a toolkit to elicit one's 'ologies' (seslink.org), i.e. theoretical and methodological commitments, and a heuristic tool to articulate and discuss individual research strategies (Hazard et al. 2020). Collaborative and participatory modelling are also useful tools to make explicit diverse and possibly contradicting viewpoints among scientists or among different stakeholders (Singer et al. 2017; Schlüter et al. 2019b).

Methods for synthesis and theory building

Social-ecological systems research has over the last two decades accumulated much in-depth, place-based knowledge and understanding of key SES features and behaviours across a diversity of contexts. Synthesising this knowledge in a way that accounts for the complex adaptive systems nature of SES, particularly context dependence, radical openness and emergence, may help to provide carefully generalised knowledge to inform SES governance (Magliocca et al. 2018). Synthesis is a research approach that draws upon many sources of data, ideas, explanations and methods in order to generalise and build theory (Magliocca et al. 2018). Efforts to synthesise existing SES knowledge are, however, complicated by a lack of approaches and methods that can deal with different types of data and the diversity of concepts and methods by which they were collected (Magliocca et al. 2018; Cox et al. 2020). Both methodological pluralism and the variability of SES dynamics across different contexts make the development of generalisable knowledge to inform middle-range theories and governance difficult (see Chapter 22 on institutional analysis; Chapter 19 on qualitative content analysis; Cox 2015; Bodin et al. 2019; De Vos, Biggs, and Preiser 2019).

Despite these challenges, recent years have seen more and more research that moves towards synthesis. Databases of variables found across empirical cases are one attempt to standardise approaches across studies that facilitate synthesis and theory building (Cox et al. 2020). However, standardisation comes at the expense of being able to adapt methodologies to specific contexts (Magliocca et al. 2018). Examples of databases that have been developed to facilitate comparison and synthesis are the thresholds database (resalliance.org/tdb-database), regime shifts database (regimeshifts.org), the SESMAD database (sesmad.dartmouth.edu) and the SES library (seslibrary.asu.edu). Similarly, there are first attempts to facilitate qualitative data sharing and synthesis (Alexander et al. 2019). Synthesis and cross-case comparison are just one approach that can be used for context-sensitive generalisation and theorising. Recent methodological developments in SES research and the social sciences include archetype analysis (Oberlack et al. 2019) and combining cross-case with within-case analysis for developing typologies (Møller and Skaaning 2017). Another methodology for theory building in SES combines the development of empirical explanations of observed phenomena with agent-based modelling to test and explore possible explanations (Magliocca et al. 2015; Schlüter et al. 2019b). Through this combination, particularly when applied in an iterative and collaborative process that involves empirical researchers and modellers, different assumptions and understandings can be made explicit and their consequences explored through modelling and field research.

Big data, machine learning and virtual/augmented reality

Significant technological and analytical developments have enhanced the generation, storage, processing and analysis of large-scale biophysical and social datasets (Franklin et al. 2017; Gorelick et al. 2017; Dong et al. 2019; see Chapter 6 on ecological field data collection; Chapter 16 on expert modelling; Chapter 17 on data mining and pattern recognition; Chapter 18 on statistical analysis; Chapter 27 on state-and-transition modelling). These advances have increased data availability and understanding of global SES, particularly land systems such as forests (Hansen et al. 2013), surface water bodies (Pekel et al. 2016), urban accessibility (Weiss et al. 2018), agriculture (Tian et al. 2019) and fisheries (Kroodsma et al. 2018). Global-scale measurements of socio-economic characteristics are generally harder to derive than biophysical land cover (Dong et al. 2019), but are also expanding through, for example, crowdsourcing of social sensing data (Fritz et al. 2017; Zulkarnain et al. 2019) and mining of location-based social media data (Di Minin, Tenkanen, and Toivonen 2015; Jendryke et al. 2017; Chapter 17 on data mining and pattern recognition).

The use and usefulness of automatically derived data (e.g. remote sensing, automatic sensors) and big data have advanced significantly through the application of artificial intelligence and machine-learning techniques, such as deep learning (e.g. Christin, Hervet, and Lecomte 2019; see also Chapter 18 on statistical analysis; Chapter 17 on data mining and pattern recognition). These techniques allow for the synergising of datasets that could previously only be used in isolation (Jendryke et al. 2017; Christin, Hervet, and Lecomte 2019; Esch et al. 2020). Deep-learning approaches can help combine different resolutions and scales of data, or integrate social and biophysical datasets for the purpose of better understanding landscape dynamics, particularly as they relate to human activities (Dong et al. 2019; Chapter 25 on historical assessment). Deep learning is also contributing significantly to the development of complex predictive and analytical models (see Chapter 17 on data mining and pattern recognition; Chapter 18 on statistical analysis), and deep learning and historical land analyses are increasingly combined with scenario development to inform strategic planning processes (Drees and Liehr 2015; Sang 2020).

Virtual and augmented reality approaches have become popular as a way to elicit human values related to (often future) ecological conditions (see Chapter 8 on participatory data collection; Paine 2016; Smithwick et al. 2018; Smithwick et al. 2019) and to create realistic future worlds in scenario development, e.g. in planning smart cities (Jamei et al. 2017; Chapter 11 on scenario development; Chapter 10 on futures analysis). 'Experiential futures' bring a future into the real world, making it an immediate, first-hand encounter (Zaidi 2019). A related 'world building' technique that is increasingly used in scenario development is the use of science fiction prototyping to depict rich, nuanced storied futures (Merrie et al. 2018). This interplay between world building and storytelling is psychologically more compelling and realistic than an abstract futurist scenario or statistical prediction (Merrie et al. 2018; Zaidi 2019). Indeed, combining virtual and augmented reality and science fiction prototyping may blur the lines between experiential futures and science fiction (Zaidi 2019).

Whereas big data, machine learning and virtual/augmented reality approaches offer exciting opportunities for advancing our understanding of cross-scale and large-scale social-ecological dynamics, and operationalising new perspectives and solutions, their application requires thoughtful reflexivity (Gulsrud et al. 2018). As discussed briefly in Chapter 3, researchers have to take into account ethical concerns about regional and demographic representation underlying big datasets, discriminatory algorithms based on narrow training data, the exclusion of certain groups (e.g. older people) in virtual/augmented reality approaches, data ownership, and privacy concerns related to where data are sourced, and what additional personal data may be collected by high-tech devices ranging from smartphones to satellites (Di Minin, Tenkanen, and Toivonen 2015; Mittelstadt and Floridi 2016; Stahl and Wright 2018). In SES, neither risks nor opportunities are fixed, but are dynamic properties of changing internal contexts and cross-scale interactions (Gulsrud et al. 2018). These uncertainties, combined with the evolving nature of technology that has been changing human–nature relationships, human agency and cross-scale interactions in SES (Ahlborg et al. 2019), mean that SES researchers should be particularly aware of unintended consequences of using high-tech methods and tools.

Advances in multi- or mixed-methods approaches

The combination of methods in multi- and mixed-methods approaches is increasingly common in SES research (see Chapter 3 for a definition of multi- and mixed methods). Methods are combined in order to study multiple scales or cross-scale interactions, collect different types of data, integrate different perspectives or triangulate findings, to name just a few. While multi- or mixed-methods approaches are useful to overcome the limitations of individual methods and to include multiple perspectives or ways of analysing a system, they need to be applied with careful consideration of possible incompatibilities of the worldviews or theoretical foundations underlying each method (Johnson and Onwuegbuzie 2004).

In the context of data collection and analysis, methods have been combined to facilitate the collection of a broad range of information and data about a situation and to better account for differences between social and ecological data. An example is the combination of household questionnaires, participatory time lines, oral histories, focus group discussions, vegetation and wild animal surveys, and analysis of remote-sensing images to inform a livelihoods assessment (Sallu, Twyman, and Stringer 2010). In this study, the different quantitative and qualitative data collected were analysed using methods such as thematic analysis and iterative reflexivity to allow for inductive interpretation of qualitative data as well as statistical analysis of quantitative data. Furthermore, method combinations can be useful to study SES across scales. An example is the combination of remotely sensed data analyses with (spatial) participatory data collection to study ecosystem change and its relation to ecosystem services (Brown et al. 2018; Delgado-Aguilar, Hinojosa, and Schmitt 2019). Finally, multi- and mixed-methods approaches allow triangulation of findings to build confidence in the results and account for limitations of individual methods (Bentley Brymer et al. 2016; Lee et al. 2019; Salomon et al. 2019).

In the context of modelling, method combinations are very common, particularly when the process of model building involves collecting and analysing empirical data, often in participatory ways (Voinov et al. 2018). Frontiers of combining empirical methods with modelling include using qualitative data/narratives to build the model structure (e.g. Lindkvist, Basurto, and Schlüter 2017), combining social network analysis (Dobson et al. 2019; Will et al. 2020) or process tracing with agent-based modelling (Orach, Duit, and Schlüter 2020). If models are constructed collaboratively, 'negotiation' processes between those who have the empirical understanding and those who develop the model can highlight gaps and differences in understanding that can then be explored with the model or further field research. Process tracing can be used to establish causal processes in SES that can be further explored through modelling. Orach, Schlüter and Österblom (2017), for example, use process tracing to identify coalition formation as a key mechanism through which environmental interest groups managed to attain their preferences in the 2013 EU Common Fisheries Policy reform. Using an agent-based model that formalised this mechanism, they could then explore how and under which conditions interest-group competition can lead to sustainable resource management (Orach, Duit, and Schlüter 2020).

Finally, the combination of different types of modelling, such as agent-based and dynamical systems modelling, allows researchers to make use of the strengths of the respective modelling approaches, such as the mathematical analysis methods available for dynamical systems modelling and the ability of agent-based modelling to represent human behaviour. An example is the combination of a system dynamics model of a lake with an agent-based model of a community to study the interplay between policy responses in the face of deteriorating lake conditions and the ecological dynamics of the lake (Martin and Schlüter 2015).

In the context of futures work, there is much experimentation with combining different tools and approaches. The Seeds of Good Anthropocenes (goodanthropocenes.net) initiative, for example, has developed a new bottom-up scenario methodology that combines a variety of futures tools, including the Mānoa method for detecting weak signals, futures wheels, the three horizons framework, and experiential futures to explore how local, potentially transformative social-ecological initiatives might grow and together create radically alternative futures (see Chapter 10 on futures analysis). Many SES scenario-development processes involve mixed-methods approaches, where narrative storylines are initially developed through various participatory approaches. These storylines may then be quantified using a variety of different models, with the outputs of some models serving as input into others. Model outputs are then typically discussed with a range of stakeholders, leading to adjustments of the storylines and models to ensure plausibility (see Chapter 11 on scenario development). In general, combining methods in knowledge co-production activities can allow for creativity and flexibility while at the same time grounding the research in biophysical and socio-economic realities. This was done, for example, by combining creative thinking and storytelling with quantitative modelling of drivers and trends to develop potential global futures of ecosystem change and human well-being during the Millennium Assessment.

Conclusion

Social-ecological systems research draws on a diverse set of approaches and methods to address real-world problems and effect change towards more sustainable and just futures. It has pioneered new ways of doing research, of doing research in a more socially just way, and of engaging with society to effect change towards more sustainable pathways. Social-ecological systems research provides opportunities to question established assumptions and fundamentally rethink the nature of reality and our ability to study and shape it. It acknowledges that the researcher is part of the SES (not just an outside observer) and poses important ethical questions. These developments reflect a fundamental shift from a mechanistic worldview towards a complexity perspective that views SES as intertwined complex adaptive systems. This shift has stimulated much exciting research and action that is visible in a proliferation of approaches and methods that at times can be bewildering.

The aim of this book is to help researchers navigate the emerging SES field by providing a comprehensive synthesis and guide to this diversity of methods, grounded in an understanding of SES as complex adaptive, intertwined systems. The book goes beyond a mere compilation of commonly used methods by reflecting on the challenges that a complexity perspective holds for how we conceptualise SES, choose and apply methods, produce knowledge and attempt to effect change within SES. We have grounded the methods in their conceptual foundations, assessed their suitability for addressing different systemic features and processes in SES, and reflected on their limitations. Together, this grounding and mapping of methods help to clarify what each method can do, how it relates to other methods, and the different approaches, knowledge types and purposes of application of each method. We hope that the book enables SES researchers to make informed choices about the method(s) to use for a particular purpose, research goal or activity in a given situation and to critically reflect on the use of a method. Furthermore, we hope that it will serve as a foundation for developing new methods or combining methods in useful and sensible ways.

No single method can by itself capture all aspects of complexity and intertwinedness. Some methods are more limited in their ability to account for complexity because of their underlying epistemology. Others are more flexible or are based on a complexity perspective such as network analysis, dynamical systems modelling, agent-based modelling and the many methods for effecting system change in co-production processes. A better understanding of the conceptual foundations, strengths and limitations of approaches and methods can support an assessment of their suitability for a problem or question of interest. Furthermore, the different perspectives that characterise SES research and the different strengths and limitations of methods call for pluralist and integrative approaches that combine or contrast different methods in order to take advantage of their differing strengths and weaknesses. However, the theoretical commitments and epistemologies underlying different methods need to be navigated with care as they may involve incompatibilities.

Social-ecological systems research diverges from the tradition of the lone genius. Instead, doing research and engaging in SES is an inherently collaborative and integrative endeavour across disciplines, knowledge systems, and science and practice. This does not mean that every research endeavour is necessarily team research; however, every researcher will, most likely, engage with various understandings and methods coming from a diversity of worldviews and epistemologies. Whereas a plurality of methods is needed to deal with the complexity and intertwinedness of SES, these processes require careful engagement and communication as well as a reflexive practice of doing research and engaging with other researchers and stakeholders. Ultimately, studying SES and affecting change towards sustainability is a continuous learning process. Social-ecological systems are continuously changing, as is our understanding of them.

Acknowledgements

We would like to thank Krisztina Jónás for helping with the analysis and synthesis of the sections on limitations and future directions of each method chapter in Part 2 of the book. Our thanks also go to Laura Elsler, Blanca González-Mon, Romina Martin, Kirill Orach and Udita Sanga for comments that greatly helped improve the manuscript. Maja Schlüter received support from the European Research Council (ERC) under the European Union's Horizon 2020 research and innovation programme (grant agreement no. 682472 – MUSES). Reinette Biggs received support from the South African Research Chairs Initiative (SARChI) (grant 98766), the Guidance for Resilience in the Anthropocene: Investments for Development (GRAID) project funded by the Swedish International Development Agency (Sida) and a Young Researchers Grant from the Vetenskapsrådet in Sweden (grant 621-2014-5137). Hayley Clements is funded by a Jennifer Ward Oppenheimer Research Grant and Kone Foundation. Alta de Vos received support from a Rhodes University Council Grant. Rika Preiser's salary was co-funded by the Swedish International Development Agency's (Sida) GRAID and Swedbio programmes.

References

Ahlborg, H., I. Ruiz-Mercado, S. Molander, and O. Masera. 2019. 'Bringing Technology into Social-Ecological Systems Research – Motivations for a Socio-Technical-Ecological Systems Approach.' *Sustainability* 11(7): 2009. doi:10.3390/su11072009.

Alexander, S.M., K. Jones, N.J. Bennett, A. Budden, M. Cox, M. Crosas, E.T. Game et al. 2019. 'Qualitative Data Sharing and Synthesis for Sustainability Science.' *Nature Sustainability*. doi:10.1038/s41893-019-0434-8.

Angeler, D.G., M. Alvarez-Cobelas, and S. Sánchez-Carrillo. 2018. 'Sonifying Social-Ecological Change: A Wetland Laments Agricultural Transformation.' *Ecology and Society* 23(2): 20. doi:10.5751/ES-10055-230220.

Angelstam, P., K. Andersson, M. Annerstedt, R. Axelsson, M. Elbakidze, P. Garrido, P. Grahn et al. 2013. 'Solving Problems in Social-Ecological Systems: Definition, Practice and Barriers of Transdisciplinary Research.' *Ambio* 42(2): 254–265. doi:10.1007/s13280-012-0372-4.

AWARD (Association for Water and Rural Development). 2017. 'Monitoring, Evaluation, Reporting and Learning for the USAID RESILIM-O Programme MERL FRAMEWORK Cooperative Agreement AID-674-A-13-00008.' Hoedspruit, South Africa.

Beach, D., and R.B. Pedersen. 2013. *Process-Tracing Methods: Foundations and Guidelines*. Cambridge: University of Michigan Press.

Bentley Brymer, A.L., J.D. Holbrook, R.J. Niemeyer, A.A. Suazo, J.D. Wulfhorst, K.T. Vierling, B.A. Newingham, T.E. Link, and J.L. Rachlow. 2016. 'A Social-Ecological Impact Assessment for Public Lands Management: Application of a Conceptual and Methodological Framework.' *Ecology and Society* 21(3): 9. doi:10.5751/ES-08569-210309.

Biesbroek, R., J. Dupuis, and A. Wellstead. 2017. 'Explaining through Causal Mechanisms: Resilience and Governance of Social-Ecological Systems.' *Current Opinion in Environmental Sustainability* 28: 64–70. doi:10.1016/j.cosust.2017.08.007.

Bodin, Ö., S.M. Alexander, J. Baggio, M.L. Barnes, R. Berardo, G.S. Cumming, L.E. Dee et al. 2019. 'Improving Network Approaches to the Study of Complex Social-Ecological Interdependencies.' *Nature Sustainability* 2(7): 551–559. doi:10.1038/s41893-019-0308-0.

Boonstra, W.J., and F.W. de Boer. 2014. 'The Historical Dynamics of Social-Ecological Traps.' *Ambio* 43(3): 260–274. doi:10.1007/s13280-013-0419-1.

Born, G., and A. Barry. 2010. 'ART-SCIENCE: From Public Understanding to Public Experiment.' *Journal of Cultural Economy* 3(1): 103–119. doi:10.1080/17530351003617610.

Boulton, J.G., P.M. Allen, and C. Bowman. 2015. *Embracing Complexity: Strategic Perspectives for an Age of Turbulence*. Oxford: Oxford University Press. doi:10.1093/acprof:oso/9780199565252.001.0001.

Bourdieu, P. 1991. *On Symbolic Power, Language and Symbolic Power*. Cambridge: Polity in association with Basil Blackwell.

Bretagnolle, V., M. Benoit, M. Bonnefond, V. Breton, J.M. Church, S. Gaba, D. Gilbert et al. 2019. 'Action-orientated Research and Framework: Insights from the French Long-Term Social-Ecological Research Network.' *Ecology and Society* 24(3): 10. doi:10.5751/ES-10989-240310.

Brown, C., B. Seo, and M. Rounsevell. 2019. 'Societal Breakdown as an Emergent Property of Large-Scale Behavioural Models of Land Use Change.' *Earth System Dynamics* 10(4): 809–845. doi:10.5194/esd-10-809-2019.

Brown, M.I., T. Pearce, J. Leon, R. Sidle, and R. Wilson. 2018. 'Using Remote Sensing and Traditional Ecological Knowledge (TEK) to Understand Mangrove Change on the Maroochy River, Queensland, Australia.' *Applied Geography* 94: 71–83. doi:10.1016/j.apgeog.2018.03.006.

Caillon, S., G. Cullman, B. Verschuuren, and E.J. Sterling. 2017. 'Moving beyond the Human–Nature Dichotomy through Biocultural Approaches: Including Ecological Well-Being in Resilience Indicators.' *Ecology and Society* 22(4): 27. doi:10.5751/ES-09746-220427.

Caniglia, G., C. Luederitz, T. von Wirth, I. Fazey, B. Martín-López, K. Hondrila, A. König et al. 2020. 'A Pluralistic and Integrated Approach to Action-oriented Knowledge for Sustainability.' *Nature Sustainability*. doi:10.1038/s41893-020-00616-z.

Carlson, A., J. Zaehringer, R. Garrett, R. Felipe Bicudo Silva, P. Furumo, A. Raya Rey, A. Torres, M. Gon Chung, Y. Li, and J. Liu. 2018. 'Toward Rigorous Telecoupling Causal Attribution: A Systematic Review and Typology.' *Sustainability* 10(12): 4426. doi:10.3390/su10124426.

Chickering, D.M. 2002. 'Learning Equivalence Classes of Bayesian-network Structures.' *Journal of Machine Learning Research* 2: 445–498.

Christin, S., É. Hervet, and N. Lecomte. 2019. 'Applications for Deep Learning in Ecology.' *Methods in Ecology and Evolution* 10(10): 1632–1644. doi:10.1111/2041-210X.13256.

Cox, M. 2015. 'A Basic Guide for Empirical Environmental Social Science.' *Ecology and Society* 20(1): 63. doi:10.5751/ES-07400-200163.

Cox, M., S. Villamayor-Tomas, N.C. Ban, G. Epstein, L. Evans, F. Fleischman, M. Nenadovic et al. 2020. 'From Concepts to Comparisons: A Resource for Diagnosis and Measurement in Social-Ecological Systems.' *Environmental Science & Policy* 107: 211–216.

Cumming, G.S., and K.A. Dobbs. 2020. 'Quantifying Social-Ecological Scale Mismatches Suggests People Should Be Managed at Broader Scales Than Ecosystems.' *One Earth* 3(2): 251–259. doi:10.1016/j.oneear.2020.07.007.

Delgado-Aguilar, M.J., L. Hinojosa, and C.B. Schmitt. 2019. 'Combining Remote Sensing Techniques and Participatory Mapping to Understand the Relations between Forest Degradation and Ecosystems Services in a Tropical Rainforest.' *Applied Geography* 104: 65–74. doi:10.1016/j.apgeog.2019.02.003.

De Vos, A., R. Biggs, and R. Preiser. 2019. 'Methods for Understanding Social-Ecological Systems: A Review of Place-Based Studies.' *Ecology and Society* 24(4): 16. doi:10.5751/ES-11236-240416.

Di Minin, E., H. Tenkanen, and T. Toivonen. 2015. 'Prospects and Challenges for Social Media Data in Conservation Science.' *Frontiers in Environmental Science* 3. doi:10.3389/fenvs.2015.00063.

Díaz, S., S. Demissew, J. Carabias, C. Joly, M. Lonsdale, N. Ash, A. Larigauderie et al. 2015. 'The IPBES Conceptual Framework – Connecting Nature and People.' *Current Opinion in Environmental Sustainability* 14: 1–16. doi:10.1016/j.cosust.2014.11.002.

Dobson, A.D.M., E. de Lange, A. Keane, H. Ibbett, and E.J. Milner-Gulland. 2019. 'Integrating Models of Human Behaviour between the Individual and Population Levels to Inform Conservation Interventions.' *Philosophical Transactions of the Royal Society B: Biological Sciences* 374(1781): 20180053. doi:10.1098/rstb.2018.0053.

Dong, J., G. Metternicht, P. Hostert, R. Fensholt, and R.R. Chowdhury. 2019. 'Remote Sensing and Geospatial Technologies in Support of a Normative Land System Science: Status and Prospects.' *Current Opinion in Environmental Sustainability* 38: 44–52.

Drees, L., and S. Liehr. 2015. 'Using Bayesian Belief Networks to Analyse Social-Ecological Conditions for Migration in the Sahel.' *Global Environmental Change* 35: 323–339. doi:10.1016/j.gloenvcha.2015.09.003.

Eigenbrode, S.D., M. O'Rourke, J.D. Wulfhorst, D.M. Althoff, C.S. Goldberg, K. Merrill, W. Morse et al. 2007. 'Employing Philosophical Dialogue in Collaborative Science.' *BioScience* 57(1): 55–64. doi:10.1641/B570109.

Eldred, S.M. 2016. 'Art–Science Collaborations: Change of Perspective.' *Nature* 537:125–126. www.nature.com/articles/nj7618-125a.

Epstein, G., T.H. Morrison, A. Lien, G.G. Gurney, D.H. Cole, M. Delaroche, S. Villamayor-Tomas, N. Ban, and M. Cox. 2020. 'Advances in Understanding the Evolution of Institutions in Complex Social-Ecological Systems.' *Current Opinion in Environmental Sustainability* 44: 58–66.

Esch, T., H. Asamer, F. Bachofer, J. Balhar, M. Boettcher, E. Boissier, P. d'Angelo et al. 2020. 'Digital World Meets Urban Planet – New Prospects for Evidence-Based Urban Studies Arising from Joint Exploitation of Big Earth Data, Information Technology and Shared Knowledge.' *International Journal of Digital Earth* 13(1): 136–157. doi:10.1080/17538947.2018.1548655.

Fernández-Giménez, M.E., L.B. Jennings, and H. Wilmer. 2019. 'Poetic Inquiry as a Research and Engagement Method in Natural Resource Science.' *Society and Natural Resources* 32(10): 1080–1091. doi:10.1080/08941920.2018.1486493.

Ferraro, P.J., J.N. Sanchirico, and M.D. Smith. 2019. 'Causal Inference in Coupled Human and Natural Systems.' *Proceedings of the National Academy of Sciences* 116(12): 5311–5318. doi:10.1073/pnas.1805563115.

Fischer, A., and A. Eastwood. 2016. 'Coproduction of Ecosystem Services as Human–Nature Interactions – An Analytical Framework.' *Land Use Policy* 52: 41–50. doi:10.1016/j.landusepol.2015.12.004.

Fischer, J., T.A. Gardner, E.M. Bennett, P. Balvanera, R. Biggs, S. Carpenter, T. Daw et al. 2015. 'Advancing Sustainability through Mainstreaming a Social-Ecological Systems Perspective.' *Current Opinion in Environmental Sustainability* 14: 144–149. doi:10.1016/j.cosust.2015.06.002.

Foucault, M. 1982. *The Archaeology of Knowledge*. New York, NY: Pantheon Books.

Franklin, J., J.M. Serra-Diaz, A.D. Syphard, and H.M. Regan. 2017. 'Big Data for Forecasting the Impacts of Global Change on Plant Communities.' *Global Ecology and Biogeography* 26(1): 6–17. doi:10.1111/geb.12501.

Fritz, S., L. See, C. Perger, I. McCallum, C. Schill, D. Schepaschenko, M. Duerauer et al. 2017. 'A Global Dataset of Crowdsourced Land Cover and Land Use Reference Data.' *Scientific Data* 4(1): 170075. doi:10.1038/sdata.2017.75.

Galafassi, D., S. Kagan, M. Milkoreit, M. Heras, C. Bilodeau, S.J. Bourke, A. Merrie, L. Guerrero, G. Pétursdóttir, and J.D. Tàbara. 2018. '"Raising the Temperature": The Arts in a Warming Planet.' *Current Opinion in Environmental Sustainability* 31: 71–79.

Gelcich, S., T.P. Hughes, P. Olsson, C. Folke, O. Defeo, M. Fernandez, S. Foale et al. 2010. 'Navigating Transformations in Governance of Chilean Marine Coastal Resources.' *Proceedings of the National Academy of Sciences* 107(39): 16794–16799. doi:10.1073/pnas.1012021107.

Giddens, A. 1984. *The Constitution of Society*. Cambridge: Polity Press.

Gómez-Baggethun, E., and M. Ruiz-Pérez. 2011. 'Economic Valuation and the Commodification of Ecosystem Services.' *Progress in Physical Geography: Earth and Environment* 35(5): 613–628. doi:10.1177/0309133311421708.

Gorelick, N., M. Hancher, M. Dixon, S. Ilyushchenko, D. Thau, and R. Moore. 2017. 'Google Earth Engine: Planetary-scale Geospatial Analysis for Everyone.' doi:10.1016/j.rse.2017.06.031.

Guerrero, A.M., N.J. Bennett, K.A. Wilson, N. Carter, D. Gill, M. Mills, C.D. Ives et al. 2018. 'Achieving the Promise of Integration in Social-Ecological Research: A Review and Prospectus.' *Ecology and Society* 23(3): 38. doi:10.5751/ES-10232-230338.

Guerrero, A.M., R.R.J. Mcallister, and K.A. Wilson. 2015. 'Achieving Cross-Scale Collaboration for Large Scale Conservation Initiatives.' *Conservation Letters* 8(2): 107–117. doi:10.1111/conl.12112.

Gulsrud, N.M., C.M. Raymond, R.L. Rutt, A.S. Olafsson, T. Plieninger, M. Sandberg, T.H. Beery, and K.I. Jönsson. 2018. '"Rage against the Machine"? The Opportunities and Risks Concerning the Automation of Urban Green Infrastructure.' *Landscape and Urban Planning* 180: 85–92. doi:10.1016/j.landurbplan.2018.08.012.

Haider, L.J., J. Hentati-Sundberg, M. Giusti, J. Goodness, M. Hamann, V.A. Masterson, M. Meacham et al. 2018. 'The Undisciplinary Journey: Early-career Perspectives in Sustainability Science.' *Sustainability Science* 13(1): 191–204. doi:10.1007/s11625-017-0445-1.

Hansen, M.C., P.V. Potapov, R. Moore, M. Hancher, S.A. Turubanova, A. Tyukavina, D. Thau et al. 2013. 'High-Resolution Global Maps of 21st-Century Forest Cover Change.' *Science* 342(6160). http://science.sciencemag.org/content/342/6160/850.

Haraway, D. 2018. 'Staying with the Trouble for Multispecies Environmental Justice.' *Dialogues in Human Geography* 8(1): 102–105. doi:10.1177/2043820617739208.

Hazard, L., M. Cerf, C. Lamine, D. Magda, and P. Steyaert. 2020. 'A Tool for Reflecting on Research Stances to Support Sustainability Transitions.' *Nature Sustainability* 3(2): 89–95. doi:10.1038/s41893-019-0440-x.

Herrfahrdt-Pähle, E., M. Schlüter, P. Olsson, C. Folke, S. Gelcich, and C. Pahl-Wostl. 2020. 'Sustainability Transformations: Socio-Political Shocks as Opportunities for Governance Transitions.' *Global Environmental Change* 63: 102097. doi:10.1016/j.gloenvcha.2020.102097.

Hertz, T., M. Mancilla García, and M. Schlüter. 2020. 'From Nouns to Verbs: How Process Ontologies Enhance Our Understanding of Social-Ecological Systems Understood as Complex Adaptive Systems.' *People and Nature* 2(2): 328–338. doi:10.1002/pan3.10079.

Jahn, T., M. Bergmann, and F. Keil. 2012. 'Transdisciplinarity: Between Mainstreaming and Marginalization.' *Ecological Economics* 79: 1–10.

Jamei, E., M. Mortimer, M. Seyedmahmoudian, B. Horan, and A. Stojcevski. 2017. 'Investigating the Role of Virtual Reality in Planning for Sustainable Smart Cities.' *Sustainability* 9(11): 2006. doi:10.3390/su9112006.

Janssen, M.A., M.L. Schoon, W. Ke, and K. Börner. 2006. 'Scholarly Networks on Resilience, Vulnerability and Adaptation within the Human Dimensions of Global Environmental Change.' *Global Environmental Change* 16(3): 240–252. doi:10.1016/j.gloenvcha.2006.04.001.

Jendryke, M., T. Balz, S.C. McClure, and M. Liao. 2017. 'Putting People in the Picture: Combining Big Location-Based Social Media Data and Remote Sensing Imagery for Enhanced Contextual Urban Information in Shanghai.' *Computers, Environment and Urban Systems* 62: 99–112. doi:10.1016/j.compenvurbsys.2016.10.004.

Johnson, R.B., and A.J. Onwuegbuzie. 2004. 'Mixed Methods Research: A Research Paradigm Whose Time Has Come.' *Educational Researcher* 33(7): 14–26. doi:10.3102/0013189X033007014.

Kaaronen, R.O. 2017. 'Affording Sustainability: Adopting a Theory of Affordances as a Guiding Heuristic for Environmental Policy.' *Frontiers in Psychology* 8. doi:10.3389/fpsyg.2017.01974.

Kok, M.T.J., K. Kok, G.D. Peterson, R. Hill, J. Agard, and S.R. Carpenter. 2017. 'Biodiversity and Ecosystem Services Require IPBES to Take Novel Approach to Scenarios.' *Sustainability Science* 12(1): 177–181. doi:10.1007/s11625-016-0354-8.

Kroodsma, D.A., J. Mayorga, T. Hochberg, N.A. Miller, K. Boerder, F. Ferretti, A. Wilson et al. 2018. 'Tracking the Global Footprint of Fisheries.' *Science* 359(6378): 904–908. doi:10.1126/science.aao5646.

Lang, D.J., A. Wiek, M. Bergmann, M. Stauffacher, P. Martens, P. Moll, M. Swilling, and C.J. Thomas. 2012. 'Transdisciplinary Research in Sustainability Science: Practice, Principles, and Challenges.' *Sustainability Science* 7(Supplement 1): 25–43. doi:10.1007/s11625-011-0149-x.

Lee, L.C., J. Thorley, J. Watson, M. Reid, and A.K. Salomon. 2019. 'Diverse Knowledge Systems Reveal Social-Ecological Dynamics That Inform Species Conservation Status.' *Conservation Letters* 12(2): e12613. doi:10.1111/conl.12613.

Lele, S., O. Springate-Baginski, R. Lakerveld, D. Deb, and P. Dash. 2013. 'Ecosystem Services: Origins, Contributions, Pitfalls, and Alternatives.' *Conservation and Society* 11(4): 343. doi:10.4103/0972-4923.125752.

Levin, S., T. Xepapadeas, A-S. Crépin, J. Norberg, A. de Zeeuw, C. Folke, T. Hughes et al. 2012. 'Social-Ecological Systems as Complex Adaptive Systems: Modeling and Policy Implications.' *Environment and Development Economics* 18(2): 111–132. doi:10.1017/S1355770X12000460.

Lindkvist, E., X. Basurto, and M. Schlüter. 2017. 'Micro-Level Explanations for Emergent Patterns of Self-Governance Arrangements in Small-Scale Fisheries – A Modeling Approach.' *PloS One* 12(4): e0175532.

Lippe, M., M. Bithell, N. Gotts, D. Natalini, P. Barbrook-Johnson, C. Giupponi, M. Hallier et al. 2019. 'Using Agent-Based Modelling to Simulate Social-Ecological Systems across Scales.' *GeoInformatica* 23(2): 269–98. doi:10.1007/s10707-018-00337-8.

Liu, J., Y. Dou, M. Batistella, E. Challies, T. Connor, C. Friis, J.D. Millington et al. 2018. 'Spillover Systems in a Telecoupled Anthropocene: Typology, Methods, and Governance for Global Sustainability.' *Current Opinion in Environmental Sustainability* 33: 58–69.

Maciejewski, K., and G.S. Cumming. 2016. 'Multi-Scale Network Analysis Shows Scale-Dependency of Significance of Individual Protected Areas for Connectivity.' *Landscape Ecology* 31(4): 761–774. doi:10.1007/s10980-015-0285-2.

Magliocca, N.R., E.C. Ellis, G.R.H. Allington, A. de Bremond, J. dell'Angelo, O. Mertz, P. Messerli, P. Meyfroidt, R. Seppelt, and P.H. Verburg. 2018. 'Closing Global Knowledge Gaps: Producing Generalized Knowledge from Case Studies of Social-Ecological Systems.' *Global Environmental Change* 50: 1–14. doi:10.1016/j.gloenvcha.2018.03.003.

Magliocca, N.R., J. van Vliet, C. Brown, T.P. Evans, T. Houet, P. Messerli, J.P. Messina et al. 2015. 'From Meta-Studies to Modeling: Using Synthesis Knowledge to Build Broadly Applicable Process-Based Land Change Models.' *Environmental Modelling & Software* 72: 10–20. doi:10.1016/j.envsoft.2015.06.009.

Mancilla García, M., T. Hertz, and M. Schlüter. 2020. 'Towards a Process Epistemology for the Analysis of Social-Ecological System.' *Environmental Values* 29(2): 221–239. doi:10.3197/096327119X15579936382608.

Martin, R., and M. Schlüter. 2015. 'Combining System Dynamics and Agent-Based Modeling to Analyze Social-Ecological Interactions – An Example from Modeling Restoration of a Shallow Lake.' *Frontiers in Environmental Science* 3. doi:10.3389/fenvs.2015.00066.

Merchant, C. 2018. *Science and Nature: Past, Present, and Future*. New York: Routledge.

Merrie, A., P. Keys, M. Metian, and H. Österblom. 2018. 'Radical Ocean Futures-Scenario Development Using Science Fiction Prototyping.' *Futures* 95: 22–32. doi:10.1016/j.futures.2017.09.005.

Meyfroidt, P. 2016. 'Approaches and Terminology for Causal Analysis in Land Systems Science.' *Journal of Land Use Science* 11(5): 501–522. doi:10.1080/1747423X.2015.1117530.

Mittelstadt, B.D., and L. Floridi. 2016. 'The Ethics of Big Data: Current and Foreseeable Issues in Biomedical Contexts.' *Science and Engineering Ethics* 22(2): 303–341.

Miyasaka, T., Q.B. Le, T. Okuro, X. Zhao, and K. Takeuchi. 2017. 'Agent-Based Modeling of Complex Social-Ecological Feedback Loops to Assess Multi-Dimensional Trade-Offs in Dryland Ecosystem Services.' *Landscape Ecology* 32(4): 707–727. doi:10.1007/s10980-017-0495-x.

Møller, J., and S-E. Skaaning. 2017. 'Explanatory Typologies as a Nested Strategy of Inquiry: Combining Cross-case and Within-case Analyses.' *Sociological Methods & Research* 46(4): 1018–10148. doi:10.1177/0049124115613778.

Moore, M-L., P. Olsson, W. Nilsson, L. Rose, and F. Westley. 2018. 'Navigating Emergence and System Reflexivity as Key Transformative Capacities: Experiences from a Global Fellowship Program.' *Ecology and Society* 23(2): 38. doi:10.5751/ES-10166-230238.

Moore, M.-L., O. Tjornbo, E. Enfors, C. Knapp, J. Hodbod, J.A. Baggio, A. Norström, P. Olsson, and D. Biggs. 2014. 'Studying the Complexity of Change: Toward an Analytical Framework for Understanding Deliberate Social-Ecological Transformations.' *Ecology and Society* 19(4): 54. doi:10.5751/ES-06966-190454.

Morris, J., and A. Lawrence. 2010. 'Learning from Monitoring & Evaluation – a Blueprint for an Adaptive Organisation.' Social & Economic Research Group, Forest Research.

Norström, A.V., C. Cvitanovic, M.F. Löf, S. West, C. Wyborn, P. Balvanera, A.T. Bednarek et al. 2020. 'Principles for Knowledge Co-Production in Sustainability Research.' *Nature Sustainability* 3(3): 182–190. doi:10.1038/s41893-019-0448-2.

Nuno, A., N. Bunnefeld, and E. Milner-Gulland. 2014. 'Managing Social-Ecological Systems under Uncertainty: Implementation in the Real World.' *Ecology and Society* 19(2): 52. doi:10.5751/ES-06490-190252.

Oberlack, C., D. Sietz, E. Bürgi Bonanomi, A. de Bremond, J. Dell'Angelo, K. Eisenack, E.C. Ellis et al. 2019. 'Archetype Analysis in Sustainability Research: Meanings, Motivations, and Evidence-Based Policy Making.' *Ecology and Society* 24(2): 26. doi:10.5751/ES-10747-240226.

Orach, K., A. Duit, and M. Schlüter. 2020. 'Sustainable Natural Resource Governance under Interest Group Competition in Policy-Making.' *Nature Human Behaviour*, May. doi:10.1038/s41562-020-0885-y.

Orach, K., M. Schlüter, and H. Österblom. 2017. 'Tracing a Pathway to Success: How Competing Interest Groups Influenced the 2013 EU Common Fisheries Policy Reform.' *Environmental Science & Policy* 76: 90–102. doi:10.1016/j.envsci.2017.06.010.

Oteros-Rozas, E., B. Martín-López, T.M. Daw, E.L. Bohensky, J.R.A. Butler, R. Hill, J. Martin-Ortega et al. 2015. 'Participatory Scenario Planning in Place-Based Social-Ecological Research: Insights and Experiences from 23 Case Studies.' *Ecology and Society* 20(4): 32. doi:10.5751/ES-07985-200432.

Paine, G. 2016. 'Ecologies of Listening and Presence: Perspectives from a Practitioner.' *Contemporary Music Review* 35(3): 362–371. doi:10.1080/07494467.2016.1239385.

Paterson, S.K., M. le Tissier, H. Whyte, L.B. Robinson, K. Thielking, M. Ingram, and J. McCord. 2020. 'Examining the Potential of Art-Science Collaborations in the Anthropocene: A Case Study of Catching a Wave.' *Frontiers in Marine Science* 7: 340. doi:10.3389/fmars.2020.00340.

Pekel, J.F., A. Cottam, N. Gorelick, and A.S. Belward. 2016. 'High-Resolution Mapping of Global Surface Water and Its Long-term Changes.' *Nature* 540(7633): 418–422. doi:10.1038/nature20584.

Polasky, S., S.R. Carpenter, C. Folke, and B. Keeler. 2011. 'Decision-Making under Great Uncertainty: Environmental Management in an Era of Global Change.' *Trends in Ecology & Evolution* 26(8): 398–404. doi:10.1016/j.tree.2011.04.007.

Popa, F., and M. Guillermin. 2017. 'Reflexive Methodological Pluralism.' *Journal of Mixed Methods Research* 11(1): 19–35. doi:10.1177/1558689815610250.

Popa, F., M. Guillermin, and T. Dedeurwaerdere. 2015. 'A Pragmatist Approach to Transdisciplinarity in Sustainability Research: From Complex Systems Theory to Reflexive Science.' *Futures* 65: 45–56. doi:10.1016/j.futures.2014.02.002.

Preiser, R., R. Biggs, A. de Vos, and C. Folke. 2018. 'Social-Ecological Systems as Complex Adaptive Systems: Organizing Principles for Advancing Research Methods and Approaches.' *Ecology and Society* 23(4): 46. doi:10.5751/ES-10558-230446.

Pricope, N.G., L. Cassidy, A.E. Gaughan, J.D. Salerno, F.R. Stevens, J. Hartter, M. Drake, and P. Mupeta-Muyamwa. 2020. 'Addressing Integration Challenges of Interdisciplinary Research in Social-Ecological Systems.' *Society & Natural Resources* 33(3): 418–431. doi:10.1080/08941920.2019.1680783.

Raymond, C.M., M. Giusti, and S. Barthel. 2018. 'An Embodied Perspective on the Co-Production of Cultural Ecosystem Services: Toward Embodied Ecosystems.' *Journal of Environmental Planning and Management* 61(5–6): 778–799. doi:10.1080/09640568.2017.1312300.

Rosa, I.M.D., H.M. Pereira, S. Ferrier, R. Alkemade, L.A. Acosta, H.R. Akcakaya, E. den Belder et al. 2017. 'Multiscale Scenarios for Nature Futures.' *Nature Ecology & Evolution* 1(10): 1416–1419. doi:10.1038/s41559-017-0273-9.

Runge, J., S. Bathiany, E. Bollt, G. Camps-Valls, D. Coumou, E. Deyle, C. Glymour et al. 2019. 'Inferring Causation from Time Series in Earth System Sciences.' *Nature Communications* 10(1): 2553. doi:10.1038/s41467-019-10105-3.

Ryan, L., and A. d'Angelo. 2018. 'Changing Times: Migrants' Social Network Analysis and the Challenges of Longitudinal Research.' *Social Networks* 53: 148–158. doi:10.1016/j.socnet.2017.03.003.

Sallu, S.M., Twyman, C., and Stringer, LC. 2010. 'Resilient or Vulnerable Livelihoods? Assessing Livelihood Dynamics and Trajectories in Rural Botswana.' *Ecology and Society* 15(4): 3. www.ecologyandsociety.org/vol15/iss4/art3.

Salomon, A.K., A.E. Quinlan, G.H. Pang, D.K. Okamoto, and L. Vazquez-Vera. 2019. 'Measuring Social-Ecological Resilience Reveals Opportunities for Transforming Environmental Governance.' *Ecology and Society* 24(3): 16. doi:10.5751/ES-11044-240316.

Sang, N., ed. 2020. *Modelling Nature-Based Solutions: Integrating Computational and Participatory Scenario Modelling for Environmental Management and Planning.* Cambridge: Cambridge University Press.

Schill, C., J.M. Anderies, T. Lindahl, C. Folke, S. Polasky, J.C. Cárdenas, A-S. Crépin, M.A. Janssen, J. Norberg, and M. Schlüter. 2019. 'A More Dynamic Understanding of Human Behaviour for the Anthropocene.' *Nature Sustainability* 2(12): 1075–10782. doi:10.1038/s41893-019-0419-7.

Schlüter, M., L. Haider, S. Lade, E. Lindkvist, R. Martin, K. Orach, N. Wijermans, and C. Folke. 2019a. 'Capturing Emergent Phenomena in Social-Ecological Systems: An Analytical Framework.' *Ecology and Society* 24(3): 11. doi:10.5751/ES-11012-240311.

Schlüter, M., K. Orach, E. Lindkvist, R. Martin, N. Wijermans, Ö. Bodin, and W.J. Boonstra. 2019b. 'Toward a Methodology for Explaining and Theorizing about Social-Ecological Phenomena.' *Current Opinion in Environmental Sustainability* 39: 44–53. doi:10.1016/j.cosust.2019.06.011.

Sieber, I.M., R. Biesbroek, and D. de Block. 2018. 'Mechanism-based Explanations of Impasses in the Governance of Ecosystem-based Adaptation.' *Regional Environmental Change.* doi:10.1007/s10113-018-1347-1.

Singer, A., S. Gray, A. Sadler, L. Schmitt Olabisi, K. Metta, R. Wallace, M.C. Lopez, J. Introne, M. Gorman, and J. Henderson. 2017. 'Translating Community Narratives into Semi-quantitative Models to Understand the Dynamics of Socio-Environmental Crises.' *Environmental Modelling & Software* 97: 46–55. doi:10.1016/j.envsoft.2017.07.010.

Smithwick, E., E. Baxter, K. Kim, S. Edel-Malizia, S. Rocco, and D. Blackstock. 2018. 'Interactive Videos Enhance Learning about Socio-Ecological Systems.' *Journal of Geography* 117(1): 40–49. doi:10.1080/00221341.2017.1374433.

Smithwick, E.A.H., C. Caldwell, A. Klippel, R.M. Scheller, N. Tuana, R.B. Bird, K. Keller et al. 2019. 'Learning about Forest Futures under Climate Change through Transdisciplinary Collaboration across Traditional and Western Knowledge Systems.' In *Collaboration Across Boundaries for Social-Ecological Systems Science: Experiences Around the World*, edited by S.G. Perz, 153–184. Cham: Palgrave Macmillan. https://link.springer.com/chapter/10.1007/978-3-030-13827-1_5.

Stahl, B.C., and D. Wright. 2018. 'Ethics and Privacy in AI and Big Data: Implementing Responsible Research and Innovation.' *IEEE Security and Privacy* 16(3): 26–33. doi:10.1109/MSP.2018.2701164.

Sugihara, G., R. May, H. Ye, C-h. Hsieh, E. Deyle, M. Fogarty, and S. Munch. 2012. 'Detecting Causality in Complex Ecosystems.' *Science* 338(6106): 496–500. doi:10.1126/science.1227079.

Tian, S., A.I.J.M. van Dijk, P. Tregoning, and L.J. Renzullo. 2019. 'Forecasting Dryland Vegetation Condition Months in Advance through Satellite Data Assimilation.' *Nature Communications* 10(1): 1–7. doi:10.1038/s41467-019-08403-x.

Voinov, A., K. Jenni, S. Gray, N. Kolagani, P.D. Glynn, P. Bommel, C. Prell et al. 2018. 'Tools and Methods in Participatory Modeling: Selecting the Right Tool for the Job.' *Environmental Modelling and Software* 26.

Weiss, D.J., A. Nelson, H.S. Gibson, W. Temperley, S. Peedell, A. Lieber, M. Hancher et al. 2018. 'A Global Map of Travel Time to Cities to Assess Inequalities in Accessibility in 2015.' *Nature* 553(7688): 333–336. doi:10.1038/nature25181.

West, S., L.J. Haider, S. Stålhammar, and S. Woroniecki. 2020. 'A Relational Turn for Sustainability Science? Relational Thinking, Leverage Points and Transformations.' *Ecosystems and People* 16(1): 304–325. doi:10.1080/26395916.2020.1814417.

Will, M., J. Groeneveld, K. Frank, and B. Müller. 2020. 'Combining Social Network Analysis and Agent-Based Modelling to Explore Dynamics of Human Interaction: A Review.' *Socio-Environmental Systems Modelling* 2: 16325. doi:10.18174/sesmo.2020a16325.

Wyborn, C., A. Datta, J. Montana, M. Ryan, P. Leith, B. Chaffin, C. Miller, and L. van Kerkhoff. 2019. 'Co-Producing Sustainability: Reordering the Governance of Science, Policy, and Practice.' *Annual Review of Environment and Resources* 44(1): 319–346. doi:10.1146/annurev-environ-101718-033103.

Yletyinen, J., J. Hentati-Sundberg, T. Blenckner, and Ö. Bodin. 2018. 'Fishing Strategy Diversification and Fishers' Ecological Dependency.' *Ecology and Society* 23(3): 28. doi:10.5751/ES-10211-230328.

Zaidi, L. 2019. 'Worldbuilding in Science Fiction, Foresight and Design.' *Journal of Futures Studies* 23(4) 15–26. doi:10.6531/JFS.201906_23(4).0003.

Zhao, Y., Y. Wei, B. Wu, Z. Lu, and L. Fu. 2018. 'A Connectivity-based Assessment Framework for River Basin Ecosystem Service Management.' *Current Opinion in Environmental Sustainability* 33: 34–41. doi:10.1016/j.cosust.2018.03.010.

Zulkarnain, F., M.D.M. Manessa, W. Suseno, A. Ardiansyah, R. Bakhtiar, A.N. Safaryanto, B.W. Widjaja, and R. Rokhmatuloh. 2019. 'People in Pixels: Developing Remote Sensing-based Geodemographic Estimation through Volunteered Geographic Information and Crowdsourcing.' *Conference Paper: Remote Sensing Technologies and Applications in Urban Environments IV.* doi:10.1117/12.2533230.

Glossary of key terms

The purpose of this glossary is to provide a clear guide to many commonly used terms in social-ecological systems (SES) research. For most of these terms, there are several definitions in the literature, as can be expected in an emerging scientific field with a high degree of methodological pluralism. The definitions given below represent the way in which the terms are used in this book. We provide expanded and alternative definitions for some terms that are particularly contested, and also cross-reference to chapters where these concepts are explained in more detail.

Adaptation, adaptive capacity: Adaptation is the process by which an SES learns, combines experiences and knowledge, and adjusts to changing external drivers and internal processes. 'Adaptive capacity' is the extent to which an SES can adjust its responses.

Adaptive governance, adaptive co-management, adaptive management: Adaptive governance encompasses a broad range of processes and interactions between public and private actors, networks, organisations and institutions to create adaptability and transformability in complex, uncertain SES. 'Governance' refers to the structures and processes by which people in societies make decisions and share power and responsibilities, and can involve both public and private actors engaging in the development, implementation and application of principles, rules, norms and institutions that guide both social and social-ecological interactions. Governance is broader than 'management', which focuses on the implementation of decisions. Adaptive governance enables adaptive co-management, which combines adaptive management (a management system that emphasises learning, treating policies as hypotheses, and management actions as experiments to test those hypotheses) with collaborative management (management that involves collaboration between diverse stakeholders). Systems of adaptive governance are considered to be more suitable for SES, as they emphasise learning and experimentation, emergence, context dependency, management of feedbacks and action in the face of uncertainty.

Anthropocene: The Anthropocene is the geological epoch we currently live in, dating from the commencement of significant human impact on the earth's geology and ecosystems from anthropogenic climate change. The Anthropocene is most commonly proposed to have started around 1950 and succeeds the Holocene, the geological epoch that followed the last ice age (approximately 12 000 years ago) which was characterised by a stable climate and favourable conditions for the development of modern human societies. The Anthropocene concept sometimes relates to discussions about the geological evidence that marks the beginning of a new epoch that is distinguished from the Holocene. More commonly, however, it refers to the pervasive impacts of people on ecosystems on the planet, across scales from local to global.

Boundary object: A boundary object is a concept, framework or model which is understood differently by different participants and serves to foster discussion about different understandings and interpretations of a situation or phenomenon. Boundary objects have different meanings for different communities, but their structure is loose and common enough that multiple worlds can interact with it. Thus, boundary objects often serve as tools of translation and to enable collaboration.

Collective action, collaborative governance: Collective action refers to action taken together by a group of people whose goal is to achieve a common objective. Collaborative governance refers to the processes and structures of public policy decision-making and management that engage people constructively across the boundaries of public agencies, levels of government, and/or the public, private and civic spheres in order to carry out a public purpose that could not otherwise be accomplished.

Common-pool resource, public good, tragedy of the commons: Common-pool resources or goods are those that are non-excludable (it is challenging to exclude potential beneficiaries) and rivalrous in consumption (one person's use of these goods detracts from another person's use). Common examples are fisheries or communal pastures. A public good is non-excludable, and also non-rivalrous, in that use by one person does not reduce availability to another person (e.g. knowledge, national security, street lights, clean air). The tragedy of the commons is a situation in a common-pool resource system where individual and group interests do not align (a social dilemma), i.e. what is optimal for an individual according to their own self-interest leads to outcomes that are not optimal for the group (hence the tragedy). Individuals need to act collectively and restrain their individual behaviour to achieve outcomes that are overall better for everyone.

Complex adaptive systems (CAS): Complex adaptive systems comprise a number of relationally constituted phenomena that interact in adaptive and non-linear ways to form emergent patterns of behaviour. Complex adaptive systems are a special instance of complex systems as they extend the definition of traditional systems theory by recognising that CAS contain adaptive components and capacities.

Diversity, redundancy: Diversity indicates the amount of variation in a system and includes three interrelated and distinct components: variety (how many different elements), balance (how many of each element) and disparity (how different the elements are from one another). Important SES elements that exhibit diversity include genes, species, landscape patches, cultural groups, livelihood strategies and governance institutions. Microdiversity is variation within a type and macrodiversity is variation between types. This concept is often linked to the notion of 'redundancy', which refers to the capacity of a system to have more (variables, numbers, connections, processes) than is required for an SES to function under current conditions.

Ecosystem services, Nature's Contributions to People (NCP): Ecosystem services, first popularised by the Millennium Assessment, are the benefits from nature that support and fulfil human life. Fields, forests and oceans, for example, provide food in the form of crops, game and fish (provisioning services). Coastal habitats such as mangroves buffer people and infrastructure along coastlines from the impacts of storms (regulating services). Natural landscapes around the world are important parts of people's cultural and spiritual identities (cultural services). Functioning, healthy ecosystems are necessary for the production of ecosystem services, but often some sort of human input or action is also required to facilitate the provision of these services and their contribution to well-being, such as forest management or cropland irrigation. Ecosystem services are therefore co-produced by people and nature.

Although the ecosystem service approach has been widely used in policy and research, it is not universally accepted by all stakeholders and researchers. Specifically, its utilitarian view of nature (particularly associated with economic valuation approaches) is incompatible with many indigenous and cultural worldviews and social science perspectives. Recently the International Panel on Biodiversity and Ecosystem Services (IPBES) proposed the concept of 'Nature's Contributions to People' (NCP) as an advance on the ecosystem service concept. NCPs are all the contributions, both positive and negative, of living nature (diversity of organisms, ecosystems, and their associated ecological and evolutionary processes) to people's quality of life. The NCP approach emphasises not only the role that culture plays in defining links between people and nature but also the role of indigenous and local knowledge in understanding nature's contributions to people. However, the NCP approach has not been universally welcomed, with some scholars arguing that it does not present a significant advance on the term 'ecosystem services', and does not capture the intertwined nature of human–nature relationships.

Emergence: Emergence occurs when phenomena are observed to have systemic properties that are different from and non-reducible to the properties of the constituent elements. Emergent properties are qualitatively different to the properties of the individual agents or elements so that the quality of the emergent property cannot be reduced to the properties of the individual parts or agents. Emergent phenomena come about as a result of complex causality.

Epistemology, ontology: Epistemology is a branch of philosophy concerned with the nature of knowledge. Epistemology concerns itself with questions about how we gain knowledge about reality. Ontology refers to philosophies and assumptions about how the world works and the essential nature of phenomena, i.e. our assumptions about the nature of reality.

Equity, social justice, epistemic justice: Equity is often conflated with equality. Equality refers to equal treatment of all, but it is only fair if all people start from the same place and have the same needs. Equity acknowledges the differential capabilities of different social groups and concerns the application of practices that enables access to resources and opportunities for everyone to successfully achieve well-being within their contexts.

To achieve equity often requires that all individuals within a community or society are treated according to their needs, rather than equally. Social justice refers to a process and goal in which the benefits and burden of society are divided between its citizens to achieve equity. Epistemic justice can be defined in negative terms: the lack of epistemic injustice is when a person or group of people are wronged in their capacity to know, or when these people or groups are unable to access knowledge that can enable having a voice or power. An example is when an illiterate person is unable to find information about their land because requesting such information requires a written form of language.

Facilitated dialogues: Facilitated dialogues are carefully designed processes aimed at supporting multi-stakeholder groups to address complex SES problems through the creation of 'safe' or 'safe enough' spaces for developing and fostering shared understanding and innovation.

Feedback: A feedback is created when a change in one part of the system creates a set of changes that eventually loop back to affect the original system component. Amplifying or reinforcing effects are called positive feedbacks and exacerbate or create more of the original change, e.g. when a forest fire grows larger and hotter because vegetation that burns releases additional energy, resulting in more and faster burning of the forest.

Dampening, balancing or negative feedbacks counteract the original change, e.g. when intensive harvesting of a certain species triggers the implementation of a rule or regulation that reduces harvesting of that species.

Framework: Frameworks identify, categorise and organise elements or variables and their linkages in a way that is relevant to understanding a particular phenomenon. Frameworks aim to guide an investigation or activity by pointing to the concepts, elements, variables, links or processes of an SES that are characteristic or critical, or that help to explain or predict particular SES outcomes. Beyond this generic aim, however, purposes and forms of frameworks vary widely. Purposes range from descriptive (conceptual frameworks), to analytical/explanatory (analytical frameworks), to serving as boundary objects for interdisciplinary collaboration or heuristics for problem solving. Some frameworks include assumptions about causal relationships between variables, whereas others only list and categorise factors that are considered most relevant for understanding a phenomenon, such as collective action of resource users. Frameworks can be represented as box-and-arrow diagrams, as different types of elements linked through lines or arrows, or as lists of tiered variables.

Human well-being: Human well-being has multiple constituents, including basic material for a good life, freedom and choice, health, good social relations and security. Well-being is at the opposite end of a continuum from poverty, which has been defined as a 'pronounced deprivation in well-being'. The constituents of well-being, as experienced and perceived by people, are situation dependent, reflecting local geography, culture and ecological circumstances.

Institutions: Institutions refer to the formal (e.g. rules, laws, constitutions, organisational entities) and informal (e.g. norms of behaviour, conventions, codes of conduct) practices that structure human interaction. Institutions play a particularly important role in decision-making and resulting social-ecological outcomes. Boundary rules, for instance, that define the eligibility of actors to harvest resources, can provide powerful incentives to invest in the management and sustainable exploitation of resources by internalising the costs and benefits of resource use.

Interdisciplinary, transdisciplinary: Teams that involve only academics from different disciplines are commonly referred to as interdisciplinary teams. In these cases, researchers work together to integrate or combine disciplinary knowledge and methods, develop and meet shared goals, and achieve a synthesis of approaches. In transdisciplinary teams, the sphere of collaboration is expanded to include relevant societal stakeholders and other non-academics with other types of knowledge (e.g. local, indigenous, practice-based), often engaging in multi-stakeholder knowledge co-production processes.

Knowledge co-production: Knowledge co-production refers to an iterative and collaborative process involving diverse expertise and knowledge actors to produce context-specific knowledge and pathways towards a sustainable future. Successful knowledge co-production is context based, explicitly recognises a range of perspectives and knowledge, considers gender, ethnicity and age, works to define a shared, meaningful, clearly defined goal, and is able to develop through an ongoing learning process.

Knowledge systems: Knowledge systems are distinct bodies of knowledge in which a particular combination of agents, culture, practices and institutions organises the production, transfer and use of knowledge. Historically, scientific knowledge has been privileged over other knowledge types in many spheres, but policymakers, scientists and managers are increasingly recognising that other knowledge systems such as practitioner and indigenous knowledge offer valuable, valid and complementary knowledge that can be used to address sustainability challenges. Since each knowledge system produces knowledge by different rules and processes, knowledge produced within one knowledge system cannot be easily validated using another knowledge system. Instead, transdisciplinary teams have to work to 'weave' different knowledge streams together.

Livelihoods: Livelihoods comprise the capabilities, material and social resources (assets), and activities required to make a living. It includes all the ways in which individuals, households and communities make a living, generate cash and non-cash incomes, and sustain themselves, their assets and networks within a given social-ecological context.

Methodological pluralism: Methodological pluralism refers to the use of more than one methodological or conceptual approach to solve a problem. In contrast to unification, in which a researcher tries to combine different approaches, pluralism underscores the autonomy of the methods or approaches used, along with their associated assumptions and theories, with respect to one another. Thus, pluralism is less integrative and less focused on unifying perspectives of various sorts across disciplinary boundaries.

Model: A model is a simplified representation of a real-world phenomenon, i.e. something (a small form) that stands for something else (the 'real thing'). Models are always developed for a certain purpose which influences how they are developed and used. Models can be conceptual (e.g. a causal loop diagram, a box-and-arrow diagram, a mental model), physical (e.g. a model of a house used by an architect), or formalised in mathematical equations or computer algorithms.

Multi-methods research, mixed-methods research, triangulation: Multi-methods research refers to the use of more than one method to understand a given phenomenon, and does not necessarily mean those methods are integrated or used for triangulation. In many cases, however, SES researchers use mixed-methods and multi-methods triangulation approaches, which seek to go beyond just using different methods within a project. A mixed-methods study can be defined as one where the researcher combines at least one quantitative method and one qualitative method to analyse data, and involves some form of integration of the two. Multi-methods triangulation (which overlaps with, but is not limited to, mixed-methods approaches) is also usually integrative, but specifically refers to approaches where insights regarding a single research problem are strategically drawn from findings generated using different methods.

Path dependency: Path dependency refers to a particular path or development trajectory of a system that is inherently difficult to change due to reinforcing feedbacks and system memory. It implies that the outcomes in a particular system are sensitive to the history of the path traced, and limited by previous events or decisions.

Policy, policy instrument, policy support tool, policy cycle: Policy refers to a definite course or method of action, selected from among alternatives and in light of given conditions, to guide and determine present and future decisions. A policy instrument is a set of mechanisms that are used to achieve a particular policy goal, whereas a policy support tool can, as the name suggests, support the implementation of a particular policy. The policy cycle refers to the process of formulating, implementing and subsequently evaluating a policy, which results in new or amended policies.

Power relations: Power relations refer to the agency someone or an institution or a set of values has over someone else or over resources (dominant or sovereign power). Power can also refer to having the agency or power to act (productive power).

Reflexivity: Reflexivity is the process of actively and continuously reflecting on the role of one's own position, subjectivity and assumptions in the research process, and how this may shape conclusions drawn from the research. Reflexive researchers recognise that they do not exist outside their system of study, and that their engagement with the system also affects the outcomes of their research. Reflexivity is not about removing the researcher from the study, or making them more 'objective', but rather making explicit the lenses through which they engage in the study.

Regime shift: Regime shift refers to large, persistent changes in the composition, structure and function of SES associated with the transgression of critical tipping points. Regime shifts have substantive impacts on ecosystem services and human well-being. They are often very costly and difficult, or even impossible, to reverse. Examples of regime shifts are the transformation of a forest to a grassland following excessive burning, or from a clear lake to an algal-dominated lake following an increase of agricultural run-off and pollutants into the lake. Regime shifts can be systemically similar to social-ecological transformations, but tend to focus on unintended, negative shifts rather than more intentional, positive shifts in SES.

Resilience: Classical definitions of resilience refer to the amount of change a system can undergo and still retain its essential structures, functions and feedbacks, as well as the capacity of the system for self-organisation, adaptation and learning. More recently, resilience has been defined as the capacity of a social-ecological system to persist in the face of disturbance and change, while continuing to adapt and develop along a pathway or transform and navigate new pathways in order to sustain human well-being. Increasingly, resilience thinking takes into account and integrates notions of governance systems, ecosystem services and human well-being, adaptive capacity and transformation.

Scale/level, cross-scale/cross-level, multi-scale/multi-level: The terms 'scale' and 'level' are sometimes used interchangeably in SES research. Although related, they have different origins. Scale typically refers to the physical dimensions, in either space or time, of phenomena or observations. This is expressed in physical units, such as metres or years. The word 'level' is mostly used to describe the discrete levels of social organisation, such as individuals, households, communities or nations. A level of organisation is not a scale, but it can have a scale. It is also important to distinguish the 'scale of observation' from the 'scale of the phenomenon'. Most ecological and human processes have characteristic scales (i.e. a typical extent or duration over which the process is expressed and has impact), which are independent of human systems of measurement. In contrast, the scale of observation is a construct based on human measurement, and has three components: extent or duration (the total area or time over which a phenomenon is observed), resolution (the interval or distance between observations) and grain (the area or duration of an individual observation). Cross-scale/cross-level refers to interactions across different scales or levels. Multiple scales (or multi-scale/multi-level) indicate the presence of more than one scale, but not necessarily of interactions cutting across scales or levels.

Self-organisation: Self-organisation is the process by which a system can develop a complex structure from unstructured beginnings. This process is not generated by a central or external agent, but comes about as a result of the interaction between the various constituents of the system and its environment.

SES intertwinedness: Contrary to the view that the separation of 'social' and 'natural' systems is arbitrary and artificial, SES are viewed as co-constituted by human and non-human agents that interact through multiple processes and relations. Nature is not merely the setting for society, and societal relations are not just external drivers that have an impact on nature. SES are not seen as merely social plus ecological systems, with their social and ecological components, but as cohesive, intertwined systems characterised by strong connections and feedbacks.

Social-ecological systems: Social-ecological systems (SES) are complex adaptive systems co-constituted by intertwined social and ecological agents and processes.

Social learning: Social learning refers to learning or changes in understanding of people that occur through social interactions and processes. New knowledge becomes situated within a group, community of practice or society and can sometimes lead to a transformation in values and worldviews.

Stewardship: A 'steward' is a custodian or carer. Ecosystem stewardship refers to the proactive shaping of physical, biological and social conditions to sustain, rather than disrupt, critical ecosystem processes that support nature and human well-being at local to planetary scales. Stewardship is driven by stewards that abide by a stewardship ethic, a philosophy that guides sustainability actions of stewards. Ecosystem stewardship aims to foster resilience to sustain desirable social-ecological conditions and enable transformation from undesirable trajectories. Stewardship has a temporal dimension – care for the future, and an equity dimension – a fairer distribution of rights, responsibilities and power across society.

Sustainability science: Sustainability science is an emerging research field dealing with social and environmental sustainability challenges and their interrelations. Sustainability is concerned with understanding how we can balance the needs of the future with achieving human well-being and healthy ecosystems in the present. Sustainability science is used very broadly: to describe disciplinary science that is focused on sustainability challenges, but more often to describe inter- and transdisciplinary science that is frequently focused on directly affecting change towards sustainability. Social-ecological systems research falls within the broader area of sustainability science, usually within the latter category.

Telecoupling: Telecoupling refers to socio-economic and environmental interactions between distant coupled human and natural systems. It builds on the concept of 'teleconnections', originating from meteorology and climate science and referring to an effect beyond the location at which a phenomenon originated. Telecoupling goes beyond 'action at a distance', also focusing on the feedbacks between social-ecological processes and outcomes in multiple interacting systems. A telecoupling arises when an action produces flows between two or more SES, which create an intended or unintended change or response in distant systems.

Transformation: Transformation refers to a fundamental change in an SES, generally towards a more sustainable or preferred outcome. It can involve systemic dynamics similar to a regime shift, but typically focuses on positive shifts, often involving radical changes in underlying worldviews, values and governance systems.

Values: The word 'value' can refer to how much something is 'worth' (how important it is, or a measure of importance), a preference someone has for a particular state of the world, or principles associated with particular worldviews or cultures. Particularly in ecosystem service research, people have traditionally distinguished instrumental value (the value that something has to a person – often associated with ecosystem services such as water provision and aesthetic beauty) and intrinsic value (something's value that originates within itself – often associated with wilderness areas and nature). More recently, however, scholars have started recognising that people also consider relational values: the preferences, principles and virtues associated with relationships with nature, and with other people in nature.

Vulnerability: People or societies are vulnerable when they are susceptible to harm from exposure to stresses associated with social-ecological change and a lack of adaptive capacity.

Index

Note: **Bold** font indicates the major references to a topic and *italic* font indicates figures.